MECHANICS OF SOLID MATERIALS

Mechanics of solid materials

JEAN LEMAITRE
Université Paris 6 – LMT Cachan

and

JEAN-LOUIS CHABOCHE
Office National d'Etudes et de Recherches

TRANSLATED BY B. SHRIVASTAVA

FOREWORD TO THE FRENCH EDITION BY
PAUL GERMAIN
Académie des Sciences, Paris

FOREWORD TO THE ENGLISH EDITION BY
FRED LECKIE
University of Illinois, Urbana–Champaign

Published by the Press Syndicate of the University of Cambridge
The Pitt Building, Trumpington Street, Cambridge CB2 1RP
40 West 20th Street, New York, NY10011–4211, USA
10 Stamford Road, Oakleigh, Melbourne 3166, Australia

Originally published in French as a
Mécanique des matériaux solides
by Dunod, Paris, 1985 and © Bordas, Paris, 1985

First published in English by Cambridge University Press 1990 as
Mechanics of solid materials

First paperback edition 1994

Printed in Great Britain at The Bath Press, Avon

British Library Cataloguing in Publication Data

Lemaitre, Jean
Mechanics of solid materials.
1. Solids. Mechanics
I. Title II. Chaboche, Jean-Louis
III. Mécanique des matériaux solides. *English*
531

Library of Congress Cataloguing in Publication Data

Lemaitre, J. (Jean), 1934-
[Mécanique des matériaux solides. English]
Mechanics of solid materials/Jean Lemaitre and Jean-Louis Chaboche:translated by B. Shrivastava; foreword to the English edition by Fred Lackie.
p. cm.
Originally published under title: Mécanique des matériaux solides.
Includes bibliographies and index.
ISBN 0-521-32853-5
1. Strength of materials. I. Chaboche, Jean-Louis. II. Title.
TA405.L3813 1990
620.1′12-dc 19 88-22913 CIP

ISBN 0 521 32853 5 hardback
ISBN 0 521 47758 1 paperback

TM

CONTENTS

FOREWORD TO FRENCH EDITION

When my young colleagues, Jean Lemaitre and Jean-Louis Chaboche invited me to write a few introductory lines on the occasion of the publication of their work *Mécanique des Matériaux Solides* (Mechanics of solids materials), I very willingly acceded to their request as an expression of trust and friendship, even though on one hand, the aim of the present work is made abundantly clear by the title and the introduction, and on the other, the well-deserved fame of the authors is quite sufficient to attract and retain the attention of readers.

The originality, I would even say the identity, of this book becomes apparent only if we place it within the evolution of the scientific subjects during the last few decades. In fact, it is part of a triple current whose recent developments it assimilates and integrates. First of all, it borrows from continuum mechanics and thermodynamics the conceptual framework and methods which, starting with a few simple concepts, allow construction of a great variety of phenomenological models necessary for describing the extremely varied behaviour of solids. Secondly, it collects, rearranges, and above all, takes advantage of the observations, schematic representations, and empirical laws which generations of engineers have used with imagination and perspicacity in guiding and accomplishing their projects. Finally, it provides an inventory of physical phenomena, especially those observed at the microscopic, molecular or atomic scale, the scale at which events determining and explaining macroscopic behaviour occur. Even if these phenomena cannot be explained and expressed in terms of formulas, they are mentioned, whenever possible, in order to clarify the results and procedures.

To my mind, this triple heritage can never be overstated. This remarkable development of continuum thermodynamics would be no more than a pure theoretical elaboration if it did not include an effective and operational understanding of the many empirical laws, patiently deduced from experiments. The knowledge resulting from these latter sources can only contribute to the advancement of the mind in as much as it helps to explain, justify, and often inspire, the theoretical developments which receive their full recognition on the basis of experiments. Even as the physics of solids penetrates further and further into the elementary physical phenomena at the microscopic scale, it can acquire its full practical importance only if it is supplemented by macroscopic disciplines which support and perfect an ever bolder and an ever more efficient technique. These disciplines, as is shown by the present day research, will become more precise and refined in their methods so long as the disciplines close to them, physics and chemistry, continue to be perfected. It appears to me that the main characteristic of this work is that it lies at the crossroads of these trends of thoughts and that it brings out their mutual relationships which, in turn, reinforce their individual interests.

To write a good book, it is not enough to have a right conception, a clear objective, and an interesting aim. It is of no interest to mention the pitfalls and the dangers inherent in the task which our authors could have encountered, because to my mind they have succeeded in avoiding them. This book is not a '*summa*' where the reader could find a presentation of the three main branches of research that I have just mentioned. But it is essentially an account, a rather complete one but without a surfeit of details that obscure the ideas, of the constitutive laws of all the so-called solid materials, metals at room or elevated temperatures, polymers exhibiting very different behaviours from a vitrous state to a rubbery state, wood, concrete, and moreover, if I may say so, in every state of their existence: damaged, cracked, or aged. It is this primary concern which is the essence of this book.

The theoretical aspects of continuum mechanics are presented, without any superfluous development, with just what is required by the reader to establish a link, if necessary, with other works; the same is true as regards the finite element method of analysis without which the work would lose much of its interest. Similar observations hold true for the other disciplinary currents mentioned above: what is presented is quite sufficient to clarify the aim of this book in describing methods which can be used to test the proposed laws, and to explain them with numerical data useful in applications. This is the reason why I do not hesitate to recommend this book to a broad category of readers who, because of the connecting threads provided herein, will easily find valuable information for their work:

theoretical researchers, who will be happy to see that their past efforts have not been useless, will find here an inspiration for the future; mechanical and civil engineers will find here not only precious data but also an enlightening and stimulating framework for their thoughts; metallurgists, chemists and physicists will discover here the knowledge of macroscopic disciplines beyond their research. Of a reasonable size, written in a clear style, free from any over-specialized language, at the level of a master's degree which remains accessible to a vast scientific public, this book should enjoy a wide and well-deserved circulation.

Finally, I cannot help but note that this book provides a new testimony of the vitality of French mechanics of solids, of the unity and the interrelation between its views and its efforts, and of the interest of the results owed to it.

The French team of researchers have indeed contributed significantly to the theoretical developments on which Jean Lemaitre and Jean-Louis Chaboche have based their work: the method of virtual power, functional analysis and methods of finite element analysis, the formulation of constitutive laws by the method of local state in plasticity and especially in viscoplasticity, and application of these methods to damage and to the mechanics of cracked media. The merit of these authors is to have been able to assimilate the above developments, so as to go further and open new fields by combining all the needs arising from physical experiments and technical applications, and thus provide to French researchers a proof of the validity of the direction of their work and of the quality of the results arising from their coherent efforts. Both of these authors have already succeeded in enlivening this productive spirit and developing it by their own personal research at ONERA and at LMT at Cachan, by their teaching, and above all by their work in the '*Large Deformations and Damage Group*' (GRECO). The present work will spread the field of their influence even further, and as a consequence, the field of French mechanics as well. This will contribute to the betterment and greater vitality of our particular approach to mechanics.

In expressing publicly to Jean Lemaitre and to Jean-Louis Chaboche my gratitude and congratulations, I hope that this book will enjoy a wide circulation and that it will contribute to an awareness, beyond our borders, of the spirit and the quality of research in the mechanics of solids which the French teams have carried out in recent years.

Paul Germain
Professor at the École Polytechnique
Secrétaire perpétuel de l'Académie des Sciences
Paris, December 1984

FOREWORD TO THE ENGLISH EDITION

It is a special pleasure for me to introduce the English translation of the book by Jean Lemaitre and Jean-Louis Chaboche. Readers will find this an ambitious book written in a bold and adventurous style. I had the good fortune to spend four months as a visitor at the Laboratoire de Mécanique et Technologie at Cachan in 1983. It is evident that the book reflects the dynamic and refreshing style of research at this laboratory. The aim of the book is to answer the important question of how the mechanical properties of materials are affected by complex loading histories and how this behaviour in turn affects the performance of engineering components. The approach is global. Theoretical formulation is combined with discriminating experiments and the computational power of the computer to define and calculate the factors which affect the life and performance of engineering components subjected to severe loading conditions. A great merit of the approach is its ability to be constantly improved. Incremental improvements in constitutive laws, for example, are easily introduced as are efforts which help to bridge the physical processes within the material and the macroscopic behaviour observed in experiment. By this means research and development effort can be well coordinated and gaps of knowledge become more clearly identified. While the properties of metals attract most attention the importance of the general approach is illustrated with reference to other materials such as concrete and polymers. A particularly attractive feature of the book is the special emphasis given to damage mechanics and in this respect it probably represents the first attempt to present a unified treatment of this rather new topic. Damage mechanics plays a crucial role when estimating the life of engineering components. The resulting procedures are of particular importance at the initial design stages or when attempts are made to predict the remaining life of existing components.

This is a book which should appeal to both academic and practising engineers who are concerned with the means of predicting the life and performance of advanced engineering components. It should also appeal to those who are involved in the demanding business of attempting to span the scales of the processes occurring at the microstructural level and the engineering properties appropriate to design.

Fred Leckie
Professor and Head of the Department of Theoretical and Applied Mechanics
University of Illinois, Urbana-Champaign, USA
March 1987

INTRODUCTION

Well done reader, you have been bold enough to open this book! and now you will be rewarded with a few explanations.

First on its title: the book deals with *mechanics*, that is the study of equilibrium and motion by considering relations between forces (or stresses) and displacements (or strains), time, and possibly temperature. It is applied to the structural *materials* used in mechanical and civil engineering: metals and alloys, natural organic materials (wood) or synthetics (plastics), and concrete. Thus, it is concerned with the study of the properties of *resistance* to *deformation* and *fracture*. These properties are intrinsic to the material, and are usually defined with reference to a volume element, independent of the geometry of the body under study.

On its spirit: the writing of the book has been organized to facilitate *transfer* of knowledge to engineering science: fundamental knowledge oriented towards practical applications, knowledge of macroscopic properties for formulating macroscopic laws of material behaviour, and a synthesis of the knowledge of theoretical and practical aspects. The recent progress in the mechanics of solid materials has resulted from simultaneous and decisive developments in all these fields:

> At the fundamental level, it is the synthesis of continuum mechanics with the method of local state in thermodynamics.
>
> At the level of applications, it is a recognition of the need to ensure higher safety and economy in the building of more sophisticated structures. It also recognizes the possibility of using faster computers to solve nonlinear problems numerically.
>
> At the level of microscopic properties, it is the theory of dislocations and the invention of electron microscopes.

At the level of macroscopic phenomena, it is the technique of identification of mathematical models from experimental results.
At the theoretical level, it is the development of functional analysis and variational formulations.
Finally, at the experimental level, it is, of course, the introduction of electronics and microcomputers in testing machines and measurement procedures.

On its content: in order to emphasize this idea of transfer of knowledge, and for the education of the young and the continued training of the 'not so young', the book includes two chapters (Chapters 1 and 2) that recall the basic knowledge necessary to understand the rest of the book, one chapter (Chapter 3) that synthesizes different material behaviours, and five chapters (Chapters 4–8) that represent original contributions to each of the five broad classes of behaviour.

The first chapter is devoted to *physical mechanisms* of deformation and fracture of metals and alloys, polymers, concrete and wood. The chapter provides a brief summary of the principal mechanisms and is written with a view to justifying the physical hypotheses used in macroscopic modelling.
The second chapter reviews the elements of *continuum mechanics* and *thermodynamics of irreversible processes* which constitute the theoretical tools used in other chapters.
The third chapter presents a schematic *classification* of the behaviour of solids based on experimentation and identification, the main methods of which are described.
The fourth chapter marks the begining of the modelling of different material behaviours. All the subsequent chapters follow the same outline, namely: domain of validity defining more or less precisely the conditions for using the models, phenomenological aspects derived experimentally, general formulation based on thermodynamics, determination and identification of particular models with examples for common materials, and a concise presentation of the associated structural analyses. Within this framework, the fourth chapter presents *linear elasticity*, *thermoelasticity*, and *viscoelasticity*.
The fifth chapter is devoted to *plasticity*. Classical isotropic plasticity is formulated starting with the dissipation potential associated with the flow criterion. Above all, we insist on plasticity with anisotropic hardening which can be used to consider the

cyclic behaviour so important for the prediction of fatigue failure.
The sixth chapter is concerned with the same questions but this time applied to metals and alloys under loads at such intermediate or high temperatures which give rise to the phenomena of viscosity: *viscoplasticity*.
The seventh chapter approaches the fracture of a volume element through the *continuum damage mechanics*. Different models have been worked out for considering and predicting the phenomena of ductile fracture, brittle fracture and fatigue fracture.
The eighth chapter (take heart, this is the last one!) deals with crack mechanics of solids. The *fracture mechanics of crack growth* is approached by energy methods which logically introduce the concept of the energy release rate. This variable, associated with stress intensity factors, is used to formulate models of fracture by instability, of ductile fracture, and of fatigue crack growth.

The book therefore covers the whole field of the mechanics of materials, but in a highly condensed fashion. For a more detailed study, it is advisable to consult the important works listed in the bibliography at the end of each chapter. This book is therefore intended for the reader, who has a good knowledge of the basic elements of continuum mechanics or of the strength of materials, but who wishes to introduce more physics into the design and manufacture of products, with or without the help of computers, or in the safety analysis of structures.

The subject matter of the book occupies a central position between what is taught in the final undergraduate years at engineering schools, at the master's degree and postgraduate levels at universities, and what is expected of professional engineers engaged in research, project engineers, and engineers engaged in the testing of materials. In order to facilitate the use of the book as a manual, each chapter has been written in a way that it can be read independently. Consequently, there is unavoidable repetition of some material. Although, most of the analysis has been carried out in intrinsic notation, the main results are expressed using index notation.

The wise reader can now appreciate the fact that the scope of this book is very wide, and in order to be able to accomplish this, the authors took full advantage of their particularly favourable human environment at the Division Résistance-Fatigue de la Direction des Structures at ONERA, created at the initiative of R. Mazet of the Laboratoire de Mécanique et Technologie at Cachan (ENS de Cachan/Université Paris 6/CNRS). Much of the material of this book is based on the notes of courses given at

Université Paris 6 (3e Cycle de Mécanique Appliquée à la Construction), at ENS de Cachan, at École Polytechnique Féminine, at École Centrale, and at various training sessions of continuing education, summer schools, and foreign universities. The authors here would like to thank publicly all those who participated in numerous and engaging discussions.

The endorsement of the GRECO CNRS '*Grandes Déformations et Endommagement*' was obtained after a critical reading of each chapter by specialists: A. Pineau, F. Sidoroff, A. Zaoui, M. Predeleanu, G. Duvaut, D. Marquis, G. Touzot, C. Oytana, K. Dang Van, D. François, H.D. Bui, R. Labourdette, and a reading in its entirety by P. Muller.

The authors wish to express their firm belief in this method. Thanks to you all for your advice. Thanks are also due to G. Combourieux who prepared the figures, and to Marie-Christine Senechal for her patience in the difficult task of reading the manuscript to ensure a neat and clear typescript.

J. Lemaitre – J. L. Chaboche
Spring, 1979–Autumn 1984

NOTATION

Operators

Operator	Meaning
X	Scalar
$\vec{x}$	Vector with components x_i
$\mathbf{X}$	Second order tensor with components X_{ij}
$\hat{X}$	Virtual quantity
$[X]$	Matrix
$\{X\}$	Column
$\dot{X}$	Time derivative of X $(= \mathrm{d}X/\mathrm{d}t)$
$\delta X/\delta N$	Pseudo derivative (as a function of the cycle number)
$\mathbf{X}^{\mathrm{T}}$	Transpose of $\mathbf{X}$
X^{+}	Laplace–Carson transform
$X \otimes Y$	Convolution product of X with Y
$\mathrm{Re}\, X$	Real part of X
$\mathrm{Im}\, X$	Imaginary part of X
f, g	Unidentified functions
$\mathscr{F}$	Unidentified functional
$Sgn(x)$	$+$ or $-$ sign of scalar x
$\langle x \rangle$	$\langle x \rangle = x$ if $x > 0$, $\langle x \rangle = 0$ if $x < 0$
$\mathbf{X}:\mathbf{Y}$	Double contracted product of $\mathbf{X}$ with $\mathbf{Y}$
X_{I}	First invariant of $\mathbf{X}$: $X_{\mathrm{I}} = Tr(\mathbf{X})$
X_{II}	Second invariant of $\mathbf{X}$: $X_{\mathrm{II}} = \frac{1}{2} Tr(\mathbf{X}^2)$
X_{III}	Third invariant of $\mathbf{X}$: $X_{\mathrm{III}} = \frac{1}{3} Tr(\mathbf{X}^3)$
$\mathbf{X}'$	Deviator of $\mathbf{X}$: $\mathbf{X}' = \mathbf{X} - \frac{1}{3} X_{\mathrm{I}} \mathbf{1}$
J_1	$J_1(\mathbf{X}) = X_{\mathrm{I}}$
J_2	$J_2(\mathbf{X}) = (3X'_{\mathrm{II}})^{1/2}$
J_3	$J_3(\mathbf{X}) = (\frac{27}{2} X'_{\mathrm{III}})^{1/3}$
$H(x)$	$H(x) = 1$ if $x \geqslant 0$, $H(x) = 0$ if $x < 0$
δ	Kronecker delta $\delta_{ij} = 1$ if $i = j$, $\delta_{ij} = 0$ if $i \neq j$
X_{Max}	Maximum value of X
X_m	Minimum value of X
$\bar{X}$	Mean value of X
ΔX	Peak to peak amplitude of X (range of X)

List of symbols used as coefficients defined every time they are used.

a	c	e	k	M	Q	S	α	η	ν
A	C	F	K	n	r	v	β	θ	τ
b	d	G	L	N	R	V	γ	λ	
B	D	H	m	P	s	w	δ	μ	

Modelling is the corruption of notation. The authors, unfortunately, have not been able to avoid using the same letter several times for denoting different quantities.

Symbols

Symbol	Meaning
a	Crack length
$\mathbf{a}$	Tensor of elastic moduli
A_{II}	Octahedral shear amplitude
A_k	Thermodynamically associated variable
$\mathbf{A}$	Tensor of elastic compliances
$\vec{b}$	Burgers' vector
c	Specific heat
C^*	Contour integral
D	Damage
D_c	Critical damage
D_u	Ductility
e	Specific internal energy
E	Internal energy
E	Young's modulus of elasticity
$\mathbf{E}$	Finite elastic transformation tensor
f	Plasticity criterion function
$\vec{f}$	Force density
f	Coefficient of Coulomb friction
$\vec{F}$	Force vector
g	Hardening function
$\vec{g}$	Vector $\vec{g} = \overrightarrow{\text{grad}}\, T$
G	Elastic shear modulus
G	Elastic energy release rate
G_s	Threshold of elastic energy release rate
G_c	Critical elastic energy release rate
G_n	Reduced elastic energy release rate
h	Hardening modulus
ΔH	Activation energy
I	Bui integral
J	Viscoelastic creep function
J	Rice integral

K	Kinetic energy
K	Elastic bulk modulus in compression
K, K_I	Stress intensity factors
K_{IC}	Toughness
L	Reference length
L_p	Useful length of a specimen
$\bar{m}$	Coefficient of friction of boundary layer
M	Material point
$\vec{n}$	Vector in the normal direction
N	Current number of cycles
N_R	Number of cycles to failure
N_F	Number of cycles to failure due to pure fatigue
p	Accumulated plastic strain
p_R	Accumulated plastic strain at fracture
P	Power
$\mathbf{P}$	Finite inelastic transformation tensor
$\{q\}$	Column of degrees of freedom
$\vec{q}$	Heat flux vector
Q	Rate of heat input
$\{Q\}$	Column of nodal forces
$\mathbf{Q}$	Tensor of reduced damage
q, Q	Hardening memory variables
r	Heat production per unit volume
r_Y	Dimension of plastic zone
R	Stiffness
R	Relaxation function in viscoelasticity
R	Resistance to ductile tearing
R	Isotropic hardening variable
s	Specific entropy
$\vec{s}$	Unit vector
S	Entropy
S	Specimen cross-section
$\mathbf{S}$	First Piola–Kirchhoff stress tensor
$\mathbf{S}^*$	Second Piola–Kirchhoff stress tensor
t	Time
t_R	Time to fracture
$\vec{t}$	Unit vector
T	Absolute temperature
T_m	Melting point
$\vec{T}$	Stress vector
u	One-dimensional displacement
u_e	Elastic displacement
u_p	Plastic displacement
$\vec{u}$	Displacement vector
$\vec{v}$	Velocity vector
V	Volume
V_k	Internal variable

$\mathscr{V}$	Potential energy
w	Strain energy density
w_e	Elastic strain energy density
w_p	Plastic strain energy density
W	Strain energy
W^*	Complementary energy
$\mathbf{X}$	Tensor variable of kinematic hardening
Y	Elastic energy density release rate
z	Complex variable
Z	Airy stress function
α	Phase of a metal
α	Coefficient of dilatation
$\boldsymbol{\alpha}$	Kinematic hardening variable
γ	Phase of a metal
γ	Density of decohesion energy
$\vec{\gamma}$	Acceleration vector
Γ	Crack front
$\boldsymbol{\Gamma}$	Couple stress tensor
δ	Crack opening displacement (COD)
$\boldsymbol{\Delta}$	Green–Lagrange finite strain tensor
ε	Uniaxial strain
ε_v	True strain
ε_e	Elastic strain
ε_p	Plastic strain
ε_R	Fracture strain
ε_D	Damage threshold strain
ε_H	Hydrostatic strain
ε_{eq}	Von Mises equivalent strain
$\boldsymbol{\varepsilon}$	Strain tensor
$\boldsymbol{\varepsilon}'$	Deviator of strain tensor
$\boldsymbol{\varepsilon}^e$	Elastic strain tensor
$\boldsymbol{\varepsilon}^p$	Plastic strain tensor
η	Exponent in the Paris crack growth law.
θ	Temperature
λ	Lame's constant of elasticity
λ	Multiplying factor
μ	Lame's constant of elasticity in shear
ν^*	Contraction coefficient
ν	Poisson's ratio
ξ	Hardening memory variable
ρ	Mass density
ρ_D	Density of dislocations
ρ	Length of plastic zone
ρ	Electric resistivity

σ	Uniaxial stress
σ_v	True stress
σ_s	Plastic threshold
σ_Y	Yield stress
σ_u	Ultimate stress at fracture
σ_H	Hydrostatic stress
σ_{eq}	Von Mises equivalent stress
$\tilde{\sigma}$	Effective stress
σ^*	Equivalent damage stress
$\boldsymbol{\sigma}$	Stress tensor
$\boldsymbol{\sigma}'$	Deviatoric stress tensor
σ_{ij}	Components of $\boldsymbol{\sigma}$
σ_i	Principal components of $\boldsymbol{\sigma}$
τ	Time
τ	Shear stress
τ_Y	Shear stress at elastic limit
τ_R	Shear stress at fracture limit
φ	Dissipation potential
φ^*	Dual potential
Φ	Specific power of dissipation
χ	Equivalent stress in creep fracture
ψ	Specific free energy
Ω	Potential of viscoplastic dissipation
$\boldsymbol{\Omega}$	Rotation rate tensor
$\boldsymbol{\Omega}$	Area reduction tensor due to damage

1

ELEMENTS OF THE PHYSICAL MECHANISMS OF DEFORMATION AND FRACTURE

Lorsqu'un théoricien trouve un résultat nouveau personne n'y croit, sauf lui! lorsqu'un expérimentateur trouve un résultat nouveau tout le monde y croit, sauf lui!

The aim of this first chapter is to give nonspecialists in physics and metallurgy a general idea of the structure and mechanisms of deformations and fracture of the principal materials used in ordinary structures. Therefore, it consists of a brief summary of the classical knowledge as found in specialized works. The main references among these are given in the bibliography. Knowledge of these basic physical mechanisms is necessary for the formulation of hypotheses upon which the macroscopic phenomenological theories of deformation and fracture can be based. These theories, presented in Chapters 4–8, must indeed integrate the phenomena associated with discrete entities such as atoms, crystals, molecules, cells, etc., to the level of homogeneous continuum models. In metals, the introduction of the concept of dislocations by Taylor and several others in 1934 marked the first decisive step in explaining the phenomenon of plastic deformation. In a more general way, the invention of transmission and scanning electron microscopes has permitted, during the 1960s, an understanding of the main mechanisms of deformation and fracture.

It must, however, be remembered that in the field of fracture where these phenomena occur at the much larger scale of crystal or molecular arrangements, there are several unresolved questions, and for this reason the schematic treatment given in this chapter remains very rudimentary.

Paradoxically, despite the large differences in the nature and structure of materials such as metals and alloys, polymers and composites, concrete, wood, there is a great unity displayed in their macroscopic behaviour. With different orders of magnitude, terms like elasticity, viscosity, plastic or permanent deformation, consolidation or hardening, brittle fracture and ductile fracture can be applied to all these materials. This is what justifies '*a priori*' the global approach to the mechanics of materials which, with the aid of concepts from continuum mechanics and thermodynamics in Chapter 2, and from rheology in Chapter 3, allows the construction of models independent of the nature of materials in their basic properties (but not in their analytical formulations).

1.1 Metals and alloys

1.1.1 *Structure*

Elements of crystallography

Atoms

Metals and alloys are made up of arrangements of atoms held together by electromagnetic forces between the electrons of neighbouring atoms. The order of magnitude of the 'radius' of an atom varies from 10^{-7} to 10^{-6} mm (1–10 Å). Stable arrangements are determined by a minimum energy condition of the atomic packing, which is a function of thermal activation. In metals, the bonds result from a sharing of electrons in the outer shells of the atoms. Metals and alloys are normally found to consist of closely packed crystals. The amorphous state can only result from complex processes requiring extremely fast cooling.

Monocrystals

The crystalline state is characterized by the regularity of the atomic arrangement; an elementary parallelpipedic pattern or lattice repeats itself periodically in all three directions.

Most of the metallic lattices belong to one of the three following systems:

Cubic centred (CC) crystals (Fig. 1.1) e.g., Feα, Cr, Mo
Face centred cubic (FCC) crystals (Fig. 1.2) e.g., Cu, Ag, Al, Ni
Close packed hexagonal (CPH) crystals (Fig. 1.3) e.g., Mg, Zn, Tiα, Coα.

Fig. 1.1. CC crystals.

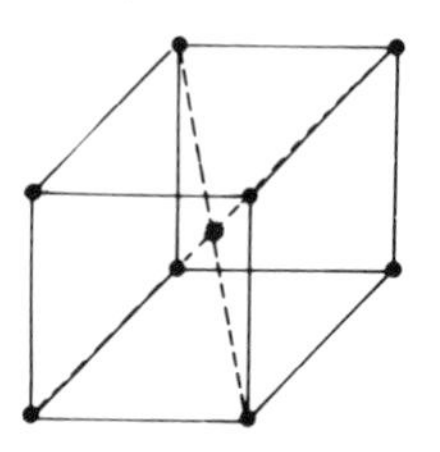

These lattices possess axes and planes of symmetry, the latter being in general the planes in which the atoms are most densely packed and which possess a lower resistance to shear. Their conventional representation is by means of Miller indices: the plane to be characterized is identified by the coordinates of its intersection with the axes of an orthogonal reference system, with the reciprocals of these numbers reduced to the three smallest integers which have the same ratios (Fig. 1.4).

Fig. 1.2. FCC crystals.

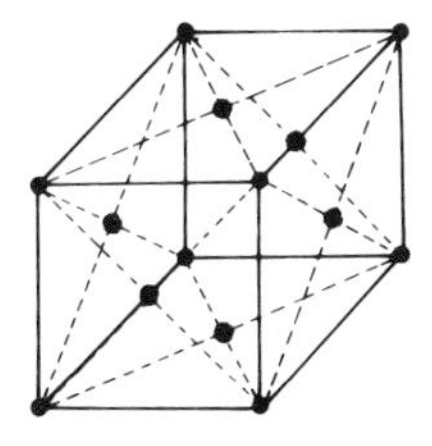

Fig. 1.3. CPH crystals.

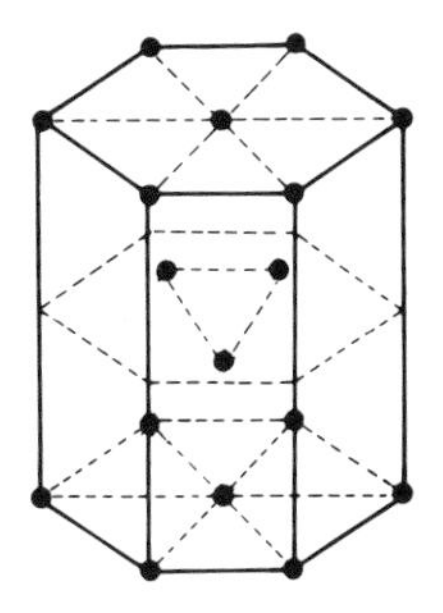

Fig. 1.4. Miller indices.

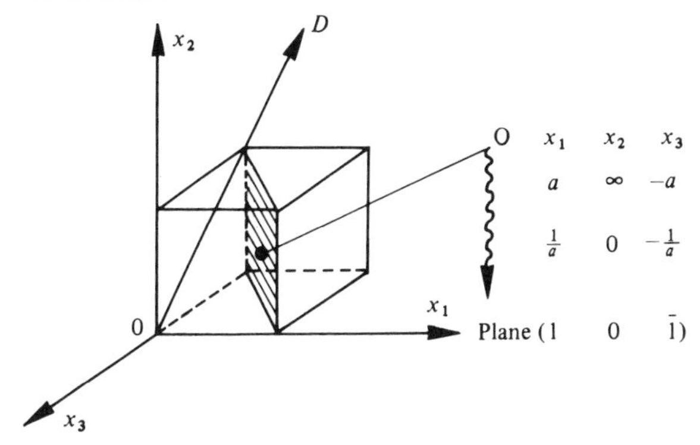

A crystallographic direction is characterized by a straight line emanating from the origin of the reference system and passing through the position of an atom. It is designated by the coordinates of that atom expressed in terms of interatomic distances. For example, the direction $0D$ in Fig. 1.4 is expressed as $\langle 01\bar{1}\rangle$.

The structure of an alloy can be either that belonging to one of the constitutive elements or an entirely different crystalline structure. In solid solutions, it is possible to have the substitution of an atom from the lattice (for example aluminium in iron) or the insertion of small atoms in the network of the solvent (for example carbon in iron for austenite (Fig. 1.5)). Metals can take different forms or phases depending on the temperature. The most common phases for steels are the γ and α phases (respectively FCC for austenite and CC for the ferrite of the carbon–iron system). It is also possible to have stable or unstable secondary phases, phases β, δ, ε, σ, and in a particular phase it is possible to distinguish the cases in which the substitute atoms occur in a disordered state (for example phase γ) or in an ordered state (for example phase γ').

Polycrystals

Metals and alloys are generally produced in a liquid state, and their structure is formed as they solidify when cooled. As the temperature of the liquid decreases, the interatomic distances become smaller; the critical distance, at which bonding occurs, is reached at several, randomly distributed sites, and these constitute the first germs or nuclei of crystal growth. The lattices are formed in the same crystalline system but in random directions. Each nucleus develops into a crystal whose growth is limited by neighbouring crystals. A polycrystal is made up of several monocrystals oriented randomly. The size of such monocrystals varies from a few microns (10^{-3} mm) to a few millimetres depending on the nature of the constituent elements of the metal, as well as on the thermal and mechanical treatments to which the polycrystal has been subjected (Fig. 1.6). It is therefore

Fig. 1.5. Solid solution of substitution (S) or insertion (I).

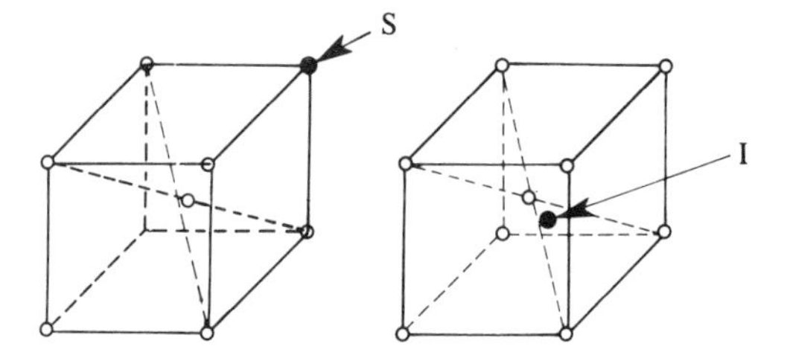

Fig. 1.6. Examples of crystals (after A. Pineau): (*a*) nickel based Inconel 718 alloy, transmission electron microscopy; (*b*) nickel Waspaloy alloy, optical microscopy; (*c*) nickel based Inconel 718 alloy, optical microscopy.

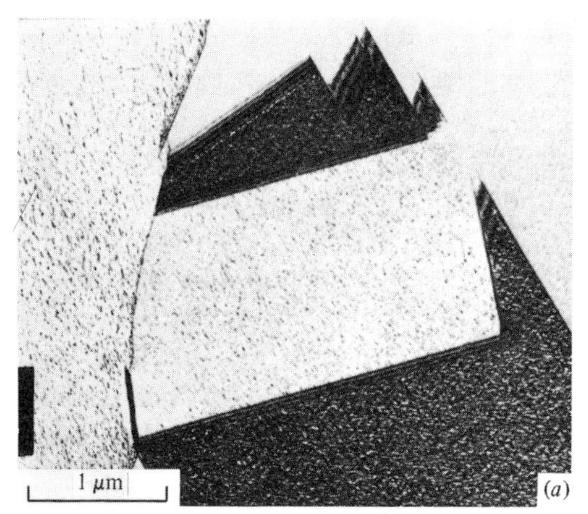

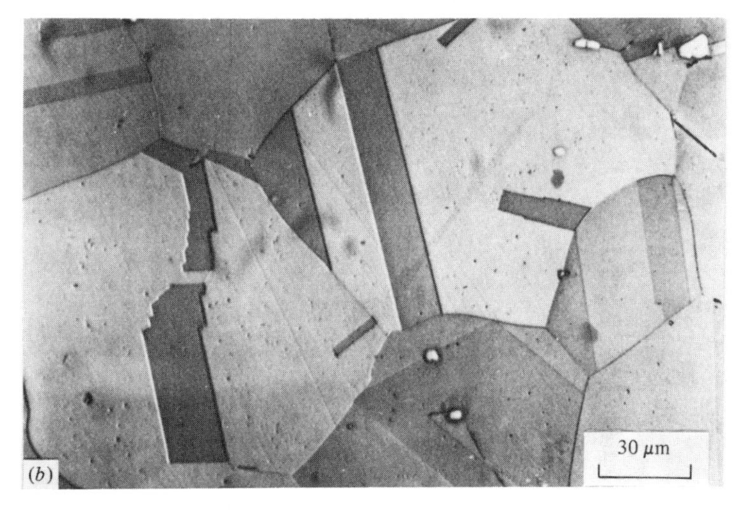

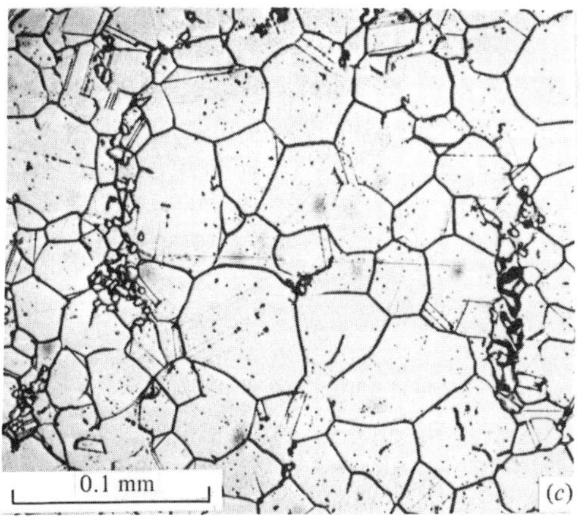

understandable that although a polycrystal consists of essentially anisotropic monocrystals, it can often be regarded as macroscopically isotropic, i.e., without any preferred direction for the properties under consideration.

Crystal defects

The structure of the perfect crystal as described above can only account for elastic deformations and for so-called brittle fracture in which the loss of cohesion occurs without noticeable macroscopic deformation. Plastic deformations and tensile fracture can only be explained by the presence of defects that disturb the crystal lattice.

Point and surface defects

Atomic point defects consist of atoms inserted or substituted in solid solutions and of vacancies (i.e. points in the lattice where atoms are missing). They result in a local distortion of the lattice.

Surface defects are the surfaces of separation between crystals or parts of a crystal where the orientations or natures of the phases are different. Their thickness is of the order of 4–5 atomic 'diameters'. For example,

grain boundaries (or crystal boundaries) in polycrystals,
dislocation loops and cells,
twin crystal boundaries,
interfaces between two phases.

Cohesion defects, consisting of the surfaces of separation in a material which lead to fracture, are micro-cracks and cavities.

Dislocations (line defects)

These are the defects which are mainly responsible for plasticity of metals. A line of dislocation is a defect in the arrangement of atoms which is repeated periodically and which represents the equilibrium state of atoms with slightly different magnetic fields. It is possible to give a schematic representation of such defects in the case of simple cubic crystals:

An edge dislocation or defect which would be created by the translation of the upper part of the crystal (Fig. 1.7).
A screw dislocation or defect which would be created by a local rotation of the upper part of the crystal (Fig. 1.8).
A dislocation loop or line which, for example, joins a pure edge dislocation to a pure screw dislocation (Fig. 1.9).

Dislocations are created during the growth of the crystals; their density, which is very high in most metals and alloys, varies from 10 km cm^{-3} for well-annealed crystals to 10^7 km cm^{-3} for heavily cold-worked samples. Fig. 1.10, which was obtained by scanning electron microscopy, shows a dislocation configuration. Prolonged heat treatments favour the arrangement of dislocations in the form of almost regular three-dimensional cells consisting of crystal planes and nodes, called *Frank* networks.

Fig. 1.7. Edge dislocation of line *DC*.

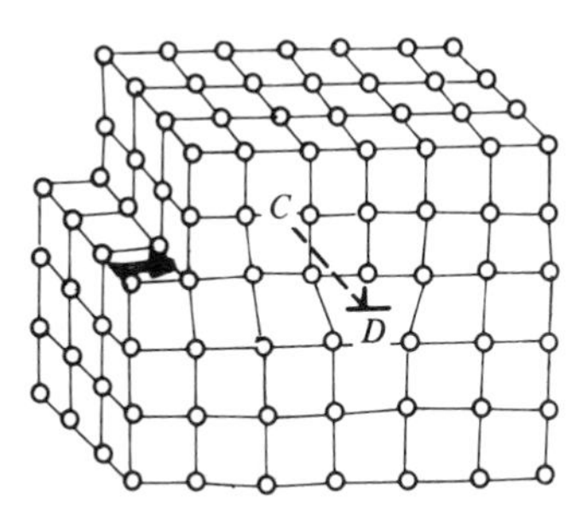

Fig. 1.8. Screw dislocation of line *DV*.

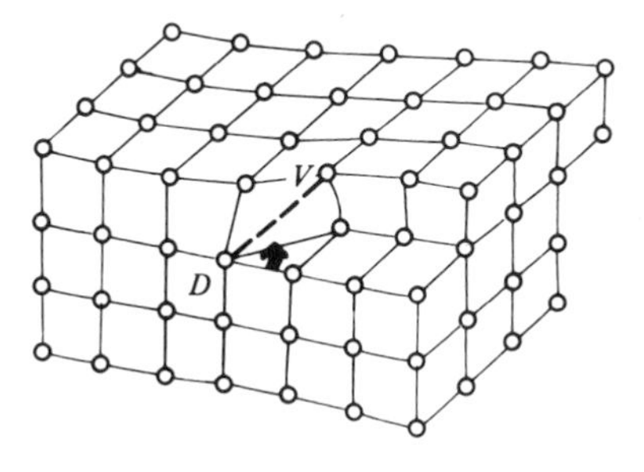

Fig. 1.9. Dislocation loop of line *BD*.

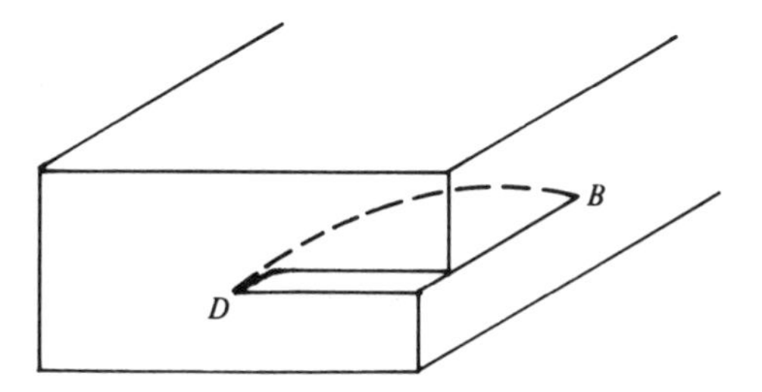

Burgers vector

A dislocation can be characterized by the 'lack of closure' vector of a contour encircling the dislocation line. Let us consider two simple cubic crystals one perfect and the other with an edge dislocation L (Fig. 1.11). In the latter, let us consider a closed loop $ABCD$ around the point L with the sense of traverse defined with respect to an orthogonal reference system. Now let us, reproduce this contour in the perfect crystal by counting the same number of interatomic distances: $A'B'C'D'E'$. To close the contour, the vector $\overrightarrow{E'A'}$ should be added. This vector by definition is the Burgers vector $\vec{b}$.

Burgers vectors have the following properties:

they are independent of the contour;
their magnitude defines the translation which would be necessary to create a dislocation;

Fig. 1.10. Dislocations in a stainless steel. Fe20 Cr–Ni–Al hardened by precipitation in the ordered phase Ni–Al (after Pineau).

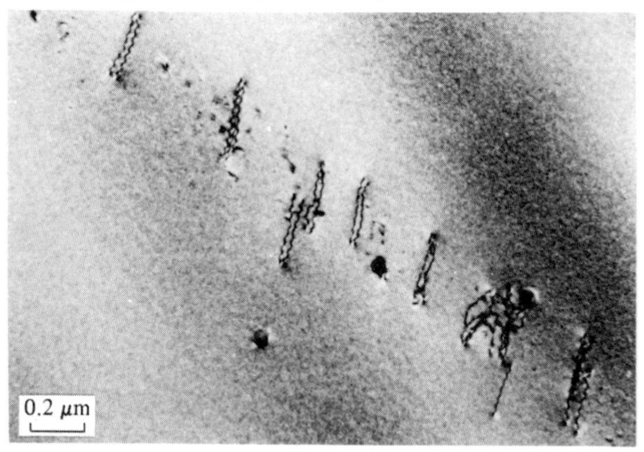

Fig. 1.11. Burgers vector.

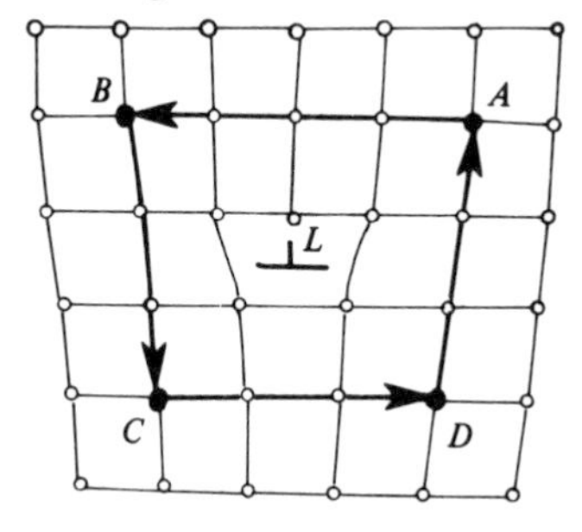

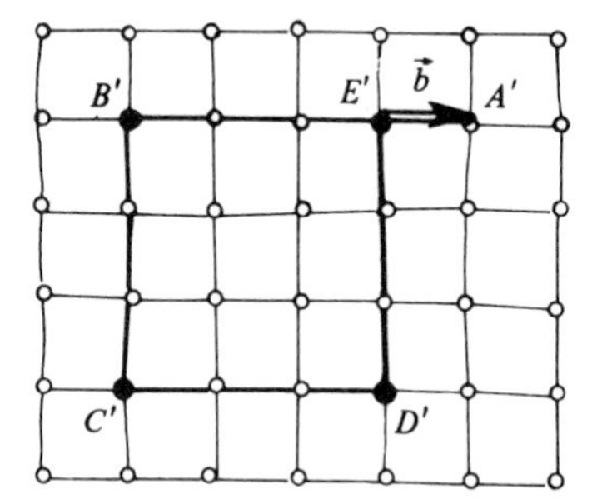

they are constant along the whole dislocation line;

if $\vec{t}$ is a unit vector on the dislocation line then:

- if $\vec{t}$ and $\vec{b}$ are perpendicular to each other, the dislocation is an edge dislocation;
- if $\vec{t}$ and $\vec{b}$ are parallel, the dislocation is a screw dislocation;
- if $\vec{t}$ and $\vec{b}$ are at some angle to each other, the dislocation is a mixed one;

the closing vector of a contour surrounding several dislocations is the sum of the Burgers vector of these dislocations.

Energy associated with dislocations

Since dislocations produce purely elastic distortions of the lattice, and since the neighbourhood of a dislocation can be considered to be a linear elastic continuum, we can calculate the elastic energy w_e stored in the material in the form of internal stresses and strains. For an edge dislocation or a screw dislocation (Fig. 1.12) of unit length and a Burgers vector of magnitude b in a medium with shear modulus μ, we find that the stored energy is of the order of

$$w_e = \mu b^2 \,\mathrm{Jm}^{-1}.$$

In the same way, it is possible to calculate the force $\vec{F}_D$ which would be exerted on a dislocation of unit vector $\vec{t}$ and Burgers vector $\vec{b}$ in a uniform stress field $\boldsymbol{\sigma}$. This is the Peach-Koehler relation

$$\vec{F}_D = (\boldsymbol{\sigma} \cdot \vec{b}) \wedge \vec{t}.$$

1.1.2 *Physical mechanisms of deformation*

Elastic deformation

Elastic deformations occur at the atomic level. The observed macroscopic effect is the result of the variations in the interatomic spacing necessary to balance the external loads, and also of the reversible movements of dislocations. These geometrical adjustments are essentially reversible. In a

Fig. 1.12. Energy associated with dislocations.

purely elastic deformation, the initial configuration of atoms is restored upon the removal of the load.

Permanent deformations

Plastic or viscoplastic permanent deformations occur at the crystal level, and are in addition to elastic deformations; they correspond to a relative displacement of atoms which remains when the load is removed. Depending on the case, the deformations are either purely intragranular (inside the grains) or involve intergranular displacements. The ratio of joint deformation to grain deformation remains small, but it generally increases with increasing temperature and also with decreasing strain rate.

Deformation by slip and twinning

Symmetry planes of the crystal lattice, which are also the reticular planes of the most densely packed atoms, form the parallel planes with the greatest distance between them. It is therefore in these planes that slip due to shear can occur in the direction of maximum shear stress, e.g., planes (1, 1, 1) in FCC crystals; and planes (1, 1, 0) and (1, 1, 2) in the BCC crystals. They occur in the form of parallel slip bands which result in steps on the exterior surface of the samples (Fig. 1.13(*a*)) or in the form of twins which consist of slips symmetric with respect to a plane (Fig. 1.13(*b*)).

Twinning is more characteristic of deformations which occur at average or room temperature. It occurs in CC and CHP crystals in conjunction with slips, but also in FCC crystals where the energy of stacking defects is low. These defects, slips or twinnings, are, in fact, heterogeneous deformations at the crystal level, but may be considered homogeneous at the macroscopic level.

Deformation by dislocation movements

The presence of dislocations considerably reduces the stability of the crystal lattice. Their mobility is the essential cause of permanent deformations, homogeneous at the macroscopic scale.

Slip displacement: when, under an external load, an edge or screw dislocation moves across a crystal, irreversible displacement occurs which is equal to the Burgers vector. Fig. 1.14 shows an edge dislocation in a cubic crystal subjected to a shear stress σ_{12}. The plane defined by the unit vector $\vec{t}$ and the Burgers vector $\vec{b}$ is called the slip plane.

This displacement mechanism requires the breaking of bonds only in the vicinity of the dislocation line, and successively from one atom to the next.

Fig. 1.13. (*a*) Slip bands in nickel based Waspaloy (after Pineau). (*b*) Twinning deformation in zinc (after Pineau).

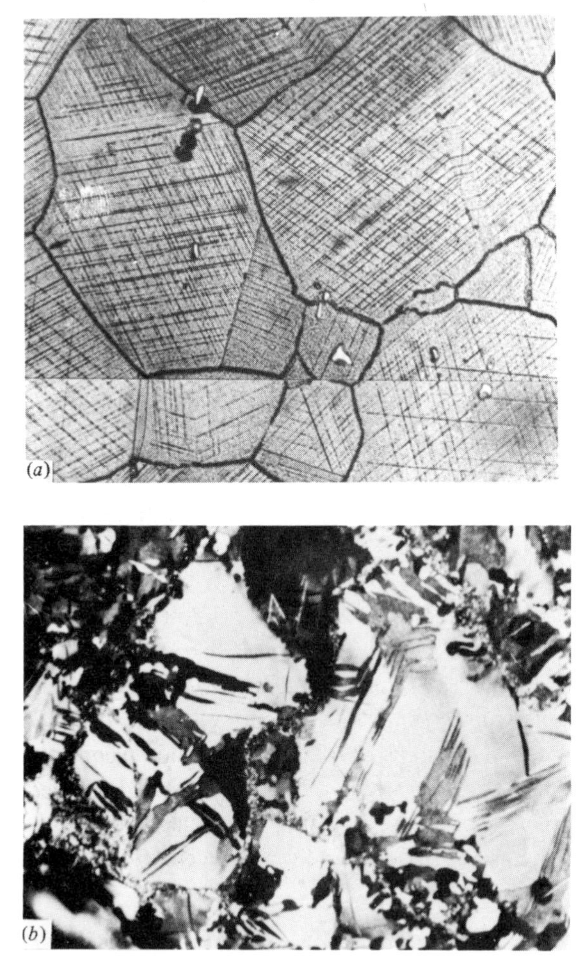

Fig. 1.14. Slip displacement of a dislocation.

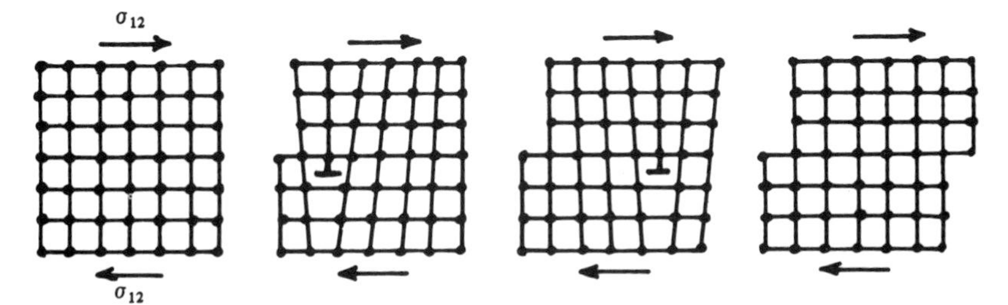

In the more complex case of a dislocation loop, the plane of the loop can digress into another plane (a perpendicular one for example) arising from its 'pure screw' point to avoid an obstacle such as an impurity; this is called a deviated slip.

Climb displacement: an edge dislocation can move perpendicularly to its slip plane with the transport of material. If a void is close to a dislocation line, then a distortion of the lattice resulting from the application of an external load can cause an atom of the lattice to jump by half a plane on the empty side of the crystal and result in the rearrangement of the whole row of atoms. Thus, in this mechanism, the dislocation climbs up by one interatomic space (Fig. 1.15). This displacement mechanism, linked to the diffusion of vacancies or foreign atoms, is favoured by thermal activation; it, therefore, occurs mostly at high temperatures ($T > \frac{1}{3} T_M$, where T_M is the melting temperature).

The rate of dislocation displacement (slip or climb) can be very low or very high depending upon the applied stress, but it cannot be higher than the speed of sound in the material under consideration.

Fig. 1.15. Climb displacement of a dislocation.

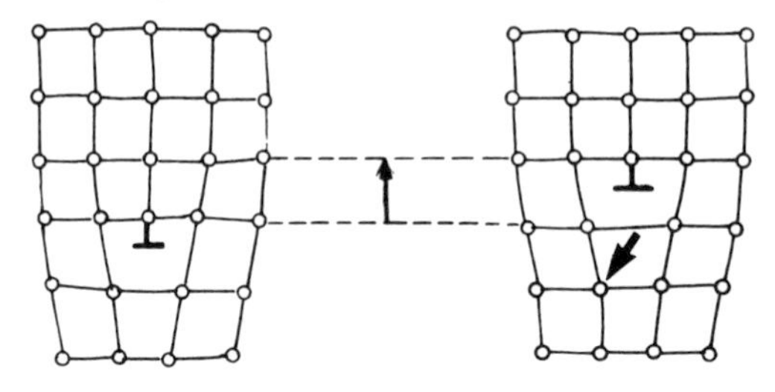

Fig. 1.16. Generation of dislocations by the Frank–Read mechanism.

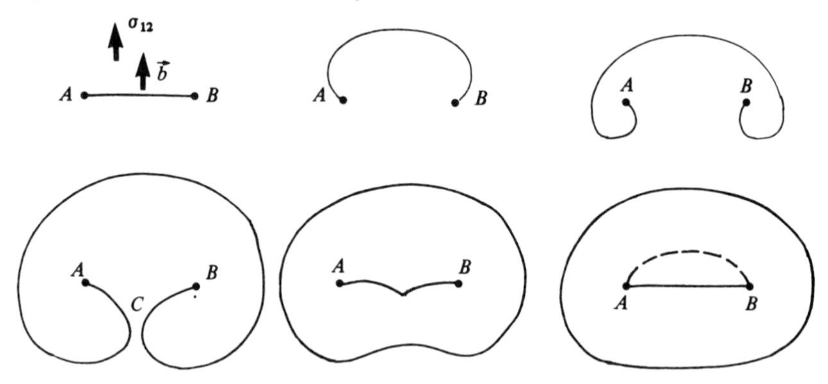

Sources of dislocations

Dislocation movements cause permanent deformations; conversely, large plastic deformations increase the density of dislocations which, in turn, increases the number of blockings and contributes to further hardening. The most important source of dislocations is that described by the Frank–Read mechanism. Let us consider a dislocation segment anchored at A and B, the points of intersection with other dislocations or impurities (Fig. 1.16), of Burgers vector $\vec{b}$ normal to AB and subjected to a shear stress σ_{12} in the same direction as $\vec{b}$. Under the applied stress, the dislocation is forced to move around the points A and B. When the two branches rejoin at C, they have the same Burgers vector but opposite unit vectors. They combine to create a large loop and a new segment AB which in turn generates another loop and so on...

Elasticity limit

It is easy to calculate the elastic limit of a perfect stack of atoms assuming that the shear stress causing one set of atoms to glide over another in a plane, is a sinusoidal function of a period equal to the interatomic distance (the Frenkel hypothesis). Assuming that in the vicinity of the equilibrium position, this shear stress is linearly related to the corresponding deformation through the shear modulus μ, the maximum τ_Y of this stress corresponding to the limit of reversibility of the movement is found to be

$$\tau_Y = \mu/2\pi.$$

This value is of the order of 100 times higher than the elastic limit in shear of common metals, but it is very close to the values observed for 'whiskers' or thin fibres which consist of perfect stacks of atoms. For example,

$$\text{Theoretical elastic limit} \begin{Bmatrix} \text{pure iron} \\ \text{whiskers of iron} \end{Bmatrix} \tau_Y = 84700/2\pi = 13550\,\text{MPa}$$

$$\text{Measured elastic limit} \begin{cases} \text{pure iron} & \tau_Y = 28\,\text{MPa} \\ \text{whiskers of iron} & \tau_Y = 11000\,\text{MPa} \end{cases}$$

The elastic limit of a real single crystal is very difficult to evaluate because of the different mechanisms which can occur:

the dislocation loops generated by the Frank–Read mechanism and which emerge at the free surface of the crystal create an irreversible deformation. The shear stress required to initiate this mechanism is of the order of

$\tau_Y = 2\mu(b/L)$ (L is the length of the source)

a value much lower than the observed elastic limit;

in fact, due to the interatomic bonds broken at the core of dislocations, the so-called Peierls–Nabaro forces prevent the slip and give rise to a higher elastic limit;

the decomposition of dislocations leads to the creation of barriers which also restrict the slip;

the Frank networks offer a resistance to shear with an order of magnitude

$$\tau_Y = \frac{\mu b}{2\pi L} \qquad (L \text{ is the length of the Frank cells});$$

in alloys, solid solutions or precipitates contribute to an increase in the resistance to dislocation movements.

Several of these mechanisms can occur simultaneously to give the values of elastic limit as measured macroscopically. In the case of polycrystals, the disorientation between the crystals and the grain boundaries, prevents the progress of irreversible deformation. Therefore, in this case we must add to the mechanisms described above a mechanism for crossing the grain boundary. By considering the equilibrium of forces due to the blocking of dislocations at the grain boundaries and those applied externally, and taking into account the fact that τ_Y, the shear elastic limit, is the stacking stress in a crystal which can activate a dislocation source at a distance r in an adjacent crystal of mean length d, we obtain the following relation known as the Petch equation

$$\tau_y = \tau_0 + \tau^*(r/d)^{1/2}.$$

It is evident from this equation that the elastic limit varies in direct proportion to the square root of the distance r, and in inverse proportion to the square root of the average crystal length d; τ_0 and τ^* are coefficients which are characteristics of the metal or the alloy.

Deformation of polycrystals

In order to define the basic assumptions of a model well, it is necessary to examine in a schematic way the sequence of mechanisms which occur in the deformation of a polycrystal during a uniaxial external loading, which first increases and then decreases (see Fig. 1.17).

Elastic deformation (0Y in Fig. 1.17)

Elastic deformations result from reversible relative movements of atoms. They are almost independent of the permanent deformations except as regards the microscopic residual stresses which result from the irreversible crystalline slips causing the macroscopic elastic deformations to be slightly different from the sum of the microscopic elastic deformations.

Elastic limit (Point Y in Fig. 1.17)

The elastic limit is characterized by the state of stress or strain which causes the first irreversible movements of dislocations. As they are difficult to detect, we define the elastic limit or the yield stress, σ_Y, by convention to be the stress for a fixed amount of permanent strain; $\varepsilon^p = 0.02\%$ or 0.05% or 0.2% depending upon the precision required (see Chapter 5).

Plastic deformation (YP in Fig. 1.17)

The first slips occur in crystals with crystallographic slip planes oriented at $\alpha = \pi/4$ to the direction of applied stress σ, where the shear stress σ_{12} is maximum. The reorientation of crystals, necessary to ensure compatibility of deformations, activates other slip systems and the deformation appears macroscopically homogeneous. It is a stable deformation and each state is that of an elastoplastic equilibrium.

Hardening or consolidation (YP in Fig. 1.17)

If the stress continues to rise, the dislocation density is increased but the number of barriers is increased even more, so that the deformation

Fig. 1.17. Deformation of a polycrystal.

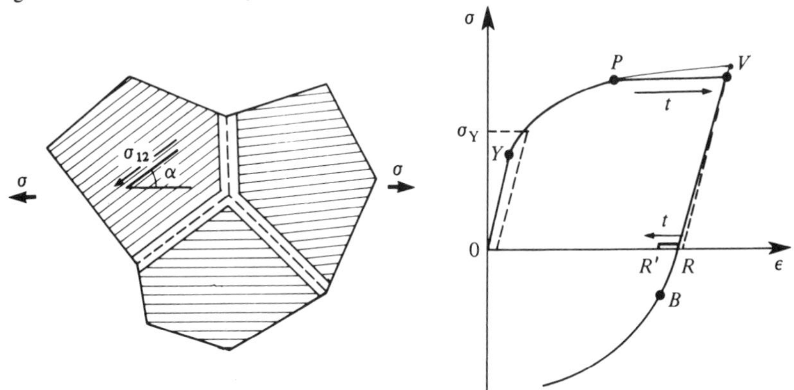

cannot progress unless the load is increased. This increased resistance to slip deformation is the phenomenon of hardening which can also result from the presence of intercrystalline microstresses induced by the incompatibility of grain to grain deformations.

Viscoplastic deformation (PV in Fig. 1.17)

If the stress continues to rise, the slips can cross and follow the grain boundaries. This phenomenon of intergranular slip is favoured by thermal activation and is especially significant at temperatures higher than one third of the absolute melting temperature. A large part of the deformation, however, remains intragranular and is comprised of slip and climb of dislocations. When the deformation can progress under constant stress with no possibility of equilibrium, we have creep flow the rate of which depends on the applied stress; this is the domain of viscoplasticity.

Restoration or recovery (RR′ in Fig. 1.17)

The modifications produced in an aggregate of polycrystals by deformation represent a deviation with regard to the thermodynamic equilibrium which tends to decrease under zero or reversed loading; the concentration of the vacancies or of the interstitial atoms decreases, dislocations of opposite signs neutralize each other, and recrystallization can occur. The recovery is a function of time and is favoured by thermal activation. On a macroscopic level, it manifests itself in a partial recovery of the deformation or in a decrease in the hardening. This description also applies to microstructurally stable materials. However, the phenomenon of ageing can occur which involves instabilities of structures or chemical compositions over a short or a very long time (e.g. ageing of some alloys after heat treatment).

Plastic or viscoplastic incompressibility

Since slip deformations do not alter the crystal structure, the total volume of an aggregate remains unchanged. Only elastic deformations produce a noticeable volume change; the change due to the increase of the density of dislocations, always remains very small. In the same way, a normal stress on the slip plane or a state of hydrostatic stress ($\sigma_1 = \sigma_2 = \sigma_3$) has no effect on permanent deformations. This has been experimentally verified up to pressures of 30 000 atmos (3000 MPa).

Anisotropy induced by permanent deformations

As permanent deformations differ from one crystal to the next, the compatibility of deformations at grain boundaries is assured only by elastic microdeformations; these remain partially locked when the load is removed resulting in self-equilibrated microscopic residual stresses. These microscopic stresses in effect load the crystal at a neutral state and can increase or decrease the external load necessary to produce new slips in a way which varies according to the direction considered: this is the anisotropy which results from permanent deformations. The Bauschinger effect, which is the lowering of the absolute value of the elastic limit in compression following a previous tensile loading, is its simplest manifestation (point B in Fig. 1.17).

Microstructural instabilities

These correspond to an instability of the metallurgic state of the microstructure and manifest themselves in the course of time by variations in strength. They are frequently induced by temperature variations (phase changes, precipitation).

Phase and structural changes generally take place during a variation of temperature. The structure of steel reveals, for example, phase δ (BCC) at very high temperatures, phase γ (FCC) at high temperatures, and phase α (BCC) at low temperatures. Precipitation (precipitate γ' for example, in a phase γ) occurs during a lowering of the temperature, after a partial dissolution of the phase γ' at the higher temperature. The new structure is extremely fine and produces significant hardening (higher resistance to deformation) but it is unstable.

Growth of precipitates occurs under constant temperature when the microstructure is initially in an unstable state. This phenomenon is thermally activated. It manifests itself by an increase in resistance to deformation in the region where dislocations can cross precipitates only by shear and then, for more important precipitate sizes, by a decrease in resistance to deformation when dislocations can go around the particles.

Ageing corresponds to an instability of microstructure which occurs at constant temperature with or without load and usually results in an increase in resistance (precipitation of carbides, for example, the precipitation of $M_{23}C_6$ at the grain boundaries of non-oxydizing austenitic steels). It is possible to demonstrate this phenomenon (as well as that of the precipitate growth) by heating a sample for some time before a creep or hardening test.

1.1.3 ***Physical mechanisms of fracture***

Elastic and permanent deformations, which take place at atomic and crystalline levels respectively maintain the cohesion of the matter. Fracture, by definition, destroys this cohesion by creating surface or volume discontinuities within the material. Therefore, fracture occurs at the larger scale of crystals: we will be concerned with microcracks or cavities of a size expressed in microns or hundredths of a millimetre, with macrocracks of the order of a millimetre, and with cracks which occur at the scale of mechanical structures, measurable in centimetres or decimetres.

The two main basic mechanisms of local fracture are brittle fracture by cleavage and ductile fracture resulting from large localized plastic deformations.

Brittle fracture

In so-called brittle fracture, only the fracture of interatomic bonds, without noticeable overall plastic deformations, is involved. These fractures occur when the local strain energy due to the external loads becomes equal to the energy necessary to pull the atom layers apart. Lattice defects or accidental geometrical imperfections result in stress concentration, and hence they play an essential part in the initiation of the fracture process. Brittle fracture by cleavage consists in a direct separation of particular crystallographic planes (for example, the plane (100) for iron). In a crystal, several parallel

Fig. 1.18. Brittle fracture by cleavage in a weak alloyed steel, scanning micrograph (after Pineau).

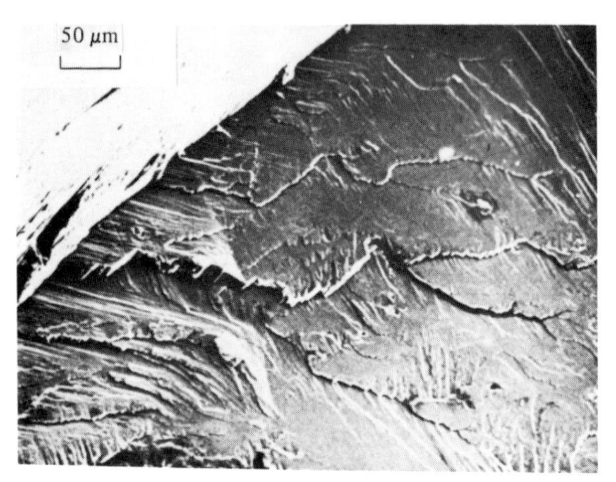

surfaces of cleavage can develop which join each other perpendicularly in the form of 'steps' arranged in 'rivers'. At grain boundaries the cleavage surfaces change direction in order to follow the crystallographic cleavage planes of the next crystal (Fig. 1.18).

FCC crystals are less vulnerable to cleavage; this occurs mostly in CC crystals such as steel with a low carbon content and in CPH crystals such as zinc and magnesium which fracture by cleavage.

Intergranular fracture is a cleavage that follows the grain boundaries. It occurs in two forms:

Brittle fracture itself which occurs at low temperature when impurities segregated at the grain boundaries lower the energy of cohesion at these boundaries.

Intergranular fracture due to creep, observed mostly at average and high temperatures (more than one third of the absolute melting temperature). It can be present with or without significant overall viscoplastic deformation, which justifies the term 'brittle' (Fig. 1.19). The defects leading to this mode of fracture are initially in the form of cavities at the boundaries where dislocations have piled up, subsequently increasing in volume and reproducing themselves, or in the form of decohesions called triple points, which result from the intersection of three crystals, following intercrystalline slips. This intergranular decohesion occurs over a period of time and can be caused, for example, by high temperature creep. When decohesion occurs at several adjacent grain boundaries, we may say that a crystalline crack has been initiated.

Ductile fracture

Ductile fracture arises from the instability which results when very large local deformations occur in the vicinity of crystalline defects. Depending upon the density of these defects, the overall macroscopic deformations may or may not be significant so that a material which shows signs of ductile fracture can have either a ductile or brittle global behaviour. The defects responsible for initiating ductile fracture are:

particles of added elements in alloys;
inclusions;
solution precipitates resulting from a heat treatment;
piling up of dislocations;
grain boundaries and triple points.

In the vicinity of the defects, the external loads create stress concentrations which lead to large plastic deformations. The particle or foreign defect, being generally less ductile than the matrix, produces an instability which results in decohesion at the interface or in fracture due to cleavage and thus initiating a microcrack or a cavity.

The growth of these cavities takes place through plastic slips with local strains that can well exceed 100% to the point where the peduncles of metal separating the holes reach a condition of instability by necking. Their brittle fracture causes coalescence of the cavities which leads to the final fracture. The fracture surface very often looks like a juxtaposition of cups having a precipitate or an inclusion at their bottoms (Fig. 1.20).

Fig. 1.19. (*a*) Intergranular brittle fracture by creep in a Cr–Mo steel (after Pineau). The small cavities on the boundaries are creep cavities. (*b*) Intergranular ductile fracture in a Nickel based Inconel 718 alloy (after Pineau).

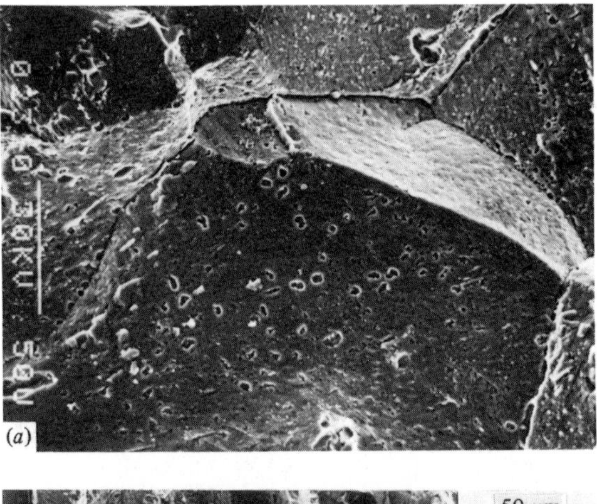

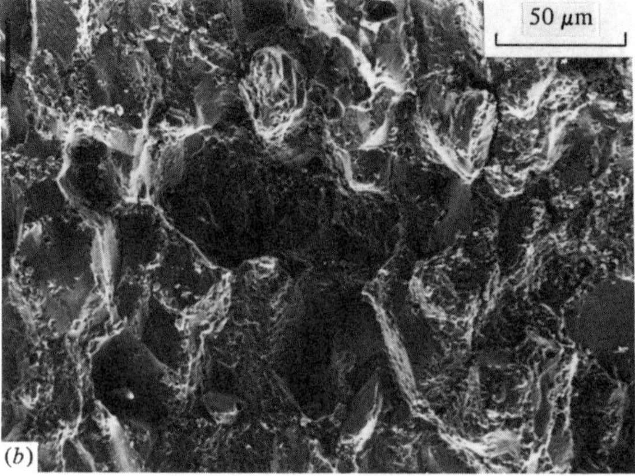

Fatigue fracture of a polycrystal

The fracture of an element under service conditions usually involves different interacting mechanisms. Fatigue failure under cyclic loads is an example worth describing because it clearly distinguishes for modelling purposes the different stages of fracture. Let us consider therefore a polycrystal subjected to a periodic load.

Accommodation stage (or nucleation)

Even if the maximum load is lower than the usual elastic limit, locally, in the vicinity of the defects the stress concentrations create cyclic plastic microdeformations, which, in turn, block further slip by virtue of multiplication of dislocation nodes. This dissipative mechanism produces a local rise in temperature which can then induce the relaxation of microstresses. Depending upon the relative importance of these two processes of hardening and relaxation, a hardening or softening of the material takes place. During this phase slip bands can form, resulting in steps on the surface of the sample.

Fig. 1.20. Transgranular ductile fracture in a 0.30 C–1 Cr–0.25 Mo steel (after Pineau).

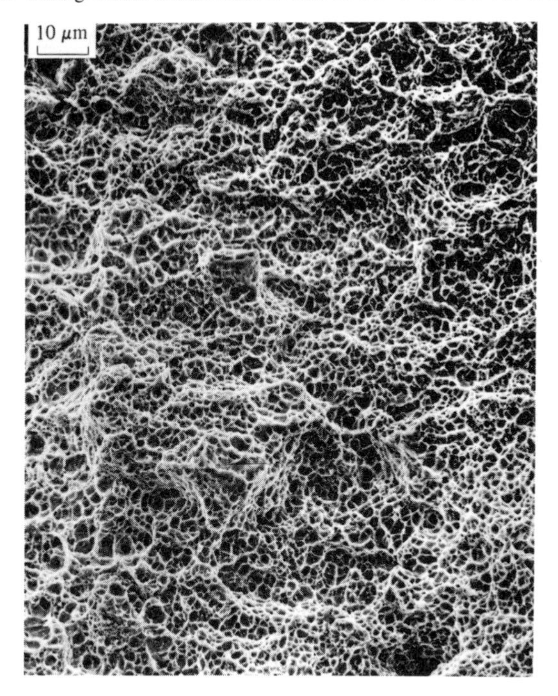

Initiation phase of microcracks (stage 1)

It is difficult to demonstrate and to study the initiation of microcracks due to fatigue. Depending upon the material and the level of the load, several mechanisms can influence the initiation:

dislocation climbs accompanied by the formation of voids;
the formation of permanent slip bands and decohesion, very often at the surface of the sample;
intrusion–extrusion mechanisms.

During this initiation phase, the defects usually follow planes with $\pm 45°$ inclination to the direction of the largest principal stress. Their presence is likely to be accelerated by a hostile environment (oxidation or corrosion).

Growth phase of microcracks (stage 2)

The initiation stage corresponds to progression from stage 1 to stage 2. Schematically, the microcracks have a tendency to orient themselves perpendicular to the direction of the maximum principal stress. This stage usually corresponds to the crossing of the first grain. Thereafter, the microcracks move through the successive grains or, less often, along the grain boundaries. When the crack size becomes significant and has acquired a well-defined direction, it grows in a preferential way, partially unloading other microcracks and generating a high stress concentration at its front. In this stage a macroscopic initiation is taking place: the material can no longer be considered as a homogeneous macroscopic medium.

Growth phase of a macrocrack

The cyclic opening and closing of cracks results in alternating plastic slips at the crack-tip, which in different crystallographic planes form a ridge of cleavage at each growth of the crack. The fractured surface reveals a succession of striations which often (but not always) permits a measurement of the crack-tip progress in each cycle (two good examples are given in Fig. 1.21).

The crack grows in this way until it reaches a critical size at which the cracked part becomes unstable. The crack then propagates rapidly, breaking the work-piece into two or more pieces.

Effects of the environment

The fracture mechanisms described above are influenced by chemical reactions generated by a hostile environment. Corrosion, more particularly stress-corrosion, affects grain boundaries and accelerates intergranular fracture. The presence of hydrogen in steel, even in small amounts, is conducive to formation of cracks by diffusion and concentration in highly stressed zones.

Fig. 1.21. Fatigue ridges (after Pineau): (*a*) stainless austenitic steel; (*b*) Inconel 718 alloy.

The most affected damage phenomenon is that of fatigue in view of the preferential localization of defects which occurs at the surfaces of areas easily accessible to corrosive agents. Generally it is observed that fatigue life in a vacuum (or in an inert gas) is longer than that in the air, the latter being higher than that in a corrosive medium (salt water, media with high sulphur content). The mechanisms involved are of an electrochemical nature resulting in the formation of passive protective layers which can break under the action of mechanical loads (stress-corrosion).

For fatigue cracks, the effect of the environment can result in either slowing down or accelerating the crack growth rate depending on whether the crack-tip has been blunted or whether there has been an increase in stress concentration by virtue of the edge effect of a partial penetration. These phenomena are not yet well known and are difficult to model.

1.2 Other materials

1.2.1 *Polymers*

Structures

Molecules

High polymer solids consist of chain-like molecules. The chain backbone is mostly made up of carbon atoms linked by carbon–carbon covalent bonds with a fracture energy which is of the order of 300 kJ mole^{-1}. Other atoms of hydrogen, oxygen, nitrogen are linked to the atoms of the chain and eventually to the neighbouring atoms by covalent links or by van der Waal type bonds; the latter have a fracture energy which is only in the order of 10 kJ mole^{-1} (Fig. 1.22).

Amorphous and semicrystalline arrangements

The flexibility of polymer molecules allows different types of arrangement. However, as a result, molecules possess a much less regular character than metal crystals.

Fig. 1.22. Terephthalate polyethylene molecule (after Ward).

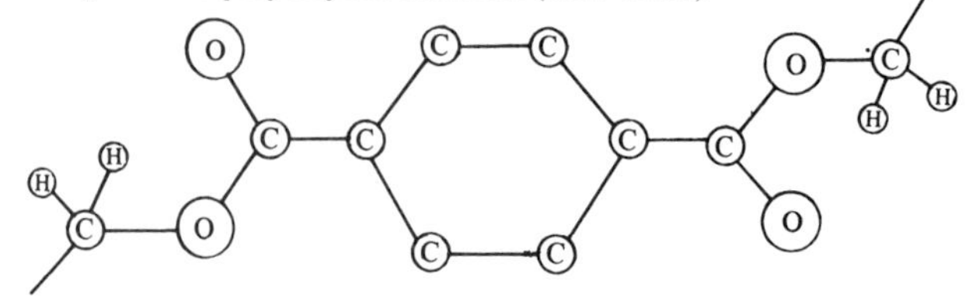

In an amorphous polymer the chains are packed randomly at the molecular scale but have a certain order at a smaller scale (Fig. 1.23). On the macroscopic scale, the amorphous materials are isotropic and often transparent. Examples of such polymers include plexiglass and polystyrene.

During polymerization and under thermomechanical effects the long-chain molecules of certain polymers tend to arrange themselves in packs to form crystallites separated by amorphous areas.

The crystalline wafers form a superstructure with a mean lattice size of 1 μm. They are more resistant than the amorphous zones and have arrangement defects. Partially crystallized polymers are

Fig. 1.23. Diagrams of (*a*) amorphous and (*b*) crystalline arrangements.

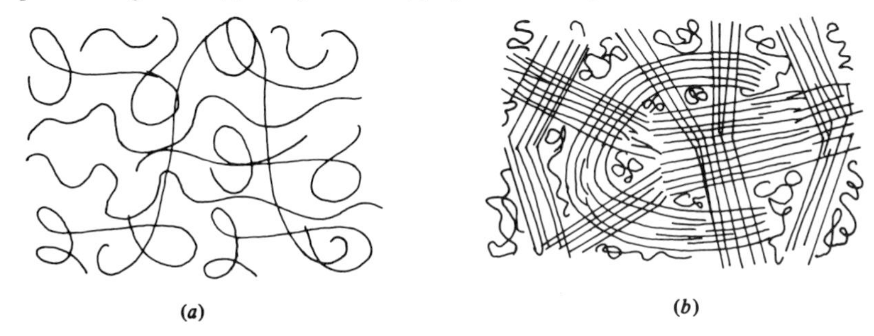

Fig. 1.24. States of polymers.

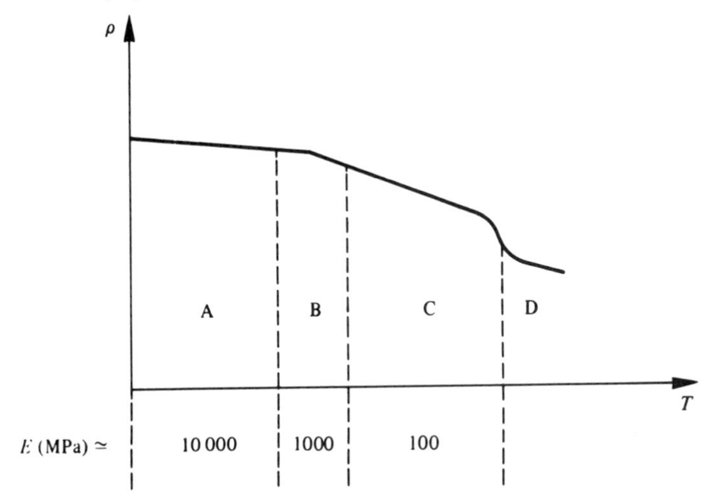

translucent or opaque. Examples of such polymers include nylon, polythene.

Different polymers

Depending upon the temperature, a polymer can occur in four different states corresponding to an increase in the free intermolecular volume with the temperature, and a decrease in the strength of the bonding forces. We may represent these states by following the graph of density ρ or of the elastic modulus E, for example, as a function of the temperature (Fig. 1.24). Room temperature could lie in any one of the four zones depending upon the polymer considered.

Glass state (A): organic glasses allow only very small deformations.

Transition state (B): in this state, we encounter linear thermoplastic polymers (cellulose, polyamides, polyesters, polyvinyls) and cross-linked polymers in which chemical decomposition occurs before melting (phenolic plastics or aminoaldehydic resins).

Rubbery state (C): elastomers consist of very long molecular chains linked together at relatively few junction points.
Fluid state (D).

Fibre–resin composites

The combination of fibres embedded in a resin results in composite materials with a specific resistance which may be higher than that of certain metals. The most commonly used fibres are of glass, Kevlar, boron, and carbon. Their diameters range from a few hundredths of a millimetre to tens of millimetres. The most frequently used resin is epoxy resin. The fibre–matrix interfaces are the weakest zones as regards resistance to deformation and to fracture. The volume of the fibres in the composite varies from 50 to 70% depending on the manner of their arrangement which can be:

unidirectional: the composite consists of parallel fibres in the form of bars or plates (Fig. 1.25(*a*));

bidirectional: by superposition of crossing unidirectional sheets or by means of tissue impregnation, plates or shells are fabricated whose stiffness and resistance vary with the direction of the load (Fig. 1.25(*b*));

tridirectional: it is also possible to manufacture composites with fibres in three directions, for example a composite with a carbon matrix and carbon fibres in three perpendicular directions (Fig. 1.25(*c*)).

By virtue of their construction, all these composites are strongly anisotropic as the stiffness and the strength of the fibres are 50–100 times higher than those of resins.

Fig. 1.25. Structures of composites: (*a*) unidirectional; (*b*) bidirectional; (*c*) tridirectional (after Vançon).

Physical mechanisms of deformation

Viscoelastic deformations

For most polymers, as long as the load remains below a certain value, deformations are elastic but involve dissipation which is globally expressed by viscosity. These deformations result from the relative movement of chain segments in which the bonds are not really destroyed but the rearrangement is thermally activated. The number of atoms free to jump per second is a function of the stress and the temperature; this gives a relation between the strain rate and the stress. This is the phenomenon of viscosity. The deformation, though reversible, stabilizes itself or disappears only after the lapse of a certain time. The limit of reversibility is of the order of a few per cent, but with a low density of bonds (as in elastomers) it can reach 50–100%.

Elastic limit

Above a characteristic load, irreversible reorientation of chain segments and crystallite wafers occurs resulting in strains which occur in addition to the elastic ones and which persist long after the removal of the load.

Permanent deformations

Permanent deformations originate both in the crystallites where dislocations are present and in amorphous regions by rotations of the bonds. Application of a high load initiates the destruction of the substructure by breaking the weakest link while new bonds can develop simply by the coming together of active elements.

> Reorientations promote the creation of new crystallites which act as reinforcement. Under constant stress, consolidation or hardening occurs with asymptotic plastic equilibrium as a function of time.
>
> As the density of the crystallites becomes stable, creep flow under constant stress can occur.
>
> Finally, as for metals, plastic deformation is accompanied by anisotropy by virtue of directional consolidation.

All these mechanisms take place in the domain of very large strains (several tens to several thousands per cent).

Physical mechanisms of fracture

In a very schematic way, we may say that fracture initiation in polymers is marked by the disappearance of molecular bonds under the combined effect of external load and thermal activation. As in the case of deformations, this phenomenon is sensitive to strain rate.

The initiation zones are mainly those zones with defects, impurities or crystalline flaws in crystallites. Fracture development can take the form of a brittle fracture for polymers in the transition or rubbery state.

1.2.2 *Granular material: concrete*

Constitution of hydraulic concrete

Concrete used in construction is composed of aggregates, cement and water. For the occasional handyman, good concrete is made with (by volume) one part cement, two parts dry sand, four parts gravel, and a 'large' amount of water.

Aggregate

Natural aggregates are obtained from rolled alluvial materials or from crushed rocks of very different natures (silicon limestone, granite, ...). An aggregate is characterized by its granular curve which gives, as a function of the mean grain size, the percentage of grains with sizes smaller than a given value.

To make mortar, only sand with a grain diameter less than 5 mm is used. For concrete, fine gravel is also used with a grain diameter up to 25 mm. In very special cases, for example in construction of dams, rocks (up to 100 mm) and rubble (160 mm and more) can be used.

The maximum size d of the aggregate to be used depends on the characteristic size of the work-piece. If h is the smaller dimension of the concrete structure under consideration and e is the distance between the reinforcing bars to be provided, then

$$d \leqslant \tfrac{1}{4}h \quad \text{or} \quad d \leqslant \tfrac{3}{4}e.$$

Cement

Portland cement essentially consists of tricalcium silicate ($\approx 60\%$), bicalcium silicate ($\approx 20\%$), gypsum ($\approx 3\%$), tricalcium aluminate ($\approx 10\%$) and tetracalcium aluminoferrite ($\approx 7\%$) ground into a powder with a grain size of the order of 10–50 μm. The process of hydration renders these constituents into a truly artificial stone.

Hydration of the cement powder

A mixture (by weight) of 66% cement and 33% water yields a paste that hardens with time. At the start of the setting, dry compounds are dissolved to the point of saturation and these then react in solution to give hydrates that precipitate coatings around the dry grains. The crystals which are formed in this process are hydrates of calcium silicate, hydroxide of lime, and various aluminates. The paste remains fluid for a few hours and then hardens with the growth of tobermorite crystals and postlandite originating from silicates. Hydration gives rise to the following phenomena:

an increase in the specific surface of the grains by a factor which might be as high as 100;

shrinkage or contraction of the order of 0.3 mm per metre;

hardening with time: the crushing strength in compression normalized to 1 on the seventh day, is 0.35 on the second day, 1.5 on the 28th day, 1.8 at the end of a year, and 2.2 at the end of a decade.

Fig. 1.26. An example of the structure of concrete (after Mazars, Lab. Audio-visuel ENS de Cachan).

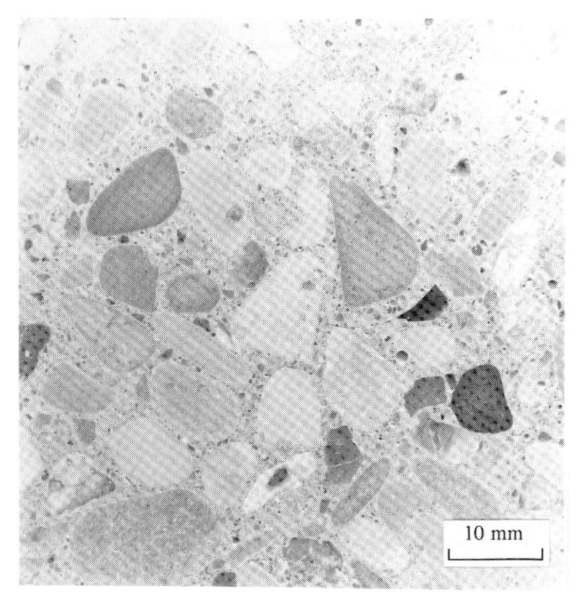

Finally the structure of hardened concrete consists of the following (Fig. 1.26):

aggregates of different sizes,

cement paste binding the aggregates,

flaws consisting of microcracks in the paste–grain bond caused by shrinkage and voids caused by air bubbles entrapped during casting.

Physical mechanisms of deformation and fracture

It is difficult to separate the phenomena of deformation and fracture, because the microcracks and the initial voids, present before any load is applied, grow through the mechanism of brittle fracture and generate permanent deformations (Fig. 1.27).

The phenomenon of permanent deformation, however, is not significant when the load is below a certain value. Deformation in this first stage is the result of almost reversible movements of atoms and can be considered elastic with a low viscosity.

Fig. 1.27. X-ray observations of the evolution of microcracks in concrete under compression (after Robinson).

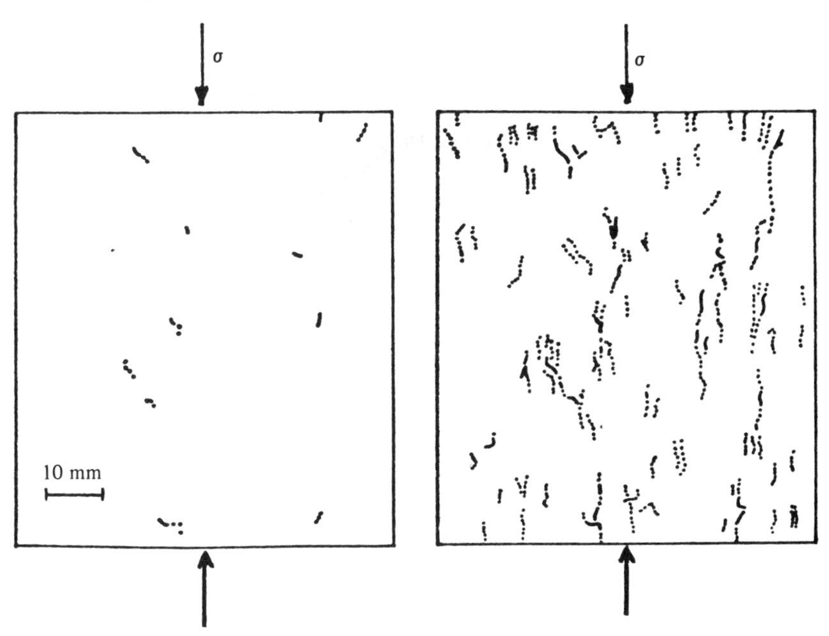

Brittle fracture which arises from a decohesion of the paste-grain is the essential phenomenon of permanent deformation and fracture and is strongly influenced by the nature of the load. The elastic limit or the fracture strength is of the order of 12 times higher in compression than in tension, which explains why concrete is used essentially in compression.

Beyond the load which corresponds to the elastic limit, microcracks in the bonds between the paste and the bigger grains begin to spread toward the periphery of the grains thus generating, at the macroscopic level, permanent deformations in conjunction with the elastic ones. For even higher loads loss of cohesion means that the microcracks reach the paste and in the case of compression loadings, assume a direction parallel to the stress. Slips appear in the crystals of the grains and thus contribute to the permanent deformation which occurs at constant volume. The microcrack damage becomes strongly anisotropic. The final stage is the fracture stage: macroscopic cracks appear, the stress necessary to produce further deformation decreases, the specific volume increases, and the final fracture occurs when microcracks join to form a surface discontinuity through the whole object.

We will see in Chapter 7 that the theory of damage coupled with that of

Fig. 1.28. Cross-section of the trunk of an Oak tree (after Lab. Audio-visuel ENS de Cachan).

elasticity is sufficient to model the nonlinear deformation behaviour of concrete.

1.2.3 ***Wood***

Anatomical structure of wood

Within a broad classification, there are two kinds of wood; resinous and foliaceous, which have different structures, but nevertheless possess common characteristics due to the mode of growth of trees.

Heterogeneity and anisotropy
The growth in diameter of trees results from the proliferation of cells in the cambium which is located under the bark of the tree, with growth activity depending on the season. In spring, the cellular tubes have a large diameter and consist of soft tissue (spring wood); at the end of summer, when the growth activity stops, the tissue becomes lighter and more fibrous (summer wood). These alternate formations result in the concentric annular rings characteristic of the macroscopic appearance of the cross-sections of trees (Fig. 1.28).

To this pseudo-periodic radial inhomogeneity is added another kind which results from the ageing of the central part (or the heartwood) with tissue in this area becoming thicker than in the periphery (the sapwood). The anisotropy of formation is therefore essentially axisymmetric, although with many imperfections which result from knots at the root of the branches, differences of exposure, internal longitudinal compressive stresses and forces and local biological accidents.

Fibres and cells
The macroscopic fibres consist of cells a few hundredths of a millimetre wide and a few millimetres long arranged parallel to the axis of the trunk. The strength of the wood increases as the cross-section of the cells decreases.

In resinous wood, the branches and the conduction tissue through which the sap circulates consist of only one type of cells. In addition, there are resinogen canals which produce the resin and the 'rays' of the wood, which consist of more dense cells arranged in the form of radial laminae (Fig. 1.29(*a*)).

Folliaceous woods have a more complex structure with three or

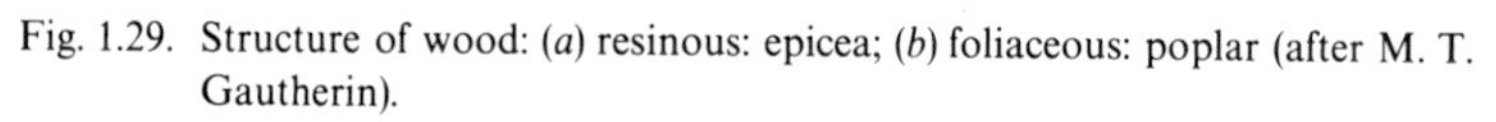

Fig. 1.29. Structure of wood: (*a*) resinous: epicea; (*b*) foliaceous: poplar (after M. T. Gautherin).

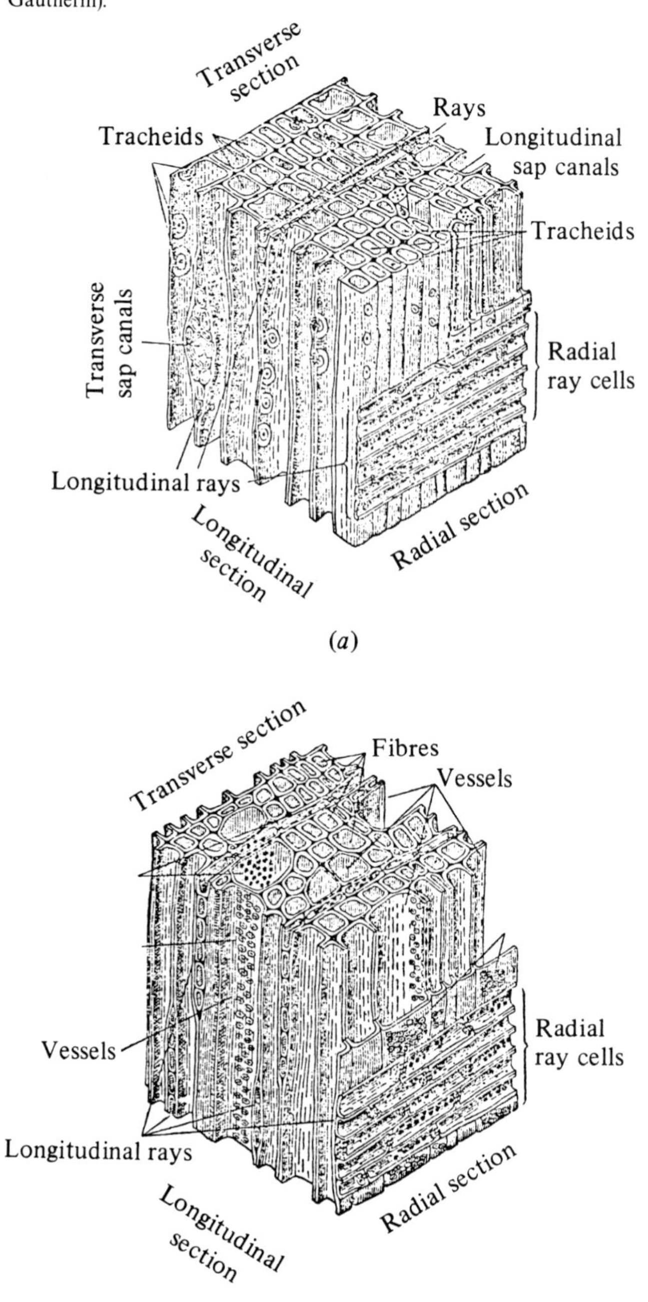

four types of cells with different functions: mechanical support, conduction, and storage (Fig. 1.29(*b*)).

Humidity

Wood cells consist mainly of cellulose (40–50% by weight) and lignin (25–30%). To these two polymers it is necessary to add water which can reach a moisture percentage of 100% of the dry weight. Water exists in chemical combination with lignin, or impregnated in cell membranes, or even in the free form in the cavities of the tissues. Moisture plays a large part in determining the mechanical properties, and therefore, drying of wood is very important. At moisture contents which are below the saturation of the fibres, the elastic characteristics and the resistance to fracture increase as the moisture content decreases.

Physical mechanisms of deformation and fracture

Elastic deformations are caused by reversible deformations of the cells, which are highly dependent upon the moisture content. They can reach 0.1–0.5%, without noticeable permanent deformation, depending upon the type of wood and the direction of the load with regard to the axis of the tree. The viscosity effect is weak.

As with concrete, it is difficult to separate the mechanisms of permanent deformation and fracture, both because macroscopic permanent deformations result from microfractures and because macroscopic fracture occurs, before noticeable deformations are produced. In any case, the principal mechanism of plastic deformation in wood consists of the slip of cells over one another, while fracture is caused by the breaking of cellulose chains, the breaking of a bond by a very high stress in the direction of fibres, and cleavage fracture at much lower stresses perpendicular to the fibres.

Bibliography

Friedel J. *Dislocations.* Pergamon, London (1964).

Hull D. *Introduction to dislocations.* Pergamon, London (1975).

McClintock, F. & Argon A. S. *Mechanical behavior of materials.* Addison-Wesley, Reading (Mass) (1966).

Yokobori T. *An interdisciplinary approach to fracture and strength of solids.* Nordhoff publication, Groningen (1968).

Garofalo F. *Déformation et rupture par fluage.* Dunod, Paris (1970).

Engel L. & Klingele H. *An atlas of metal damage.* Wolf Science Books, New York (1981).

Ward I. M. *Mechanical properties of solid polymers.* Wiley Interscience, New York (1971).
Kausch H. H. *Polymer fracture.* Springer Verlag, Berlin (1978).
Neville A. M. *Properties of concrete.* J. Wiley, New York (1973).
Baron J. & Sautgrey R. *Le béton hydraulique.* Presses de l'ENPC, Paris (1982).
Kollman F. F. P. & Cote W. A., *Principles of wood science and technology.* Springer Verlag, Berlin (1968).

2

ELEMENTS OF CONTINUUM MECHANICS AND THERMODYNAMICS

A chaque phénomène sa variable
A chaque variable sa loi d'évolution

To model the physical phenomena of deformation and fracture, described briefly in Chapter 1, a method based upon general principles which govern the variables representative of the state of the material medium is needed. The objective of this chapter is to present in a condensed form all the basic concepts that will be used in the following chapters. Two types of modelling are necessary: one, the so-called kinematical or mechanical modelling is concerned with the motions and forces in the continuum, and the second, the so-called phenomenological or physical modelling introduces the variables characterizing the phenomena under study.

The presentation given here is the result of the work of Germain as presented in a post graduate course and subsequently incorporated by one of the authors (Lemaitre) in his course, and also contained in *Cours de mécanique des milieux continus* cited in the bibliography of this chapter. This reference contains the details necessary for a deeper understanding of the concepts. Here, only the essential results are given, and the mathematical derivations have been largely omitted. The notation used is almost identical to that in the work referred to above which should help the reader who is anxious to go deeper into the subject.

The framework of mechanics presented here is based on the principle of virtual power. Although the basic idea was presented by D'Alembert as long ago as 1750, it is only, with the development of the variational methods of functional analysis in around 1970 (Duvaut, Lions, Nayroles), that systematic use of this principle has been made. The choice of a particular virtual movement for a given medium leads naturally to consistent definitions of stresses and strains, and to equations of equilibrium with corresponding boundary conditions.

The state of a continuum material depends, in general, on the whole history of its mechanical variables, and the modelling of its behaviour may be based on hereditary or integral laws. In order to obtain a formalism which is directly accessible to the methods of functional analysis, we prefer to adopt the approach of the thermodynamics of irreversible processes by introducing state variables. This approach was initiated first by chemists, and was applied to continuum mechanics by Eckart and Biot around 1950. The thermodynamic potential allows us to define associated variables from observable variables and internal variables chosen for the study of the phenomenon. This then naturally leads to the state laws. The pseudo-potential of dissipation furnishes the complementary laws of evolution for the variables which describe irreversible processes according to the Moreau (1970) formalism.

It should be mentioned, however, that the elementary thermodynamics presented here will be used without consideration of thermal or true dynamic effects for most of the phenomena studied. Nevertheless, this thermodynamic framework will be very useful to guide and limit the possible choices in phenomenological modelling.

2.1 Statement of the principle of virtual power

2.1.1 *Motion and virtual power*

In order to obtain a schematic representation of the forces involved in the phenomenon under study, it is convenient to imagine fictitious or virtual motions and to analyse the resulting work or power. For example:

to analyse the gravity forces acting on a car, we can imagine that it is being lifted (virtual motion from bottom to top);
to analyse the friction forces involved when the car skids on the road, we can imagine that the car is being pulled along with its brakes on (horizontal virtual motion);
to analyse the rigidity of the suspension, we can imagine that the body of the car is being moved with respect to the wheels (relative virtual motion of one point with respect to another).

More generally, a virtual motion of a material medium with respect to a frame of reference is defined at any instant by a velocity vector field dependent on the material point (M):

$$\vec{\hat{v}}(M).$$

By a suitable choice of $\vec{\hat{v}}(M)$ the mechanical phenomenon can be represented more or less precisely.

The virtual power of a system of forces in a given virtual motion is, by definition, a linear continuous function of the scalar value of $\vec{\hat{v}}(M)$ equal to the work done per unit time in the considered phenomenon:

$$\hat{P}(\vec{\hat{v}}(M)).$$

2.1.2 *Frames of reference and material derivatives*

Motion may be described in terms of either Eulerian variables, or, alternatively, in terms of Lagrangian variables.

> The Eulerian variables are the current time t and the coordinates of the current position of the material point M; these variables identify the current configuration. The velocity of the point is expressed by:
>
> $$\vec{v}(M) = \vec{v}(x_1(t), x_2(t), x_3(t), t).$$
>
> The Lagrangian description is based on the current time t and the coordinates of the initial position M_0 of the material point M. This identification is done with respect to the initial configuration and the velocity is expressed by:
>
> $$\vec{v}(M) = \vec{v}(x_1^0, x_2^0, x_3^0, t).$$

The concept of a derivative following the particle, called the material derivative, is applicable to any quantity defined on a set whose motion is being followed.

> In terms of Lagrangian variables, the material derivative is identical to the partial derivative with respect to time. For example, the acceleration of a point M is expressible as:
>
> $$\vec{\gamma} = \mathrm{d}\vec{v}/\mathrm{d}t = \partial\vec{v}/\partial t.$$
>
> In contrast, in terms of Eulerian variables, there is no such identity because the current coordinates of the point M depend on time. One then has:
>
> $$\vec{\gamma} = \frac{\mathrm{d}\vec{v}}{\mathrm{d}t} = \frac{\partial\vec{v}}{\partial t} + \frac{\partial\vec{v}}{\partial\vec{x}}\frac{\partial\vec{x}}{\partial t} = \frac{\partial\vec{v}}{\partial t} + \vec{v}\cdot\overrightarrow{\mathrm{grad}}\ \vec{v}$$

or in index notation,

$$\gamma_i = \frac{\mathrm{d}v_i}{\mathrm{d}t} = \frac{\partial v_i}{\partial t} + v_{i,j}v_j$$

where we have used the Einstein summation convention:

$$v_{i,j}v_j = v_{i,1}v_1 + v_{i,2}v_2 + v_{i,3}v_3.$$

2.1.3 *Principle of virtual power (Germain, 1972)*

It is possible in an isolated material medium to distinguish between the external forces acting on the medium and the internal forces which represent the bonds existing between all possible parts of the medium.

Axiom of objectivity

The virtual power of internal forces in any rigid movement of the medium is zero.

Axiom of equilibrium (static or dynamic)

For any material identified in an absolute frame of reference, at every instant and for any virtual movement, the virtual power of the acceleration quantities (i.e., inertia forces) $\hat{P}_{(a)}$ is equal to the sum of the virtual power of internal forces $\hat{P}_{(i)}$ and of external forces $\hat{P}_{(x)}$

- $$\hat{P}_{(i)} + \hat{P}_{(x)} = \hat{P}_{(a)}.$$

These two axioms embody the fundamental law of dynamics.

Fig. 2.1. Isolated material medium.

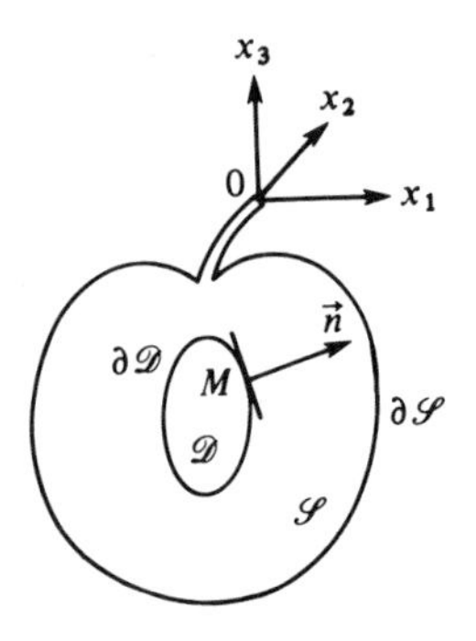

2.2 Method of virtual power

2.2.1 *Rates of strain and stress*

The application of the principle of virtual power to an isolated region for a particular virtual motion and a choice of the linear forms which appear in the expression for virtual power lead directly to the mechanical equations of continua.

Let there be a region $\mathscr{D}$ with boundary $\partial\mathscr{D}$ in the interior of volume $\mathscr{S}$ with boundary $\partial\mathscr{S}$, and let $\vec{n}$ be the outward normal at a point M of $\partial\mathscr{D}$ in an orthogonal frame of reference $(0, x_1, x_2, x_3)$ (see Fig. 2.1).

Choice of a virtual motion, theory of first gradient

The first choice consists in defining the space of virtual velocities. Here a simple vector field of displacement rate $\vec{\hat{v}}(M)$ is sufficient as long as micro-rotations are not taken into account (micropolar media excluded). The second choice is concerned with the terms to be included in the virtual power calculations; a single term in $\vec{\hat{v}}(M)$ is insufficient to describe the behaviour of deformable bodies. The simplest idea consists in adding to it a field of velocity gradients, $\overrightarrow{\text{grad}}\ \vec{\hat{v}}$ or $\hat{v}_{i,j}$. Thus, for the theory of first gradient, the virtual motion is given by:

$$\vec{\hat{v}}(M) \text{ and } \overrightarrow{\text{grad}}\ \vec{\hat{v}}.$$

By decomposition of $\overrightarrow{\text{grad}}\ \vec{\hat{v}}$ into its symmetric and antisymmetric parts, one obtains the rate of deformation tensor $\hat{\mathbf{D}}$ (of second order) and the rate of rotation tensor $\hat{\mathbf{\Omega}}$, as follows (the symbol $(\ )^{\mathrm{T}}$ denotes a transpose):

$$\overrightarrow{\text{grad}}\ \vec{\hat{v}} = \tfrac{1}{2}[\overrightarrow{\text{grad}}\ \vec{\hat{v}} + (\overrightarrow{\text{grad}}\ \vec{\hat{v}})^{\mathrm{T}}] + \tfrac{1}{2}[\overrightarrow{\text{grad}}\ \vec{\hat{v}} - (\overrightarrow{\text{grad}}\ \vec{\hat{v}})^{\mathrm{T}}]$$

$$\hat{\mathbf{\Omega}} = \tfrac{1}{2}[\overrightarrow{\text{grad}}\ \vec{\hat{v}} - (\overrightarrow{\text{grad}}\ \vec{\hat{v}})^{\mathrm{T}}]$$

$$\hat{\mathbf{D}} = \tfrac{1}{2}[\overrightarrow{\text{grad}}\ \vec{\hat{v}} + (\overrightarrow{\text{grad}}\ \vec{\hat{v}})^{\mathrm{T}}]$$

or

$$\hat{D}_{ij} = \tfrac{1}{2}(\hat{v}_{i,j} + \hat{v}_{j,i}).$$

Virtual power of internal forces

The virtual power of internal forces is defined by the integral over the whole domain $\mathscr{D}$ of a volume density which is supposed to contain *a priori* three terms in $\vec{\hat{v}}$, $\hat{\mathbf{D}}$ and $\hat{\Omega}$ associated respectively to a vector $\vec{f}^{\,*}$, and two second order tensors, $\boldsymbol{\sigma}$ which is symmetric, and $\boldsymbol{\Gamma}$ which is antisymmetric:

$$\hat{P}_{(i)} = -\int_{\mathscr{D}} (\vec{f}^* \cdot \vec{\hat{v}} + \boldsymbol{\sigma} : \hat{\mathbf{D}} + \boldsymbol{\Gamma} : \hat{\boldsymbol{\Omega}})\, \mathrm{d}V$$

where the symbol : denotes the tensorial product contracted on two indices and the minus sign is in accordance with the convention used in thermodynamics.

The first axiom of the principle of virtual power requires that for a rigid motion of the solid:

in translation $\vec{v} \neq 0$, $\mathbf{D} = 0$, $\boldsymbol{\Omega} = 0$;
in rotation $\vec{v} = 0$, $\mathbf{D} = 0$, $\boldsymbol{\Omega} \neq 0$

the power $\hat{P}_{(i)}$ is zero.

Application of the fundamental lemma of the physics of continuum media to any function $f(M)$ defined and continuous in $\mathscr{D}$ gives

$$\int_{\mathscr{D}} f(M)\, \mathrm{d}V = 0 \quad \forall \mathscr{D} \quad \text{in} \quad \mathscr{S} \Rightarrow f(M) = 0 \quad \text{in} \quad \mathscr{D},$$

which leads to $\vec{f}^* \cdot \vec{\hat{v}} = 0$ and $\boldsymbol{\Gamma} : \boldsymbol{\Omega} = 0$ regardless of $\vec{\hat{v}}$ and $\boldsymbol{\Omega}$ so that

$$\vec{f}^* = 0, \qquad \boldsymbol{\Gamma} = 0$$

and

$$\hat{P}_{(i)} = -\int_{\mathscr{D}} \boldsymbol{\sigma} : \hat{\mathbf{D}}\, \mathrm{d}V.$$

It will be seen that the tensor $\boldsymbol{\sigma}$ introduced above is the Cauchy stress tensor. The first axiom of virtual power and the choice of virtual motion according to the theory of first gradient are together equivalent to the hypothesis generally made to introduce the stress tensor. In this hypothesis, it is supposed that the internal forces can be represented schematically by a surface density of cohesive forces $\vec{T}$ which represent action at very short distance. By enlarging upon this hypothesis, it is possible to define a stress vector dependent on the material point M under consideration, on time t, and linearly on the normal $\vec{n}$ to $\partial\mathscr{D}$ at M:

$$\vec{T}(M, t, \vec{n}).$$

The stress tensor $\boldsymbol{\sigma}$ can be shown to be given by:

$$\vec{T}(M, t, \vec{n}) = \boldsymbol{\sigma}(M, t) \cdot \vec{n}.$$

The tensor is symmetric and the components σ_{ij} of its matrix representation in the frame of reference $(0, x_1, x_2, x_3)$ are given by

$$T_i = \sigma_{ij} n_j.$$

These components are shown schematically in Fig. 2.2. On each face, the stress vector is decomposed into one normal component and two shear components.

The method of virtual power enables us to bring more precision to the hypotheses. Moreover, it allows us, if necessary, to construct theories based on other hypotheses: for example, the introduction of a rotation field in the space of virtual velocities for a micropolar medium, the introduction of a second gradient in the axioms of virtual power to describe the deformations more precisely.

Virtual power of external forces

The external forces consist of:

Fig. 2.2. Components of the stress vector and of the representative matrix of the stress tensor (after Germain-Muller).

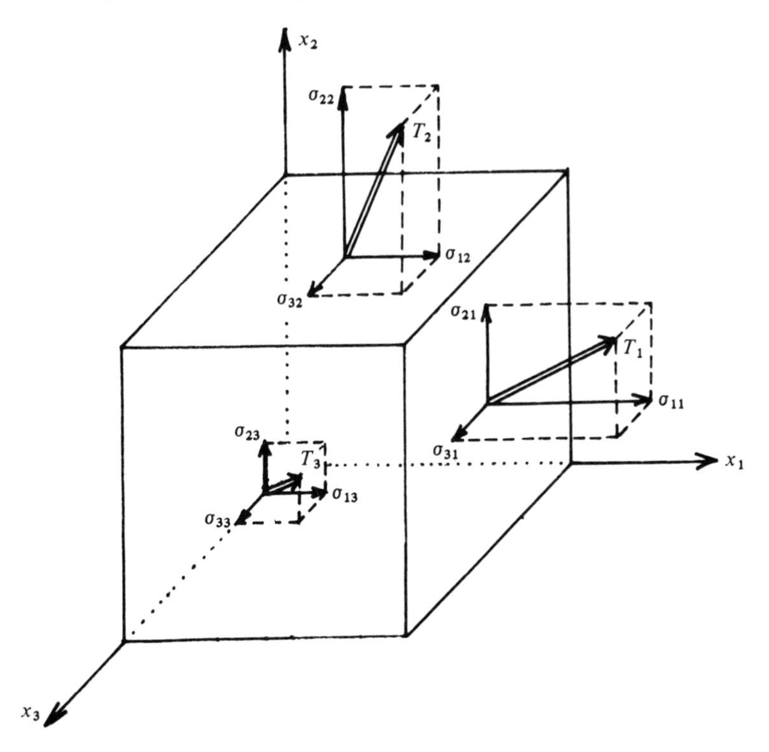

the forces exerted at a distance by systems external to $\mathscr{S}$, supposed to be defined by a volume density of force $\vec{f}$ (not of a tensor or of a couple);
the contact forces schematically represented by a surface density $\vec{T}$ (Cauchy's stress hypothesis)

$$\hat{P}_{(x)} = \int_{\mathscr{D}} \vec{f} \cdot \vec{\hat{v}} \, dV + \int_{\partial\mathscr{D}} \vec{T} \cdot \vec{\hat{v}} \, dS.$$

We may add a term involving $\overrightarrow{\text{grad}}\ \vec{v}$, but this can rigorously be shown to be zero.

Virtual work of acceleration quantities

If $\vec{\gamma}$ is the acceleration vector of a particle M, and ρ is the mass density (the values for some materials are given in Table 2.4), the power of the acceleration quantities is expressed by:

$$\hat{P}_{(a)} = \int_{\mathscr{D}} \vec{\gamma} \cdot \vec{v} \rho \, dV.$$

2.2.2 ***Equations of equilibrium***

By applying the axiom of equilibrium of the principle of virtual power to the region $\mathscr{D}$ we obtain the following equilibrium equations:

$$\hat{P}_{(i)} + \hat{P}_{(x)} = \hat{P}_{(a)} \quad \forall \vec{\hat{v}}$$

- $$-\int_{\mathscr{D}} \boldsymbol{\sigma} : \hat{\mathbf{D}} \, dV + \int_{\mathscr{D}} \vec{f} \cdot \vec{\hat{v}} \, dV + \int_{\partial\mathscr{D}} \vec{T} \cdot \vec{\hat{v}} \, dS = \int_{\mathscr{D}} \vec{\gamma} \cdot \vec{\hat{v}} \rho \, dV$$

or

$$-\int_{\mathscr{D}} \sigma_{ij} \hat{D}_{ij} \, dV + \int_{\mathscr{D}} f_i \hat{v}_i \, dV + \int_{\partial\mathscr{D}} T_i \hat{v}_i \, dS = \int_{\mathscr{D}} \gamma_i \hat{v}_i \rho \, dV.$$

In order to take advantage of the fact that this identity holds for arbitrary virtual motions $\vec{\hat{v}}$, it is essential to have $\vec{\hat{v}}$ itself in the first term. This is achieved by applying the divergence theorem (or integration by parts) as follows:

$$-\int_{\mathscr{D}} \boldsymbol{\sigma} : \hat{\mathbf{D}} \, dV = -\int_{\mathscr{D}} \boldsymbol{\sigma} : \overrightarrow{\text{grad}}\ \vec{v} \, dV$$

$$= -\int_{\partial\mathscr{D}} \boldsymbol{\sigma} \cdot \vec{\hat{v}} . \vec{n} \, dS + \int_{\mathscr{D}} \text{div}\, \boldsymbol{\sigma} \cdot \vec{\hat{v}} \, dV$$

where $\vec{n}$ is the outward unit normal at the boundary $\partial\mathscr{D}$ of the region $\mathscr{D}$. Then, the equation corresponding to the second axiom, may be written as:

$$-\int_{\partial\mathscr{D}} \boldsymbol{\sigma}\cdot\vec{\hat{v}}\cdot\vec{n}\,\mathrm{d}S + \int_{\mathscr{D}} \operatorname{div}\boldsymbol{\sigma}\cdot\vec{\hat{v}}\,\mathrm{d}V + \int_{\mathscr{D}} \vec{f}\cdot\vec{\hat{v}}\,\mathrm{d}V + \int_{\partial\mathscr{D}} \vec{T}\cdot\vec{\hat{v}}\,\mathrm{d}S = \int_{\mathscr{D}} \vec{\gamma}\cdot\vec{\hat{v}}\rho\,\mathrm{d}V$$

or

$$\int_{\mathscr{D}} (\operatorname{div}\boldsymbol{\sigma} + \vec{f} - \rho\vec{\gamma})\cdot\vec{\hat{v}}\,\mathrm{d}V + \int_{\partial\mathscr{D}} (\vec{T} - \boldsymbol{\sigma}\cdot\vec{n})\cdot\vec{\hat{v}}\,\mathrm{d}S = 0 \quad \forall\vec{\hat{v}}.$$

This identity, in accordance with the fundamental lemma, can be satisfied for any field $\vec{\hat{v}}$ if and only if:

- $$\operatorname{div}\boldsymbol{\sigma} + \vec{f} - \rho\vec{\gamma} = 0 \quad \text{in} \quad \mathscr{D}$$

or

$$\sigma_{ij,j} + f_i - \rho\gamma_i = 0 \quad \text{in} \quad \mathscr{D},$$

and

$$\vec{T} = \boldsymbol{\sigma}\cdot\vec{n} \quad \text{or} \quad T_i = \sigma_{ij}n_j \quad \text{on} \quad \partial\mathscr{D}.$$

The first equation in its two forms expresses the local static or dynamic equilibrium. The second one defines the stress vector as the surface density of the forces introduced. It shows that $\boldsymbol{\sigma}$ is really the Cauchy stress tensor: a second order symmetric tensor. It also yields the boundary conditions on forces if the axiom of equilibrium of virtual power is applied to the whole region $\mathscr{S}$ under consideration.

If $\vec{T}^{\mathrm{d}}$ represents the density of the applied forces on the boundary $\partial\mathscr{S}$ of the region $\mathscr{S}$, the same argument as the preceding one leads to:

$$\vec{T}^{\mathrm{d}} = \sigma\cdot\vec{n}$$

$$\text{or } T_i^{\mathrm{d}} = \sigma_{ij}n_j \quad \text{on} \quad \partial\mathscr{S}.$$

2.2.3 *Strains and displacements*

The method of virtual power introduces in a natural way the kinematical variables: the rate of displacement $\vec{v}$ and the rate of deformation $\mathbf{D}$, defined at every instant t. Beyond this description of motion, the study of a solid continuum requires the characterization of its current state of deformation with respect to a reference configuration which may be chosen to be the initial configuration, or the relaxed configuration if elastic and inelastic phenomena are present simultaneously. It is then necessary to introduce the displacement vector and the strain tensor for finite deformations, or for infinitesimal deformations when these quantities are small.

Assumptions of small strains and small displacements

In the case of infinitesimal deformations, the distinction between the Eulerian and Lagrangian variables can be ignored. We may then neglect second order terms in the analysis and pass from the latter to the former set of variables.

Let $\vec{u}$ then be the displacement vector defined by:

$$\vec{u} = \int_0^t \vec{v}\,\mathrm{d}t \qquad \text{or} \qquad \vec{v} = \dot{\vec{u}},$$

where $t = 0$ corresponds to the reference configuration. Let $\boldsymbol{\varepsilon}$ be the strain tensor defined by:

$$\boldsymbol{\varepsilon} = \int_0^t \mathbf{D}\,\mathrm{d}t \qquad \text{or} \qquad \mathbf{D} = \dot{\boldsymbol{\varepsilon}}.$$

Invoking the definition of the tensor $\mathbf{D}$, we may then write:

$$\boldsymbol{\varepsilon} = \int_0^t \tfrac{1}{2}[\overrightarrow{\text{grad}}\ \vec{v} + (\overrightarrow{\text{grad}}\ \vec{v})^{\mathrm{T}}]\,\mathrm{d}t.$$

The assumption of small deformations results in the equivalence of the Eulerian and Lagrangian variables, and thus the material derivative is a partial derivative and the operations of gradient and integration become commutative:

$$\int_0^t \overrightarrow{\text{grad}}\ \vec{v}\,\mathrm{d}t = \overrightarrow{\text{grad}}\left(\int_0^t \vec{v}\,\mathrm{d}t\right) = \overrightarrow{\text{grad}}\ \vec{u}$$

whence

$$\boldsymbol{\varepsilon} = \tfrac{1}{2}[\overrightarrow{\text{grad}}\ \vec{u} + (\overrightarrow{\text{grad}}\ \vec{u})^{\mathrm{T}}].$$

The strain tensor $\boldsymbol{\varepsilon}$ defined from the deformation rate tensor $\mathbf{D}$ is therefore the symmetric part of the gradient of displacement $\vec{u}$ defined from the velocity $\vec{v}$.

The equations of continuum mechanics can be summarized in the case of small strains and small displacements by the following equations:

- $$\operatorname{div}\boldsymbol{\sigma} + \vec{f} = \rho\vec{\gamma}$$

or $\sigma_{ij,j} + f_i = \rho\ddot{u}_i$ and

- $$\boldsymbol{\varepsilon} = \tfrac{1}{2}[\overrightarrow{\text{grad}}\ \vec{u} + (\overrightarrow{\text{grad}}\ \vec{u})^{\mathrm{T}}]$$

or $\varepsilon_{ij} = \frac{1}{2}(u_{i,j} + u_{j,i})$.

To solve a problem we need to add to the above equations:

the boundary conditions, generally given in terms of tractions, $\boldsymbol{\sigma}\cdot\vec{n} = \vec{T}^{\mathrm{d}}$, on part $\partial\mathcal{S}_{\mathrm{F}}$ of the surface, and in terms of displacements, $\vec{u} = \vec{u}^{\mathrm{d}}$, on the complementary part $\partial\mathcal{S}_{\mathrm{u}}$ of the surface.

the constitutive laws which characterize the physics of the medium in terms of the relations between stress and strain; these will be studied in Chapters 4, 5 and 6.

In practice, the assumption of small strains can be applied as long as the strain modulus remains smaller than the order of magnitude of the precision with which the computations are to be done, i.e. when:

$$|\varepsilon| < (2\text{–}5) \times 10^{-2}.$$

This will be the case for most of the topics treated in this book.

Finite deformations with large strains and large displacements

When the assumption of small strains and small displacements is untenable it is necessary to introduce new concepts regarding the geometry of deformation, which are described in detail in the works cited in the references of this chapter; here, they are only summarized.

Let M be a material point in the current configuration of the body, at a time t, represented by Lagrangian variables with respect to the initial configuration M_0 at t_0 (Fig. 2.3):

$$M = M(M_0, t_0, t)$$

and let $\mathcal{F}$ be the transformation which carries M_0 to M by a vector $\vec{x}$:

$$\vec{x}(M) = \mathcal{F}(\vec{x}(M_0), t_0, t).$$

The deformation may be expressed by means of the linear tangential transformation which expresses the transformation in the vicinity of M_0 and M:

$$\mathrm{d}\vec{x}(M) = \mathbf{F}\cdot\mathrm{d}\vec{x}(M_0)$$

where $\mathbf{F}$ is the deformation gradient tensor of $\mathcal{F}$

$$\mathbf{F} = \partial\mathcal{F}/\partial\vec{x}(M_0).$$

If we chose the reference state as the current configuration in M, we are led to characterize the deformation by the Almansi–Euler strain tensor. If we choose the reference state to be the initial configuration in M_0, we may then introduce the Green–Lagrange tensor $\mathbf{\Delta}$ by means of the following definition:

$$\mathbf{\Delta} = \tfrac{1}{2}(\mathbf{F}^{\mathrm{T}} \cdot \mathbf{F} - \mathbf{1})$$

where a dot denotes the contracted product of the tensors on one index, and where $\mathbf{1}$ is the unit second order tensor.

An important aspect of large deformations, which concerns the constitutive laws for elastoplastic (or viscoelastic) behaviour, is linked to the separation of (total) strain into elastic (reversible) and inelastic (irreversible) strains. The rule of partition used for small strain $\boldsymbol{\varepsilon} = \boldsymbol{\varepsilon}^{\mathrm{e}} + \boldsymbol{\varepsilon}^{\mathrm{p}}$ (see Section 4.1) can only be generalized for large deformations through the introduction of a relaxed intermediate configuration. Fig. 2.3 gives a schematic representation of the decomposition of $\mathbf{F}$ into an inelastic deformation $\mathbf{P}$ between the initial and intermediate configuration, and an elastic deformation $\mathbf{E}$ between the intermediate configuration and the current one:

$$\mathbf{F} = \mathbf{E} \cdot \mathbf{P}.$$

The rate of deformation is defined with respect to the current configuration:

$$\mathbf{D} = \tfrac{1}{2}[\overrightarrow{\mathrm{grad}}\ \vec{v} + (\overrightarrow{\mathrm{grad}}\ \vec{v})^{\mathrm{T}}]$$

where $\vec{v}$ is the velocity vector of the material point. It can be decomposed in

Fig. 2.3. Initial (M_0), current (M) and relaxed (M_r) configurations in finite deformation.

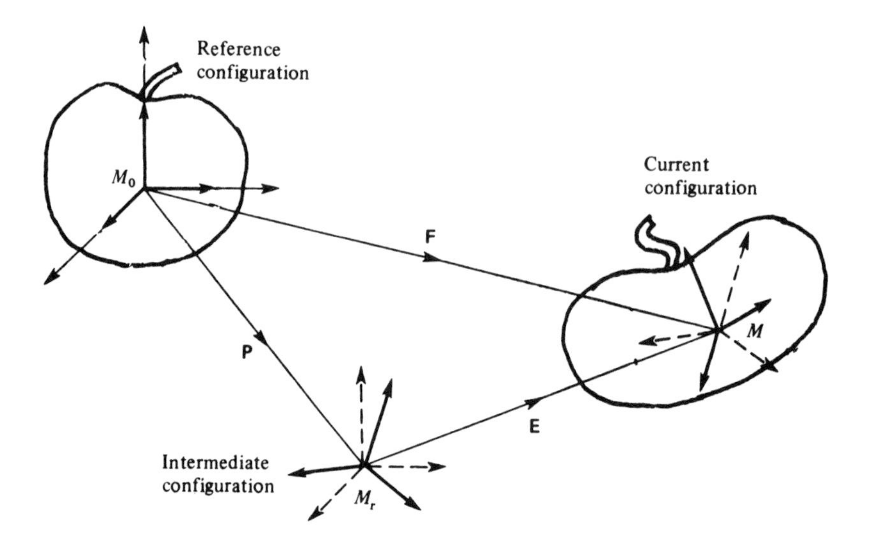

the additive fashion:

$$\mathbf{D} = \mathbf{D}^{e} + \mathbf{D}^{p}.$$

In contrast, the total strain tensor $\boldsymbol{\Delta}$ cannot be as simply decomposed as the sum of an elastic strain tensor (defined with respect to the relaxed configuration):

$$\boldsymbol{\Delta}^{e} = \tfrac{1}{2}(\mathbf{E}^{T}\mathbf{E} - \mathbf{1})$$

and an inelastic strain tensor (defined with respect to the initial configuration):

$$\boldsymbol{\Delta}^{p} = \tfrac{1}{2}(\mathbf{P}^{T}\mathbf{P} - \mathbf{1}).$$

The geometrical aspects related to large deformations will not be discussed in this book. However, it should be remembered that the laws developed here can be extended to large deformations, but only at the cost of a very complicated presentation. Table 2.1 indicates schematically the correspondences which exist between the variables and the equations in terms of the velocity vector in the current configuration and in terms of small and large strains. The great similarity between the first and the second columns should be noted: the quantities in the first column are effectively the derivatives with respect to time of those in the second one. As indicated in the table, the treatment of large deformations often requires the use of stress tensors different from the Cauchy stress tensor $\boldsymbol{\sigma}$ defined with respect to the current configuration. The equations of motion (and of equilibrium) given in the preceding paragraph in terms of Eulerian variables (current configuration) remain valid. They can, however, be replaced by those in Lagrangian variables, by introducing the second Piola–Kirchoff stress tensor and using the initial mass density.

In the one-dimensional case, we often use the true strain ε_v and the true stress σ_v, defined as Eulerian variables in the current configuration. Assuming homogeneous strain, the true strain can be expressed as:

$$d\varepsilon_v = dl/l = dx(M)/x(M), \quad \text{or} \quad \varepsilon_v = \int_{x_0(M_0)}^{x(M)} (dx/x),$$

and by introducing the expression for small strains with respect to the initial configuration $\varepsilon = (x - x_0)/x_0$, we obtain:

- $$\varepsilon_v = \ln(l/l_0) = \ln(x/x_0) = \ln(1 + \varepsilon).$$

The Cauchy stress σ_v used until now is the true stress. We may only consider it as identical to the nominal stress $\sigma = F/S_0$ (which happens to be the first

Table 2.1. *Strain and displacement variables*

Description in terms of velocity	Description in terms of small deformations	Description in terms of finite deformations
Eulerian variables $M = M(x,t)$	Eulerian variables = Lagrangian variables	Lagrangian variables $M = M(M_0, t_0, t)$
Current configuration	Current configuration = initial configuration	Initial configuration
Velocity $\vec{v}$	Displacement $\vec{u}$	Deformation $x(M) = \mathscr{F}(x(M_0), t_0, t)$
Velocity gradient $\overrightarrow{\text{grad}}\ \vec{v}$	Displacement gradient $\overrightarrow{\text{grad}}\ \vec{u}$	(Linear tangent) Deformation gradient $\mathbf{F} = \partial\mathscr{F}/\partial x(M_0)$
Deformation rate tensor $\mathbf{D} = \frac{1}{2}[\overrightarrow{\text{grad}}\ \vec{v} + (\overrightarrow{\text{grad}}\ \vec{v})^T]$	Strain tensor $\boldsymbol{\varepsilon} = \frac{1}{2}[\overrightarrow{\text{grad}}\ \vec{u} + (\overrightarrow{\text{grad}}\ \vec{u})^T]$	Green–Lagrange tensor $\boldsymbol{\Delta} = \frac{1}{2}(\mathbf{F}^T\mathbf{F} - \mathbf{1})$
Elastic strain rate tensor $\mathbf{D}^e$	Elastic strain tensor $\boldsymbol{\varepsilon}^e$	Elastic deformation $\mathbf{E}$ with reference to a relaxed configuration Elastic strain tensor $\boldsymbol{\Delta}^e = \frac{1}{2}(\mathbf{E}^T\mathbf{E} - \mathbf{1})$
Inelastic strain rate tensor $\mathbf{D}^p$	Inelastic strain tensor $\boldsymbol{\varepsilon}^p$	Inelastic deformation $\mathbf{P}$ Inelastic strain tensor $\boldsymbol{\Delta}^p = \frac{1}{2}(\mathbf{P}^T\mathbf{P} - \mathbf{1})$
Partition $\mathbf{D} = \mathbf{D}^e + \mathbf{D}^p$	Partition $\boldsymbol{\varepsilon} = \boldsymbol{\varepsilon}^e + \boldsymbol{\varepsilon}^p$	Decomposition $\mathbf{F} = \mathbf{E}\cdot\mathbf{P}$ $\boldsymbol{\Delta} \neq \boldsymbol{\Delta}^e + \boldsymbol{\Delta}^p$)
	$\dot{\boldsymbol{\varepsilon}} = \mathbf{D}$, $\dot{\boldsymbol{\varepsilon}}^e = \mathbf{D}^e$, $\dot{\boldsymbol{\varepsilon}}^p = \mathbf{D}^p$	$\dot{\boldsymbol{\Delta}} \neq \mathbf{D}$, $\dot{\boldsymbol{\Delta}}^e \neq \mathbf{D}^e$, $\dot{\boldsymbol{\Delta}}^p \neq \mathbf{D}^p$
Cauchy stress tensor $\boldsymbol{\sigma}$	Cauchy stress tensor $\boldsymbol{\sigma}$	First Piola–Kirchhoff stress tensor $\mathbf{S} = \det(\mathbf{F})\boldsymbol{\sigma}\cdot\mathbf{F}^{T-1}$ Second Piola–Kirchhoff stress tensor $\mathbf{S}^* = \mathbf{F}^{-1}\cdot\mathbf{S}$

Piola–Kirchoff stress) in the small strain case. Otherwise, it is necessary to take into account the change of the cross-sectional area:

$$S = S_0(1 - v^*\varepsilon)^2 \approx S_0(1 - 2v^*\varepsilon)$$

where v^* is the coefficient of elastic or elastoplastic contraction. As a first approximation, we may take $v^* = \frac{1}{2}$ for elastoplastic cases, and write:

- $$\sigma_v = \frac{F}{S_0(1 - v^*\varepsilon)^2} \approx \sigma(1 + \varepsilon) \simeq \sigma e^{\varepsilon_v}$$

2.2.4 *Tensorial representation: invariants*

In writing three-dimensional constitutive laws, the assumption of isotropic behaviour leads to the use of invariants of the stress tensor $\boldsymbol{\sigma}$ and the strain tensor $\boldsymbol{\varepsilon}$ and also of their deviators defined by:

$$\boldsymbol{\sigma}' = \boldsymbol{\sigma} - \tfrac{1}{3}\mathrm{Tr}(\boldsymbol{\sigma})\mathbf{1} \qquad \text{or} \qquad \sigma'_{ij} = \sigma_{ij} - \tfrac{1}{3}\sigma_{kk}\delta_{ij}$$
$$\boldsymbol{\varepsilon}' = \boldsymbol{\varepsilon} - \tfrac{1}{3}\mathrm{Tr}(\boldsymbol{\varepsilon})\mathbf{1} \qquad \text{or} \qquad \varepsilon'_{ij} = \varepsilon_{ij} - \tfrac{1}{3}\varepsilon_{kk}\delta_{ij}.$$

These second order tensors possess three basic invariants defined by three independent scalar functions. Mathematically, we may define the three invariants as the three coefficients of the following cubic equation in x:

$$\det(\boldsymbol{\sigma} - x\mathbf{1}) = 0.$$

This equation is called the characteristic equation. The three invariants are:

$$\Sigma_{\mathrm{I}} = \mathrm{Tr}(\boldsymbol{\sigma}) = \sigma_{ii} = \sigma_{11} + \sigma_{22} + \sigma_{33}$$
$$\Sigma_{\mathrm{II}} = \tfrac{1}{2}[(\mathrm{Tr}(\boldsymbol{\sigma}))^2 - \mathrm{Tr}(\boldsymbol{\sigma}^2)] = \tfrac{1}{2}(\sigma_{ii}^2 - \sigma_{ij}\sigma_{ij})$$
$$= \sigma_{11}\sigma_{22} + \sigma_{22}\sigma_{33} + \sigma_{33}\sigma_{11} - \sigma_{23}^2 - \sigma_{31}^2 - \sigma_{12}^2$$
$$\Sigma_{\mathrm{III}} = \det(\boldsymbol{\sigma}).$$

In practice, we may prefer to use the following invariants:

$$\sigma_{\mathrm{I}} = \Sigma_{\mathrm{I}} = \mathrm{Tr}(\boldsymbol{\sigma})$$
$$\sigma_{\mathrm{II}} = \tfrac{1}{2}\mathrm{Tr}(\boldsymbol{\sigma}^2) = \tfrac{1}{2}\sigma_{ij}\sigma_{ij}$$
$$\sigma_{\mathrm{III}} = \tfrac{1}{3}\mathrm{Tr}(\boldsymbol{\sigma}^3) = \tfrac{1}{3}\sigma_{ij}\sigma_{jk}\sigma_{ki}.$$

In the same way, we can define the invariants of the strain tensor:

$$\varepsilon_{\mathrm{I}} = \mathrm{Tr}(\boldsymbol{\varepsilon}) = \varepsilon_{ii}$$
$$\varepsilon_{\mathrm{II}} = \tfrac{1}{2}\mathrm{Tr}(\boldsymbol{\varepsilon}^2) = \tfrac{1}{2}\varepsilon_{ij}\varepsilon_{ij}$$
$$\varepsilon_{\mathrm{III}} = \tfrac{1}{3}\mathrm{Tr}(\boldsymbol{\varepsilon}^3) = \tfrac{1}{3}\varepsilon_{ij}\varepsilon_{jk}\varepsilon_{ki}.$$

In plasticity, we often use the invariants of the deviatoric stress and deviatoric plastic strain, defined in the same fashion as above: $s_{\mathrm{II}}, s_{\mathrm{III}}, e_{\mathrm{pII}}, e_{\mathrm{pIII}}$ (since $s_{\mathrm{I}} = e_{\mathrm{pI}} = 0$). For convenience we will use functions of these invariants which represent concrete, physical quantities. It is practical to choose homogeneous invariants that can be identified in the simple tension case with the first component of the tensor. We thus define:

For stresses:

the hydrostatic stress

$$J_1(\boldsymbol{\sigma}) = \sigma_1 = 3\sigma_{\mathrm{H}}$$

the equivalent stress in the sense of von Mises

$$J_2(\boldsymbol{\sigma}) = (3s_{\text{II}})^{1/2} = \sigma_{\text{eq}}$$

the third homogeneous invariant of a uniaxial state of stress

$$J_3(\boldsymbol{\sigma}) = (\tfrac{27}{2}s_{\text{III}})^{1/3}.$$

For strains:

the volumetric strain

$$I_1(\boldsymbol{\varepsilon}) = \varepsilon_{\text{I}} = 3\varepsilon_{\text{H}}$$

the equivalent inelastic strain in the sense of von Mises

$$I_2(\boldsymbol{\varepsilon}^{\text{p}}) = (\tfrac{4}{3}e_{\text{pII}})^{1/2} = \varepsilon_{\text{peq}}.$$

The invariants defined in this way are sufficient for writing constitutive laws of isotropic materials. In the anisotropic case, the tensorial arguments, characteristic of the anisotropy must also be used. For example, let $\mathbf{f}$ be an anisotropic function relating the stress tensor $\boldsymbol{\sigma}$ to the strain tensor $\boldsymbol{\varepsilon}$ for a general anisotropic behaviour:

$$\boldsymbol{\varepsilon} = \mathbf{f}(\boldsymbol{\sigma}).$$

If $\vec{a}_1, \vec{a}_2, \vec{a}_3$ are the base vectors of an orthonormal set of axes in a reference configuration, we may write:

$$\boldsymbol{\varepsilon} = \mathbf{f}(\boldsymbol{\sigma}) = \mathbf{g}(\boldsymbol{\sigma}, \vec{a}_1, \vec{a}_2, \vec{a}_3).$$

To each rotation $\vec{q}\,\vec{a}_1$, $\vec{q}\,\vec{a}_2$, $\vec{q}\,\vec{a}_3$ of the reference axes, we may associate the transformations:

$$\vec{q}\cdot\boldsymbol{\sigma}\cdot\vec{q}^{\text{T}} \quad \text{and} \quad \vec{q}\cdot\boldsymbol{\varepsilon}\cdot\vec{q}^{\text{T}}$$

which shows that $\mathbf{g}$ is an isotropic function which can then be written as:

$$\boldsymbol{\varepsilon} = C_{\text{p}}\mathbf{G}_{\text{p}}.$$

$\mathbf{G}_{\text{p}}$ are the generating tensors. They number 21 in the general case, but can be expressed as linear functions of a smaller number of independent generators. C_{p} are coefficients which are functions of the 21 invariants associated with the tensor $\boldsymbol{\sigma}$ and with the three vectors $\vec{a}_1$, $\vec{a}_2$, $\vec{a}_3$. These invariant coefficients can also be expressed as functions of a smaller number of independent invariants.

2.3 Fundamental statements of thermodynamics

2.3.1 *Conservation laws; first principle*

The equilibrium equation established in the preceding section can be interpreted as a form of the law of conservation of momentum:

$$(\mathrm{d}/\mathrm{d}t)\int_{\mathscr{D}} \rho\vec{v}\,\mathrm{d}V - \int_{\partial\mathscr{D}} \vec{T}\,\mathrm{d}S = \int_{\mathscr{D}} \vec{f}\,\mathrm{d}V$$

as it can be deduced by applying the divergence theorem with:

$$\vec{T} = \boldsymbol{\sigma}\cdot\vec{n} \quad \text{and} \quad \mathrm{d}\vec{v}/\mathrm{d}t = \vec{\gamma}$$

(where $\mathrm{d}/\mathrm{d}t$ denotes the material derivative).

In the same way, the symmetry of the stress tensor, $\sigma_{ij} = \sigma_{ji}$, which results directly from the principle of virtual power, can be considered as a consequence of the equation of dynamic moment balance.

The second law of conservation is that of conservation of mass, which with the notation introduced, can be expressed as:

$$(\mathrm{d}/\mathrm{d}t)\int_{\mathscr{D}} \rho\,\mathrm{d}V = 0.$$

First principle of thermodynamics

This constitutes the third important law of conservation: the conservation of energy. Let us again consider a domain $\mathscr{D}$ with a boundary $\partial\mathscr{D}$ in the interior of the material medium $\mathscr{S}$. Let E be its internal energy, and e the specific internal energy:

$$E = \int_{\mathscr{D}} \rho e\,\mathrm{d}V.$$

Let K be its kinetic energy:

$$K = \tfrac{1}{2}\int_{\mathscr{D}} \rho\vec{v}\cdot\vec{v}\,\mathrm{d}V.$$

Let Q be the rate at which heat is received by the region $\mathscr{D}$; it consists of two terms, the heat generated within the volume $\mathscr{D}$ by the external agencies (inductive heating, for example), and the heat received by conduction through the boundary $\partial\mathscr{D}$ of $\mathscr{D}$:

$$Q = \int_{\mathscr{D}} r\,\mathrm{d}V - \int_{\partial\mathscr{D}} \vec{q}\cdot\vec{n}\,\mathrm{d}S$$

where r is the volumetric density of the internal heat production, $\vec{q}$ is the heat flux vector and $\vec{n}$ is the outward unit normal to $\partial\mathscr{D}$. Let $P_{(x)}$ be the actual power of the external forces:

$$P_{(x)} = \int_{\mathscr{D}} \vec{f}\cdot\vec{v}\,\mathrm{d}V + \int_{\partial\mathscr{D}} \vec{T}\cdot\vec{v}\,\mathrm{d}S.$$

The first principle of thermodynamics is expressed by:

- $$(\mathrm{d}/\mathrm{d}t)(E+K) = P_{(x)} + Q \quad \forall\mathscr{D}$$

or

$$(\mathrm{d}/\mathrm{d}t)\int_{\mathscr{D}} \rho(e + \tfrac{1}{2}\vec{v}\cdot\vec{v})\,\mathrm{d}V = \int_{\mathscr{D}} (\vec{f}\cdot\vec{v} + r)\,\mathrm{d}V + \int_{\partial\mathscr{D}} (\vec{T}\cdot\vec{u} - \vec{q}\cdot\vec{n})\,\mathrm{d}S.$$

From this it is possible to derive a local expression which involves only the power of internal forces and the heat received. To do so, we write the second axiom of the principle of virtual power by treating the real motion as the virtual one:

$$P_{(x)} = P_{(a)} - P_{(i)}.$$

Now, noting that

$$P_{(a)} = \int \rho\vec{\gamma}\cdot\vec{v}\,\mathrm{d}V = \frac{\mathrm{d}}{\mathrm{d}t}\frac{1}{2}\int \rho\vec{v}\cdot\vec{v}\,\mathrm{d}V = \frac{\mathrm{d}K}{\mathrm{d}t}$$

we obtain

$$\frac{\mathrm{d}E}{\mathrm{d}t} + \frac{\mathrm{d}K}{\mathrm{d}t} = \frac{\mathrm{d}K}{\mathrm{d}t} - P_{(i)} + Q.$$

Since

$$Q = \int_{\mathscr{D}} r\,\mathrm{d}V - \int_{\partial\mathscr{D}} \vec{q}\cdot\vec{n}\,\mathrm{d}S = \int_{\mathscr{D}} r\,\mathrm{d}V - \int_{\mathscr{D}} \mathrm{div}\,\vec{q}\,\mathrm{d}V$$

we have

$$(\mathrm{d}/\mathrm{d}t)\int_{\mathscr{D}} \rho e\,\mathrm{d}V = \int_{\mathscr{D}} \boldsymbol{\sigma}:\mathbf{D}\,\mathrm{d}V + \int_{\mathscr{D}} r\,\mathrm{d}V - \int_{\mathscr{D}} \mathrm{div}\,\vec{q}\,\mathrm{d}V.$$

This identity is valid for any region $\mathscr{D}$. Hence, according to the fundamental lemma it implies that:

$$\rho(\mathrm{d}e/\mathrm{d}t) = \boldsymbol{\sigma}:\mathbf{D} + r - \mathrm{div}\,\vec{q}$$

or, under the assumption of small strains:

- $$\rho \dot{e} = \boldsymbol{\sigma} : \dot{\boldsymbol{\varepsilon}} + r - \operatorname{div} \vec{q}$$

or

$$\rho \dot{e} = \sigma_{ij} \dot{\varepsilon}_{ij} + r - q_{i,i}.$$

2.3.2 *Entropy; second principle*

In addition to the internal energy and the rate of heating it is necessary to introduce two more new variables: temperature and entropy. We assume that it is possible to represent the temperature by a scalar field of positive values defined at each instant t and at all points of the domain $\mathscr{D}$ under study : $T(M, t)$. Entropy expresses a variation of energy associated with a variation in the temperature. It is defined for a domain $\mathscr{D}$ by means of specific entropy s per unit mass:

$$S = \int_{\mathscr{D}} \rho s \, \mathrm{d}V.$$

The second principle

The second principle postulates that the rate of entropy production is always greater than or equal to the rate of heating divided by the temperature:

- $$\frac{\mathrm{d}S}{\mathrm{d}t} \geqslant \int_{\mathscr{D}} \frac{r}{T} \mathrm{d}V - \int_{\partial \mathscr{D}} \frac{\vec{q} \cdot \vec{n}}{T} \mathrm{d}S \quad \Bigg| \forall \mathscr{D}$$

where $\mathrm{d}/\mathrm{d}t$ denotes the material derivative. Use of the divergence theorem leads to

$$\int_{\mathscr{D}} \left(\rho \frac{\mathrm{d}s}{\mathrm{d}t} + \operatorname{div} \frac{\vec{q}}{T} - \frac{r}{T} \right) \mathrm{d}V \geqslant 0.$$

This inequality is valid for any region $\mathscr{D}$ of the body and implies the following local form of the irreversibility of the entropy production rate:

$$\rho \frac{\mathrm{d}s}{\mathrm{d}t} + \operatorname{div} \frac{\vec{q}}{T} - \frac{r}{T} \geqslant 0.$$

The fundamental inequality containing the first and second principles is obtained by replacing r with the expression resulting from the equation of conservation of energy:

$$\rho \frac{\mathrm{d}s}{\mathrm{d}t} + \operatorname{div} \frac{\vec{q}}{T} - \frac{1}{T} \left(\rho \frac{\mathrm{d}e}{\mathrm{d}t} - \boldsymbol{\sigma} : \mathbf{D} + \operatorname{div} \vec{q} \right) \geqslant 0.$$

Noting that:

$$\operatorname{div}\frac{\vec{q}}{T} = \frac{\operatorname{div}\vec{q}}{T} - \frac{\vec{q}\cdot\overrightarrow{\operatorname{grad}}\, T}{T^2}$$

and multiplying by $T > 0$, we obtain:

$$\rho\left(T\frac{\mathrm{d}s}{\mathrm{d}t} - \frac{\mathrm{d}e}{\mathrm{d}t}\right) + \boldsymbol{\sigma}:\mathbf{D} - \vec{q}\cdot\frac{\overrightarrow{\operatorname{grad}}\, T}{T} \geqslant 0.$$

The Clausius–Duhem inequality is obtained by introducing a new variable, the specific free energy Ψ defined by:

$$\Psi = e - Ts.$$

Differentiating this, we obtain:

$$\frac{\mathrm{d}\Psi}{\mathrm{d}t} = \frac{\mathrm{d}e}{\mathrm{d}t} - T\frac{\mathrm{d}s}{\mathrm{d}t} - s\frac{\mathrm{d}T}{\mathrm{d}t} \quad \text{or} \quad T\frac{\mathrm{d}s}{\mathrm{d}t} - \frac{\mathrm{d}e}{\mathrm{d}t} = -\left(\frac{\mathrm{d}\Psi}{\mathrm{d}t} + s\frac{\mathrm{d}T}{\mathrm{d}t}\right)$$

which, when substituted in the first term of the fundamental inequality, yields:

$$\boldsymbol{\sigma}:\mathbf{D} - \rho\left(\frac{\mathrm{d}\Psi}{\mathrm{d}t} + s\frac{\mathrm{d}T}{\mathrm{d}t}\right) - \vec{q}\cdot\frac{\overrightarrow{\operatorname{grad}}\, T}{T} \geqslant 0.$$

For small perturbations, the above may be written as:

- $$\boldsymbol{\sigma}:\dot{\boldsymbol{\varepsilon}} - \rho(\dot{\Psi} + s\dot{T}) - \vec{q}\cdot\frac{\overrightarrow{\operatorname{grad}}\, T}{T} \geqslant 0$$

or

$$\sigma_{ij}\dot{\varepsilon}_{ij} - \rho(\dot{\Psi} + s\dot{T}) - q_i\frac{T_{,i}}{T} \geqslant 0.$$

2.4 Method of local state

In order to avoid any confusion in the reader's mind, it is time to summarize the variables used to describe the thermomechanical behaviour of solids. The concept of thermodynamic potential will clarify everything! However, before giving its definition, a choice must be made with regard to the nature of the variables. In this choice lie both the weakness and the richness of the method of phenomenological thermodynamics: weakness because the choice is partly subjective and results in different models depending on the inclination of the authors, and richness because it allows the formulation of the theories to be adopted to the study of one or more phenomena, either coupled or uncoupled, depending on the intended use.

2.4.1 *State variables*

The method of local state postulates that the thermodynamic state of a material medium at a given point and instant is completely defined by the knowledge of the values of a certain number of variables at that instant, which depend only upon the point considered. Since the time derivatives of these variables are not involved in the definition of the state, this hypothesis implies that any evolution can be considered as a succession of equilibrium states. Therefore, ultrarapid phenomena for which the time scales of the evolutions are of the same order as the relaxation time for a return to thermodynamic equilibrium (atomic vibrations) are excluded from this theory's field of application. Physical phenomena can be described with a precision which depends on the choice of the nature and the number of state variables. The processes defined in this way will be thermodynamically admissible if, at any instant of the evolution, the Clausius–Duhem inequality is satisfied. The state variables, also called thermodynamic or independent variables, are the observable variables and the internal variables.

Observable variables

The formalism of continuum mechanics and thermodynamics as developed above requires the existence of a certain number of state variables; these are the observable variables:

the temperature T
the total strain $\boldsymbol{\varepsilon}$ (assuming small strains).

We limit ourselves to the two observable variables as they are the only ones which occur in elasticity, viscoelasticity, plasticity, viscoplasticity, damage and fracture phenomena. For reversible (or elastic) phenomena, at every instant of time, the state depends uniquely on these variables. For example, the reversible power is defined with the help of the associated stress $\boldsymbol{\sigma}$ as:

$$\Phi_e = \boldsymbol{\sigma} : \dot{\boldsymbol{\varepsilon}}.$$

Internal variables

For dissipative phenomena, the current state also depends on the past history which is represented, in the method of local state, by the values at each instant of other variables called internal variables.

Plasticity and viscoplasticity require the introduction of the plastic (or viscoplastic) strain as a variable. For small strains, the plastic strain $\boldsymbol{\varepsilon}^{\mathrm{p}}$ is the permanent strain associated with the relaxed configuration. This configuration is obtained by 'elastic unloading', leading to the additive strain decomposition:

$$\boldsymbol{\varepsilon} = \boldsymbol{\varepsilon}^{\mathrm{p}} + \boldsymbol{\varepsilon}^{\mathrm{e}}.$$

The two internal variables related to the above decomposition may formally be defined as: the plastic strain $\boldsymbol{\varepsilon}^{\mathrm{p}}$, and the thermoelastic strain $\boldsymbol{\varepsilon}^{\mathrm{e}}$ (including, as well, the possibility of thermal dilatation).

Other phenomena such as hardening, damage, fracture, require the introduction of other internal variables of a less obvious nature. These represent the internal state of matter (density of dislocations, crystalline microstructure, configuration of microcracks and cavities, etc.) and there are no means of measuring them by direct observation. They do not appear explicitly either in the conservation laws or in the statement of the second principle of thermodynamics. They are called internal variables, but in fact, they are state variables which will be treated as observable ones.

There is no objective way of choosing the nature of the internal variables best suited to the study of a phenomenon. The choice is dictated by experience, physical feeling and very often by the type of application. They will be defined in the different chapters as the need arises. For their general study, they will be denoted by $V_1, V_2, \ldots, V_k \ldots$; V_k representing either a scalar or a tensorial variable.

2.4.2 *Thermodynamic potential, state laws*

Once the state variables have been defined, we postulate the existence of a thermodynamic potential from which the state laws can be derived. Without entering into the details, let us say that the specification of a function with a scalar value, concave with respect to T, and convex with respect to other variables, allows us to satisfy *a priori* the conditions of thermodynamic stability imposed by the inequalities that can be derived from the second principle. It is possible to work in an equivalent way with different potentials. Here we choose the free specific energy potential Ψ, which depends on observable state variables and internal variables:

$$\Psi = \Psi(\boldsymbol{\varepsilon}, T, \boldsymbol{\varepsilon}^{\mathrm{e}}, \boldsymbol{\varepsilon}^{\mathrm{p}}, V_k).$$

In elastoplasticity (or viscoplasticity) the strains appear only in the form of their additive decomposition $\boldsymbol{\varepsilon} - \boldsymbol{\varepsilon}^{\mathrm{p}} = \boldsymbol{\varepsilon}^{\mathrm{e}}$, so that:

$$\Psi = \Psi((\boldsymbol{\varepsilon} - \boldsymbol{\varepsilon}^{\mathrm{p}}), T, V_k) = \Psi(\boldsymbol{\varepsilon}^{\mathrm{e}}, T, V_k)$$

which shows that:

$$\partial\Psi/\partial\boldsymbol{\varepsilon}^{\mathrm{e}} = \partial\Psi/\partial\boldsymbol{\varepsilon} = -\,\partial\Psi/\partial\boldsymbol{\varepsilon}^{\mathrm{p}}.$$

We now use the Clausius–Duhem inequality with:

$$\dot{\Psi} = \frac{\partial\Psi}{\partial\boldsymbol{\varepsilon}^{\mathrm{e}}}:\dot{\boldsymbol{\varepsilon}}^{\mathrm{e}} + \frac{\partial\Psi}{\partial T}\dot{T} + \frac{\partial\Psi}{\partial V_k}\dot{V}_k.$$

to obtain:

$$\left(\boldsymbol{\sigma} - \rho\frac{\partial\Psi}{\partial\boldsymbol{\varepsilon}^{\mathrm{e}}}\right):\dot{\boldsymbol{\varepsilon}}^{\mathrm{e}} + \boldsymbol{\sigma}:\dot{\boldsymbol{\varepsilon}}^{\mathrm{p}} - \rho\left(s + \frac{\partial\Psi}{\partial T}\right)\dot{T} - \rho\frac{\partial\Psi}{\partial V_k}\dot{V}_k - \frac{\vec{q}}{T}\cdot\overrightarrow{\operatorname{grad}}\,T \geqslant 0.$$

A classical hypothesis permits us to cancel some terms in this inequality independently. We may imagine, first of all, an elastic deformation taking place at constant ($\dot{T}=0$) and uniform ($\overrightarrow{\operatorname{grad}}\,T = 0$) temperature which alters neither the plastic strain ($\dot{\boldsymbol{\varepsilon}}^{\mathrm{p}} = 0$) nor the internal variables ($\dot{V}_k = 0$). For this to happen, it is necessary to consider that the elastic deformations can occur at a time scale higher that those which would question the validity of the hypothesis of local state, and lower than those of dissipative phenomena. Since the Clausius–Duhem inequality holds regardless of any particular $\dot{\boldsymbol{\varepsilon}}^{\mathrm{e}}$, it necessarily follows that:

$$\boldsymbol{\sigma} - \rho(\partial\Psi/\partial\boldsymbol{\varepsilon}^{\mathrm{e}}) = 0.$$

Assuming this equality to hold, we now imagine a thermal deformation in which $\dot{\boldsymbol{\varepsilon}}^{\mathrm{p}} = 0$, $\dot{V}_k = 0$, $\overrightarrow{\operatorname{grad}}\,T = 0$. Then, since T is arbitrary, it follows that:

$$s + \partial\Psi/\partial T = 0.$$

These expressions define the thermoelastic laws:

- $\boldsymbol{\sigma} = \rho(\partial\Psi/\partial\boldsymbol{\varepsilon}^{\mathrm{e}})$,
- $s = -\,\partial\Psi/\partial T$.

We note that:

$$\boldsymbol{\sigma} = \rho(\partial\Psi/\partial\boldsymbol{\varepsilon}^{\mathrm{e}}) = \rho(\partial\Psi/\partial\boldsymbol{\varepsilon}) = -\,\rho(\partial\Psi/\partial\boldsymbol{\varepsilon}^{\mathrm{p}})$$

which shows that the stress is a variable associated with the elastic strain, with the total strain, and with the plastic strain (with a minus sign).

In an analogous manner, we define the thermodynamic forces associated with the internal variables by:

$$A_k = \rho(\partial\Psi/\partial V_k).$$

These relations constitute the state laws:

the entropy s and the stress tensor $\boldsymbol{\sigma}$ having been defined elsewhere, the specification of the thermodynamic potential $\Psi(\boldsymbol{\varepsilon}^e, T, V_k)$ furnishes the coupled or uncoupled theories of thermo-elasticity;

in contrast, the variables A_k associated with the internal variables, which have not yet been introduced, are defined by the specification of the thermodynamic potential $\Psi(\ldots, V_k)$;

s, $\boldsymbol{\sigma}$ and $A_1, A_2, \ldots, A_k$ constitute the associated variables. The vector formed by these variables is the gradient of the function Ψ in the space of the variables T, $\boldsymbol{\varepsilon}^e$, V_k. This vector is normal to the surface $\Psi = \text{constant}$.

The associated variables form a set of normal variables in duality with the observable and internal state variables. Table 2.2 summarizes the set of variables introduced in this way.

2.4.3 *Dissipation, complementary laws*

As we have seen, the thermodynamic potential allows us to write relations between observable state variables and associated variables. However, for internal variables it allows only the definition of their associated variables. In order to describe the dissipation process, mainly the evolution of the internal variables, a complementary formalism is needed. This is precisely the objective of the dissipation potentials.

Intrinsic dissipation, thermal dissipation

Taking into account the state laws and putting $\vec{g} = \overrightarrow{\text{grad}}\ T$, the Clausius–Duhem inequality can be reduced to express the fact that dissipation is necessarily positive:

- $$\Phi = \boldsymbol{\sigma}:\dot{\boldsymbol{\varepsilon}}^p - A_k\dot{V}_k - \vec{g}\cdot\vec{q}/T \geqslant 0.$$

We note that Φ is a sum of the products of the force variables or dual

Table 2.2. *Thermodynamic variables*

State variables		
Observable	Internal	Associated variables
ε		$\boldsymbol{\sigma}$
T		s
	ε^{e}	$\boldsymbol{\sigma}$
	ε^{p}	$-\boldsymbol{\sigma}$
	V_k	A_k

variables $\boldsymbol{\sigma}$, A_k, $\vec{g}$ with the respective flux variables $\dot{\boldsymbol{\varepsilon}}^{p}$, $-\dot{V}_k$, $-\vec{q}/T$. The sum of the first two terms:

$$\Phi_1 = \boldsymbol{\sigma}:\dot{\boldsymbol{\varepsilon}}^{p} - A_k \dot{V}_k$$

is called the intrinsic dissipation (or mechanical dissipation). It consists of plastic dissipation plus the dissipation associated with the evolution of the other internal variables; it is generally dissipated by the volume element in the form of heat. The last term:

$$\Phi_2 = -\vec{g}\cdot\frac{\vec{q}}{T} = -\frac{\vec{q}}{T}\cdot\overrightarrow{\mathrm{grad}}\ T$$

is the thermal dissipation due to the conduction of heat.

Dissipation potential

In order to define the complementary laws related to the dissipation process, we postulate the existence of a dissipation potential (or pseudo-potential) expressed as a continuous and convex scalar valued function of the flux variables, wherein the state variables may appear as parameters:

$$\varphi(\dot{\boldsymbol{\varepsilon}}^{p}, \dot{V}_k, \vec{q}/T).$$

This potential is a positive convex function with a zero value at the origin of the space of the flux variables, $\dot{\boldsymbol{\varepsilon}}^{p}$, $\dot{V}_k$, $\vec{q}/T$. The complementary laws are then expressed by the normality property (or normal dissipativity):

$$\boldsymbol{\sigma} = \frac{\partial \varphi}{\partial \dot{\boldsymbol{\varepsilon}}^{p}} \quad A_k = -\frac{\partial \varphi}{\partial \dot{V}_k} \quad \vec{g} = -\frac{\partial \varphi}{\partial(\vec{q}/T)}.$$

The thermodynamic forces are the components of the vector $\overrightarrow{\mathrm{grad}}\ \varphi$ normal to the $\varphi = \text{constant}$ surfaces in the space of the flux variables.

In fact, the complementary laws are more easily expressed in the form of the evolution laws of flux variables as functions of dual variables. The Legendre–Fenchel transformation enables us to define the corresponding potential $\varphi^*(\boldsymbol{\sigma}, A_k, \vec{g})$, the dual of φ with respect to the variables $\dot{\boldsymbol{\varepsilon}}^{\mathrm{p}}$, $\dot{V}_k$ and $\vec{q}/T$. By definition:

$$\varphi^*(\boldsymbol{\sigma}, A_k, \vec{g}) = \operatorname*{Sup}_{(\dot{\boldsymbol{\varepsilon}}^{\mathrm{p}}, \dot{V}_k, \vec{q}/T)} ((\boldsymbol{\sigma}:\dot{\boldsymbol{\varepsilon}}^{\mathrm{p}} - A_k \dot{V}_k - \vec{g}\cdot\vec{q}/T) - \varphi(\dot{\boldsymbol{\varepsilon}}^{\mathrm{p}}, \dot{V}_k, \vec{q}/T)).$$

The transformation, written in a slightly clumsy form, is illustrated graphically in Fig. 2.4 in which only one variable has been retained.

It can be shown that, if the function φ^* is differentiable, the normality property is preserved for the variables $\dot{\boldsymbol{\varepsilon}}^{\mathrm{p}}$, $-\dot{V}_k$, $-\vec{q}/T$, and the complementary laws of evolution can then be written as:

- $\dot{\boldsymbol{\varepsilon}}^{\mathrm{p}} = \partial\varphi^*/\partial\boldsymbol{\sigma}$,
- $-\dot{V}_k = \partial\varphi^*/\partial A_k$,
- $-\dfrac{\vec{q}}{T} = \partial\varphi^*/\partial\vec{g}$.

Let us note once more the properties that the potentials φ and φ^* must possess for the automatic satisfaction of the second principle of thermodynamics: they must be nonnegative, convex functions, zero at the origin: $(\boldsymbol{\sigma} = A_k = g = 0)$. Later, we will generally use the potential φ^* and the rela-

Fig. 2.4. Construction of the graph of a potential $\varphi^*(\sigma)$, dual of $\varphi(\dot{\varepsilon}^{\mathrm{p}})$, by the Legendre–Fenchel transformation.

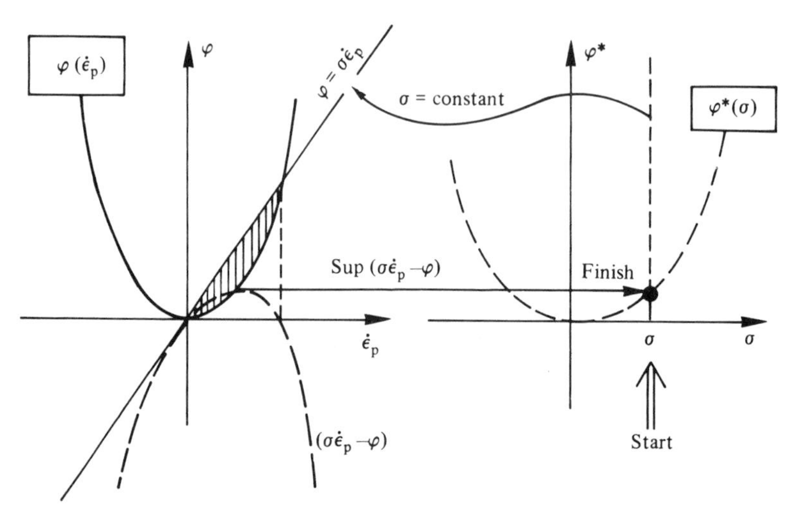

Table 2.3. *Dissipation variables*

Flux variables	Dual variables
$\dot{\boldsymbol{\varepsilon}}^{\mathrm{p}}$	$\boldsymbol{\sigma}$
$-\dot{V}_{\mathrm{k}}$	A_{k}
$-\vec{q}/T$	$\vec{g} = \overrightarrow{\mathrm{grad}}\ T$

tions expressing the evolution of the flux variables. It should be noted that the normality rule is sufficient to ensure the satisfaction of the second principle of thermodynamics, but it is not a necessary condition. This rule applies to generalized standard materials. A standard material is defined as that for which only the first of the above three rules, $\dot{\boldsymbol{\varepsilon}}^{\mathrm{p}} = \partial\varphi^*/\partial\boldsymbol{\sigma}$, applies. This first relation yields the plasticity or viscoplasticity laws. The second equation expresses the evolution laws of the internal variables, and the third one leads to the Fourier law of thermostatics. Table 2.3 provides a summary of the dissipation variables.

The whole problem of modelling a phenomenon lies in the determination of the analytical expressions for the thermodynamic potential Ψ and for the dissipation potential φ or its dual φ^*, and their identification in characteristic experiments. In fact the values of φ or φ^* are almost impossible to measure as they represent an energy usually dissipated as heat. The flux variables and the dual variables are quite easy to measure and it is on their values that the modelling and identification are based. The complementary laws of evolution are therefore directly identified but the dissipation potential is used as a guideline for writing their analytical expression.

It should be noted that one can generalize the dissipation potentials by including the state variables themselves as parameters. The above development is not modified at all. The dissipation potentials are then written as:

$$\varphi(\dot{\boldsymbol{\varepsilon}}^{\mathrm{p}}, \dot{V}_k, \quad \vec{q}/T; \boldsymbol{\varepsilon}^{\mathrm{e}}, T, V_k)$$

$$\varphi^*(\underbrace{\boldsymbol{\sigma}, A_k, \vec{g}}_{\text{variable}}; \underbrace{\boldsymbol{\varepsilon}^{\mathrm{e}}, T, V_k}_{\text{parameters}}).$$

Onsager's symmetry relations

A first simplification consists in assuming that the function φ^* is of a positive-definite quadratic form in terms of the dual variables. Then denoting the flux variables by $\dot{V}_\alpha$ and the dual variables by A_α the potential

φ^* may be written

$$\varphi^* = \tfrac{1}{2}C_{\alpha\beta}(\boldsymbol{\varepsilon}^e, T, V_k)A_\alpha A_\beta.$$

Consequently, under such circumstances, every complementary law of evolution is linear with respect to the corresponding dual variables:

$$\dot{V}_\alpha = C_{\alpha\beta}A_\beta.$$

The matrix $C_{\alpha\beta}$ is symmetric. This property is known as Onsager's symmetry relation.

Decoupling of intrinsic and thermal dissipation

A second simplification consists in assuming a decoupling of intrinsic and thermal dissipations. This does not mean that the corresponding physical mechanisms are decoupled. This assumption amounts to considering the dissipation potential as the sum of two terms, one dependent on the dual variables $\boldsymbol{\sigma}$, A_k, and the other on the variable $\vec{g}$:

$$\varphi^* = \varphi_1^*(\boldsymbol{\sigma}, A_k) + \varphi_2^*(\vec{g})$$

and the second principle of thermodynamics is satisfied by the following inequalities respectively:

$$\Phi_1 = \boldsymbol{\sigma}:\dot{\boldsymbol{\varepsilon}}^p - A_k\dot{V}_k = \boldsymbol{\sigma}:\frac{\partial\varphi_1^*}{\partial\boldsymbol{\sigma}} + A_k\frac{\partial\varphi_1^*}{\partial A_k} \geqslant 0$$

$$\Phi_2 = -\vec{g}\cdot\frac{\vec{q}}{T} = \vec{g}\cdot\frac{\partial\varphi_2^*}{\partial\vec{g}} \geqslant 0.$$

Note that, since φ^* is convex in $\boldsymbol{\sigma}$ and A_k and passes through the origin, the first of the above inequalities is automatically satisfied. We then have:

$$\boldsymbol{\sigma}:\frac{\partial\varphi_1^*}{\partial\boldsymbol{\sigma}} + A_k\frac{\partial\varphi_1^*}{\partial A_k} \geqslant \varphi_1^* \geqslant 0.$$

The phenomena of instantaneous dissipation

When the behaviour is independent of the velocities, the function $\varphi(\dot{\boldsymbol{\varepsilon}}^p, \dot{V}_k)$ is a positive, homogeneous function of degree 1 and its dual function φ^* is nondifferentiable. By extension, we write that $\dot{\boldsymbol{\varepsilon}}^p$ belongs to the subdifferential of φ^* defined by:

$$\partial\varphi^*_{(\boldsymbol{\sigma}_0)} = \{\dot{\boldsymbol{\varepsilon}}^p / \varphi^*_{(\boldsymbol{\sigma})} \geqslant \varphi^*_{(\boldsymbol{\sigma}_0)} + \dot{\boldsymbol{\varepsilon}}^p:(\boldsymbol{\sigma} - \boldsymbol{\sigma}_0), \forall\boldsymbol{\sigma}\}.$$

In addition, we take the convex function of the criterion $f(\boldsymbol{\sigma}, A_k)$ whose convex, $f=0$, has φ^* as an indicator function.

$$\varphi^* = 0 \text{ if } f < 0 \rightarrow \dot{\boldsymbol{\varepsilon}}^{\mathrm{p}} = 0$$

$$\varphi^* = +\infty \text{ if } f = 0 \rightarrow \dot{\boldsymbol{\varepsilon}}^{\mathrm{p}} \neq 0.$$

A proof, not given here, allows us to assert that it is equivalent to write:

$$\dot{\boldsymbol{\varepsilon}}^{\mathrm{p}} \in \partial\varphi^*_{(\boldsymbol{\sigma})} \text{ and } \dot{\boldsymbol{\varepsilon}}^{\mathrm{p}} = \frac{\partial F}{\partial \boldsymbol{\sigma}} \dot{\lambda} \quad \text{if } \begin{cases} f = 0 \\ \dot{f} = 0 \end{cases}$$

where F is a potential function equal to f in the case of 'associated' theories and $\dot{\lambda}$ is a multiplier determined by the consistency condition $\dot{f} = 0$.

The equations describing normality have to be replaced by:

$$\dot{\boldsymbol{\varepsilon}}^{\mathrm{p}} = \dot{\lambda}(\partial F/\partial \boldsymbol{\sigma}), \quad -\dot{V}_k = \dot{\lambda}(\partial F/\partial A_k)$$

or

- $$\boldsymbol{\varepsilon}^{\mathrm{p}} = \dot{\lambda}\partial f/\partial \boldsymbol{\sigma}, \quad -V_k = \dot{\lambda}\partial f/\partial A_k$$

2.5 Elements of heat

2.5.1 *Fourier's law*

The law of heat diffusion, or Fourier's law, expresses a linear relation between the heat flux vector $\vec{q}$ and its dual variable $\vec{g}$. This is a direct consequence of the two simplifications introduced regarding dissipation potentials. In fact, we let:

$$\varphi_2^* = \tfrac{1}{2}\mathbf{C}\cdot\vec{g}\cdot\vec{g}$$

and

$$-\frac{\vec{q}}{T} = \frac{\partial \varphi_2^*}{\partial \vec{g}} = \mathbf{C}\cdot\vec{g} = \mathbf{C}\cdot\overrightarrow{\text{grad}}\, T.$$

If we now make the hypothesis that diffusion properties are isotropic for the material under consideration, then the tensor $\mathbf{C}$ is reduced to a scalar tensor. Moreover, this scalar is considered to vary inversely with respect to the temperature so that it is possible to write:

- $$\vec{q} = -k\,\overrightarrow{\text{grad}}\, T$$

or

$$q_i = -kT_{,i}$$

Table 2.4. *Order of magnitudes of the thermodynamic properties of some materials*

	Melting temperature T_M °C	Density ρ kg m^{-3}	Thermal conductivity k Wm^{-1} °C^{-1}	Specific heat C_ε J kg^{-1} °C^{-1}
Aluminium alloys	600	2800	115 at 20 °C 150 at 200 °C	900 at 20 °C
Steels	1500	7800	46 at 20 °C 29 at 900 °C	460 at 20 °C 625 at 900 °C
Brass	900	8600	128 at 400 °C	450 at 400 °C
Polymers	200	1300	—	1500 at 29 °C
Concrete	2200	2000	1.2 at 20 °C	880 at 20 °C
Wood	—	500	0.15 at 20 °C	2000 at 20 °C

where k is the coefficient of thermal conductivity, a characteristic property of the material (see Table 2.4 for values pertinent to some particular materials).

2.5.2 *Heat equation*

Let us return to the equation of the conservation of energy (the first principle of thermodynamics):

$$\rho \dot{e} = \boldsymbol{\sigma}:\dot{\boldsymbol{\varepsilon}} + r - \operatorname{div} \vec{q}$$

and replace $\rho\dot{e}$ by the expression derived from $e = \Psi + Ts$:

$$\rho \dot{e} = \rho \dot{\Psi} + \rho T \dot{s} + \rho \dot{T} s$$

and $\dot{\Psi}$ by its expression as a function of the state variables, so that:

$$\rho \dot{e} = \rho \left(\frac{1}{\rho} \boldsymbol{\sigma}:\dot{\boldsymbol{\varepsilon}}^e - s\dot{T} + \frac{1}{\rho} A_k \dot{V}_k \right) + \rho T \dot{s} + \rho \dot{T} s$$

and we obtain:

$$\boldsymbol{\sigma}:\dot{\boldsymbol{\varepsilon}}^e + A_k \dot{V}_k + \rho T \dot{s} = \boldsymbol{\sigma}:\dot{\boldsymbol{\varepsilon}} + r - \operatorname{div} \vec{q}.$$

Now, with $s = (-\partial\bar{\Psi}/\partial T)(\boldsymbol{\varepsilon}^e, T, V_k)$, we may express $\dot{s}$ by:

$$\begin{aligned} \dot{s} &= -\frac{\partial^2 \Psi}{\partial \boldsymbol{\varepsilon}^e \partial T}:\dot{\boldsymbol{\varepsilon}}^e - \frac{\partial^2 \Psi}{\partial T^2}\dot{T} - \frac{\partial^2 \Psi}{\partial V_k \partial T}\dot{V}_k \\ &= -\frac{1}{\rho}\frac{\partial \boldsymbol{\sigma}}{\partial T}:\dot{\boldsymbol{\varepsilon}}^e + \frac{\partial s}{\partial T}\dot{T} - \frac{1}{\rho}\frac{\partial A_k}{\partial T}\dot{V}_k. \end{aligned}$$

By introducing the specific heat defined by:

$$C = T\,\partial s/\partial T$$

and taking into account Fourier's law $\vec{q} = -k\,\overrightarrow{\text{grad}}\,T$ or

$$\text{div}\,\vec{q} = -k\,\text{div}(\overrightarrow{\text{grad}}\,T) = -k\,\Delta T$$

we obtain, using $\dot{\boldsymbol{\varepsilon}}^{\text{p}} = \dot{\boldsymbol{\varepsilon}} - \dot{\boldsymbol{\varepsilon}}^{\text{e}}$,

- $$k\,\Delta T = \rho C\dot{T} - \boldsymbol{\sigma}:\dot{\boldsymbol{\varepsilon}}^{\text{p}} + A_k\dot{V}_k - r - T\left(\frac{\partial\boldsymbol{\sigma}}{\partial T}:\dot{\boldsymbol{\varepsilon}}^{\text{e}} + \frac{\partial A_k}{\partial T}\dot{V}_k\right)$$

which is the complete heat equation and where Δ denotes the Laplacian operator.

Heat propagation

The classical heat equation corresponds to a process:

without variation in inelastic strains, $\boldsymbol{\sigma}:\dot{\boldsymbol{\varepsilon}}^{\text{p}} = 0$;
without variation in internal variables, $A_k\dot{V}_k = 0$;
without internal generation of heat created by the external sources, $r = 0$;
without thermomechanical coupling, $(\partial\boldsymbol{\sigma}/\partial T):\dot{\boldsymbol{\varepsilon}}^{\text{e}} = 0$, $(\partial A_k/\partial T)\dot{V}_k = 0$.

The values of these last two terms are actually negligible in most applications. The specific heat is then the specific heat at constant strain:

$$C = C_\varepsilon = T(\partial s/\partial T)_\varepsilon.$$

(Table 2.4 gives some specific values of C_ε.) Under such conditions we may write:

- $$k\Delta T = \rho C_\varepsilon\dot{T}.$$

Adiabatic overheating

The complete heat equation also allows us to find the rise in temperature of a medium subjected to mechanical dissipation. To do this it is sufficient to solve:

$$\rho C\dot{T} = \boldsymbol{\sigma}:\dot{\boldsymbol{\varepsilon}}^{\text{p}} - A_k\dot{V}_k + r + k\Delta T + T\left(\frac{\partial\boldsymbol{\sigma}}{\partial T}:\dot{\boldsymbol{\varepsilon}}^{\text{e}} + \frac{\partial A_k}{\partial T}\dot{V}_k\right).$$

The classical equation used to calculate the overheating of metallic materials, for example during the forming process, corresponds to:

$k\Delta T = 0$: adiabatic evolution
$r = 0$: no internal heat production generated by external sources
$(\partial \boldsymbol{\sigma}/\partial T)\dot{\boldsymbol{\varepsilon}}^{e} = 0$, $(\partial A_k/\partial T)\dot{V}_k = 0$: no thermomechanical coupling.

$A_k \dot{V}_k$ represents the nonrecoverable energy stored in the material. For metals, this is the energy of the field of the residual microstresses which accompany the increase in the dislocation density. It represents only 5–10% of the term $\boldsymbol{\sigma}:\dot{\boldsymbol{\varepsilon}}^{p}$ and is often neglible:

$$A_k \dot{V}_k \approx 0.$$

The equation of adiabatic overheating is given by:

$$\bullet \qquad \rho C_\varepsilon \dot{T} = \boldsymbol{\sigma}:\dot{\boldsymbol{\varepsilon}}^{p}.$$

Bibliography

Germain P. & Muller P. *Introduction à la mécanique des milieux continus.* Masson, Paris (1980).
Germain P. *Cours de mécanique des milieux continus.* Masson, Paris (1973).
Valid R. *La mécanique des milieux continus et le calcul des structures.* Eyrolles, Paris (1977).
Bamberger Y. *Mécanique de l'ingénieur, II: Milieux déformables.* Hermann (1981).
Nemat-Nasser S. *Mechanics today.* Vol. I, II, III, IV, V. Pergamon Press, New York (1972–1980).
Sanchez-Palencia E. *Non homogeneous media and vibration theory.* Springer Verlag, Berlin (1980).
Truesdell C. *The elements of continuum mechanics.* Springer Verlag, Berlin (1966).
Eringen A. C. *Mechanics of continua.* J. Wiley, New York (1967).
Kestin J. & Rice J. R. *A critical review of thermodynamics.* Stuart Ed. Mono book (1970).

3

IDENTIFICATION AND RHEOLOGICAL CLASSIFICATION OF REAL SOLIDS

L' expérience (d'un Laboratoire), c'est l'ensemble des erreurs qu'on ne recommencera plus.

Continuum mechanics and thermodynamics (Chapter 2) constitute the basic theoretical tools for the formulation of the physical phenomena of deformation and fracture. For fundamental and practical reasons, we model each broad class of phenomena separately. The aim of this chapter is to differentiate from a qualitative point of view and identify the most common types of material behaviour. The phenomenological method used is based on observed experimental results. We, therefore, present some basic elements on the types of tests, the machines and the modern measurement techniques likely to be used. Progress in electronics, automatic controls, digital measurements, and more recently in microprocessors has resulted in a radical transformation, especially during the 1970s, of the laboratories engaged in characterization of materials. We no longer have to be content with approximate measurements of a few quantities; we are now in a position to measure the evolution of any mathematically well defined variable precisely. The identification of complex models has thus become possible, but it requires numerical methods for the identification of nonlinear processes which still belong to the domain of 'heuristic' techniques.

The resulting schematic classification allows us to associate, *a priori*, to each material, a theory of only the dominant phenomena in a limited domain of state variables. Beyond its fundamental interest, this offers a guide for the choice of material in the design stage and helps to simplify the estimation of the resistance of a structure under service loads. The method used is that of rheology initiated by Bingham around 1930, with decisive developments taking place during the 1950s.

3.1 The global phenomenological method

We generally distinguish three broad methods for formulating the constitutive laws of materials.

> The microscopic approach attempts to model the mechanics of deformation and fracture at the atomic, molecular and crystalline levels as described in Chapter 1; the macroscopic behaviour being the result of an integration or of an averaging process performed over the microscopic variables at the scale of the volume element of continuum mechanics.
>
> The thermodynamic approach introduces a homogeneous continuous medium equivalent to the real medium and represents microscopic physical phenomena by means of macroscopic 'internal variables'.
>
> The functional approach leads to the hereditary laws of the integral type which appear as characteristic functions of materials, and which are themselves expressed in terms of macroscopic variables. This approach will be partly used in the section on viscoelasticity in Chapter 4.

None of these three approaches allows direct identification of materials. Microscopic variables (density of dislocation, density of cavities, texture...) are difficult to measure, and moreover are difficult to use in practical computations. Thermodynamic potentials are practically inaccessible to measurement, and internal variables, by definition, cannot be measured directly. As the hereditary functions need the knowledge of the whole history of the observable variables, they pose theoretical as well as experimental problems.

The global phenomenological method consists in studying the volume element of matter through the relations between cause and effect which exist between the physically accessible variables constituting the input and output of the process under study. In this way, we are able to determine the material response to a specific input. These responses are sufficient to characterize the materials qualitatively, but they do not constitute (except for linear phenomena) the constitutive laws.

By a volume element, in the sense of solid mechanics, we mean a volume of a size large enough with respect to the material heterogeneities, and small enough for the partial derivatives of the equations of the continuum to be meaningful. Table 3.1 gives the order of magnitude of the reasonable sizes

Table 3.1. *Orders of magnitude of representative volume elements*

Materials	Inhomogeneities	Volume element
Metals and alloys	Crystals 1 μm–0.1 m	$0.5 \times 0.5 \times 0.5$ mm
Polymers	Molecules 10 μm–0.05 mm	$1 \times 1 \times 1$ mm
Wood	Fibres 0.1 mm–1 mm	$1 \times 1 \times 1$ cm
Concrete	Granulates $\simeq$ 1 cm	$10 \times 10 \times 10$ cm

of representative volume elements, below which size it becomes illusive, to give a physical meaning to stress or strain other than as an average over the 'homogenized' volume in question.

The physically accessible variables of the volume element are those that can simply be deduced from the four classical and measurable magnitudes of mechanics: displacement, force, time, temperature.

The strains and their rates:
total three-dimensional strain $\boldsymbol{\varepsilon}$, or one-dimensional ε, with its large deformation expression $\varepsilon_v = \ln(1 + \varepsilon)$,
reversible elastic strain $\boldsymbol{\varepsilon}^e$ or ε^e,
permanent strain $\boldsymbol{\varepsilon}^p$ or ε^p.
The three-dimensional stress $\boldsymbol{\sigma}$, or one-dimensional stress σ, with the approximate expression for large deformations $\sigma_v \simeq \sigma(1 + \varepsilon)$.
The temperature T.
The time t or the number of cycles to rupture t_R or N_R.

The classification resulting from the global phenomenological method should not be considered intrinsic. It provides, in fact, only a frame of reference for general characteristics. The behaviour of a given material can be represented by a schematic model only in relation to the envisaged usage and the desired precision of the predictions.

A given piece of steel at room temperature can be considered to be:

Linear, elastic for structural analysis,
viscoelastic for problems of vibration damping,
rigid, perfectly plastic for calculation of the limit loads,
hardening elastoplastic for an accurate calculation of the permanent deformations,
elastoviscoplastic for problems of stress relaxation,
damageable by ductility for calculation of the forming limits,
damageable by fatigue for calculation of the life-time,
etc.

3.2 Elements of experimental techniques and identification process

3.2.1 *Characteristic tests*

The classical characteristic tests are essentially conducted in simple tension, or tension–compression at constant temperature. The classification of real solids is therefore based upon these tests. The specimen is subjected to an axial load (force or displacement) which produces a uniform state of stress or strain within the whole useful volume of the specimen which can be considered as one volume element. What follows applies to isotropic materials; some complementary remarks concerning anisotropic materials are given at the end of this section.

The uniaxial state is defined by a one-dimensional state of stress and a two-dimensional state of strain:

$$\sigma = \begin{bmatrix} \sigma & 0 & 0 \\ 0 & 0 & 0 \\ 0 & 0 & 0 \end{bmatrix} \qquad \varepsilon = \begin{bmatrix} \varepsilon & 0 & 0 \\ 0 & -\nu^*\varepsilon & 0 \\ 0 & 0 & -\nu^*\varepsilon \end{bmatrix}$$

where ν^* is the coefficient of contraction, equal to Poisson's ratio ν in elasticity.

In elastoplasticity or elastoviscoplasticity the hypotheses of the decoupling of elastic and plastic strains ($\boldsymbol{\varepsilon} = \boldsymbol{\varepsilon}^e + \boldsymbol{\varepsilon}^p$) and of plastic incompressibility ($\mathrm{Tr}(\boldsymbol{\varepsilon}^p) = 0$) allow us to express the contraction coefficient ν^* simply as:

$$\begin{bmatrix} \varepsilon & 0 & 0 \\ 0 & -\nu^*\varepsilon & 0 \\ 0 & 0 & -\nu^*\varepsilon \end{bmatrix} = \begin{bmatrix} \varepsilon_e & 0 & 0 \\ 0 & -\nu\varepsilon_e & 0 \\ 0 & 0 & -\nu\varepsilon_e \end{bmatrix} + \begin{bmatrix} \varepsilon_p & 0 & 0 \\ 0 & -\frac{1}{2}\varepsilon_p & 0 \\ 0 & 0 & -\frac{1}{2}\varepsilon_p \end{bmatrix}$$

$$\nu^* = \nu\frac{\varepsilon_e}{\varepsilon} + \frac{1}{2}\frac{\varepsilon_p}{\varepsilon}$$

or with the help of the linear-elastic law: $\varepsilon_e = \sigma/E$, as:

$$\bullet \qquad \nu^* = \frac{1}{2} - \frac{\sigma}{E\varepsilon}\left(\frac{1}{2} - \nu\right).$$

The graph of this function, compared with some experimental results for an aluminium alloy with hardening characteristics similar to the curve in Fig. 3.2, is given in Fig. 3.1.

Hardening test in simple tension or compression

This is the most common test. The specimen is subjected to a deformation at constant speed. The response consists of a variation in the stress σ as a function of the strain ε showing the hardening of the material (Fig. 3.2).

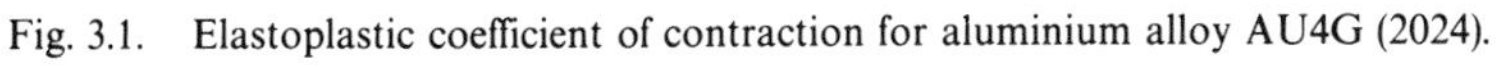

Fig. 3.1. Elastoplastic coefficient of contraction for aluminium alloy AU4G (2024).

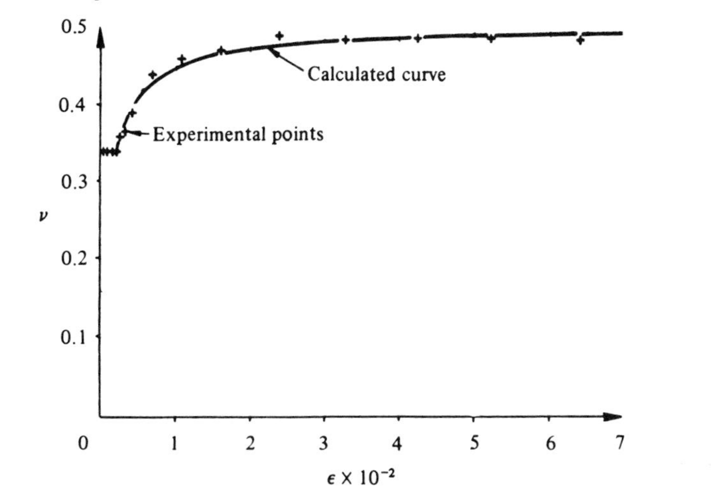

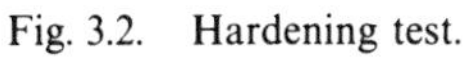

Fig. 3.2. Hardening test.

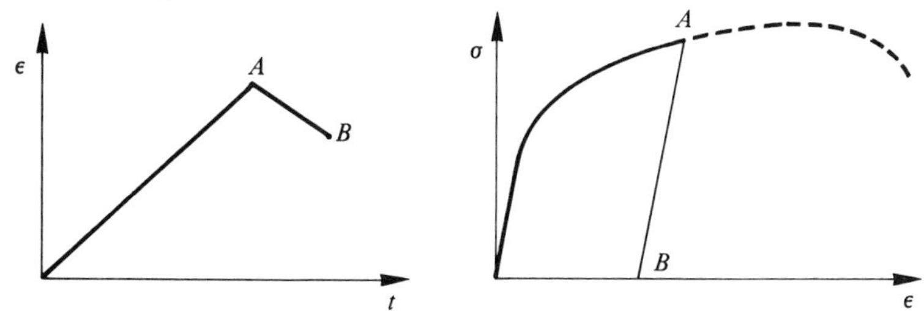

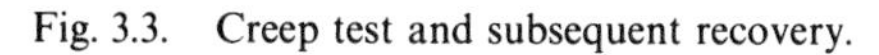

Fig. 3.3. Creep test and subsequent recovery.

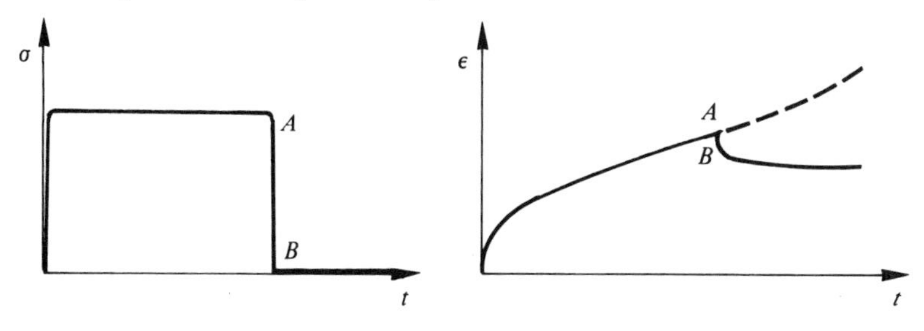

Creep test in simple tension or compression

In this test the specimen is subjected to a constant state of stress (generally apparent stress) and the resulting variation in strain ε as a function of time t is determined. This also characterizes the hardening and the viscosity of the material (Fig. 3.3). The strain variation after the stress is removed (point B) corresponds to the recovery test. The partial strain recovery is indicated on the right hand side of Fig. 3.3.

Relaxation in simple tension or compression

This test is complementary to the preceding one in that the stress response to a constant state of strain is determined. It is governed mainly by the

Fig. 3.4. Relaxation test.

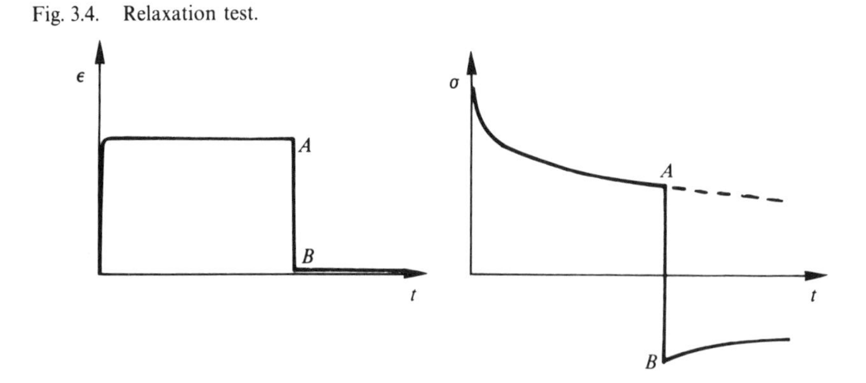

Fig. 3.5. Multiple hardening-relaxation test.

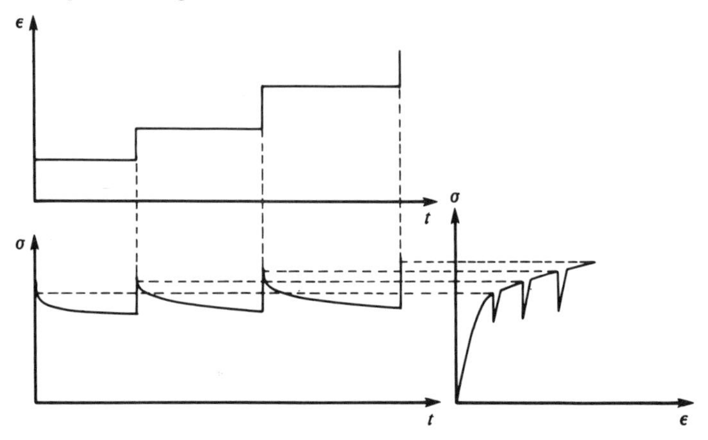

viscosity but also depends on the hardening induced by the initial load (Fig. 3.4).

Multiple hardening-relaxation test

As the name indicates, this test combines the two types of tests and allows us to determine from a single test on a single specimen the hardening characteristics, and also the viscosity, by means of successive relaxations each at different strain levels (Fig. 3.5).

Fig. 3.6. Cyclic test under prescribed strain.

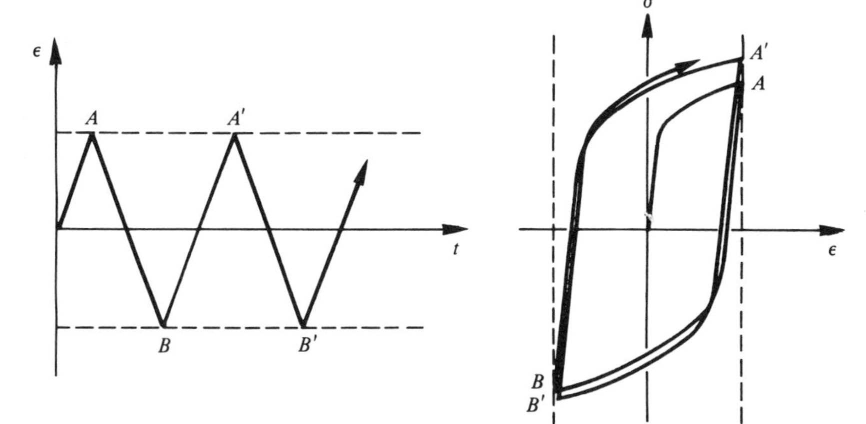

Fig. 3.7. Cyclic hardening curve.

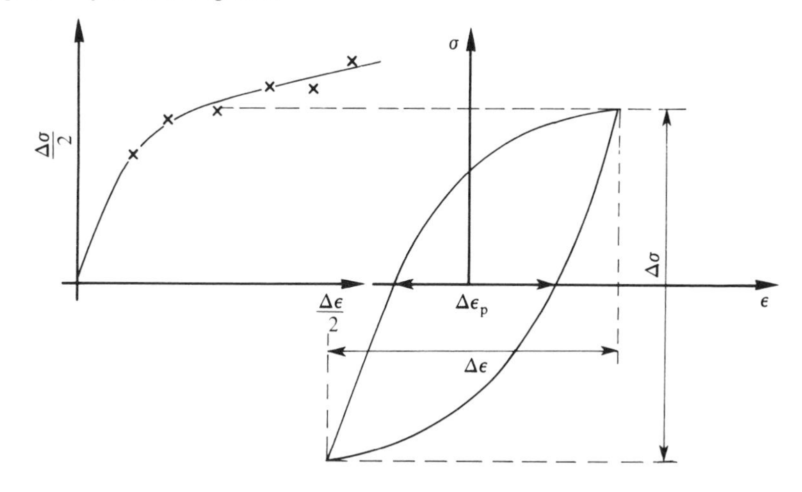

Cyclic test

In this test the specimen is subjected to a periodic load (stress or strain) and the evolution of the cyclic responses is studied by obtaining a (σ, ε) graph (Fig. 3.6). In general the response tends to be stabilized after a certain number of cycles. We can then obtain the cyclic curve which represents a relation between the peaks of the stabilized loops, corresponding to different stress or strain levels (Fig. 3.7). In the case of polymeric materials, or, more generally, in viscoelastic materials the peaks of the loops are difficult to define. We then speak of 'harmonic' tests.

Fracture tests

The four preceding types of tests are also used to determine the corresponding fracture conditions:

stress and strain at fracture;
time or number of cycles to fracture;
energy dissipated in fracture.

Certain characteristics require the use of specimens of a particular geometry (notched specimens for measuring impact strength in the Charpy test, cracked specimens for measuring toughness, etc.).

Multiaxial tests

Unfortunately, these tests are rarely conducted because of the experimental difficulties they present. Among the possible tests are: tension–shear (or compression–shear), biaxial tension and triaxial compression. The tension–shear test performed on circular tubes subjected to tension and torsion is the most interesting one for the characterization of anisotropy. The tension force and the twisting couple are applied either simultaneously or successively and the corresponding rotation and length changes are recorded.

Problems of anisotropic materials

There are even greater difficulties in the case of anisotropic materials, that is when preferred directions exist in the material.

The tension or compression test is still easy to interpret when done along a principal direction of anisotropy.

The torsion test on hollow specimens is more tricky. For a specimen cut from an anisotropic metal sheet, for example, shear deformations are no longer uniform around the circumference (because the shear moduli are different in the longitudinal and transverse directions).

Generally, multiaxial tests on anisotropic materials cannot be interpreted without making reference to a particular modelling scheme. Only the biaxial test along the principal directions of anisotropy is free from this defect.

3.2.2 ***Experimental techniques***

Testing machines

The phenomenological method requires the experiments to be made on a volume element of matter. As far as possible the specimens should be subjected to a uniform field of stress, strain and temperature. This is one of the difficulties of mechanical testing for the characterization of materials.

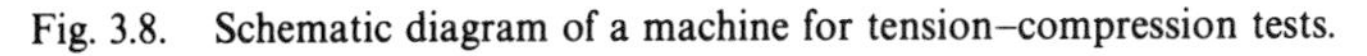

Fig. 3.8. Schematic diagram of a machine for tension–compression tests.

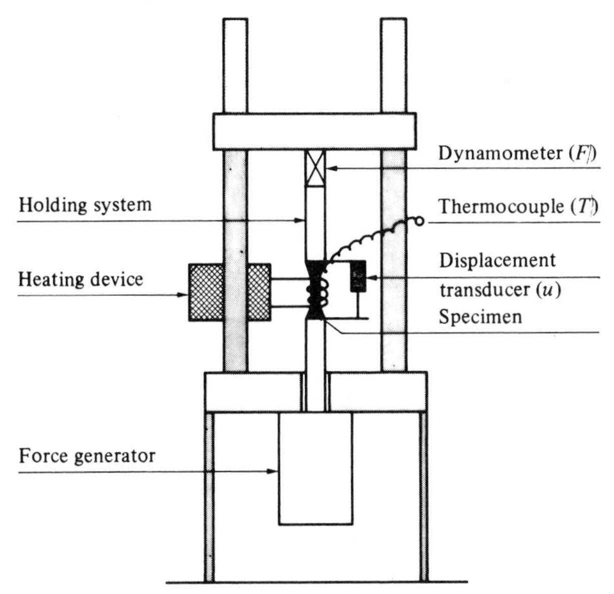

Uniaxial monotonic and cyclic tests

The most common test is the simple tension test in which the useful part of the specimen is subjected to a uniform uniaxial stress field. Fig. 3.8 gives a schematic drawing of a modern set-up, the essential elements of which are as follows:

> the system holding the heads of the specimen,
> the dynamometer measuring the force applied to the specimen,
> the transducer measuring the variation of length of the specimen,
> the frame of the machine, which should be as stiff as possible,
> the heating device,
> the device for applying the forces.

Depending on the type of loading required, the stresses are applied differently: a system of weights for creep testing machines, a continuous screw-nut system driven by an electric motor, a servo-controlled hydraulic system for more sophisticated set-ups. The specimen itself includes a useful part consisting of a cylindrical shaft (Fig. 3.9), end grips, and between the two, shoulders designed to minimize the stress concentration.

For tension tests, the useful part can be very long but is limited by the restriction due to machining and heating devices. In contrast, compression tests require much more compact specimens in order to avoid buckling problems. Actual examples are given later.

For tests at high temperatures either resistance furnaces are used, or the Joule effect (due to the electrical resistance of the sample) or high frequency induction are used to heat the specimen; the latter method is a better one for tests of short or medium duration. The choice of the technique depends on the type of test (monotonic or cyclic) and on the temperatures to be reached (uniform, constant or variable temperature).

Multiaxial tests

In this case tests are conducted to obtain complex stress states in a volume element which could be considered as isostatic. It should be possible to

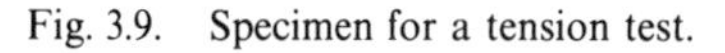

Fig. 3.9. Specimen for a tension test.

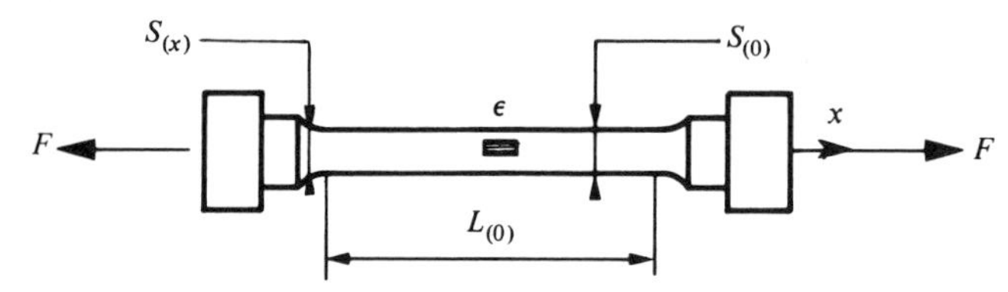

determine the stress state solely by the global equilibrium equations, independent of the material behaviour. Different techniques are used.

Test on cylindrical tubular specimens in tension–compression and internal or external pressure

This is an interesting technique as it does not require special machines. The hydraulic pressurization device (at room temperature) or the gas device (at high temperature) can be adapted to most tension–compression machines. If the wall-thickness e of the specimen is small enough with respect to the radius R, ordinary biaxial states are obtained. Even a pure shear can be simulated ($\sigma_2 = -\sigma_1$) but the principal directions of stress remain fixed (Fig. 3.10). In soil and rock mechanics, this type of test is done with solid (not hollow) samples and with external pressure produced by a pressure cell.

Tension–compression–torsion tests on hollow cylindrical specimens

The possible states of stress are more limited here, but it is possible to study the behaviour for a load path in which the principal directions of

Fig. 3.10. Case of tension–compression – internal pressure and the domain of possible stress states.

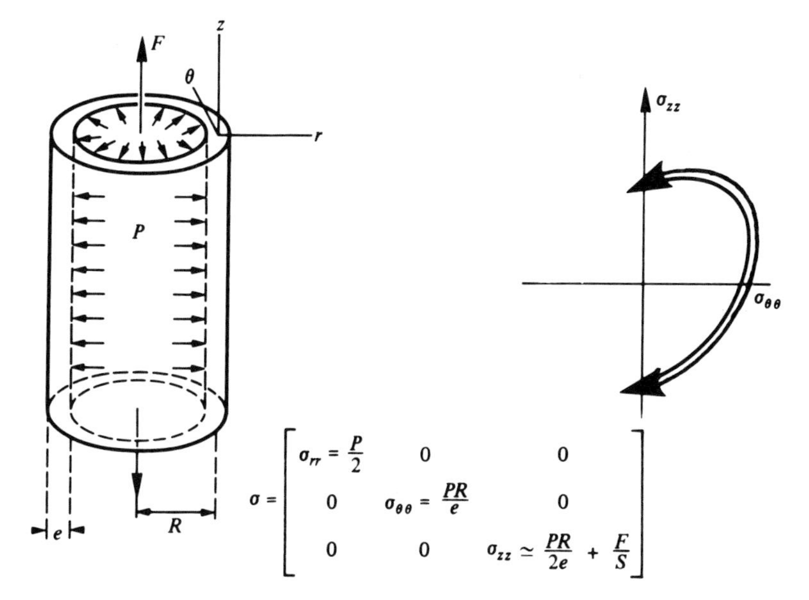

stress are changing (Fig. 3.11). The machine consists of a linear jack and a screw jack with a decoupling device. Modern servo-controlled hydraulic machines are computer driven.

Biaxial tests on cruciform specimens

For this test the machine requires two or four linear jacks placed 90° apart. The transverse bending stiffness of the arms of the cross should be

Fig. 3.11. Case of tension–compression–torsion and the domain of possible stress (C is the twisting couple, J is the torsional moment of inertia of the specimen).

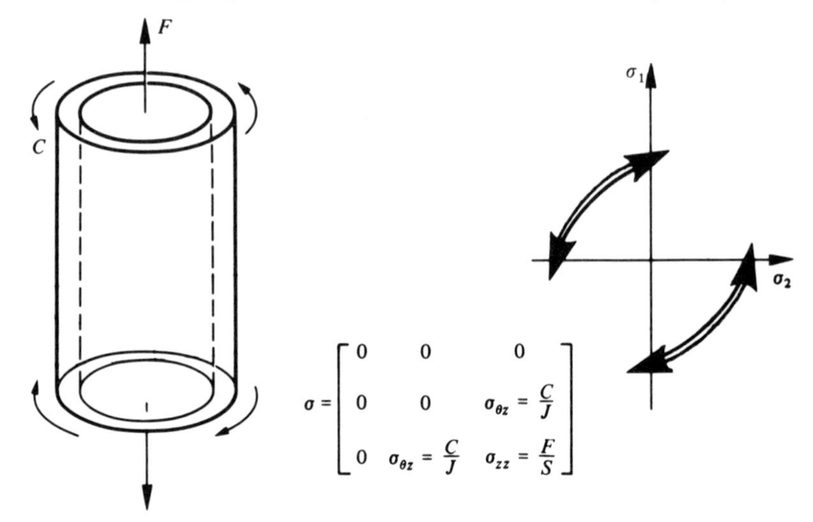

Fig. 3.12. Case of biaxial tension and the domain of possible stress states.

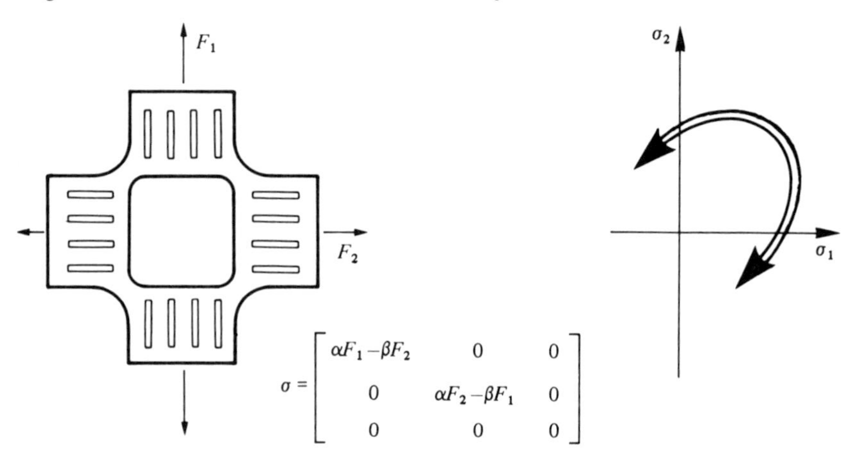

low enough to obtain uniform stress fields. One can, for example, use hollow specimens (Fig. 3.12). The possible loads are such that the principal stresses are positive and their directions remain constant. We may carry out biaxial tension tests on specimens made from plane metal sheets or on specimens in the shape of spherical caps loaded by internal pressure, but then only proportional loading paths can be obtained.

Tests in triaxial compression

This type of test is mostly used in soil and rock mechanics. It requires machines which are delicate to set up.

Techniques of measuring characteristic variables

Forces and stresses

Stresses can be deduced from forces by static equations for isostatic cases. Three examples are given in Figs. 3.10–3.12.

The forces (or couples) applied to the specimens are measured by dynamometers mounted in series with the specimens. The dynamometers, equipped with strain gauges, have a relative precision of 10^{-3} which is generally sufficient for the characterization of materials.

Displacements and strains

The measurement of relative displacement between two points on a specimen is a much more delicate operation than the measurement of the loading force. To measure relative displacement it is necessary to connect a strain gauge in 'parallel' with the sample and not in series. Essentially two techniques are used: local measurements by electrical resistance strain gauges, and global displacement measurements.

Wire strain gauges

A resistance wire, well glued to the specimen, experiences the same strain as the specimen. It therefore follows that the variation in electrical resistance will be proportional to the strain ε. A precise measurement of this variation in resistance is obtained by employing a Wheatstone bridge; this can detect strains of the order of $\varepsilon = 10^{-7}$. However, the qualification 'well glued' mentioned above limits the applicability of this particularly useful technique. Adhesives do not withstand high temperatures, and hence, these strain gauges are only used at, or below, room temperature, or at slightly higher temperatures (not more than 200–400 °C).

Displacement transducers

The strain can also be deduced from the equations of continuum mechanics and knowledge of the displacement between two material points of the specimen. The axial displacement can be measured by an extensometer attached either to the cylindrical part of the specimen through knife edges, or to the gripping heads of the specimen. The variation in the diameter of the specimen can also be measured but without using the knife edges which very often initiate a fracture. The evaluation, however, must take into account the variation of Poisson's coefficient in the presence of plastic strains (see Section 3.2.1). With inductive extensometers or with strain gauges, an absolute precision of 1 μm can be attained. With optical extensometers this precision can be improved to 0.2 μm, but these are expensive and delicate instruments.

Plastic deformation. Useful lengths of specimens

The useful length of a sample is the length L_0 which becomes L under the load, and which can be used to calculate the longitudinal strain by the simple relation:

$$\varepsilon = (L - L_0)/L_0.$$

The plastic strain can be obtained by subtracting the elastic strain from the total strain ε (measured, for example, by an electric resistance strain gauge). In tension (compression) we then have:

$$\varepsilon_p = \varepsilon - \varepsilon_e = \varepsilon - \sigma/E = \varepsilon - F/ES$$

where S is the current cross-sectional area of the specimen.

The modulus of elasticity is obtained from data at the beginning of the test when the force is sufficiently low for the specimen to remain elastic. We then obtain:

$$E = F/S\varepsilon_e.$$

When the extensometer cannot be connected directly onto the cylindrical part of the specimen, we take an equivalent useful length L_p corrected to take into account the plastic deformations in the connection areas. Employing the notation of Fig. 3.9, this is defined by:

$$L_p = \frac{u_p}{\varepsilon_{p(0)}} = 2\int_0^{L/2} \frac{\varepsilon_{p(x)}}{\varepsilon_{p(0)}}\,\mathrm{d}x.$$

It will be seen in Chapter 5 that the plastic strain can be expressed as a power function of the stress $\varepsilon_p = (\sigma/K)^{1/M}$. By a method of slicing, and neglecting the triaxiality effect in the connection zones, we can easily show that:

• $$L_p = 2\int_0^{L/2} \left(\frac{S_{(0)}}{S_{(x)}}\right)^M \mathrm{d}x.$$

The useful plastic length is thus defined from the geometry of the specimen (law of variation of section $S_{(x)}$) and the hardening exponent. The procedure used to measure the plastic strain is therefore the following:

determine the force–displacement relation by direct readings in a tension test (for example),

subtract the initial displacement u_j and the elastic displacement to obtain the plastic displacement. The stiffness R of the specimen and the grips at the level of the extensometer is obtained from the initial readings in the elastic domain (Fig. 3.13). Then:

$$u_p = u - u_j - F/R.$$

The relation $F(u_p)$ furnishes an approximation to the hardening

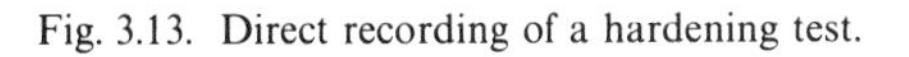
Fig. 3.13. Direct recording of a hardening test.

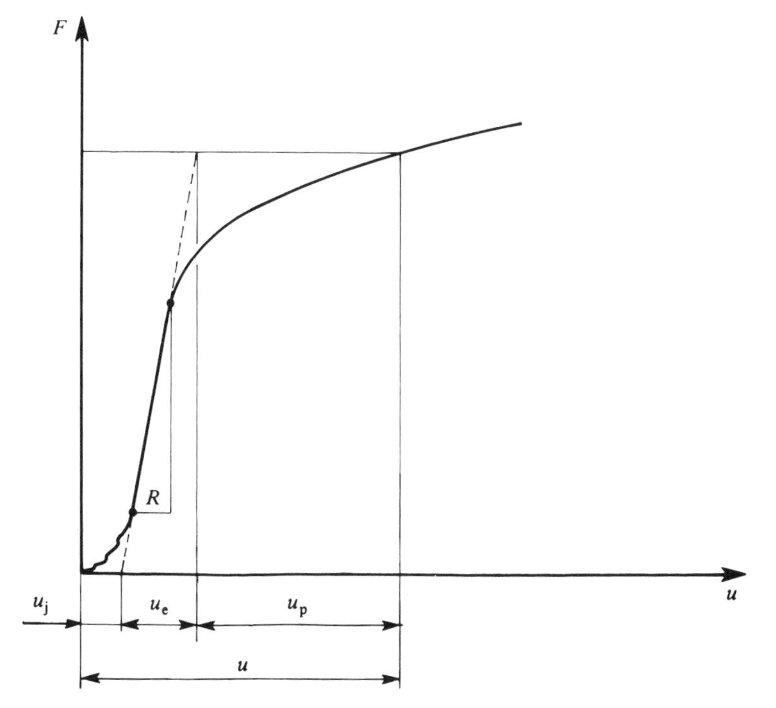

exponent M. In fact, an identity exists between the exponents of the power laws:

$$\sigma = K\varepsilon_{\mathrm{p}}^{M} \quad \text{and} \quad F = K'u_{\mathrm{p}}^{M}$$

since

$$\sigma = F/S \quad \text{and} \quad \varepsilon_{\mathrm{p}} = u_{\mathrm{p}}/L_{\mathrm{p}}.$$

The equivalent useful length L_{p} can be derived from these equations, and hence,

$$\varepsilon_{\mathrm{p}} = u_{\mathrm{p}}/L_{\mathrm{p}}.$$

It should be noted that this useful length can be defined by a preliminary test for a given specimen, material, and temperature. Even if the hypotheses used to establish it are not completely rigorous, they are sufficient since the correction is, in fact, small with respect to the length of the cylindrical zone.

Temperature

The most commonly used technique for measuring temperature involves the use of thermocouples: these measure the electromotive force which arises due to the Pelletier effect between a 'hot' soldering of two wires, such as chromium and aluminium ones for example, with both ends at the same reference temperature.

When equipment sensitive to measurements in microvolts is available, the theoretical precision of a chromium–aluminium thermocouple is 2.5×10^{-2} °C. This precision is quite adequate for characterization tests of materials for which an absolute precision of the order of ± 0.1–0.5 °C can be satisfactory. The most critical condition occurs for viscoplastic behaviour. A calculation based on the Dorn relation presented in Chapter 6 expresses the rate of the plastic strain as a function of temperature:

$$\dot{\varepsilon}_{\mathrm{p}} = f(\sigma, \ldots) \exp(-\Delta H/kT).$$

This relation can be used to show that the desired relative precision of temperature measurements must be of the order of 15–70 times more than that for the measurement of strains:

$$\delta\varepsilon_{\mathrm{p}}/\varepsilon_{\mathrm{p}} \approx 15 \text{ to } 70(\delta T/T).$$

A strain measurement with a level of precision of 2% at 500 °C requires a uniform temperature throughout the specimen within a tolerance of approximately 0.5 °C.

The thermocouples must be welded onto the specimen but this weld constitutes a site for fracture initiation. Consequently, thermocouples can be used only for tests designed for the study of deformation behaviour. In the case of static fracture tests or fatigue fracture tests, we may employ optical transducers which measure infrared emission from the sample. This, however, has a drawback in that the emission power of the specimen varies with its state of damage.

Damage

It will be seen in Chapter 7 that global measures of damage involve only the evaluation of stresses and strains as functions of time or the number of cycles.

Crack length or crack surface

The measurement of crack length or of crack surface is often a delicate task as the crack-tip or the crack-front is not always well defined physically, and corresponds to highly damaged zones full of microcracks.

Optical measurement

This measurement is performed by direct visual means – more precisely, by a photographic process: a camera with an automatic tripping mechanism sends an image at regular intervals. After processing and enlarging the images (reference marks are needed in the working zone to provide a scale) the crack length is obtained by direct measurement. The precision of the result depends on the illumination, enlargement, quality of the reference marks and good definition of the crack-tip. The precision varies between 0.05 and 0.01 mm.

Gauges with cut wires

These consist of wires glued on a support at a spacing of 0.5–1 mm. The support itself is then glued in front of the tip of the crack whose length is to be measured. As the crack grows, it cuts the wires one after the other. This results in a variation of the resistance of the gauge or in an incremental signal that activates a recording device. The precision of this system without interpolation is of the order of the spacing between the wires.

Method of potential drop

The specimen or cracked element, which should be conducting but insulated, is powered continuously, at a point far from the crack, by an electric current of high intensity (≈ 20 A), preferably a direct current. A potential plug with its two points on either side of, and close to the crack, measures a potential difference V which is a function of the dimensions of the specimen, of the location of the current supply, of the potential plug, and of the crack length a. A measurement of V therefore permits the determination of a, if the function $a(V)$ is known either through a preliminary calibration done with another means of measurement, or by calculation. A diagram of this measurement technique is given in Fig 3.14. Even though the potential difference to be measured is very low, of the order of tenths of a microvolt, the precision of the measurement can reach a few hundredths of a millimetre for crack lengths measured in thin metal plates, provided that the measurement is synchronized with the maximum opening of the crack.

This method can also be applied to measure the depths of noncrossing crack-fronts in massive parts. The calibration, although more complex, can be done using a rheo-electrical analogy tank; in this case, the precision is of the order of tenths of a millimetre in measuring the coordinates of each point of the crack-tip.

Specimens

The preparation of the specimens is always a long and delicate task as, unless care is taken, fracture can occur prematurely and not precisely where it is expected! It always involves a compromise between the implications of the machine characteristics (loading type, maximum force, stiffness, heating

Fig. 3.14. Diagram showing fatigue crack length measurement by the potential drop technique (after Baudin and Policella).

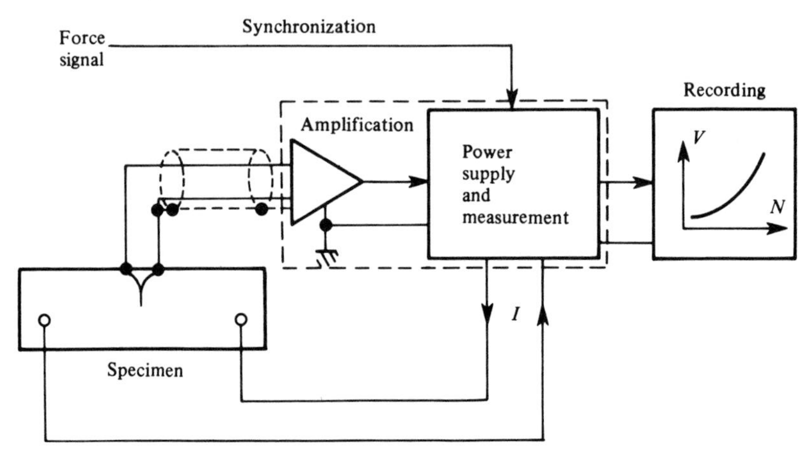

system) and the implications of the measurements to be carried out (sensitivity of the dynamometers, elongation measurements by extensometers placed locally on the useful part of the specimen or externally on the gripping heads). A few examples of dimensioned drawings which have

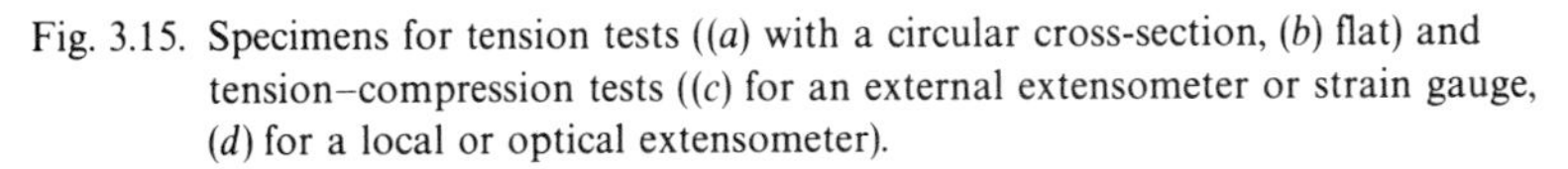

Fig. 3.15. Specimens for tension tests ((*a*) with a circular cross-section, (*b*) flat) and tension–compression tests ((*c*) for an external extensometer or strain gauge, (*d*) for a local or optical extensometer).

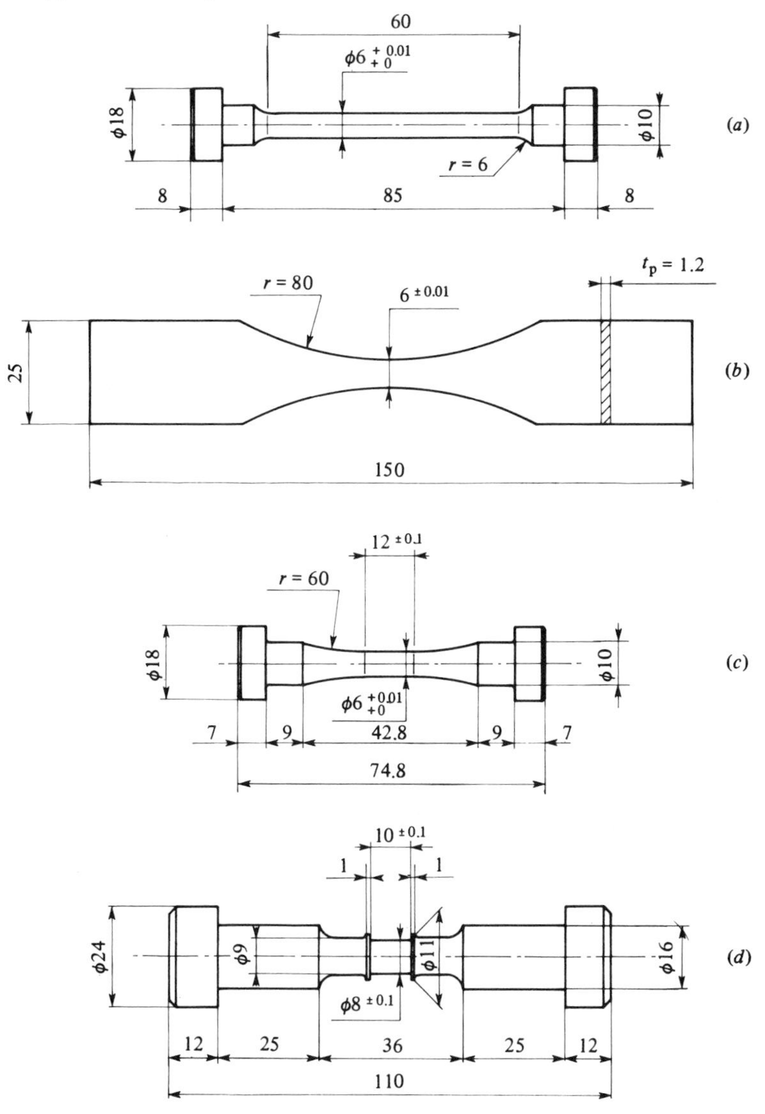

proved to be reliable are given in Figs. 3.15–3.19 (after Dauzou, LMT Cachan).

3.2.3 *Identification methods*

Thermodynamics gives the general formulation of models without specifying their analytical form (except, however, for linear behaviour) or numerical values. On the other hand, experiments provide, for each material, the quantitative relations to be verified by the models constructed to represent the phenomena under study. Identification can be defined as all the work which consists in specifying the functions which appear in the model and in finding the numerical values of the coefficients which define the functions for each material.

Fig. 3.16. (*a*) Specimen for tension. (*b*) Specimen for pure shear in compression.

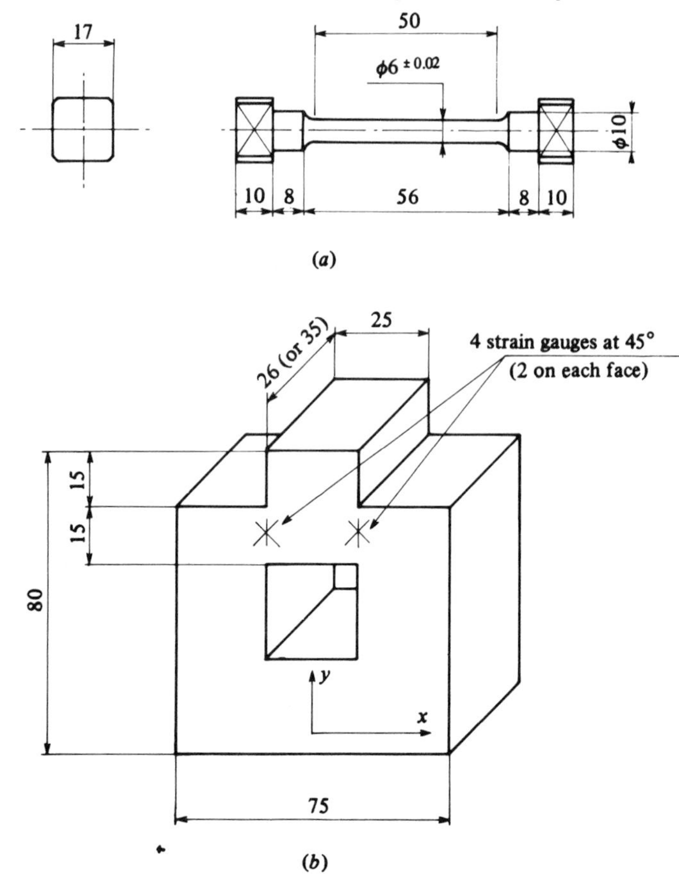

This represents a difficult task which does not follow any rigorous rules and in which experience and 'art of model construction' play a major role in steering between theory and experiment.

Analytical formulation of models

The ratio of quality versus price

Given a set of experimental results, it is always possible to find a function which represents it with an error not greater than the margin of uncertainty in measurements. This is called fitting. Since the number of experimental

Fig. 3.17. Specimen for tension–torsion. (*a*) Fixation by pins. (*b*) Fixation by clamping.

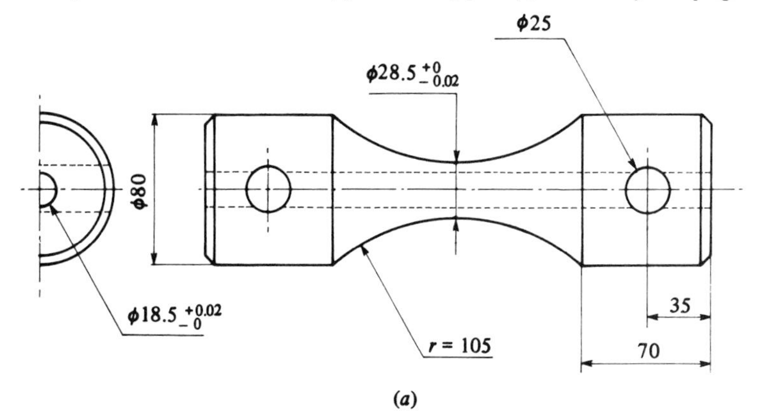

(*a*)

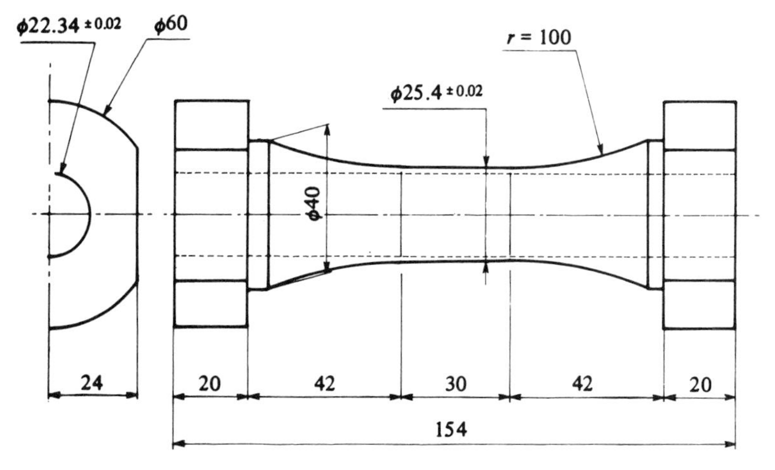

(*b*)

points can be large, the fitting may require functions with a large number of representative coefficients.

On the other hand, a model which has the ambition of becoming a law, must possess a general character, so that while identified only by a restricted number of experiments, it is representative of other types of experiments with a predictive capability. The totality of the situations verified by a model is its field of validity. This characterizes the goodness of the model and is expressed qualitatively by the set of all possible variation histories of the variables and quantitatively by the range of the variations within which the model agrees with the physics. For example, the Hencky–Mises plasticity law (Chapter 5) is valid for an isotropic material under radial loading up to strains lower than the damage threshold (Chapter 6).

The number of coefficients represents the price to pay since the difficulties of identification essentially lie in this number. It may be easy to identify say two coefficients by a 'hand procedure', but the identification of say five coefficients in a model will involve considerable numerical work; and the identification of say ten of them really belongs to 'computer-aided art'.

Fig. 3.18. (*a*) Specimen for ductile damage test. (*b*) Specimen for fatigue test.

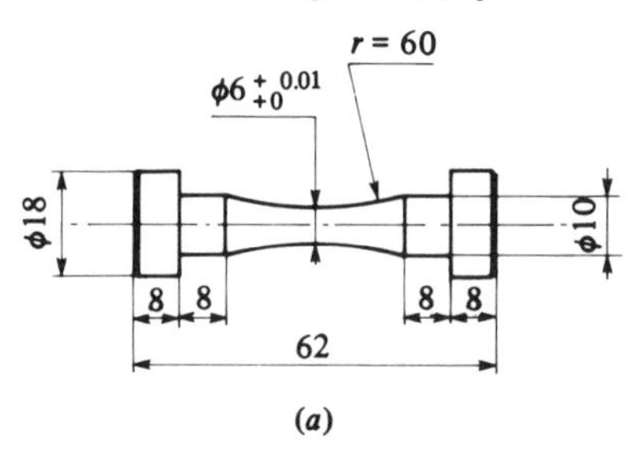

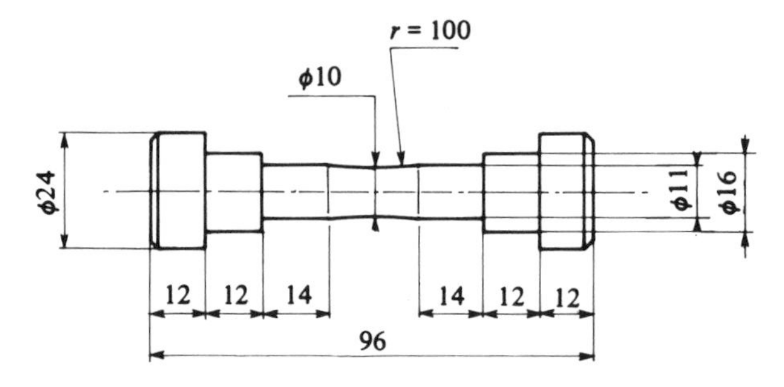

Therefore, to evaluate a model it is necessary to examine this relation of quality/price = domain of validity/number of coefficients clearly.

Modelling of nonlinearities

Most of the phenomena studied in this book (plasticity, viscoplasticity, damage, cracking) are 'very strongly' nonlinear, in the sense that a linearization limited to the first order Taylor expansion represents the phenomenon only for a very small change in the variables.

Among all the analytical possibilities for representing nonlinearities, the following will be considered:

the exponential: $\exp(aX)$, or the logarithm $\ln(aX)$,
the power function aX^N where N is a coefficient which may take the

Fig. 3.19. Crack specimens. (*a*) ASTM CT specimens. (*b*) Irwin specimen.

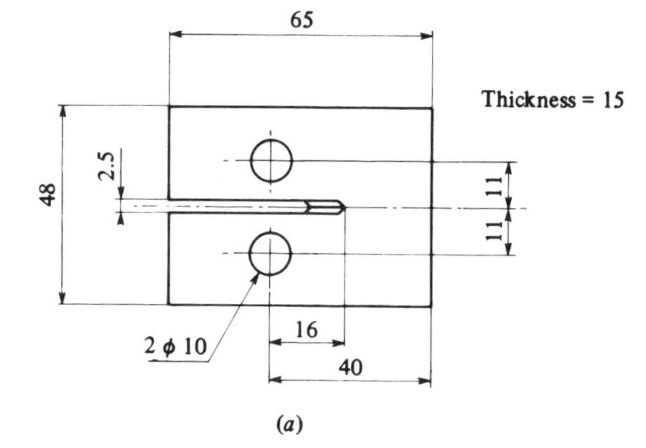

(*a*)

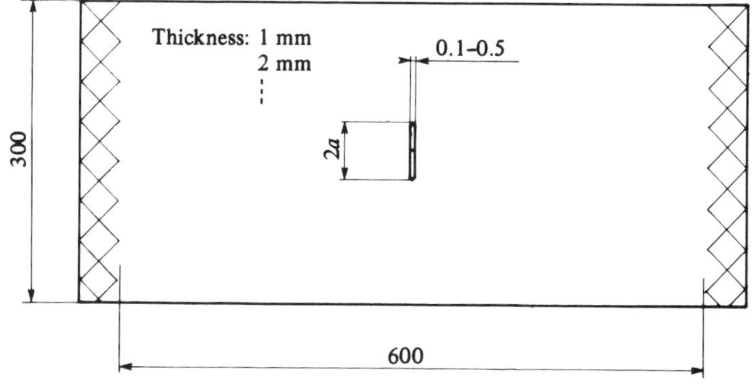

(*b*)

values, 2, 5, 10, 20 or even 100. In some cases it is necessary to use the function $aX^{N(X)}$ in which N itself is a function of X.

The reasons for this are essentially twofold:

the easy identification by logarithmic transformation which yields linear relations,
the possibility of obtaining analytical solutions of models for studying their properties.

Rule of linear accumulation
An important property, whether or not satisfied in a model, is the rule of linear accumulation (for example, the Robinson rule for creep and the Palmgreen–Miner rule for fatigue) which, in fact, is a property common to any differential equation, linear or nonlinear, which admits separation of variables.

Models resulting from dissipation potentials (see Chapter 2) always have the general form:

$$\dot{X} = f(X, V(t))$$

where X is the evolution variable of the phenomenon and is a function of the causal variables V (taken here to be just one variable for simplicity).

Let us show that if this differential equation admits separation of variables:

$$\dot{X} = g(X)h(V)$$

(where g and h may be nonlinear functions), then it implies the rule of linear accumulation. For the initial condition $X = 0$ at $t = 0$, the solution of this equation can be expressed as:

$$[g(X)]^{-1}\,\mathrm{d}X = h(V)\,\mathrm{d}t$$

or

$$\int_0^X [g(x)]^{-1}\,\mathrm{d}x = \int_0^t h(V)\,\mathrm{d}t.$$

Consider first the case where V is constant equal to V_0. The corresponding solution $t(X, V_0)$ is:

$$t(X, V_0) = [h(V_0)]^{-1} \int_0^X [g(x)]^{-1}\,\mathrm{d}x.$$

Consider now the case where V varies with time and let:

$$\tau(X, V) = [h(V)]^{-1} \int_0^X [g(x)]^{-1}\,\mathrm{d}x$$

be a function of X and the time derived from $t(X, V_0)$ by replacing V_0 with $V(t)$. Since $\mathrm{d}t = [h(V)]^{-1}[g(X)]^{-1}\,\mathrm{d}X$, let us calculate:

$$\int_0^{t^*} \frac{\mathrm{d}t}{\tau(X^*, V(t))}$$

where t^* is the time which corresponds to X^*, i.e., $X^* = X(t^*)$. We have:

$$\int_0^{t^*} \frac{\mathrm{d}t}{\tau(X^*, V(t))} = \int_0^{t^*} \frac{[h(V)]^{-1}[g(x)]^{-1}\,\mathrm{d}x}{[h(V)]^{-1}\displaystyle\int_0^{X^*} [g(x)]^{-1}\,\mathrm{d}x}$$

$$= \int_0^{X^*} \frac{[g(x)]^{-1}\,\mathrm{d}x}{\displaystyle\int_0^{X^*} [g(x)]^{-1}\,\mathrm{d}x} = 1.$$

Hence,

- $$\int_0^{t^*} \frac{\mathrm{d}t}{\tau(X^*, V(t))} = 1.$$

This relation expresses the fact that the accumulation of the time ratio $\mathrm{d}t/\tau$ is equal to 1. Here τ is the solution obtained by considering a constant process instead of the variable process signified by $V(t)$.

This is the linear accumulation rule which applies to all separable differential equations. For example, in fatigue, Miner's rule for a variable loading amplitude of stress $\Delta\sigma(N)$ is:

$$\int_0^{N_R} \frac{\mathrm{d}N}{N_F(\Delta\sigma(N))} = 1$$

where N_R is the number of cycles to fracture for the type of loading considered and N_F is the reference number of cycles to fracture for a periodic loading of range $\Delta\sigma$.

Scatter and randomness in coefficients

Another important aspect is the scattering of the experimental results. This can be due to:

> the nature of the phenomenon, inhomogeneities, random growth processes, load uncertainty,

the inherent scattering in each sample arising from the production or casting process and also from the heat treatment, and so on....

Usually, scattering is of the order of 1–5% for elastic strains, 10–50% for plastic or viscoplastic strains, 50–100% (factor of 2) of the number of cycles for low cycle fatigue crack initiation or growth, and 1000% (factor of 10) for fatigue failure in the high cycle regime.

These orders of magnitude of scattering indicate at once the pronounced influence of nonlinearities of the phenomena. The simplest and most efficient way of including this scattering in a model consists in giving a statistical definition to the multiplying coefficients present in the model, for example the coefficient A in:

$$\dot{X} = AX^m V^n.$$

With sufficient experimental results, it is possible to define A by its probability density, or by the probability distribution curve, or often just by its mean value and standard deviation.

Fig. 3.20 shows the viscoplastic behaviour of the alloy INCO 718 at 550 °C, observed through the viscosity and hardening functions (see Chapter 6). The scattering from one sample has an effect on the ordinates, i.e. on the stress, but not on the nonlinearity of the two functions.

Using $M = 13$ and $N = 140$ for a larger number of tests, it has been possible to derive the statistical distribution of the strength coefficient K of the viscoplastic law:

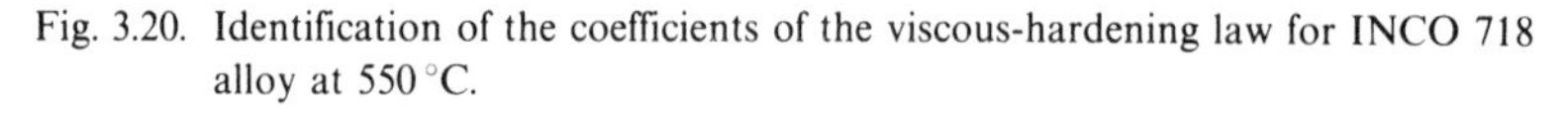

$$\sigma = K\varepsilon_p^{1/M}\dot{\varepsilon}_p^{1/N}.$$

Fig. 3.20. Identification of the coefficients of the viscous-hardening law for INCO 718 alloy at 550 °C.

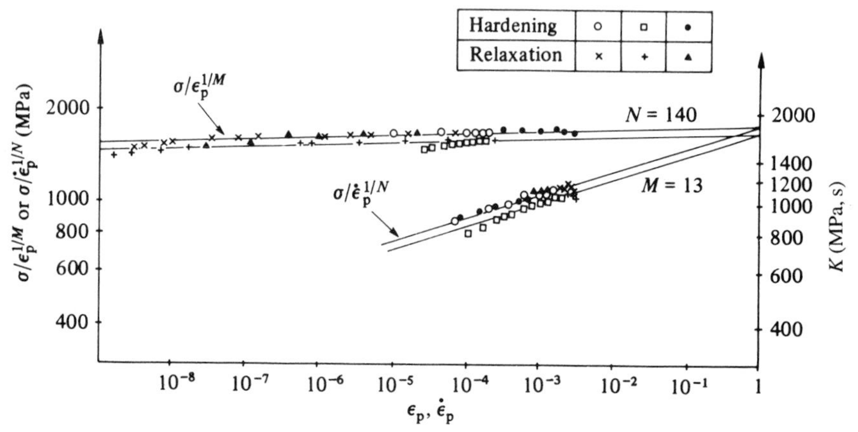

Fig. 3.21 shows that this distribution is very close to the distribution observed for the conventional yield stress (at 0.2% permanent strain).

Numerical methods of identification

Knowing a mathematical model by its analytical expression, and obtaining a set of experimental results in which all the variables of the model have been activated, we are in a position to calculate the unknown coefficients which give the best representation of the experimental results.

Generalities

Let us mention at once that problems arise differently depending on the test data available and the model under study. Two cases have to be distinguished.

The constitutive law is directly identified, i.e., we seek the entity (or the transfer function) representative of the material. One example of this is a law in which a variable X and its derivative $\dot{X}$ are present:

$$H(X, \dot{X}, Y) = 0$$

Fig. 3.21. Probability densities for the $\sigma_{0.2}$ yield stress and for the strength coefficient K of INCO 718 alloy at 550°C.

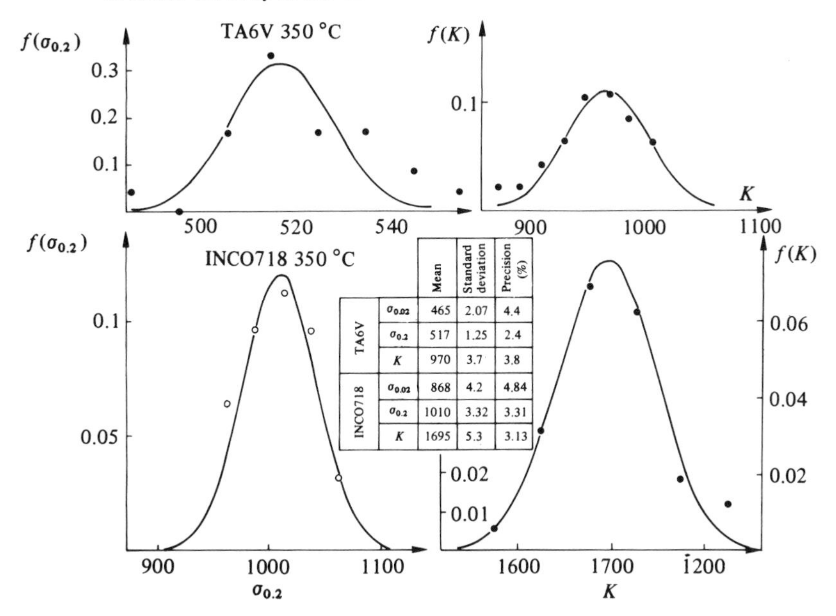

		Mean	Standard deviation	Precision (%)
TA6V	$\sigma_{0.02}$	465	2.07	4.4
	$\sigma_{0.2}$	517	1.25	2.4
	K	970	3.7	3.8
INCO718	$\sigma_{0.02}$	868	4.2	4.84
	$\sigma_{0.2}$	1010	3.32	3.31
	K	1695	5.3	3.13

Once a series of triplets of values $(X, \dot{X}, Y)$ have been measured (at the same instant), the coefficients which define the above function H can be obtained by direct fitting of the experimental points. Depending upon the particular case, we may use a linear or nonlinear least-squares method to obtain these coefficients (see below).

The response is identified, i.e., the coefficients of the function H are determined by adjusting, in the best possible way, the response values obtained from the assumed constitutive equation for a given load (input) against values obtained experimentally under the same load. In this case we use a nonlinear least-squares method, since the response cannot usually be stated explicitly in a simple analytic way.

Remarks

A given model cannot be identified correctly unless a sufficient number of test results are available which embrace a significant range of variation of each of the parameters (for example, the parameters X, $\dot{X}$ and Y in the above function H). Otherwise, we run the risk of not determining one or more coefficients well enough.

For an identical material and model, it may be necessary to define several sets of coefficients, each better suited to a domain of variation or to a load type, for example, rapid transient loads, short-term loads, long-term loads and stationary loads.

Linear least-squares method

Our objective is to minimize the discrepancy between the experimental data and the theoretically calculated values from a model. Different norms may be used, and generally speaking, we have the problem of minimizing an error function:

$$h(a)$$

which depends on the unknown coefficients a_i $(i = 1, 2, \dots, n)$. Different methods of minimization can be used. They have different efficiencies; however, in general, convergence may be difficult to achieve, due to large nonlinearities of the phenomena generally studied. We limit ourselves here to presenting the linear least-squares and Gauss–Newton methods when the error function is chosen as the sum of the squares of the discrepancies:

$$h(a) = \tfrac{1}{2} \sum_{j=1}^{m} E_j^2 = \tfrac{1}{2} \sum_{j=1}^{m} [y_j^{\mathrm{C}}(a, x_j) - y_j^{\mathrm{E}}(x_j)]^2.$$

y_j^E and y_j^C represent respectively the experimental and the calculated values of one of the parameters as a function of the others denoted by x_j (jth experimental point). Fig. 3.22 illustrates in a schematic way how this function is defined.

The linear least-squares method is used when the expression for y^C is a linear function of the coefficients a_i:

$$y_j^C = A_0(x_j) + A_1(x_j)a_1 + \cdots + A_n(x_j)a_n.$$

To minimize h it is sufficient to require that:

$$\partial h/\partial a_k = 0 \quad \forall k = 1, 2, \ldots, n$$

Then, since h is quadratic we immediately find, with $A_{ij} = A_i(x_j)$:

$$\sum_i \left(\sum_j A_{jk} A_{ji} \right) a_i = \sum_j A_{jk}(y_j^E - A_0).$$

The solution of this system of linear equations, whose matrix is symmetric, easily furnishes the unknown a_is.

Gauss–Newton method

This is a generalization of the preceding method. For a nonlinear expression of a function, y^C, we introduce the linearization:

$$y_j^C(a) = y_j^C(a^0) + \sum_{i=1}^{n} \frac{\partial y_j^C}{\partial a_i}(a^0)\Delta a_i$$

Fig. 3.22. Identification graph.

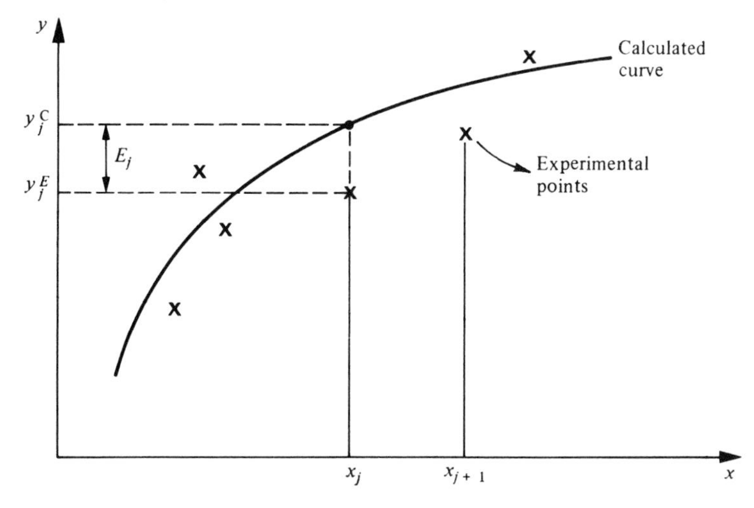

For a given set of coefficients a^0, the error function at each point y_j^C and its derivative with respect to each of the coefficients is calculated (these gradients can rarely be determined explicitly and it is generally necessary to proceed with small finite increments of each coefficient). This expression is formally identical to the expression in the previous section. To obtain a set of values Δa_i, that generally leads to an acceptable solution, it is sufficient to replace a^1 by:

$$a_i^1 = a_i^0 + \Delta a_i.$$

From this new solution, we again proceed in the same way. Note that, the convergence of this iterative method is not always certain. However, when it does converge, it does so rapidly. This means that the method is easy to use and relatively inexpensive if the number of coefficients remains small.

Fig. 3.23 shows the example of identification of the plasticity law for 316 L steel by a tension test conducted at 20 °C. The chosen equation contains three coefficients:

$$\sigma = \sigma_Y + K_Y \varepsilon_p^{1/M_Y}$$

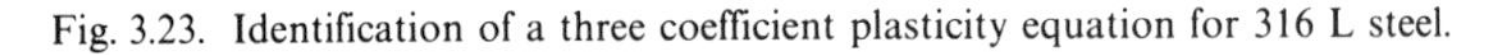

Fig. 3.23. Identification of a three coefficient plasticity equation for 316 L steel.

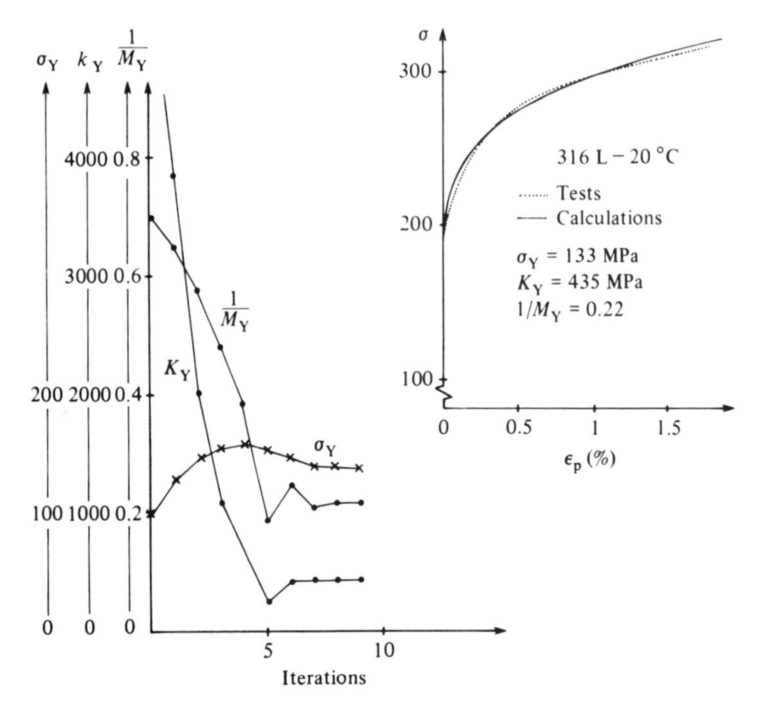

and we find that for different starting solutions, the convergence is sufficiently fast and accurate.

Variation of characteristic parameters as a function of temperature
Notwithstanding the exceptions, the characteristic parameters of a material that define each constitutive model depend on the temperature. A general method which allows us to introduce this dependency consists in writing these coefficients in the form of a parabolic function which is defined piecewise. For example, for the parameter α, let:

$$\alpha(T) = a_i(T - T_i)^2 + b_i(T - T_i) + c_i \quad \text{for} \quad T_i < T < T_{i+1}.$$

The coefficients a_i, b_i, c_i are chosen in such a way as to ensure the continuity of α and $\mathrm{d}\alpha/\mathrm{d}T$:

$$\begin{aligned} c_{i+1} &= a_i(T_{i+1} - T_i)^2 + b_i(T_{i+1} - T_i) + c_i \\ b_{i+1} &= 2a_i(T_{i+1} - T_i) + b_i. \end{aligned}$$

They are determined step by step, by the measured values of the parameter α, and by eventually introducing intermediate values to ensure a good fit.

3.3 Schematic representation of real behaviour

The qualitative aspect of the response of materials to characteristic tests allows us to classify them by the following adjectives: rigid, elastic, viscous, plastic and perfectly plastic. To each particular type there corresponds a mathematical theory developed in Chapters 4–8.

3.3.1 *Analogical models*

Analogical models are assemblies of mechanical elements with responses similar to those expected in the real material. They are used, very often for didactive purposes, to provide a concrete illustration of the constitutive equations. The analogy stops there and never concerns itself with the physical mechanisms themselves.

The most commonly used elements are the following:

the spring to represent linear elasticity

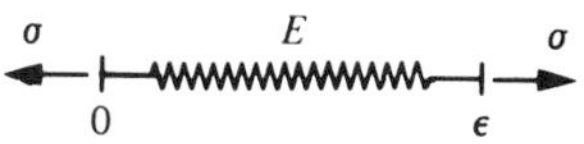

σ E σ 0 ϵ

$$\sigma = E\epsilon$$

the damper to represent linear or nonlinear viscosity

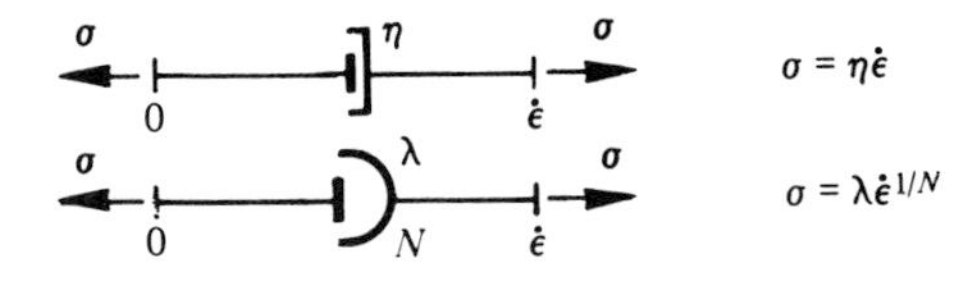

the skidding block to represent a stress threshold

$$-\sigma_s < \sigma < \sigma_s$$

the stopping block to represent a strain threshold

$$-\epsilon_s \leqslant \epsilon \leqslant \epsilon_s$$

These different elements (index i) may be assembled:

either in series:

$$\varepsilon = \sum_i \varepsilon_i \qquad \sigma = \sigma_i$$

or in parallel:

$$\sigma = \sum_i \sigma_i \qquad \varepsilon = \varepsilon_i$$

or in mixed groupings.

Using electro-mechanical analogies these models can also be realized as electrical networks but numerical calculations have replaced these analog devices.

3.3.2 *Rigid solid and perfect fluid*

These are only mentioned as a matter of interest as they do not belong to the mechanics of deformable solids. In practice, the distinction between solids and fluids is a subjective one, and it can only be linked to the choice of a time scale:

a solid admits a state of equilibrium under load;
a fluid undergoes a flow for any load even for a very small one.

It is not easy to differentiate between an infinitely slow flow and an equilibrium reached in an infinite time! The ambiguity can be resolved by using a time scale linked to the phenomenon under consideration. But then, the notions of fluid and solid lose their objective meaning: it is possible to consider a given polymer as a solid for shock problems and as a fluid for problems of long-term stability.

3.3.3 *Viscous fluid*

A body is called a 'viscous fluid' if its response to characteristic tests is of the type shown in Fig. 3.24. Sometimes such a body is also called a viscoplastic solid (see Chapter 6).

There is flow for any value of stress $\dot{\varepsilon} = f(\sigma)$.

A simple analogical model is the Maxwell model which consists of a linear spring and a damper in series. Its constitutive equation is:

$$\dot{\varepsilon} = \frac{\dot{\sigma}}{E} + \frac{\sigma}{\eta}$$

Its response to a relaxation test $\varepsilon = \varepsilon_0 H(t)$ where $H(t)$ is the Heaviside step function, $H = 0$ if $t < 0$ and $H = 1$ if $t \geqslant 0$, is:

$$\sigma = E\varepsilon_0 \exp\left(-\frac{E}{\eta}t\right).$$

Application to 'soft' solids: thermoplastic polymers in the vicinity of their melting temperature, fresh concrete (neglecting its ageing), numerous metals at a temperature close to their melting point.

3.3.4 *Elastic solids*

The deformation of elastic solids is essentially reversible. A detailed study of these solids is presented in Chapter 4.

Fig. 3.24. Viscous fluids.

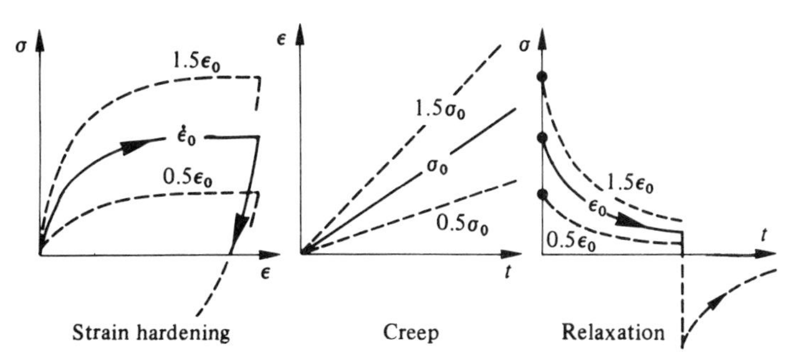

Perfectly elastic solid

The qualitative responses of a perfectly elastic solid to three characteristic tests are as shown in Fig. 3.25.

The reversibility is instantaneous. The stress–strain relation is:

$$\sigma = f(\varepsilon).$$

The analogical model of the linear elasticity is the spring:

$$\sigma = E\varepsilon.$$

Applications: metals, concrete, rocks loaded below the elastic limit.

Viscoelastic solid (Fig. 3.26)

The recovery of deformation is 'delayed' and is achieved only after an infinite time:

$$\sigma = f(\varepsilon, \dot{\varepsilon}).$$

Fig. 3.25. Perfectly elastic solid.

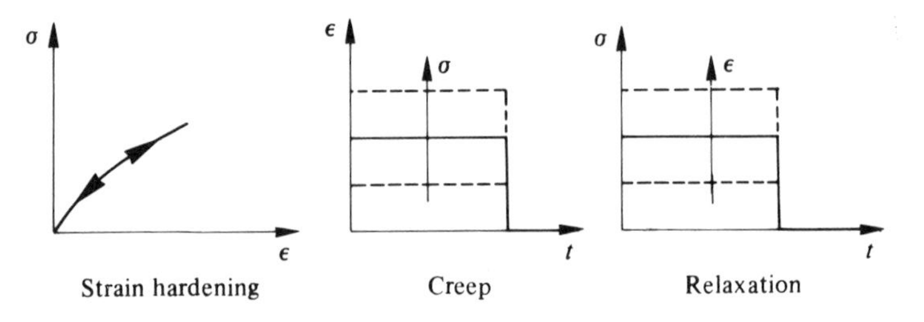

Fig. 3.26. Viscoelastic solids.

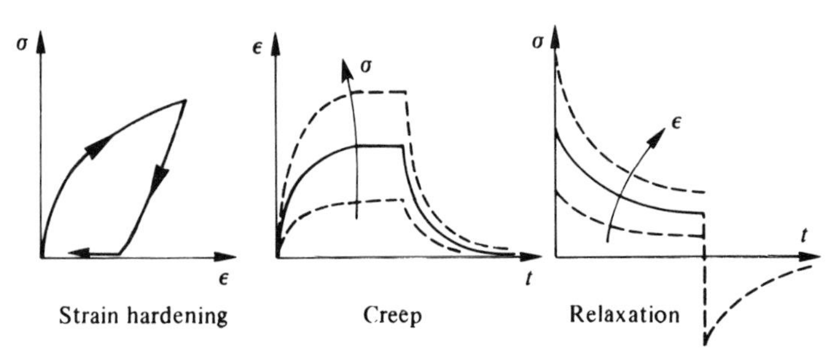

A simple analogical model is the Kelvin–Voigt model which consists of a linear spring and a damper connected in parallel:

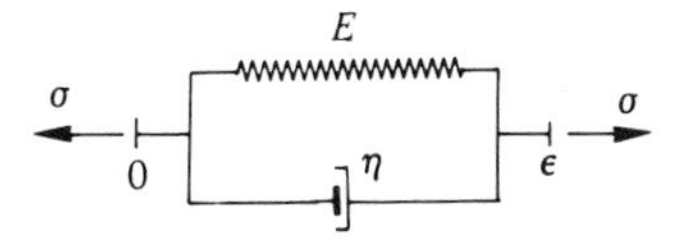

Its response to a creep test $\sigma = \sigma_0 H(t)$ is:

$$\varepsilon = \frac{\sigma_0}{E}\left[1 - \exp\left(-\frac{E}{\eta}t\right)\right].$$

The generalized analogical models of Kelvin–Voigt and Maxwell for solids are described in Chapter 4. Viscoelastic solids often exhibit the phenomenon of ageing; this point is discussed in Section 3.3.8.

Applications: organic polymers, rubber, wood when the load is not too high.

3.3.5 ***Plastic solids***

Plastic solids are those which after the application of a load reveal instantaneously stable permanent deformations and which are in equilibrium with the load. Their behaviour is not explicitly related to time.

By definition, the plastic strain ε_p is that which corresponds to the relaxed configuration $\varepsilon_p = \varepsilon(\sigma = 0)$. Chapter 5 is devoted to a detailed study of plastic solids.

Rigid perfectly plastic solid (Fig. 3.27)

The strain is either zero or negligible until a stress threshold σ_s is reached and is arbitrary at this value of stress, independent of the strain rate $\dot{\varepsilon}$ in a

Fig. 3.27. Rigid perfectly plastic solid.

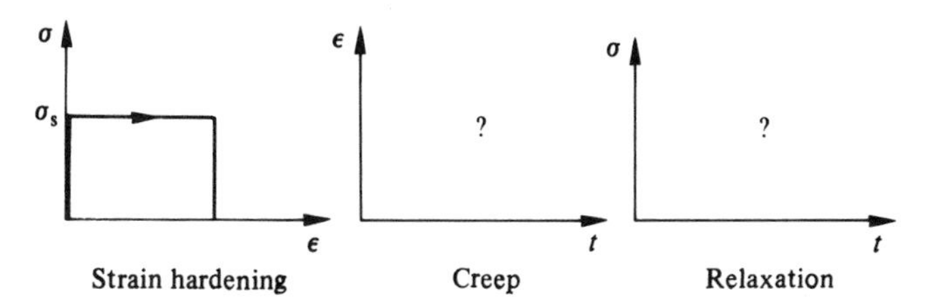

hardening test or time in a creep or a relaxation test:

$$|\sigma| < \sigma_s \rightarrow \varepsilon = 0$$

$$\sigma = \sigma_s \operatorname{Sgn}(\dot{\varepsilon}) \rightarrow \varepsilon = \varepsilon_p \qquad \text{(arbitrary)}$$

The analogical model is a skidding block.
Applications: soil mechanics, analysis of metal forming (cf. Chapter 5).

Elastic perfectly plastic solid (Fig. 3.28)

The strain is linear elastic before the threshold σ_s is reached. Thereafter, the strain is arbitrary and independent of the strain rate for this value of stress:

$$|\sigma| < \sigma_s \rightarrow \varepsilon = \varepsilon_e = \sigma/E$$

$$\sigma = \sigma_s \operatorname{Sgn}(\dot{\varepsilon}) \rightarrow \varepsilon = \varepsilon_p \qquad \text{(arbitrary).}$$

The analogical model for an elastic, perfectly plastic solid is the Saint–Venant model consisting of a linear spring in series with a skidding block:

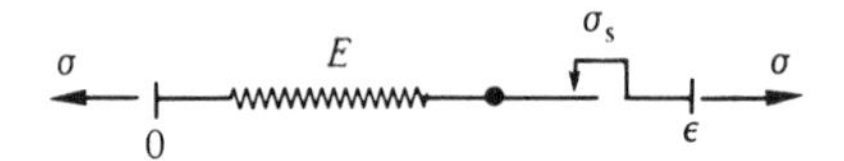

Applications: steels with a low carbon content (i.e., those exhibiting a plateau) for $\varepsilon < 2 \times 10^{-2}$ and $T < \frac{1}{4}T_M$ (where T_M is the melting temperature in K), limit analysis (cf. Chapter 5).

Elastoplastic hardening solid (Fig. 3.29)

The total strain is the sum of a linear elastic part and a plastic part which is zero before the threshold σ_s:

$$|\sigma| < \sigma_s \rightarrow \varepsilon = \varepsilon_e = \sigma/E$$

$$|\sigma| \geqslant \sigma_s \rightarrow \varepsilon = \varepsilon_e + \varepsilon_p = \sigma/E + g(\sigma).$$

Fig. 3.28. Elastic perfectly plastic solid.

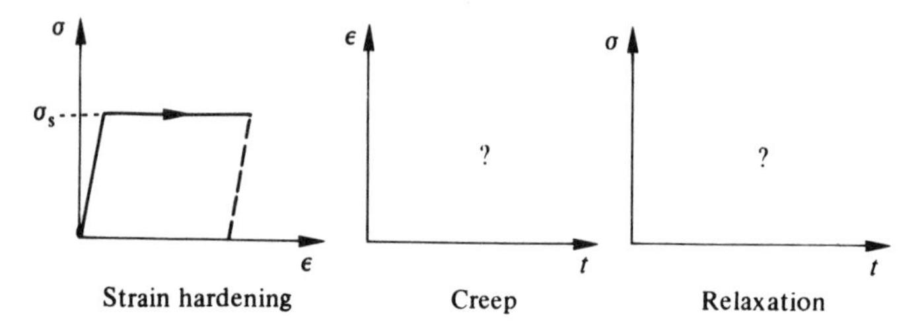

The corresponding analogical model is the generalized Saint–Venant model. Assuming that the thresholds σ_{s_i} are in increasing order and that the threshold has just been reached for the element j, its constitutive equation is the following:

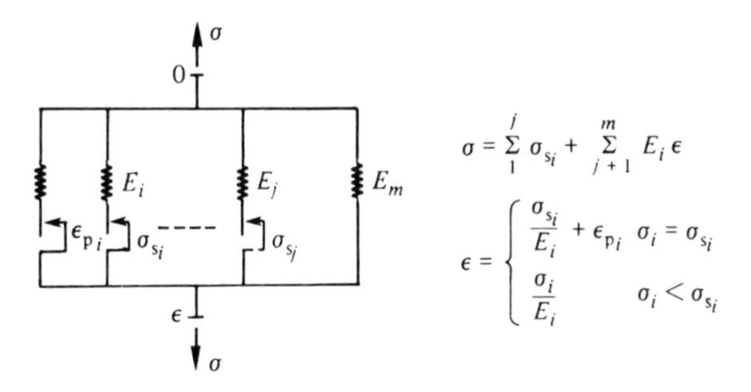

The stress–strain curve is piecewise linear, the threshold stresses are given by:

$$\sigma'_{s_j} = \sum_1^{j-1} \sigma_{s_i} + \frac{\sigma_{s_j}}{E_j} \sum_j^m E_i.$$

Its hardening curve after unloading from tension can be derived from its hardening curve in compression by a homothetic transformation with ratio 2 and a centre P' symmetrical to the unloading point P with respect to the origin 0 (Fig. 3.30).

Applications: metals and alloys at temperatures lower than one quarter of their absolute melting temperature (expressed in K).

3.3.6 *Viscoplastic solids*

Viscoplastic solids are those which exhibit permanent deformations after the application of loads (like plastic solids) but which continue to undergo a

Fig. 3.29. Elastoplastic hardening solid.

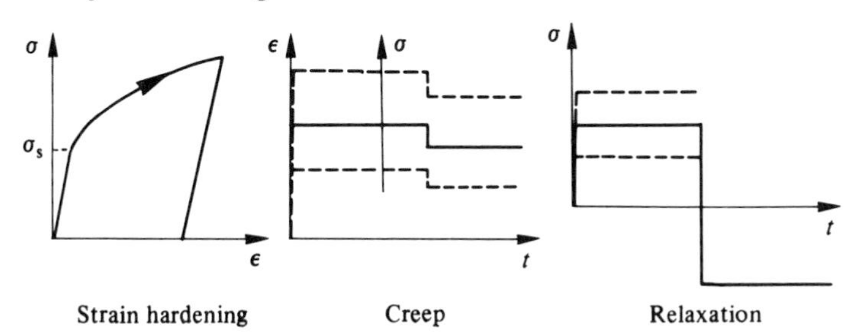

creep flow as a function of time under the influence of the applied load (equilibrium is impossible). They are studied in Chapter 6.

Perfectly viscoplastic solid (Fig. 3.31)

The rate of permanent strain (as for viscous fluids) is a function of the stress: $\sigma(\dot{\varepsilon})$.

Norton model

$$\sigma = \lambda \dot{\varepsilon}^{1/N}.$$

Fig. 3.30. Stress–strain curve for the generalized Saint–Venant model.

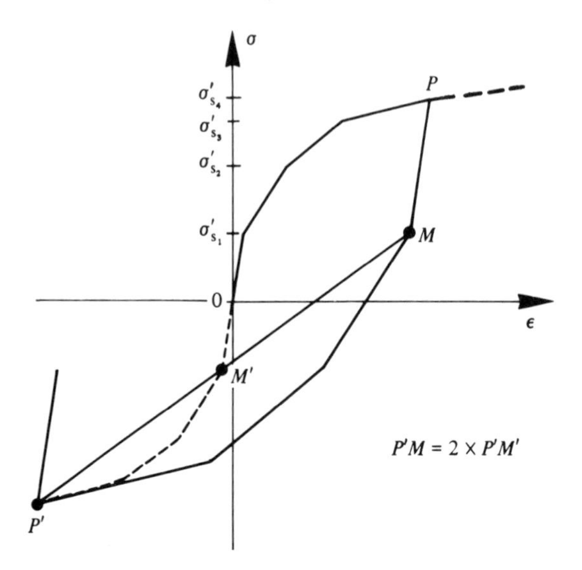

Fig. 3.31. Perfectly viscoplastic solid.

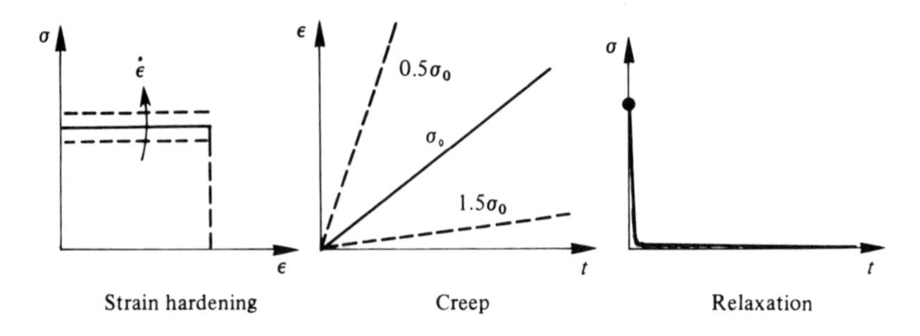

Applications: rough schematic representation of metals and alloys at temperatures higher than one third of their absolute melting point (in K).

Elastic perfectly viscoplastic solid (Fig. 3.32)

The elasticity is no longer considered negligible but the rate of plastic strain is only a function of the stress. There is no influence of hardening.

$$|\sigma| < \sigma_s \rightarrow \varepsilon = \varepsilon_e = \sigma/E$$

$$|\sigma| \geqslant \sigma_s \rightarrow \varepsilon = \varepsilon_e + \varepsilon_p \quad \dot{\varepsilon} = \dot{\sigma}/E + f(\sigma).$$

Bingham–Norton Model:

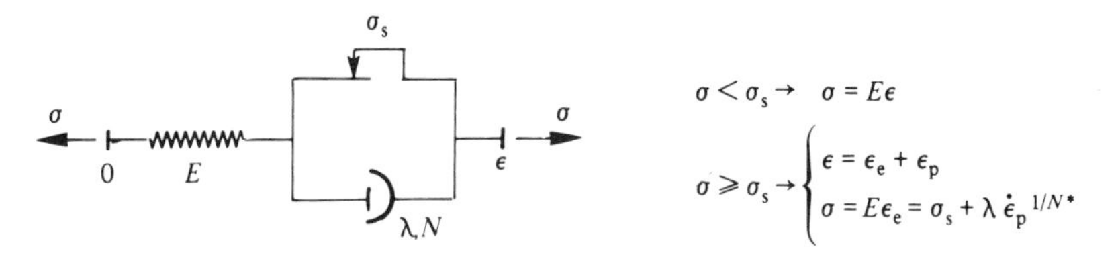

Its response to a relaxation test is:

$$\varepsilon = \varepsilon_0 H(t) \rightarrow \sigma = \sigma_s + \frac{E\varepsilon_0 - \sigma_s}{\left[1 + \frac{(N-1)E}{\lambda^N}(E\varepsilon_0 - \sigma_s)^{N-1} t\right]^{1/(N-1)}}.$$

Elastoviscoplastic hardening solid (Fig. 3.33)

This is the most complex schematic representation because the stress depends on the plastic strain rate and on the plastic strain itself or on some

Fig. 3.32. Elastic perfectly viscoplastic solid.

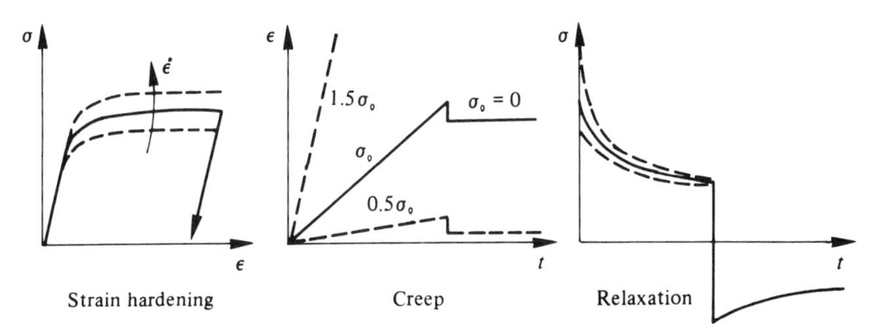

other hardening variable.

$$|\sigma| < \sigma_s \rightarrow \varepsilon = \varepsilon_e = \sigma/E$$

$$|\sigma| \geqslant \sigma_s \rightarrow \varepsilon = \varepsilon_e + \varepsilon_p$$

$$\sigma = E\varepsilon_e = f(\varepsilon_p, \dot{\varepsilon}_p).$$

Applications: metals and alloys at medium to high temperatures, wood under high loads.

3.3.7 ***Characterization of work-hardening***

There are different ways of schematically representing the hardening of materials induced by deformations.

Isotropic hardening

Even though most materials exhibit a strong, hardening induced anisotropy, the isotropic hardening representation is very often used. This is because of its simplicity and because it is a good representation in the case of proportional loading, i.e. when the representative stress vector maintains a constant direction in the stress space.

In order to demonstrate the presence of this isotropy, or lack of it (i.e. anisotropy), we must examine the results of experiments conducted with different load directions: biaxial tension tests, tension (or compression) and torsion tests. The initial anisotropy is detected by tests conducted on samples taken in different directions with respect to the material. A material will be considered to justify the hypothesis of isotropic hardening if the boundary of the elastic domain is found to depend only on a scalar parameter (Fig. 3.34).

Fig. 3.33. Elastoviscoplastic solid.

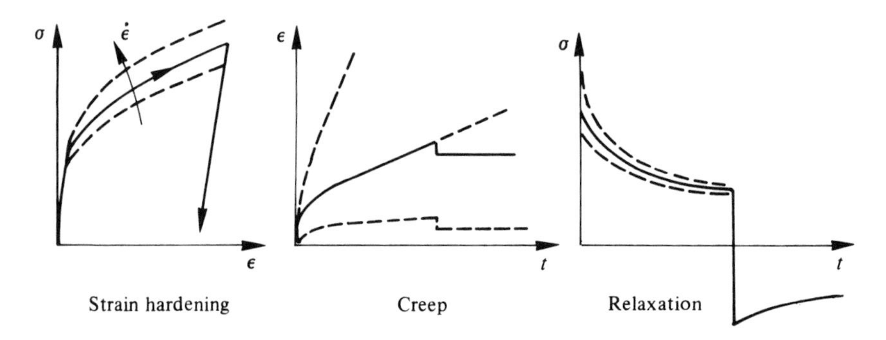

The compression curve subsequent to the initial loading in tension in a work-hardening test can be derived from the monotonic tensile curve by a homothetic transformation with ratio 1 and with centre at the point of the zero stress (point B in Fig. 3.34).

The loading curves, which represent points corresponding to the limit of elasticity in a two-dimensional stress space, of normal stress and shear stress (tension–torsion tests on thin tubes at different states of hardening), are derived from one another by homothetic transformation about the centre 0 (Fig. 3.34).

Kinematic hardening

A very useful schematic representation of anisotropic hardening is that of linear kinematic hardening in which the elastic domain retains a constant size but moves about in the stress space by translation. The 'centre of the elastic domain' (C in Fig. 3.35) represents the internal stress of the neutral state (or back stress).

The one-dimensional compression curve can be derived from the new tension curve by a homothetic transformation with ratio -1 and centre C.

In the tension–torsion test the loading curves corresponding to different hardening states can be derived from one another by translation of the vectors such as $\overrightarrow{OC}$.

Fig. 3.34. Isotropic hardening: (*a*) tension–compression test; (*b*) tension–torsion test.

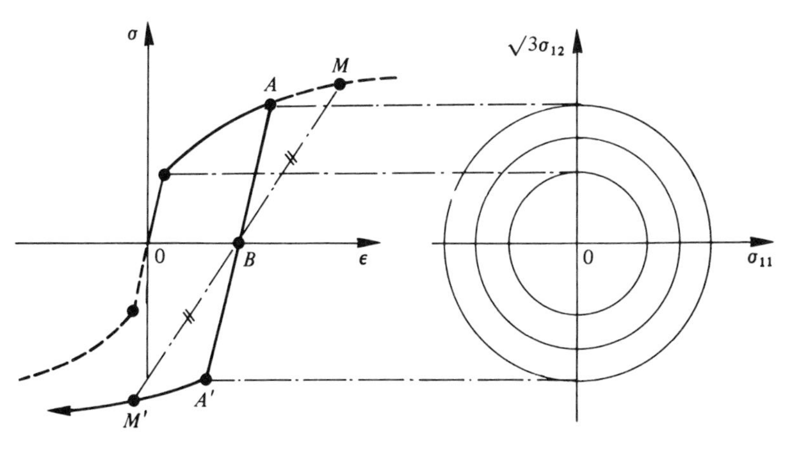

Bauschinger effect

The Bauschinger effect manifests itself when a specimen is subjected to a tension test followed by a compression test; it is often found that since the tension test was carried out first, the material has hardened in tension (increased yield stress) but has softened in compression. Fig. 3.36 shows that the yield stress in compression is lower than that if the test were carried out in compression first.

Of the two simple schematic representations mentioned above, the kinematic hardening one is closer to the real case and represents a first approximation to the Bauschinger effect.

Effect of cyclic loadings

In tension–compression cyclic loadings, most metals and alloys experience a variation in their hardening properties during the cycles. They may soften

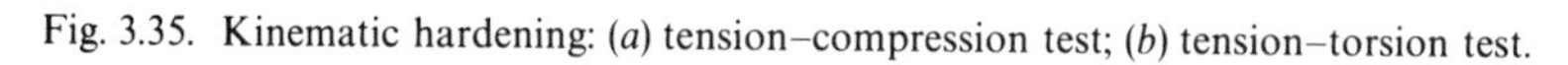
Fig. 3.35. Kinematic hardening: (*a*) tension–compression test; (*b*) tension–torsion test.

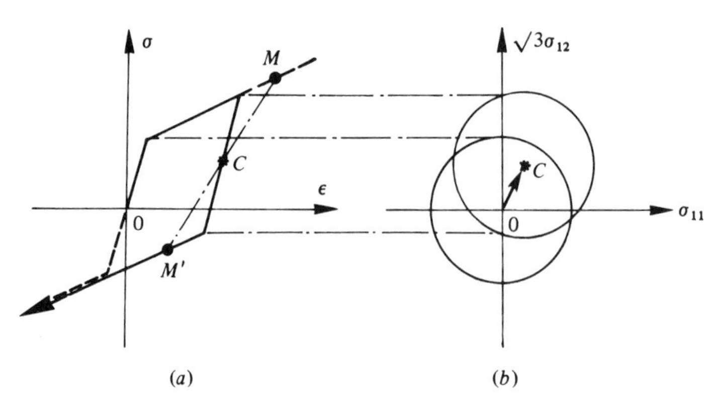

Fig. 3.36. Bauschinger effect.

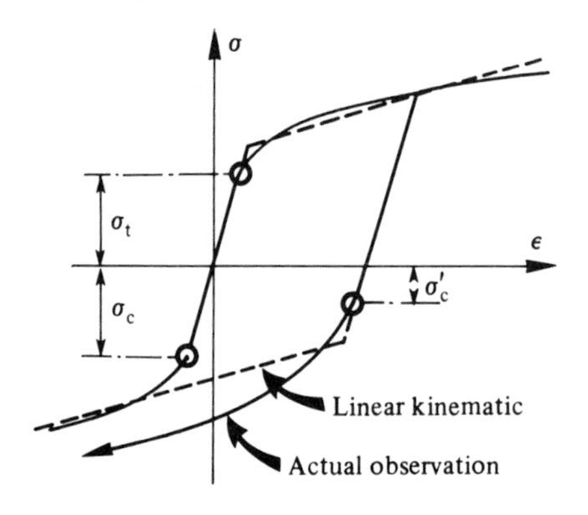

Fig. 3.37. A stress–strain cycle.

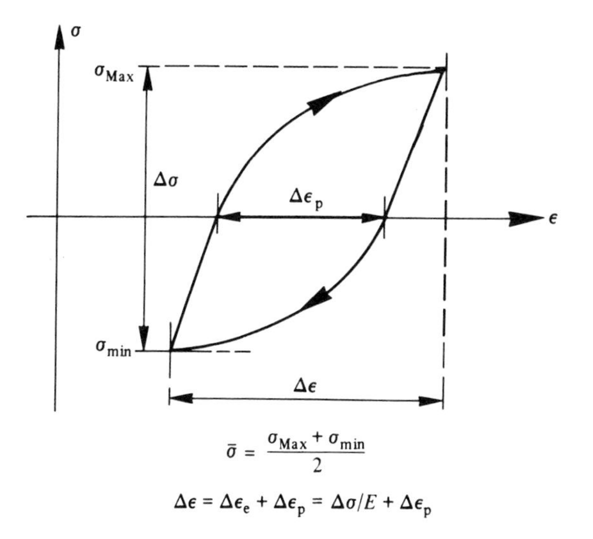

Fig. 3.38. Phenomenon of cyclic softening: (*a*) controlled strain; (*b*) controlled stress

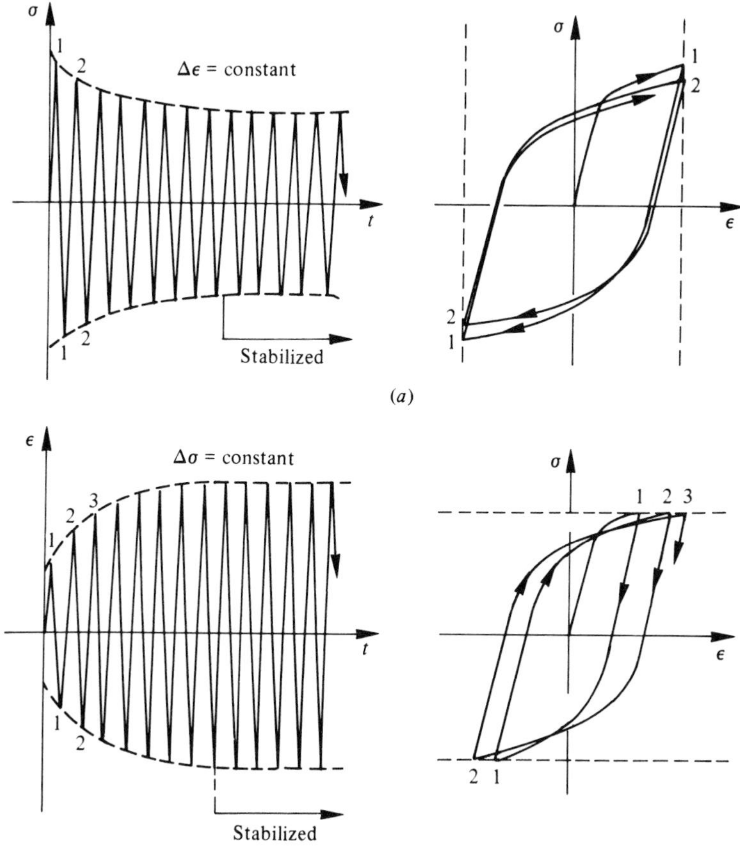

or harden depending on the material, temperature, and the initial state. The quantities generally used in describing the results of cyclic tests (with stabilized cycles) are defined in Fig. 3.37.

The softening is said to occur when the stress range $\Delta\sigma$ decreases during successive cycles under controlled strain (Fig. 3.38(*a*)), or when the strain range $\Delta\varepsilon$ increases in a stress controlled test (Fig. 3.38(*b*)). On the other hand, a cyclic hardening corresponds to a rise in the stress range $\Delta\sigma$ when strain is controlled (Fig. 3.39(*a*)) or to a fall in the strain range $\Delta\varepsilon$ when the

Fig. 3.39. Phenomenon of cyclic hardening: (*a*) controlled strain; (*b*) controlled stress.

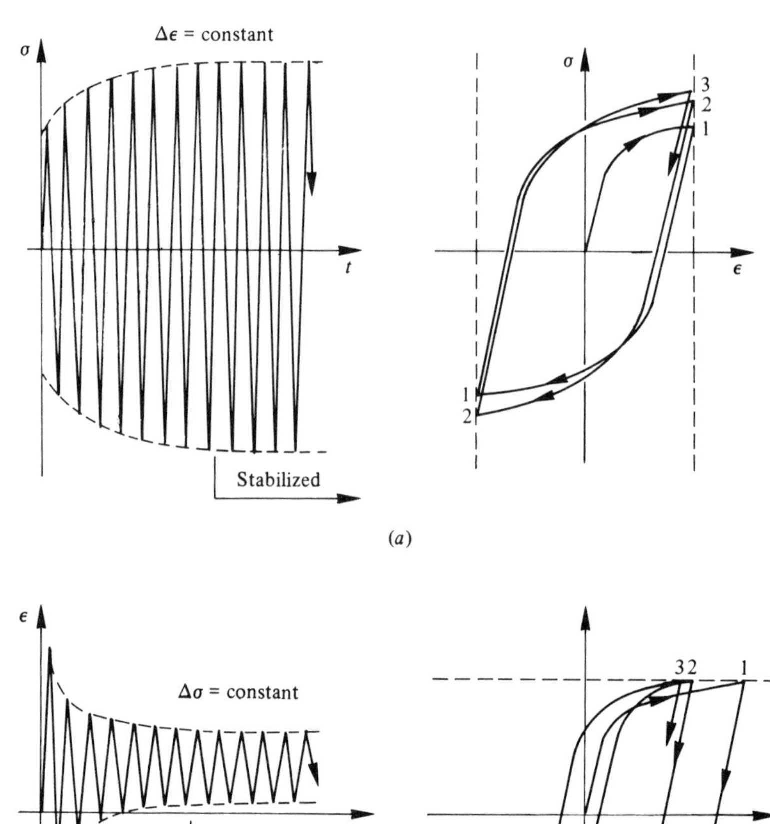

test is stress-controlled (Fig. 3.39(*b*)). When a periodic load induces a periodic response, we have stabilization in the sense that there exists a stabilized cycle (see Section 3.2.1).

If the load is not purely alternating, additional effects can occur (Fig. 3.40(*a*) and (*b*)). In nonsymmetric stress-controlled tests, either shakedown may occur or, more often, a ratchetting effect may be induced. In the case of ratchetting, there is a progressive increase in strain at each cycle, even in a stabilized regime (so that we have a periodic strain

Fig. 3.40. Phenomena of (*a*) shakedown, (*b*) ratchetting, (*c*) non-relaxation and (*d*) relaxation of the mean stress.

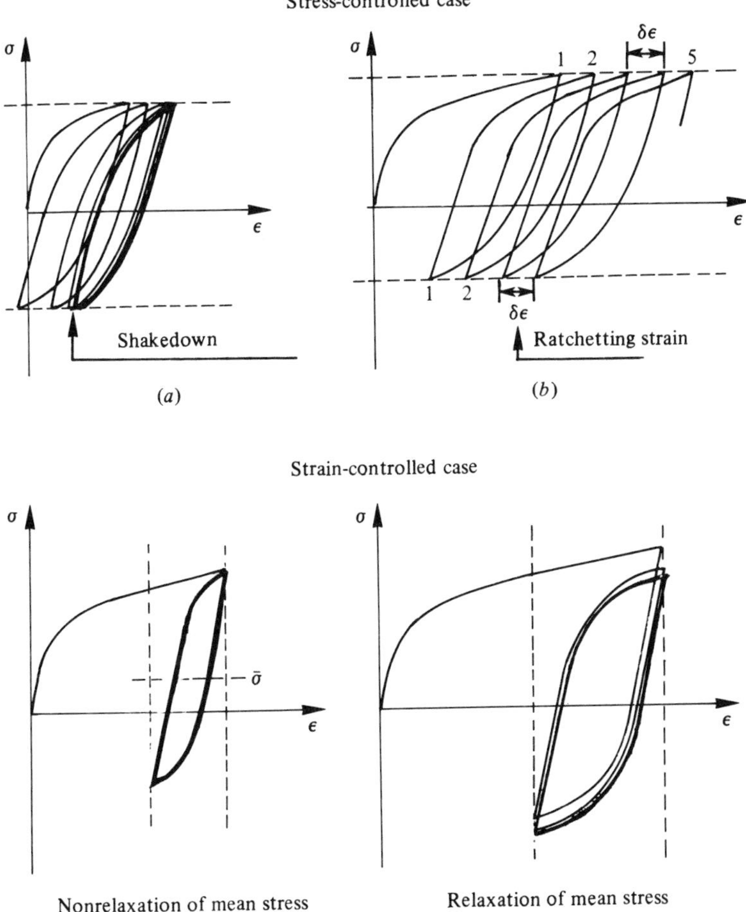

superimposed on a secular term representing progressive strain). Correspondingly, in a strain-controlled test we have the phenomena of relaxation or nonrelaxation of the mean stress (Fig. 3.40(*c*) and (*d*)).

3.3.8 *Ageing*

The materials considered in the previous sections are assumed to be stable in the sense that their characteristic properties do not change with time. The effects of ageing on a material are determined by comparing responses to characteristic tests (for example Fig. 3.41).

(*a*) The reference test is conducted before any ageing of the material (tensile test, for example).

(*b*) The same test is conducted on a specimen, initially identical, after waiting for a long or short period.

If the responses to the tests (*a*) and (*b*) are different, then ageing is said to have occurred. If the response is identical in both tests the material is stable. Materials which age include not only concrete and polymers, but also metal alloys under certain temperature conditions.

3.4 Schematic representation of fracture

Characterization of fracture remains a delicate task as it is partly founded on the physical examination of the morphology of the fractured parts and depends on the scale at which the analysis is performed: the microscopic scale of the fracture mechanisms, the macroscopic scale of volume element, and the scale of the structure. The last two lead to the modelling described in Chapters 7 and 8.

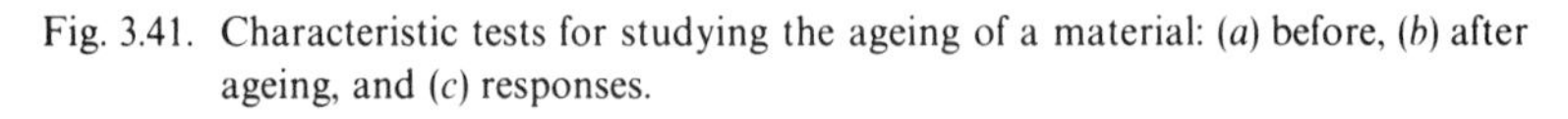

Fig. 3.41. Characteristic tests for studying the ageing of a material: (*a*) before, (*b*) after ageing, and (*c*) responses.

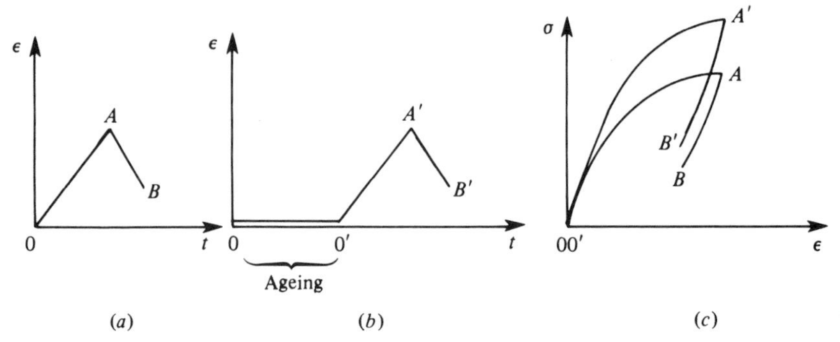

The observation of failure patterns in metals requires the preparation of samples by the use of metallographic techniques. Visual examination is limited to a resolution of the order 0.1 mm. With a binocular magnifying glass (with a maximum magnification of the order of 100 ×) it is already possible to identify the breaking zones which correspond to microbrittle fracture (shiny appearance), to fatigue fracture (dull appearance) or to a macroductile fracture (change in geometry). With an optical microscope (with a maximum magnification of the order of 1000 ×) it is possible to begin to observe the loss of cohesion associated with microbrittle fracture, the striations associated with fatigue fracture, and the cavities which lead to microductile fracture; but it is mostly with the use of the electron microscope (with magnification of 10 000 – 100 000 × and even up to 10^6 ×) that these defects can be finely analyzed so as to be able to deduce the kinematics of fracture mechanisms described in Chapter 1.

3.4.1 *Fracture by damage of a volume element*

At the level of the volume element, brittle microdecohesion and microductility induce very significant local plastic deformations. They are always present but in varying proportions, and the schematic representation depends on this proportion.

The volume element is considered as totally damaged, i.e., broken following the initiation of a macroscopic crack. In a structure, a conventional crack size of 1 mm^2 or a characteristic dimension of 1 mm is acceptable with respect to the crystalline or molecular nature of metals and polymers and with respect to analyses based on continuum mechanics. This dimension is of the order of 1 cm for wood and of 10 cm for concrete.

Macrobrittle damage

This is damage through decohesion experienced under monotonic loading by materials such as concrete in the absence of irreversible macroscopic strains. The only irreversible deformations which occur are due to the arrangement of the cracks constituting the damage. In a hardening test the fracture occurs without appreciable necking (Fig. 3.42(*a*)).

Ductile plastic damage

Schematically this is damage associated with large plastic strains which leads to the growth of cavities through the mechanism of microductile instability.

The accumulated plastic strain is more than $(10\text{–}50) \times 10^{-2}$.
The hardening curve possesses a large necking phase (Fig. 3.42(*b*)).

Brittle viscoplastic damage

This is the damage which occurs in conjunction with creep strain. Therefore it is concerned with metals at moderate and high temperatures at which the decohesion of microbrittle fracture at grain boundaries is dominant.

Fig. 3.42. (*a*) Hardening curve of a brittle material. (*b*) Hardening curve of a ductile material.

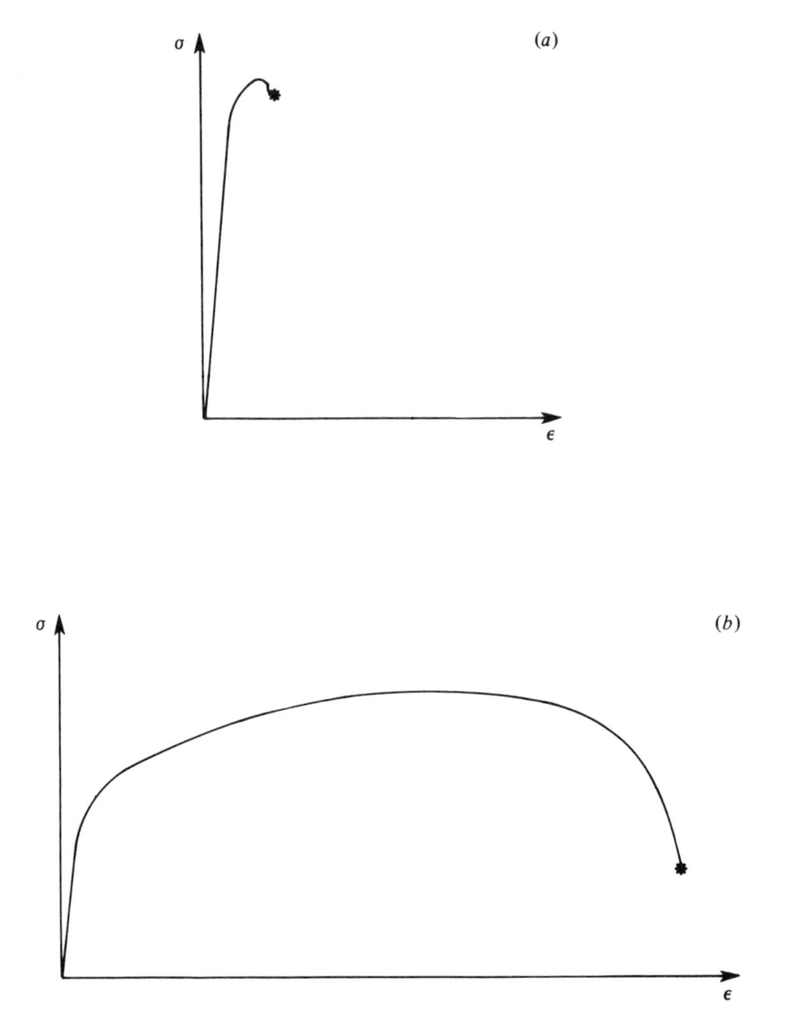

Irreversible deformations can be significant but they have no direct influence on the mechanism of decohesion. It is essentially a function of time. As the two mechanisms of decohesion and deformation are present simultaneously, this type of damage cannot be characterized by global tests. It can be detected only through a micrographic study of the fractured surfaces. An example is given in Fig. 1.19.

Fatigue damage

Under the action of repeated loads, which may or may not be periodic, microbrittle fractures occur in a discontinuous fashion with respect to time. They are often intracrystalline and appear as striations on the fractured surface. An example is given in Fig. 1.21. This damage occurs in the small strain regime and is essentially a function of the number of load cycles.

3.4.2 ***Fracture by crack propagation in a structure***

When a crack is initiated in a structure (1 mm, 1 cm, 10 cm) then under the action of the load, it may grow and cause the fracture of that structure, i.e., breaking into several pieces. Here again, different mechanisms can occur, depending on the material and the type of load.

Crack propagation by brittle fracture

In this type of fracture, crack propagation results from an instability phenomenon. It spreads with great speed without appreciable plastic deformations. The only energies which come into play in creating the surface discontinuity of the crack are the stored elastic energy and the fracture energy. The graph of force versus length or area of the crack is like that shown in Fig. 3.43(*a*).

Crack propagation by ductile fracture

In this type of fracture, the crack growth is stable. It propagates with a speed which is a function of the rate of loading of the structure. In the energy balance equation, the energy dissipated in plastic deformation in the vicinity of the crack-tip must be taken into consideration. The corresponding graph is like that shown in Fig. 3.43(*b*).

Creep crack growth

In metals at high temperature, the crack can grow under a constant load at a speed which depends on the level of the load.

Fatigue crack growth

By the same mechanism as for the volume element, periodic or nonperiodic repeated loads generate a discontinuous propagation of the crack for which the number of cycles is the appropriate kinematic variable. Striations can be observed on the crack surface.

Fig. 3.43. Curve of force versus crack area: (*a*) brittle fracture, (*b*) ductile fracture.

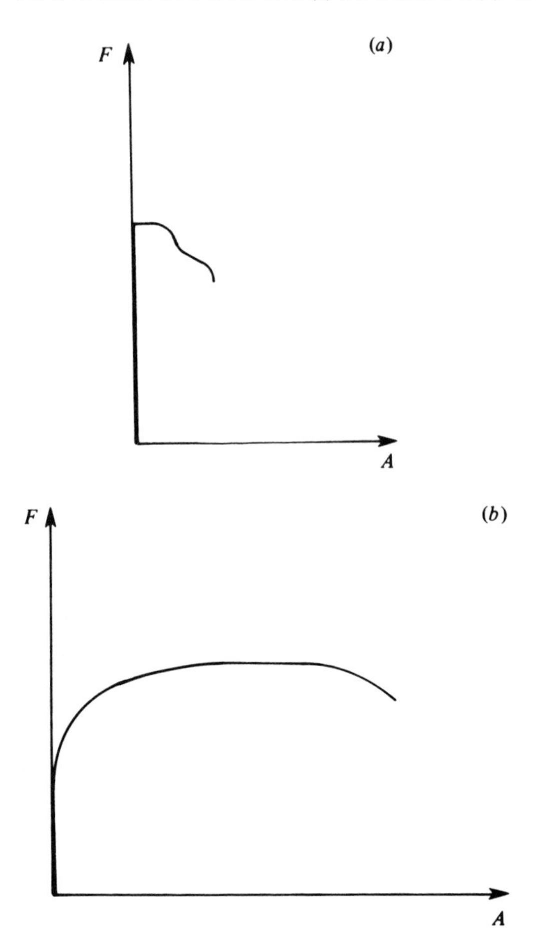

3.5 Schematic representation of friction

In problems related to forming, it becomes necessary to model frictional forces at the points of contact between the deformed material and the tool (as, for example, in the extrusion process). The interactions between the surfaces are dependent upon the roughness of the surfaces in contact as this reduces the area of real contact to a fraction of its apparent value. As a consequence, any normal force induces local stresses beyond the elastic limit.

3.5.1 *Coulomb model*

This model is derived from the theory of friction by adhesion. The coefficient of friction f is defined as the ratio of the tangential force to the normal applied force. The tangential force is expressed as the product of the real area of contact and the fracture shear stress τ_R of the less resistant of the two materials. Assuming that the normal force can be expressed by the product of real contact area and the Brinel hardness H of the materials, it follows that the coefficient of friction is: $f = \tau_R/H$ which is a quantity characteristic of the two materials in contact.

The Coulomb model expresses the shear stress σ_{12} opposing the relative translation of the two materials subjected to a normal contact stress σ_{11} in the following form:

- $\sigma_{12} = f\sigma_{11}$ if $f\sigma_{11} < \tau_R$,
- $\sigma_{12} = \tau_R$ if $f\sigma_{11} \geqslant \tau_R$.

Table 3.2 gives values of f for some materials.

3.5.2 *Boundary layer model*

A thin layer of oxides or of lubricants providing perfect adhesion between the two solids is often formed at an interface. Assuming that the relative movement between the two bodies can only take place through the shear of this layer by a shear stress equal to a fraction $\bar{m}$ of the fracture shear stress or the plastic yield stress in shear of the less resistant material, we have:

- $\sigma_{12} = \bar{m}\tau_R$

where $\bar{m}$ is the coefficient of friction of the boundary layer $0 \leqslant \bar{m} \leqslant 1$, $\bar{m} = 0$ characterizes contact without friction and $\bar{m} = 1$ characterizes the 'sticking'

Table 3.2. *Approximate values of coefficients of friction*

Metal	Temperature °C	Lubrication	f	$\bar{m}$
Mild steel	20	Dry	0.25	0.5
Mild steel	950	Dry	0.45	0.3–1
Mild steel	20	Oil	0.03	0.05
Steel	20	Soaps	0.05	0.07
Aluminium	20	Oil	0.05	0.15
Copper	20	Dry	0.1	0.9
Copper	850	Graphite	0.25	0.2

contact of materials at high temperatures. Table 3.2 gives several characteristic values of $\bar{m}$.

Bibliography

Persoz B. *Introduction à l'étude de la rhéologie.* Dunod, Paris (1960).

Malvern L. E. *Introduction to the mechanics of a continuous medium.* Prentice Hall New Jersey (1969).

Mandel J. *Propriétés mécaniques des matériaux.* Eyrolles, Paris (1978).

Lieurade H. P. *La pratique des essais de fatigue.* P.Y.C. Edition, Paris (1982).

Broyden C. G., Fletcher R., Murray W., Powel M. J. D. & Swann W. M. *Numerical methods for unconstrained optimization.* Murray Ed. (1972).

Loveday M. S., Day M. F., Dyson B. F. *Measurement of high temperature mechanical properties of materials.* National Physical Laboratory. Her Majesty's Stationery Office, London (1982).

ASTM. *Metals Handbook*: Vols. 1–12. American Society for Metals, Philadelphia.

4

LINEAR ELASTICITY, THERMOELASTICITY AND VISCOELASTICITY

Une loi est un modèle qui n'est plus (et pas encore!) contesté.

Hooke (1676), Young (1807), Cauchy (1822) and Timoshenko (1934) have said almost everything about the linear-elastic behaviour of materials. In this field, we will therefore limit ourselves to giving a summary in the form of formulae. However, we will deal with anisotropic elasticity, which is so important for composite materials, and the identification of the coefficients. The term elasticity is taken here to mean the reversible deformations mentioned in Chapter 3. Neither thermal dissipation (thermoelasticity) nor mechanical dissipation (viscoelasticity) is excluded.

Thermoelasticity introduces several additional coefficients into the constitutive law including the dilatation coefficient, and permits the treatment of problems involving temperature variations, such as thermal stress analysis problems.

The theory of viscoelasticity was developed considerably with rheology during the 1960s. We will restrict ourselves in this chapter to linear viscoelasticity which is sufficient to deal with the mechanical behaviour of certain polymers.

These theories are continuum theories; the inhomogeneities are assumed to be small enough with respect to the size of the volume element that the results of experiments conducted on the latter are really the characteristics of average macroscopic behaviour. This is also true for composite materials and concrete whose behaviour can be more accurately described by homogenization techniques.

4.1 Elasticity

4.1.1 *Domain of validity and use*

All solid materials possess a domain in the stress space within which a load variation results only in a variation of elastic strains. As discussed in

Chapter 1, these elastic strains consist of reversible movements of atoms, molecules, or cells, corresponding to strains not exceeding $(0.2\text{–}0.5) \times 10^{-2}$ for metals, composites, concrete and wood. This fact, therefore, justifies the hypothesis of small strain elastic behaviour. The dissipation is zero.

The limit of the elastic domain, the exact study of which is presented in Chapter 5, depends on the temperature and on the previous loadings, experienced during the production process or in service as a mechanical part. The temperature, while conducive to irreversible movements, at the same time decreases the elastic domain until it becomes zero at the melting point. On the other hand, mechanical loads can considerably increase this domain as a result of the hardening phenomenon.

Thus, before applying the theory of linear elasticity, it is necessary to make sure that the order of magnitude of the stresses is indeed compatible with the elastic domain of the material under consideration as defined in Chapter 5. It may not always be possible to do so '*a priori*' but it can always be done '*a posteriori*'.

The hypothesis of elastic isotropy has been amply verified for metallic materials and a little less so for concrete. For composites and wood, however, it is necessary to use special anisotropic theories of elasticity.

4.1.2 *Formulation*

The method of local state presented in Chapter 2 is followed. The schematic representation of an elastic solid, as defined in Chapter 3, shows that the state variables to be used are temperature and elastic strain. If we restrict ourselves to isothermal deformations, only the elastic strain equal to the total strain appears as a state variable; the temperature acts as a parameter on which the elastic moduli depend. Since only small deformations are considered, there is no need to distinguish between the initial and final configurations of the body.

Thermodynamic potential

In order to obtain a linear theory, it is sufficient to choose as a convex thermodynamic potential a positive definite quadratic function in the components of the strain tensor.

Let

$$\Psi = (1/2\rho)\mathbf{a}:\boldsymbol{\varepsilon}:\boldsymbol{\varepsilon}$$

where the symbol : denotes the contraction of the tensorial product on two indices and where ρ is the density and $\mathbf{a}$ is a fourth order tensor whose components are the elastic moduli which may depend on the temperature.

By definition of the associated variable, the stress tensor $\boldsymbol{\sigma}$ is derived from the potential Ψ to give the law of state:

$$\boldsymbol{\sigma} = \rho(\partial \Psi / \partial \boldsymbol{\varepsilon}) = \mathbf{a} : \boldsymbol{\varepsilon}.$$

The tensor $\mathbf{a}$ satisfies certain symmetry properties. Since

$$\sigma_{ij} = \rho(\partial \Psi / \partial \varepsilon_{ij}) = a_{ijkl}\varepsilon_{kl} \quad \text{and} \quad \sigma_{kl} = a_{klij}\varepsilon_{ij}$$

we prove the following symmetry of the moduli by differentiating the above:

$$a_{ijkl} = \frac{\partial \sigma_{ij}}{\partial \varepsilon_{kl}} = \rho \frac{\partial^2 \Psi}{\partial \varepsilon_{ij} \partial \varepsilon_{kl}} = \rho \frac{\partial^2 \Psi}{\partial \varepsilon_{kl} \partial \varepsilon_{ij}} = \frac{\partial \sigma_{kl}}{\partial \varepsilon_{ij}} = a_{klij}.$$

The relation

- $$\boldsymbol{\sigma} = \mathbf{a} : \boldsymbol{\varepsilon}$$

is the generalized Hooke's law for a linear, isotropic, elastic continuum which depends on 21 independent coefficients of the symmetric tensor of the elastic moduli. By inverting the above relations, or by defining a dual potential:

$$\Psi^* = (1/2\rho)\mathbf{A} : \boldsymbol{\sigma} : \boldsymbol{\sigma}$$

where $\mathbf{A}$ is a symmetric fourth order tensor of elastic compliances, we have:

$$\boldsymbol{\varepsilon} = \mathbf{A} : \boldsymbol{\sigma}.$$

In the case of nonlinear elasticity, one chooses other forms of the potential function Ψ which, of course, possess the convexity property.

Linear isotropic elasticity

Isotropy and linearity require that the potential Ψ be a quadratic invariant of the strain tensor, i.e. a linear combination of the square of the first invariant $\varepsilon_{\mathrm{I}}^2 = [\mathrm{Tr}(\boldsymbol{\varepsilon})]^2$ and the second invariant $\varepsilon_{\mathrm{II}}^2 = \frac{1}{2}\mathrm{Tr}(\boldsymbol{\varepsilon}^2)$:

$$\Psi = (1/2\rho)(\lambda \varepsilon_{\mathrm{I}}^2 + 4\mu\varepsilon_{\mathrm{II}})$$

and

- $$\boldsymbol{\sigma} = \rho(\partial \Psi / \partial \boldsymbol{\varepsilon}) = \lambda \,\mathrm{Tr}(\boldsymbol{\varepsilon})\mathbf{1} + 2\mu\boldsymbol{\varepsilon}$$

where $\mathbf{1}$ is the unit tensor, or

$$\sigma_{ij} = \lambda \varepsilon_{kk}\delta_{ij} + 2\mu\varepsilon_{ij}$$

where λ and μ are Lame's two coefficients. Employing the dual form:

$$\boldsymbol{\varepsilon} = \mathbf{A} : \boldsymbol{\sigma}$$

we find that

- $$\boldsymbol{\varepsilon} = \frac{1+\nu}{E}\boldsymbol{\sigma} - \frac{\nu}{E}\mathrm{Tr}(\boldsymbol{\sigma})\mathbf{1}$$

or $$\varepsilon_{ij} = \frac{1+\nu}{E}\sigma_{ij} - \frac{\nu}{E}\sigma_{kk}\delta_{ij}$$

where E is Young's modulus and ν is Poisson's ratio. Their values for different materials are given in the table in Section 4.1.4.

Alternatively, the relations between the spherical parts:

$$\sigma_H = \tfrac{1}{3}\operatorname{Tr}(\boldsymbol{\sigma})$$

$$\varepsilon_H = \tfrac{1}{3}\operatorname{Tr}(\boldsymbol{\varepsilon})$$

and the deviatoric parts:

$$\boldsymbol{\sigma}' = \boldsymbol{\sigma} - \sigma_H \mathbf{1}$$

$$\boldsymbol{\varepsilon}' = \boldsymbol{\varepsilon} - \varepsilon_H \mathbf{1}$$

are also useful, namely:

- $$\varepsilon_H = \frac{1-2\nu}{E}\sigma_H,$$
- $$\boldsymbol{\varepsilon}' = \frac{1+\nu}{E}\boldsymbol{\sigma}'.$$

Finally, we list some well-known relations between Lame's coefficients, Young's modulus, Poisson's ratio, shear modulus and the bulk modulus in hydrostatic compression:

$$\lambda = \frac{\nu E}{(1+\nu)(1-2\nu)}, \qquad \mu = \frac{E}{2(1+\nu)}$$

$$E = \mu\frac{3\lambda + 2\mu}{\lambda + \mu}, \qquad \nu = \frac{\lambda}{2(\lambda + \mu)}$$

$$G = \frac{E}{2(1+\nu)} = \mu, \qquad K = \frac{E}{3(1-2\nu)} = \frac{3\lambda + 2\mu}{3}.$$

In the isotropic elasticity law derived from a quadratic potential, the only condition which must be satisfied by the coefficients (modulus E is assumed positive) is:

$$-1 < \nu < \tfrac{1}{2}.$$

This condition follows immediately from the general case of orthotropic elasticity (see Section 4.1.2).

Plane stress assumption

This is used, for example, for thin plates and is:

$$\sigma_{33} = \sigma_{13} = \sigma_{23} = 0.$$

Using $\sigma_{33}=0$, the two-dimensional stress–strain relation can be easily obtained as:

$$\sigma_{ij}=\lambda'\varepsilon_{kk}\delta_{ij}+2\mu\varepsilon_{ij} \qquad (i \text{ and } j=1,2)$$

where

$$\lambda'=\frac{2\lambda\mu}{\lambda+2\mu}=\frac{\nu E}{1-\nu^2}.$$

The relation between $\boldsymbol{\varepsilon}$ and $\boldsymbol{\sigma}$ becomes:

$$\begin{Bmatrix}\varepsilon_{11}\\ \varepsilon_{22}\\ \varepsilon_{12}\end{Bmatrix}=\begin{bmatrix}\dfrac{1}{E} & -\dfrac{\nu}{E} & 0\\ -\dfrac{\nu}{E} & \dfrac{1}{E} & 0\\ 0 & 0 & \dfrac{1+\nu}{E}\end{bmatrix}\begin{Bmatrix}\sigma_{11}\\ \sigma_{22}\\ \sigma_{12}\end{Bmatrix}$$

$$\varepsilon_{33}=-\frac{\nu}{1-\nu}(\varepsilon_{11}+\varepsilon_{22})$$

$$\varepsilon_{13}=0$$

$$\varepsilon_{23}=0.$$

Plane strain assumption

This is used for, example, for thick plates and is:

$$\varepsilon_{33}=\varepsilon_{13}=\varepsilon_{23}=0$$

$$\begin{Bmatrix}\sigma_{11}\\ \sigma_{22}\\ \sigma_{12}\end{Bmatrix}=\begin{bmatrix}\lambda+2\mu & \lambda & 0\\ \lambda & \lambda+2\mu & 0\\ 0 & 0 & 2\mu\end{bmatrix}\begin{Bmatrix}\varepsilon_{11}\\ \varepsilon_{22}\\ \varepsilon_{12}\end{Bmatrix}$$

$$\sigma_{33}=\frac{\lambda}{2(\lambda+\mu)}(\sigma_{11}+\sigma_{22})=\nu(\sigma_{11}+\sigma_{22})$$

$$\sigma_{13}=0$$

$$\sigma_{23}=0.$$

Orthotropic elasticity

A medium is said to be orthotropic for a given property if this property is invariant with respect to a change in direction obtained by symmetry relative to two orthogonal planes (the symmetry with respect to the third orthogonal plane then necessarily follows). In other words, if the elastic characteristics of specimens cut along the axes of symmetry with respect to a particular plane P_1 and along those with respect to another orthogonal plane P_2 are identical, then the material is orthotropic with P_1 and P_2 as the planes of orthotropy. The intersections of the three symmetry planes define the principal axes of orthotropy. This model provides a good representation of the properties of not only unidirectional composites and wood, but also of laminated metallic products.

The compliance matrix of an orthotropic material contains only nine independent coefficients. Indeed it is possible to write the relations between stress and strain in a system of coordinates in which axis 1 forms an angle α or $-\alpha$ with one of the principal axes of orthotropy:

$$\{\varepsilon\}^{(\alpha)} = [A]^{(\alpha)}\{\sigma\}^{(\alpha)}$$
$$\{\varepsilon\}^{(-\alpha)} = [A]^{(-\alpha)}\{\sigma\}^{(-\alpha)}.$$

The matrices $[A]^{(\alpha)}$ and $[A]^{(-\alpha)}$ differ only by elementary rotations due to the change of axes. By definition, orthotropy means that $\{\varepsilon\}^{(\alpha)} = \{\varepsilon\}^{(-\alpha)}$ if identical stresses are applied along the α and $-\alpha$ sets of axes, i.e., $\{\sigma\}^{(\alpha)} = \{\sigma\}^{(-\alpha)}$. This remains true for arbitrary α and for the six independent components of strain. We, therefore, obtain six relations between the elastic compliance coefficients. Another six relations are obtained by similar considerations with respect to the second principal axis of orthotropy (those with respect to the third axis are automatically satisfied). Thus there remain $21 - 6 - 6 = 9$ independent coefficients.

With the principal axes of orthotropy as the reference axes, the elasticity law can be expressed in the following form:

$$\begin{Bmatrix} \varepsilon_{11} \\ \varepsilon_{22} \\ \varepsilon_{33} \\ \varepsilon_{23} \\ \varepsilon_{31} \\ \varepsilon_{12} \end{Bmatrix} = \begin{bmatrix} \dfrac{1}{E_1} & -\dfrac{\nu_{12}}{E_1} & -\dfrac{\nu_{13}}{E_1} & 0 & 0 & 0 \\ -\dfrac{\nu_{21}}{E_2} & \dfrac{1}{E_2} & -\dfrac{\nu_{23}}{E_2} & 0 & 0 & 0 \\ -\dfrac{\nu_{31}}{E_3} & -\dfrac{\nu_{32}}{E_3} & \dfrac{1}{E_3} & 0 & 0 & 0 \\ 0 & 0 & 0 & \dfrac{1}{2G_{23}} & 0 & 0 \\ 0 & 0 & 0 & 0 & \dfrac{1}{2G_{31}} & 0 \\ 0 & 0 & 0 & 0 & 0 & \dfrac{1}{2G_{12}} \end{bmatrix} \begin{Bmatrix} \sigma_{11} \\ \sigma_{22} \\ \sigma_{33} \\ \sigma_{23} \\ \sigma_{31} \\ \sigma_{12} \end{Bmatrix}$$

with the symmetry conditions:

$$\frac{\nu_{12}}{E_1}=\frac{\nu_{21}}{E_2},\quad \frac{\nu_{13}}{E_1}=\frac{\nu_{31}}{E_3},\quad \frac{\nu_{23}}{E_2}=\frac{\nu_{32}}{E_3}.$$

The nine independent coefficients are:

three moduli of extension E_1, E_2, E_3 in the directions of orthotropy
three shear moduli G_{12}, G_{23}, G_{31}
three coefficients of contraction for example, ν_{12}, ν_{23}, ν_{31}.

These coefficients must satisfy certain inequalities to ensure that the thermodynamic potential used in the derivation of the above laws is indeed a positive definite quadratic function:

$$\rho\Psi=\tfrac{1}{2}:\boldsymbol{\varepsilon}:\boldsymbol{\varepsilon}=\tfrac{1}{2}\boldsymbol{\sigma}:\boldsymbol{\varepsilon}=\tfrac{1}{2}\boldsymbol{\sigma}:\mathbf{A}:\boldsymbol{\sigma}.$$

If we consider the extensional and shear moduli to be positive quantities, then Ψ will be a positive, definite quadratic function if the eigenvalues of the matrix of the contraction coefficients are all strictly positive. This condition requires that the following inequalities are satisfied:

$$1-\nu_{12}\nu_{21}>0,\quad 1-\nu_{23}\nu_{31}>0,\quad 1-\nu_{13}\nu_{31}>0$$

$$1-\nu_{12}\nu_{23}\nu_{31}-\nu_{21}\nu_{13}\nu_{32}-\nu_{21}\nu_{12}-\nu_{31}\nu_{13}-\nu_{32}\nu_{23}>0.$$

Examples of the values of orthotropic elasticity coefficients for some materials are given in the table in Section 4.1.4.

Transversely isotropic elasticity

An elastic medium is transversely isotropic if its elastic characteristics remain invariant for all pairs of directions symmetric with respect to an axis. Accordingly, if $\vec{x}_3$ is the axis of transverse isotropy then the material is 'isotropic' in planes normal to $\vec{x}_3$. A laminated composite obtained by the bonding of layers formed from unidirectional fibres and stacked in 'all' directions, can be considered as a transversely isotropic material. The same is true for alloys obtained from oriented solidification.

A proof, analogous to that for orthotropic material, shows that in the case of transverse isotropy only five elastic coefficients can be independent. More precisely, the property of isotropy perpendicular to the x_3 direction implies that:

$$\begin{Bmatrix} \varepsilon_{11} \\ \varepsilon_{22} \\ \varepsilon_{33} \\ \varepsilon_{12} \\ \varepsilon_{23} \\ \varepsilon_{13} \end{Bmatrix} = \begin{bmatrix} \frac{1}{E_1} & -\frac{\nu_{12}}{E_1} & -\frac{\nu_{13}}{E_1} & 0 & 0 & 0 \\ -\frac{\nu_{12}}{E_1} & \frac{1}{E_1} & -\frac{\nu_{13}}{E_1} & 0 & 0 & 0 \\ -\frac{\nu_{13}}{E_1} & -\frac{\nu_{13}}{E_1} & \frac{1}{E_3} & 0 & 0 & 0 \\ 0 & 0 & 0 & \frac{1+\nu_{12}}{E_1} & 0 & 0 \\ 0 & 0 & 0 & 0 & \frac{1}{2G_{13}} & 0 \\ 0 & 0 & 0 & 0 & 0 & \frac{1}{2G_{13}} \end{bmatrix} \begin{Bmatrix} \sigma_{11} \\ \sigma_{22} \\ \sigma_{33} \\ \sigma_{12} \\ \sigma_{23} \\ \sigma_{13} \end{Bmatrix}.$$

With reference to the orthotropic case, we have the following equalities:

$E_1 = E_2$, same extensional moduli in directions 1 and 2;
$\nu_{13}/E_1 = \nu_{23}/E_2$, same contraction coefficient in directions 1 and 2 for a tension applied in direction 3;
$G_{13} = G_{23}$, same shear moduli for shear in planes perpendicular to axis 2 and axis 1;
$2G_{12} = E_1/(1 + \nu_{12})$, shear modulus around axis 3.

The five coefficients which characterize the material are:

two extensional moduli E_1 and E_3,
one shear modulus G_{13},
two coefficients of contraction ν_{12} and ν_{13}.

Values of these coefficients for some materials are listed in the table in Section 4.1.4.

4.1.3 *Identification*

Identification of the characteristic coefficients of elasticity consists in evaluating stiffnesses in tension–compression static tests, in vibration tests, or in wave propagation tests. These three broad types of tests do not yield exactly the same results for the same material because dynamic methods do not take into account certain viscous movements of atoms or molecules, and accordingly, give slightly higher values of the moduli.

Static methods

These methods are recommended for the calculation of the deformations and the static resistance of structures. Characterization of materials at room temperature or at temperatures below 200–300 °C is done by uniaxial tension or compression tests on standard specimens in which the axial strain ε_{11} and the transverse strains ε_{22}, ε_{33} are measured by electric strain gauges.

Isotropic elastic case

The axial stress σ_{11} is determined by measuring the loading force on the specimen (see Chapter 3) and we obtain:

- $E = \sigma_{11}/\varepsilon_{11}$
- $\nu = -\varepsilon_{22}/\varepsilon_{11}$.

The relative accuracy generally obtained is of the order of 1%.

Orthotropic elastic case

The nine characteristic coefficients of orthotropic elasticity can be obtained by tension or compression tests on specimens cut out along the three directions of orthotropy and in three directions at $\pi/4$ (Fig. 4.1). To determine the coefficients from these tests, it is necessary to express by means of the orthotropic constitutive law, the relation between the normal stress $\sigma^{(n)}$ and the normal strain $\varepsilon^{(n)}$ in any direction $\vec{n}$ with respect to the axes of orthotropy (Fig. 4.1).

$$\vec{n} = \vec{x}_1 \sin\alpha \cos\Psi + \vec{x}_2 \sin\alpha \sin\Psi + \vec{x}_3 \cos\alpha.$$

A uniaxial stress σ^* in a reference system with $\vec{n}$ as a base vector has the

Fig. 4.1. Directions for the characterization of elastic orthotropy.

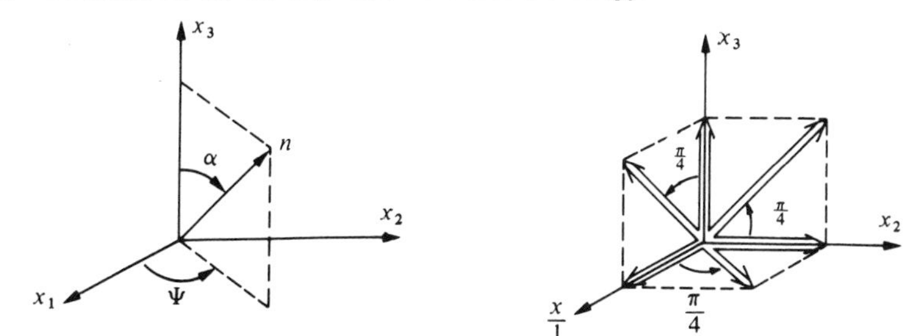

following components with respect to the orthotropy axes (x_1, x_2, x_3).

$$[\sigma] = \sigma^* \begin{bmatrix} \sin^2\alpha\cos^2\Psi & \sin^2\alpha\sin\Psi\cos\Psi & \sin\alpha\cos\alpha\cos\Psi \\ \sin^2\alpha\sin\Psi\cos\Psi & \sin^2\alpha\sin^2\Psi & \sin\alpha\cos\alpha\sin\Psi \\ \sin\alpha\cos\alpha\cos\Psi & \sin\alpha\cos\alpha\sin\Psi & \cos^2\alpha \end{bmatrix}$$

The corresponding strain components are deduced from the orthotropic constitutive law:

$$\boldsymbol{\varepsilon} = \mathbf{A} : \boldsymbol{\sigma}.$$

We can then compute the following quantities depending on the measurements of $\sigma^{(n)}$, $\varepsilon_{11}^{(n)}$, $\varepsilon_{22}^{(n)}$, $\varepsilon_{33}^{(n)}$ for the six tension tests in the six particular directions:

$$\vec{n} = \vec{x}_1 \rightarrow E_1 = \frac{\sigma^{(n)}}{\varepsilon_{11}^{(n)}}, \quad \nu_{12} = -\frac{\varepsilon_{22}^{(n)}}{\varepsilon_{11}^{(n)}}, \quad \nu_{13} = -\frac{\varepsilon_{33}^{(n)}}{\varepsilon_{11}^{(n)}}$$

$$\vec{n} = \vec{x}_2 \rightarrow E_2 = \frac{\sigma^{(n)}}{\varepsilon_{22}^{(n)}}, \quad \nu_{21} = -\frac{\varepsilon_{11}^{(n)}}{\varepsilon_{22}^{(n)}}, \quad \nu_{23} = -\frac{\varepsilon_{33}^{(n)}}{\varepsilon_{22}^{(n)}}$$

$$\vec{n} = \vec{x}_3 \rightarrow E_3 = \frac{\sigma^{(n)}}{\varepsilon_{33}^{(n)}}, \quad \nu_{31} = -\frac{\varepsilon_{11}^{(n)}}{\varepsilon_{33}^{(n)}}, \quad \nu_{32} = -\frac{\varepsilon_{22}^{(n)}}{\varepsilon_{33}^{(n)}}$$

$$\vec{n} = \frac{\sqrt{2}}{2}(\vec{x}_1 + \vec{x}_2) \rightarrow \frac{1}{G_{12}} = 4\frac{\varepsilon^{(n)}}{\sigma^{(n)}} - \frac{1-\nu_{21}}{E_2} - \frac{1-\nu_{12}}{E_1}$$

$$\vec{n} = \frac{\sqrt{2}}{2}(\vec{x}_2 + \vec{x}_3) \rightarrow \frac{1}{G_{23}} = 4\frac{\varepsilon^{(n)}}{\sigma^{(n)}} - \frac{1-\nu_{23}}{E_2} - \frac{1-\nu_{32}}{E_3}$$

$$\vec{n} = \frac{\sqrt{2}}{2}(\vec{x}_3 + \vec{x}_1) \rightarrow \frac{1}{G_{31}} = 4\frac{\varepsilon^{(n)}}{\sigma^{(n)}} - \frac{1-\nu_{13}}{E_1} - \frac{1-\nu_{31}}{E_3}.$$

Tests in pure shear or in shear by the torsion of tubes allow interesting cross-checking. The relative precision obtained is of the order of 1–5%.

Dynamic methods

In the low or average frequency range (10–100 Hz), we may determine Young's modulus by measuring the first natural frequencies of flexural vibrations of a beam. For example, for a cantilever beam fixed at one end and free at the other, the first natural frequency is given by:

$$f = \frac{\omega}{2\pi} = \frac{3.5156}{2\pi L^2}\left(\frac{EI}{\rho S}\right)^{1/2}$$

from which

$$E = \omega^2 \frac{\rho S L^4}{12.359\,6 I}$$

where S is the (constant) cross-sectional area of the beam, L is its length, I is its flexural moment of inertia, ρ is the mass density of the material, and ω is the first natural frequency in radians obtained from the test measurements.

In the high frequency range (> 20 000 Hz) the measurement of the period of traverse of ultrasonic waves can be used to determine the elastic constants. It can be shown that in a linear isotropic medium of elastic constants E and ν and of density ρ, the rate of the propagation v_L of the longitudinal waves, and v_T that of the transverse waves, are given by:

$$v_L^2 = \frac{E}{\rho} \frac{1-\nu}{(1+\nu)(1-2\nu)}, \quad v_T^2 = \frac{E}{\rho} \frac{1}{2(1+\nu)}$$

from which it follows that:

$$E = \rho v_T^2 \frac{3v_L^2 - 4v_T^2}{v_L^2 - v_T^2}$$

$$\nu = \frac{1}{2} \frac{v_L^2 - 2v_T^2}{v_L^2 - v_T^2}.$$

With ultrasonic techniques, it is possible to measure the time of traverse and therefore the rates with a relative precision of the order of 10^{-4} which yields a relative precision of 3×10^{-3} for Young's modulus E and Poisson's ratio ν.

4.1.4 *Table of elastic properties of common materials*

Material	Temperature °C	Young's modulus MPa	Poisson's ratio
Aluminium alloy 2024	20	72 000	0.32
	200	66 000	0.325
	500	50 000	0.35
Titanium alloy Ti 4Al 4Mn	20	115 000	0.34
	200	103 400	
	315	95 000	
XC10 carbon steel	20	216 000	0.29
	200	205 000	0.30
	600	170 000	0.315
Grey cast iron	20	100 000	0.29

4.1.4 *Table of elastic properties of common materials*

Material	Temperature °C	Young's modulus MPa	Poisson's ratio
A316 stainless steel	20	196 000	0.3
	200	170 000	
	700	131 000	
A5 aluminimum	20	68 000	0.33
Bronze	180	61 000	
	20	130 000	0.34
Plexiglass	20	2 900	0.4
Araldite	20	3 000	0.4
Rubber	20	2	0.5
Unidirectional composites glass-epoxy (in long. direction)	20	19 000	0.30
Unidirectional composite carbon–epoxy (in long. direction)	20	87 600	0.32
Concrete	20	30 000	0.2
Granite	20	60 000	0.27
Wood: pinewood			
fibre direction	20	17 000	0.45
transverse direction	20	1 000	0.79

4.1.5 *Concepts of the finite element method*

We outline here the basis of the finite element method which is considered as a standard numerical method of solving field problems in elasticity. Other methods exist. In particular, those based on integral equation formulation, but we will not discuss them here.

Generalities

The objective of approximate numerical methods of structural analysis is to find in a class of 'reasonable' displacement (or stress) fields those which

> minimize the potential energy $\mathscr{V}$ for kinematically admissible fields,
> maximize the complementary potential energy $\mathscr{V}^*$ for statically admissible fields.

The finite element method consists in calculating the potential energy $\mathscr{V}$ or $\mathscr{V}^*$ as the sum of the energies of all the elements of the structure divided into finite elements (these can be line, surface or volume elements depending on the schematic representation of the structure).

In each element the unknown fields are represented by a linear combination of functions of spatial coordinates. The coefficients of this combination depend on the displacements of the nodes which belong to the element (its vertices for example). These latter quantities are the ones that constitute the unknowns of the problem (nodal unknowns or degrees of freedom). In practice, the finite element method is equivalent to using the principle of virtual work as will be used later.

After imposing equality of nodal displacements at common nodes of elements, minimization of the potential energy with respect to these displacements leads to a system of linear algebraic equations. Thus, a problem of solving linear partial differential equations is replaced by that of solving a set of linear algebraic equations. Having solved the system, we can find the displacement, strains and stresses at points within the elements.

Formulation of the element properties

Numerous types of elements can be developed. A general methodology exists for the derivation of the characteristics of an element with any number of degrees of freedom. Our purpose here will only be to illustrate the principles of the finite element method in the simplest case, that of a plane triangular element with three nodes (Fig. 4.2).

The displacements of a point M identified by the coordinates x_1, x_2 are

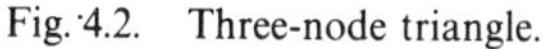
Fig. 4.2. Three-node triangle.

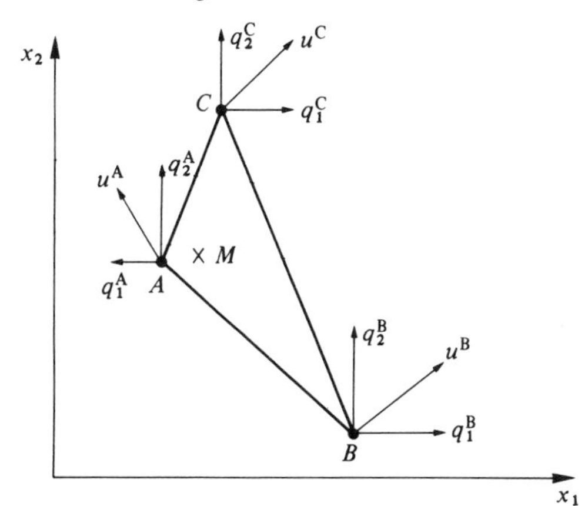

written in the form of a column vector:[†]

$$U = \begin{bmatrix} U_1(x_1, x_2) \\ U_2(x_1, x_2) \end{bmatrix}.$$

The degrees of freedom of the element are the six displacement components belonging to nodes A, B, C:

$$q = \begin{Bmatrix} q_1^A \\ q_2^A \\ q_1^B \\ q_2^B \\ q_1^C \\ q_2^C \end{Bmatrix}.$$

The unknown displacements u are represented by a combination of linear functions:

$$u = \begin{Bmatrix} u_1 \\ u_2 \end{Bmatrix} = \begin{Bmatrix} C_1 + C_2 x_1 + C_3 x_2 \\ C_4 + C_5 x_1 + C_6 x_2 \end{Bmatrix}.$$

Since at a node the displacement functions u must assume values identical to the nodal displacements q, the six coefficients $C_1, \ldots, C_6$ can be determined as linear functions of the six nodal displacements. The displacement function can then be expressed as:

$$u = \begin{Bmatrix} u_1 \\ u_2 \end{Bmatrix} = \begin{Bmatrix} N^{BC}(x_1, x_2) q_1^A + N^{CA}(x_1, x_2) q_1^B + N^{AB}(x_1, x_2) q_1^C \\ N^{BC}(x_1, x_2) q_2^A + N^{CA}(x_1, x_2) q_2^B + N^{AB}(x_1, x_2) q_2^C \end{Bmatrix}.$$

In the present case, the functions $N(x_1, x_2)$, called shape functions, are linear. In abridged notation, these relations may be expressed as:

$$u = Nq.$$

The strain components, expressed as a column vector, are:

$$\varepsilon = \begin{Bmatrix} \varepsilon_{11} \\ \varepsilon_{22} \\ 2\varepsilon_{12} \end{Bmatrix} = \begin{Bmatrix} u_{1,1} \\ u_{2,2} \\ u_{1,2} + u_{2,1} \end{Bmatrix} = Bq$$

where B is a 3×6 matrix derived from N by differentiation. The factor 2 in the component $2\varepsilon_{12}$ is used in order to write the strain energy simply as $\sigma^{\mathrm{T}}\varepsilon$

[†] The matrix symbol [ ] is omitted for all variables in this section.

where the stress components are represented by the column vector;[†]

$$\sigma = \begin{Bmatrix} \sigma_{11} \\ \sigma_{22} \\ \sigma_{12} \end{Bmatrix} = \frac{E}{1-\nu^2} \begin{bmatrix} 1 & \nu & 0 \\ \nu & 1 & 0 \\ 0 & 0 & \dfrac{1-\nu}{2} \end{bmatrix} \begin{Bmatrix} \varepsilon_{11} \\ \varepsilon_{22} \\ 2\varepsilon_{12} \end{Bmatrix} = a\varepsilon$$

in the case of isotropic elasticity and a state of plane stress.

The equilibrium of an element is expressed by invoking the principle of virtual work (see Chapter 2) with respect to an arbitrary kinematically admissible displacement field $u'(M)$ and the associated strain field $\varepsilon'(M)$:

$$\int_{\mathscr{D}} f^{\mathrm{T}} u' \,\mathrm{d}V + \int_{\partial\mathscr{D}} F^{\mathrm{T}} u' \,\mathrm{d}S - \int_{\mathscr{D}} \sigma^{\mathrm{T}} \varepsilon' \,\mathrm{d}V = 0 \qquad \forall u'$$

where f and F are respectively the externally applied body forces acting in $\mathscr{D}$ and the surface forces acting on $\partial\mathscr{D}$. Taking into account the relation between strains and nodal displacements, we may write:

$$\sigma^{\mathrm{T}} \varepsilon' = \{a\varepsilon\}^{\mathrm{T}} \varepsilon' = q^{\mathrm{T}} B^{\mathrm{T}} a B q'$$

where q' represents any choice for the values of the nodal displacements. By substituting the above into the principle of virtual work, we easily obtain:

$$\left(\int_{\mathscr{D}} f^{\mathrm{T}} N \,\mathrm{d}V + \int_{\partial\mathscr{D}} F^{\mathrm{T}} N \,\mathrm{d}S - q^{\mathrm{T}} \int_{\mathscr{D}} B^{\mathrm{T}} a B \,\mathrm{d}V \right) q' = 0$$

or

$$[-q^{\mathrm{T}} K + Q^{\mathrm{T}}] q' = 0 \qquad \forall q'.$$

Since this equation is satisfied for arbitrary q', we conclude that:

$$Kq = Q$$

where K is the stiffness matrix of the element. It is symmetric since:

$$K = \int_{\mathscr{D}} B^{\mathrm{T}} a B \,\mathrm{d}V$$

and a is symmetric. Q is the column vector of generalized, externally

[†] The superscript T denotes the transpose.

applied, nodal forces.

$$Q = \int_{\mathscr{D}} N^{\mathrm{T}} f \, \mathrm{d}V + \int_{\partial\mathscr{D}} N^{\mathrm{T}} F \, \mathrm{d}S.$$

If the external forces are known, the nodal displacements q are given by the solution of the set of linear equations $Kq = Q$. Knowing q we can determine within the element, the displacements $u = Nq$, the strains $\varepsilon = Bq$, and the stresses $\sigma = a\varepsilon$.

Assembly

To solve the problem of the whole structure (the calculation of q with respect to all elements) the virtual work of the whole structure has to be written by summing on all the elements. At the same time the condition of continuity of structure has to be imposed, which means that the displacements of a node common to several elements are the same for all these elements. Note that if the equality of nodal displacements ensures the equality of displacements at the interelement boundary, then the element is said to be a conforming one. In practice, this assembly operation consists in adding the stiffness contributions (and external forces) from each of the elements which meet at that node. More precisely, letting ${}^{E}q$ be the total degrees of freedom of all elements without considering their interconnections, and q be the total degrees of freedom when nodal compatibility is enforced, we may write:

$${}^{E}q = Aq$$

where A is a rectangular matrix since the dimension of q is less than that of ${}^{E}q$. Similarly, the matrix ${}^{E}K$ and the column vector ${}^{E}Q$ obtained by simple juxtaposition of the individual element stiffness matrices and the force vectors are:

$${}^{E}K = \begin{bmatrix} {}^{1}K & & & \\ & {}^{2}K & & \\ & & \cdot & \\ & & & \cdot \\ & & & & \cdot \end{bmatrix} \qquad {}^{E}Q = \begin{Bmatrix} {}^{1}Q \\ {}^{2}Q \\ \cdot \\ \cdot \\ \cdot \end{Bmatrix}$$

In terms of these matrices, the principle of virtual work can be written as:

$$[-{}^{E}q^{\mathrm{T}}\,{}^{E}K + {}^{E}Q^{\mathrm{T}}]\,{}^{E}q' = 0$$

where for ${}^{E}q'$ to be kinematically admissible, it must satisfy the relation ${}^{E}q' = Aq'$. For the assembled equation this means that:

$$[-(Aq)^{\mathrm{T}}\,{}^{E}K + {}^{E}Q^{\mathrm{T}}]Aq' = 0 \qquad \forall q'$$

which, in view of the arbitrariness of q', leads to:

$$-A^{\mathrm{T}}\,{}^{E}KAq + A^{\mathrm{T}}\,{}^{E}Q = 0$$

which may be written as:

$$Kq = Q$$

where K is the assembled stiffness matrix $K = A^{\mathrm{T}}\,{}^{E}KA$, and Q and q are the column vectors which represent the externally applied nodal forces and nodal displacements respectively.

Solution of an elasticity problem by the finite element method thus requires the establishment of the stiffness matrix K and the load vector Q and then the numerical solution of the above set of linear equations. It is often necessary to divide a structure into a large number of elements with numerous degrees of freedom (from 1000 to 10 000 or even more). The matrix K can therefore be very large. However, it possesses a banded structure with nonzero terms around the main diagonal. Outside this band all the coefficients are zero which makes it easier to decompose K. The band width of the assembled stiffness matrix depends on the scheme used to number the nodes of the structures. A large difference in the node numbers of an element results in a large band width and the numbering scheme should preferably minimize this difference.

4.2 Thermoelasticity

Thermoelasticity involves one more variable than elasticity: the temperature. This induces the additional phenomenon of dilatation. The intrinsic dissipation is always zero but thermal dissipation does take place. The theory of thermoleasticity must be used for elastic materials whenever the loads involve variations in temperature. We limit ourselves here to aspects concerned with the constitutive law itself and with the determination of its characteristic coefficients as functions of temperature.

4.2.1 *Formulation*

Let T_0 be a reference temperature of a volume element of an elastic material and T its temperature at a subsequent instant so that the temperature difference, $\theta = T - T_0$, is small with respect to T_0. Under such conditions the elastic properties and the density ρ can be considered to be constant.

The law of classical linear thermoelasticity is obtained by again selecting as the thermodynamic potential a positive definite quadratic form which, in addition to strain components, has temperature as one of the variables. Assuming isotropy of responses in both the deformation and dilatation phenomena, and taking cognizance of the classically used law, we are led to write:

$$\Psi = \frac{1}{\rho}\left[\tfrac{1}{2}(\lambda\varepsilon_{\mathrm{I}}^2 + 4\mu\varepsilon_{\mathrm{II}}) - (3\lambda + 2\mu)\alpha\theta\varepsilon_{\mathrm{I}}\right] - \frac{C_\varepsilon}{2T_0}\theta^2$$

where α and C_ε are two coefficients with a meaning given by the state laws derived from Ψ:

$$\boldsymbol{\sigma} = \rho(\partial\Psi/\partial\boldsymbol{\varepsilon}) = \lambda\,\mathrm{Tr}\,(\boldsymbol{\varepsilon})\mathbf{1} + 2\mu\boldsymbol{\varepsilon} - (3\lambda + 2\mu)\alpha\theta\mathbf{1}$$

or

$$\sigma_{ij} = \lambda\varepsilon_{kk}\delta_{ij} + 2\mu\varepsilon_{ij} - (3\lambda + 2\mu)\alpha\theta\delta_{ij}$$

and

$$s = -\frac{\partial\Psi}{\partial T} = -\frac{\partial\Psi}{\partial\theta} = \frac{1}{\rho}(3\lambda + 2\mu)\alpha\,\mathrm{Tr}\,(\boldsymbol{\varepsilon}) + \frac{C_\varepsilon}{T_0}\theta.$$

This last equation allows us to calculate

$$(\partial s/\partial T)_{\varepsilon = \mathrm{constant}} = (\partial s/\partial\theta)_{\varepsilon = \mathrm{constant}} = C_\varepsilon/T_0.$$

which shows that $C_\varepsilon = T_0(\partial s/\partial T)$ is the specific heat at constant strain.

By inverting the first law of state or by employing the dual potential $\Psi^*(\boldsymbol{\sigma}, \theta)$, we may express strains as functions of stresses and temperature:

● $$\boldsymbol{\varepsilon} = \frac{1+\nu}{E}\boldsymbol{\sigma} - \frac{\nu}{E}\mathrm{Tr}\,(\boldsymbol{\sigma})\mathbf{1} + \alpha\theta\mathbf{1}$$

or

$$\varepsilon_{ij} = \frac{1+\nu}{E}\sigma_{ij} - \frac{\nu}{E}\sigma_{kk}\delta_{ij} + \alpha\theta\delta_{ij}$$

which constitutes the law of linear isotropic thermoelasticity. By considering a variation of temperature with zero stress we obtain:

$$\varepsilon_{ij} = \alpha\theta\delta_{ij}$$

which defines the isotropic dilatation coefficient α.

The coefficients λ, μ, α and C_ε are taken to be constant for small temperature variations ($\theta/T_0 \ll 1$). In practice, they vary with the temperature but 'weakly' as shown by Fig. 4.5.

The thermoelastic constitutive equation enables us to calculate the strains and stresses due to the thermal effect as soon as the temperature field is known. For the treatment of a complete thermoelastic problem in which the flux (and not the temperature itself) is given, it is necessary to use Fourier's law and the heat equation mentioned in Chapter 2.

4.2.2 ***Identification***

The thermoelastic constitutive equation is written under the assumption of small temperature variations. If such is not the case, it is used step by step with the variable coefficients E, ν and α; this implies that their evolution as a function of temperature is known.

We are therefore confronted with a major experimental difficulty: the almost impossible task of measuring, with sufficient precision, the local deformation as soon as the temperature exceeds the limit for using electric resistance strain gauges. This limit varies from 200 to 300 °C. On the other hand, it is easy to obtain the dilatation coefficient by measuring the elongation of long wires heated in ovens; some values are given in Fig. 4.5.

For the elastic modulus we may also use rather long wires with an extensometer attached to the part of the wire outside the oven. However, this method raises the problem of the reliability of the fixture used for applying the forces (slipping, wrong definition of the useful length . . .). The two methods described below represent the means to obtain relatively good values from relatively short samples.

Stiffness comparison method

It is possible to overcome the problem of local measurement by employing two tension test specimens of the same material with identical heads but

Fig. 4.3. Tension test specimens for the stiffness comparison method.

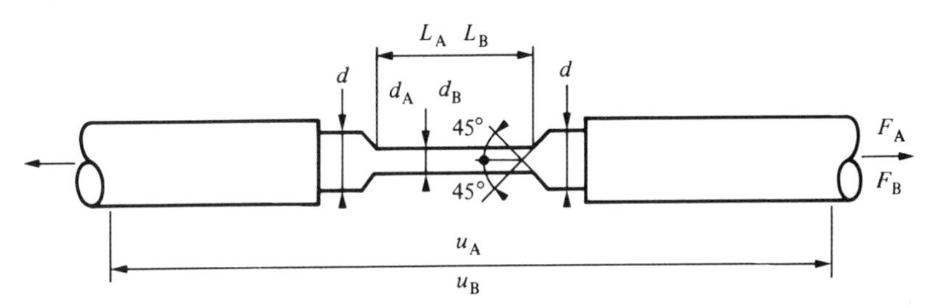

different useful lengths. The only measurements to be taken consist of recording the force F and the displacement u, which comprises the deformation of the useful part of the specimen, of its head, and of the part outside the heating zone and the heating device where the displacement transducer (or extensometer) is attached (Fig. 4.3). This latter part will be called the fixing-line part.

Let there be two samples A and B with dimensions as defined in Fig. 4.3. Let R_A and R_B be the measured stiffnesses depending on the geometry of the specimens and on their unknown Young's modulus E:

$$R_A = \delta F/\delta u_A, \quad R_B = \delta F/\delta u_B.$$

The stiffnesses of the useful parts of the two specimens are:

$$R_A^u = ES_A/L_A, \quad R_B^u = ES_B/L_B \quad \text{with } S = \pi d^2/4.$$

The second unknown is the stiffness R_m of the identical heads of the two specimens and of the fixing-line which is included in the measurement done with the extensometer. As a first approximation, it is permissible to consider that the measurement includes contributions from the conical grips (the difference between the two grips being very minimal). Moreover, the effects of stress concentration are considered negligible. We, therefore, obtain two equations in two unknowns:

$$\frac{1}{R_A} = \frac{1}{R_A^u} + \frac{1}{R_m}$$

$$\frac{1}{R_B} = \frac{1}{R_B^u} + \frac{1}{R_m}$$

which can easily be solved to obtain E:

- $$E = \frac{L_A/S_A - L_B/S_B}{1/R_A - 1/R_B}.$$

The precision of the method depends on the relative orders of magnitude of the different terms which appear in the expression for E. Generally $L_A/S_A = 2$–4 times L_B/S_B represents a good compromise between the numerical aspect of precision and the experimental aspect which requires the machine to load the specimen A without exceeding the elastic limit of the material. This method gives a relative precision of the order of $\pm 2\%$. The same principle used for torsion tests provides values of the shear modulus G, and when E is known, of Poisson's ratio, v. Denoting the torsional stiffness per unit angle of twist by R^* it can be shown that:

$$\bullet \qquad G = \frac{32}{\pi}\left(\frac{L_A/d_A^4 - L_B/d_B^4}{1/R_A^* - 1/R_B^*}\right).$$

Thermal stress method

This method leads directly to the evolution of Young's modulus E and the dilatation coefficient α as functions of temperature, without displacement measurements. The load is compressive, which is of interest in view of the fact that in applications the hottest parts of the structure are usually loaded in compression. Only one specimen is used in this method of testing, but it requires a very stiff testing machine or even a special rig.

A cylindrical compression specimen ($L/d \approx 1.5$) in series with a dynamometer is placed between two thick plates fixed so as to permit almost zero relative displacement between them (Fig. 4.4). Heating of the specimen, by induction for example, induces in the sample a force measured by the dynamometer. Its expression given by the equations of thermoelasticity contains two unknowns E and α. A second piece of information is furnished by the existence of a preload at the reference temperature.

The specimen with cross-sectional area S and length L is raised to a temperature T_1 which is the state of reference for strains.

Fig. 4.4. Device for thermal compression without displacement (after Tichkiewitch).

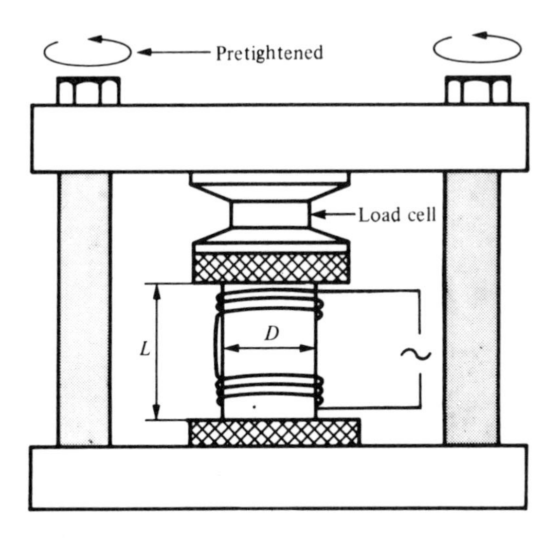

A preload F_1 is applied by the movement of one of the plates (indicated schematically in Fig. 4.4 by tightening of bolts). If E_1 is Young's modulus at the reference temperature T_1, the strain in the specimen is:

$$\varepsilon_1 = F_1/E_1 S.$$

The temperature of the specimen is then raised to T_2 without significant movement of the plates (since the stiffness of the clamps is very large). Denoting the stiffness of the dynamometer-mounted assembly by R_m, the strain in the specimen is given by:

$$\varepsilon_2 = \varepsilon_1 - (F_2 - F_1)/R_m L.$$

The linear thermoelastic law $\varepsilon = \sigma/E + \alpha\theta$ applied between the states $(T_1, F_1 = 0)$ and (T_2, F_2) gives:

$$\varepsilon_2 = \frac{F_2}{E_2 S} + \alpha_{21}(T_2 - T_1)$$

where E_2 and α_{21} denote the Young's modulus at temperature T_2 and the 'secant' coefficient of dilatation between the temperatures T_2 and T_1. Equating the two expressions for ε_2 we obtain the expression:

$$F_2 = F_1 \frac{1/E_1 S + 1/R_m L}{1/E_2 S + 1/R_m L} - \alpha_{21} \frac{T_2 - T_1}{1/E_2 S + 1/R_m L}$$

which shows that F_2 is linearly related to F_1 in the form:

$$F_2 = aF_1 + b.$$

By repeating the same operation several times with different values of preload F_1 but the same temperatures T_1 and T_2, constants a and b can be determined from the measurements of F_1 and F_2. We may then determine E_2 and α_{21} from:

- $$E_2 = \frac{E_1 R_m L a}{R_m L + E_1 S(1 - a)}$$
- $$\alpha_{21} = -\frac{b(E_2 S + R_m L)}{E_2 S R_m L(T_2 - T_1)}.$$

Knowing E_1, at a temperature T_1 (for example room temperature), a first experiment gives E_2 at T_2 and also α_{21}. Knowing E_2 at T_2, a

Fig. 4.5. Evolution of (*a*) the dilatation coefficient α, (*b*) Young's modulus E and (*c*) Poisson's ratio ν, as functions of the temperature T (after Ben Cheikh).

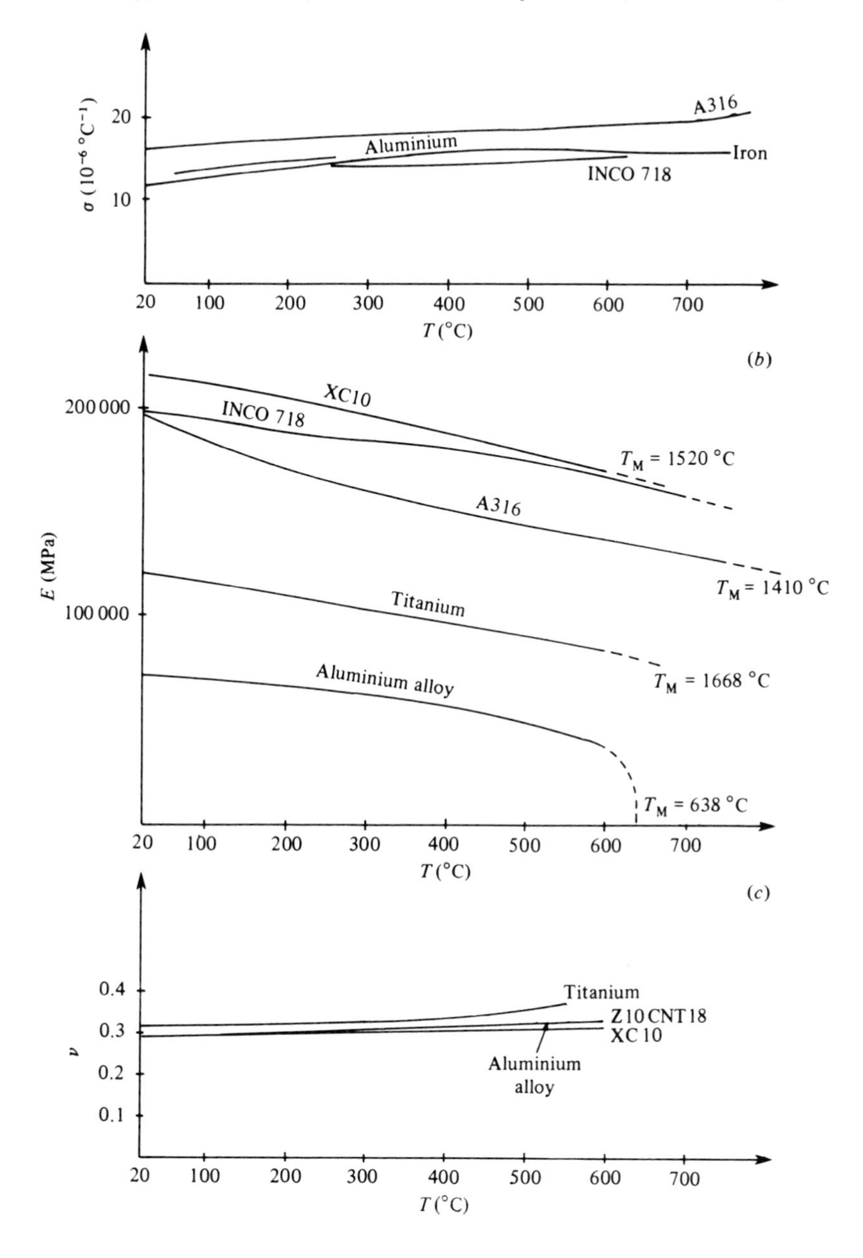

second experiment gives E_3 and α_{23} at T_3 and so on. The stiffness R_m of the assembly is first determined by using instead of the specimen a test bar with a much higher stiffness.

In following the above procedure, we can proceed in steps of 100 °C or 50 °C with a precision that is, of course, decreasing by cumulative effects, but which remains, however, in the order of 2% until over 1000 °C. It is, moreover, necessary to make sure that the elastic limit is not exceeded at the force F_2 for the temperature T_2 and that the effects of stress relaxation are negligible. The properties of A316 stainless steel given in Section 4.2.3 were obtained by this method.

4.2.3 *Thermoelastic properties of common materials*

Graphs showing the evolution of the dilatation coefficient α, Young's modulus E and Poisson's ratio ν, as functions of the temperature T, for different materials are given in Fig. 4.5.

4.3 Viscoelasticity

4.3.1 *Domain of validity and use*

As discussed in Chapter 1, polymers, and, to a lesser extent, concrete and wood exhibit dissipative phenomena associated with elasticity that can be expressed globally by viscosity. Limiting ourselves to isothermal viscoelasticity we have the situation in which thermal dissipation is zero but intrinsic dissipation is not. The strain rate now becomes an independent, thermodynamic variable.

The theory of viscoelasticity can be used to deal with the reversible evolutions of viscoelastic solids as functions of time as defined in Chapter 3. In dynamic situations, this results in a very low damping (10^{-4}) at room temperature, but it can reach 10^{-2}–1 for polymers.

The Kelvin–Voigt and Maxwell models are derived here from a direct application of thermodynamics to materials with internal variables. For the particular case of linear viscoelasticity, the above functional approach is simpler to formulate and use. We should, however, note that these formulations are valid only for stable materials. The ageing of concrete and polymers requires special treatment which is not described here.

4.3.2 *Thermodynamic formulation*

Kelvin–Voigt solid

This model corresponds to the following choice of variables:

the observable variable is the total strain $\boldsymbol{\varepsilon}$, its associated variable is the stress $\boldsymbol{\sigma}$;
in order to define the reversible power and the dissipated power, it is necessary to divide the stress into an 'elastic' part $\boldsymbol{\sigma}^{e}$ and an anelastic one $\boldsymbol{\sigma}^{an}$. This division corresponds to the analogical Kelvin–Voigt model of Chapter 3 with a spring and a damper in parallel. Accordingly, we have:

$$\boldsymbol{\sigma} = \boldsymbol{\sigma}^{e} + \boldsymbol{\sigma}^{an}$$

where $\boldsymbol{\sigma}^{e}:\dot{\boldsymbol{\varepsilon}}$ is the reversible power per unit volume and $\boldsymbol{\sigma}^{an}:\dot{\boldsymbol{\varepsilon}}$ is the intrinsic dissipation per unit volume.

The linear isotropic theory is obtained by choosing the same thermodynamic potential as for linear isotropic elasticity:

$$\Psi = (1/2\rho)(\lambda\varepsilon_{I}^{2} + 4\mu\varepsilon_{II}).$$

To express the anelastic stress $\boldsymbol{\sigma}^{an}$ (and not the strain rate $\dot{\boldsymbol{\varepsilon}}$) in order to add it to the elastic stress $\boldsymbol{\sigma}^{e}$, it is necessary to use the dissipation potential $\varphi(\dot{\boldsymbol{\varepsilon}})$. For the same reasons as before, a positive definite quadratic form of strain rate components is chosen:

$$\varphi = \tfrac{1}{2}(\lambda\theta_{\lambda}\dot{\varepsilon}_{I}^{2} + 4\mu\theta_{\mu}\dot{\varepsilon}_{II})$$

where $\dot{\varepsilon}_{I}$ and $\dot{\varepsilon}_{II}$ are the first and second principal invariants of the strain rate tensor $\dot{\varepsilon}$, and where θ_{λ} and θ_{μ} are two additional coefficients which characterize the viscosity of the material.

According to the method of local state described in Chapter 2, the state law is obtained as:

$$\boldsymbol{\sigma}^{e} = \rho(\partial\Psi/\partial\boldsymbol{\varepsilon}) = \lambda\,\mathrm{Tr}\,(\boldsymbol{\varepsilon})\mathbf{1} + 2\mu\boldsymbol{\varepsilon}$$

and the complementary law as:

$$\boldsymbol{\sigma}^{an} = \partial\varphi/\partial\dot{\boldsymbol{\varepsilon}} = \lambda\theta_{\lambda}\,\mathrm{Tr}\,(\dot{\boldsymbol{\varepsilon}})\mathbf{1} + 2\mu\theta_{\mu}\dot{\boldsymbol{\varepsilon}}.$$

The mechanical Kelvin–Voigt model is then obtained by adding the two stresses:

- $$\boldsymbol{\sigma} = \lambda(\operatorname{Tr}(\boldsymbol{\varepsilon}) + \theta_\lambda \operatorname{Tr}(\dot{\boldsymbol{\varepsilon}}))\mathbf{1} + 2\mu(\boldsymbol{\varepsilon} + \theta_\mu \dot{\boldsymbol{\varepsilon}})$$

or

$$\sigma_{ij} = \lambda(\varepsilon_{kk} + \theta_\lambda \dot{\varepsilon}_{kk})\delta_{ij} + 2\mu(\varepsilon_{ij} + \theta_\mu \dot{\varepsilon}_{ij}).$$

θ_λ and θ_μ are the characteristic retardation times; they can be identified from the results of a uniaxial tensile test and shear test (torsion of tubes for example). The shear test is governed by the relation:

$$\sigma_{12} = 2\mu(\varepsilon_{12} + \theta_\mu \dot{\varepsilon}_{12}).$$

The response $\varepsilon_{12}(t)$ in a creep experiment in shear ($\sigma_{12} = \text{constant}$), is given by the solution of the above differential equation. With the initial condition $\varepsilon = 0$ at $t = 0$, we find the solution as:

$$\varepsilon_{12} = \frac{\sigma_{12}}{2\mu}\left[1 - \exp\left(-\frac{t}{\theta_\mu}\right)\right]$$

which enables the identification of the two coefficients μ and θ_μ.

Remembering the definition of Poisson's ratio v, the tension test is governed by:

$$\sigma_{11} = [\lambda(1 - 2v) + 2\mu]\varepsilon_{11} + [\lambda(1 - 2v)\theta_\lambda + 2\mu\theta_\mu]\dot{\varepsilon}_{11}.$$

Letting:

$$E = \lambda(1 - 2v) + 2\mu$$

as in elasticity and

$$\eta = \lambda(1 - 2v)\theta_\lambda + 2\mu\theta_\mu,$$

the above relation can be written as:

$$\sigma_{11} = E\varepsilon_{11} + \eta\dot{\varepsilon}_{11}.$$

Hence, the response $\varepsilon_{11}(t)$ to a creep test in tension with $\sigma_{11} = \text{constant}$ (and the initial condition $\varepsilon = 0$ at $t = 0$) is:

$$\varepsilon_{11} = \frac{\sigma_{11}}{E}\left[1 - \exp\left(-\frac{E}{\eta}t\right)\right]$$

which, having determined μ and θ_μ, allows identification of the coefficients E and η, and then λ and θ_λ; v is measured from the lateral strain: $\varepsilon_{22} = -v\varepsilon_{11}$.

Maxwell body

The Maxwell model corresponds to the following choice of variables:

the observable variable is always the total strain $\boldsymbol{\varepsilon}$ which is associated to the stress $\boldsymbol{\sigma}$,
to define the reversible power and the dissipated power the strain is now divided into elastic and anelastic parts, so that:

$$\boldsymbol{\varepsilon} = \boldsymbol{\varepsilon}^{\mathrm{e}} + \boldsymbol{\varepsilon}^{\mathrm{an}}$$

where $\boldsymbol{\sigma}:\dot{\boldsymbol{\varepsilon}}^{\mathrm{e}}$ defines the reversible power, and $\boldsymbol{\sigma}:\dot{\boldsymbol{\varepsilon}}^{\mathrm{an}}$ is the intrinsic dissipation.

Then, in accordance with the method of local state, the law of state derived from the thermodynamic potential $\Psi(\boldsymbol{\varepsilon} - \boldsymbol{\varepsilon}^{\mathrm{an}})$ is:

$$\boldsymbol{\sigma} = \rho(\partial\Psi/\partial\boldsymbol{\varepsilon}) = \rho(\partial\Psi/\partial\boldsymbol{\varepsilon}^{\mathrm{e}})$$

and the complementary law derived from the dual dissipation potential $\varphi^*(\boldsymbol{\sigma})$ is:

$$\dot{\boldsymbol{\varepsilon}}^{\mathrm{an}} = \partial\varphi^*/\partial\boldsymbol{\sigma}.$$

In order to express the total strain rate as the sum of $\dot{\boldsymbol{\varepsilon}}^{\mathrm{e}}$ and $\dot{\boldsymbol{\varepsilon}}^{\mathrm{an}}$ it is better to work with the dual thermodynamic potential $\Psi^*(\boldsymbol{\sigma})$. Moreover, in order to obtain a linear theory, the same potential as for elasticity is chosen:

$$\Psi^* = \frac{1}{2\rho}\left\{\frac{1+\nu}{E}\mathrm{Tr}(\boldsymbol{\sigma}^2) - \frac{\nu}{E}[\mathrm{Tr}(\boldsymbol{\sigma})]^2\right\}$$

with

$$\boldsymbol{\varepsilon}^{\mathrm{e}} = \rho\frac{\partial\Psi^*}{\partial\boldsymbol{\sigma}} = \frac{1+\nu}{E}\boldsymbol{\sigma} - \frac{\nu}{E}\mathrm{Tr}(\boldsymbol{\sigma})\mathbf{1}.$$

In the same manner, the dissipation potential φ is defined by a positive definite quadratic form which now includes two new coefficients τ_1 and τ_2 which are characteristic of the viscosity and which may be determined by uniaxial tension tests and shear tests:

$$\varphi = \frac{1}{2}\left(\frac{1+\nu}{E\tau_1}\mathrm{Tr}(\boldsymbol{\sigma}^2) - \frac{\nu}{E\tau_2}[\mathrm{Tr}(\boldsymbol{\sigma})]^2\right).$$

By writing $\dot{\boldsymbol{\varepsilon}} = \dot{\boldsymbol{\varepsilon}}^{\mathrm{e}} + \dot{\boldsymbol{\varepsilon}}^{\mathrm{an}} = \dot{\boldsymbol{\varepsilon}}^{\mathrm{e}} + \partial\varphi^*/\partial\boldsymbol{\sigma}$ we obtain the following constitutive relation for the Maxwell body:

- $$\dot{\boldsymbol{\varepsilon}} = \frac{1+\nu}{E}\left(\dot{\boldsymbol{\sigma}} + \frac{\boldsymbol{\sigma}}{\tau_1}\right) - \frac{\nu}{E}\left(\mathrm{Tr}(\dot{\boldsymbol{\sigma}}) + \frac{\mathrm{Tr}(\boldsymbol{\sigma})}{\tau_2}\right)\mathbf{1}$$

or

$$\varepsilon_{ij} = \frac{1+\nu}{E}\left(\dot{\sigma}_{ij} + \frac{\sigma_{ij}}{\tau_1}\right) - \frac{\nu}{E}\left(\dot{\sigma}_{kk} + \frac{\sigma_{kk}}{\tau_2}\right)\delta_{ij}.$$

Applied to the uniaxial case, which corresponds to the analogical model of a spring with a damper in series described in Chapter 3, the above relations give:

$$\dot{\varepsilon} = \frac{\dot{\sigma}}{E} + \left(\frac{1+\nu}{E\tau_1} - \frac{\nu}{E\tau_2} \right) \sigma$$

or, by writing

$$\frac{1+\nu}{E\tau_1} - \frac{\nu}{E\tau_2} = \frac{1}{\eta}$$

we obtain

$$\dot{\varepsilon} = \frac{\dot{\sigma}}{E} + \frac{\sigma}{\eta}.$$

This is a fluid model because the possibility of equilibrium at constant stress does not exist.

Generalized Maxwell model for solids

It is possible to obtain an equilibrium by adding a spring in parallel to the analogical Maxwell model. In the same way, it becomes possible to represent more complex behaviour of solids by adding other simple elements in parallel (Fig. 4.6). This also constitutes a generalization of the Kelvin–Voigt model.

The uniaxial constitutive law of such a model is expressed as:

$$\sigma = \sigma_0 + \sigma_1 + \cdots + \sigma_j + \cdots$$

$$\varepsilon = \sigma_0 / E_\infty$$

$$\dot{\varepsilon} = \frac{\dot{\sigma}_1}{E_1} + \frac{\sigma_1}{\eta_1}$$

$$\dot{\varepsilon} = \frac{\dot{\sigma}_j}{E_j} + \frac{\sigma_j}{\eta_j}.$$

It is also possible to introduce internal strains in the model:

$$\dot{\varepsilon} = \dot{\varepsilon}_j^{\mathrm{e}} + \dot{\varepsilon}_j^{\mathrm{an}}.$$

By virtue of equality of stress in the spring and in the damper of an elementary model, the evolution equation for internal anelastic strains is obtained:

$$\dot{\varepsilon}_j^{\mathrm{an}} = \frac{E_j}{\eta_j}(\varepsilon - \varepsilon_j^{\mathrm{an}}).$$

The constitutive equation of the model may then be expressed in another form as:

$$\sigma = E_\infty \varepsilon + \sum_{j=1}^{n} \eta_j \dot{\varepsilon}_j^{\mathrm{an}}$$

$$\dot{\varepsilon}_1^{\mathrm{an}} = \frac{E_1}{\eta_1}(\varepsilon - \varepsilon_1^{\mathrm{an}})$$

$$\dot{\varepsilon}_j^{\mathrm{an}} = \frac{E_j}{\eta_j}(\varepsilon - \varepsilon_j^{\mathrm{an}}).$$

The strains $\varepsilon_j^{\mathrm{an}}$ play the role of internal variables in the sense of thermodynamics ($\sum_{j=1}^{n} \sigma_j \dot{\varepsilon}_j^{\mathrm{an}}$ is the intrinsic dissipation). The constitutive equation consists of a relation which contains internal variables and as many differential equations of evolution as the internal variables introduced. Note that this formulation can also be considered as a particular case of viscoplasticity (linear viscosity without yield limit) which will be studied in Chapter 6.

In order to represent complex viscoelastic behaviour, it is possible to generalize the association of springs and dampers in this way but soon the identification problem becomes unsolvable as it is necessary to determine at

Fig. 4.6. Generalized Maxwell model.

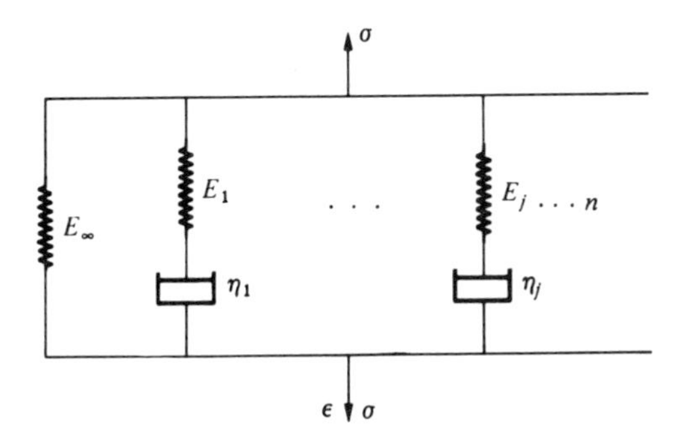

least as many coefficients as the number of elements introduced. The passage to an infinite number of elements or a continuum model is achieved by the concepts of creep or relaxation spectra. However, for linear viscoelasticity, the functional or hereditary approach offers undeniable advantages over that of discrete state variables.

4.3.3 *Functional formulation*

Creep function

Let us attempt to establish formally the response $\varepsilon(t)$ to a uniaxial load $\sigma(t)$ of a viscoelastic material under the following assumptions:

the strain $\varepsilon(t)$ is a functional of the whole history of the stress

$$\varepsilon(t) = \mathscr{F}(\sigma(\tau)), \qquad -\infty < \tau \leqslant t;$$

the material is nonageing and obeys the principle of objectivity; the functional $\mathscr{F}$ is linear.

Now consider a creep loading:

$$\sigma = \sigma_0 H(t - \tau)$$

where σ_0 is a constant, τ a reference instant and H the Heaviside step function: $H = 0$ if $t - \tau < 0$ and $H = 1$ if $t - \tau \geqslant 0$.

In view of the assumptions made, the response in terms of strain is:

$$\varepsilon = J(t - \tau)\sigma_0.$$

By definition $J(t - \tau)\sigma_0$ is the creep function, a characteristic uniaxial viscoelastic property of the material (Fig. 4.7). It is independent of the initial

Fig. 4.7. Creep function.

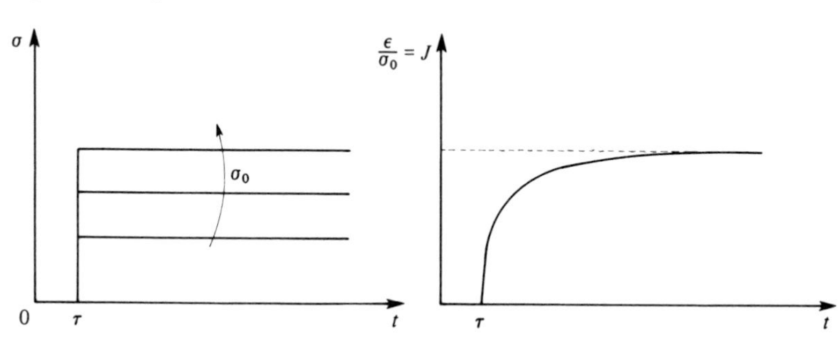

instant τ. Now consider a function represented by a series of steps (Fig. 4.8):

$$\sigma = \sum_{j=1}^{n} \Delta\sigma_j H(t - t_j).$$

The linearity or Boltzmann's principle of superposition can be used to write the strain response in the form:

$$\varepsilon(t) = \sum_{j=1}^{n} J(t - t_j)\Delta\sigma_j.$$

Finally, consider a load represented by a differentiable continuous or piecewise continuous function. Such a function can be considered as the limit of a series of step functions. The response $\varepsilon(t)$ can then be expressed in the form of a Riemann integral plus terms due to any discontinuities present in the loading.

● $$\varepsilon(t) = \int_0^t J(t - \tau)\frac{d\sigma}{d\tau}(\tau)\,d\tau + \sum_{j=1}^{n} J(t - t_j)\Delta\sigma_j.$$

This expression can be written using the notation of a convolution product and the definition of a derivative in the sense of distribution, $D\sigma/D\tau$, for the discontinuities:

$$\varepsilon(t) = J \otimes \frac{D\sigma}{D\tau}.$$

Relaxation function

The dual formulation for seeking the response $\sigma(t)$ to a uniaxial strain loading $\varepsilon(t)$ leads, under the same assumptions as those for the creep

Fig. 4.8. Creep response to a series of step loadings.

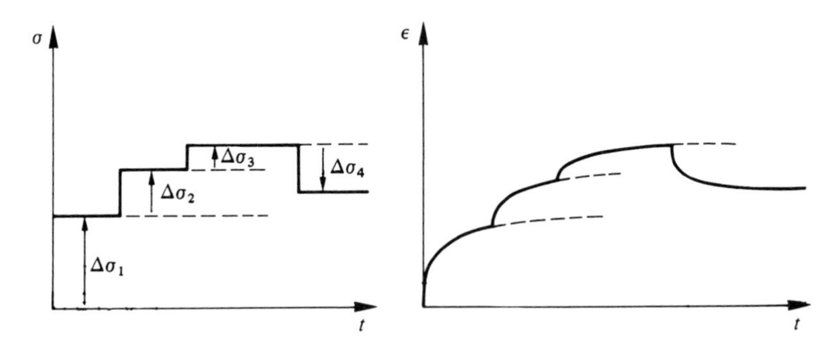

function, to

the definition of a relaxation function $R(t-\tau)$ as the response:

$$\sigma(t) = R(t-\tau)\varepsilon_0$$

to a step relaxation loading $\varepsilon = \varepsilon_0 H(t-\tau)$ (Fig. 4.9);
the response $\sigma(t)$ to any loading $\varepsilon(t)$:

$$\bullet \quad \sigma(t) = \int_0^t R(t-\tau)\frac{d\varepsilon}{d\tau}(\tau)\,d\tau + \sum_{j=1}^{n} R(t-t_j)\Delta\varepsilon_j$$

or

$$\sigma(t) = R \otimes \frac{D\varepsilon}{D\tau}.$$

For uniaxial loading, the material is characterized by the creep or by the relaxation function, one deducible from the other. For the creep test we have:

$$\sigma = \sigma_0 H \rightarrow \varepsilon = J\sigma_0$$

but

$$\sigma = \sigma_0 H = R \otimes \frac{d\varepsilon}{d\tau} = R \otimes \frac{dJ}{d\tau}\sigma_0.$$

In the same way, for the relaxation test:

$$\varepsilon = \varepsilon_0 H = J \otimes \frac{dR}{d\tau}\varepsilon_0$$

from which we conclude that:

$$R \otimes \frac{dJ}{d\tau} = J \otimes \frac{dR}{d\tau} = H.$$

Fig. 4.9. Relaxation function.

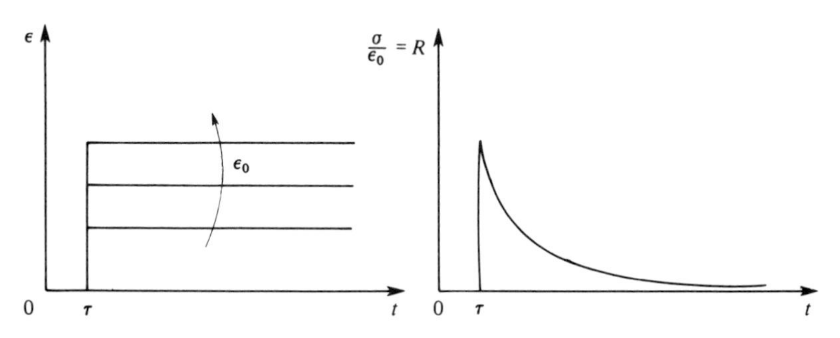

Generalized constitutive law

The assumptions made in connection with the definition of the creep function, when applied to the case of a general load lead us to define:

a creep function matrix $[J]$ such that:

$\sigma_{ij} = \text{constant} = \sigma_{ij}^0 \rightarrow \varepsilon_{ij} = J_{ijkl}(t-\tau)\sigma_{kl}^0$

a relaxation function matrix $[R]$ such that:

$\varepsilon_{ij} = \text{constant} = \varepsilon_{ij}^0 \rightarrow \sigma_{ij} = R_{ijkl}(t-\tau)\varepsilon_{kl}^0$

the responses to arbitrary loadings:

$\boldsymbol{\sigma}(t)$ or $\boldsymbol{\varepsilon}(t)$

$$\varepsilon_{ij}(t) = J_{ijkl} \otimes \frac{D\sigma_{kl}}{D\tau} \quad \text{or} \quad \sigma_{ij}(t) = R_{ijkl} \otimes \frac{D\varepsilon_{kl}}{D\tau}.$$

If we assume isotropy of behaviour, then with the same reasoning as used for elasticity in Section 4.1.2 it is possible to reduce to two the number of independent functions which define the components of each of the matrices $[J]$ and $[R]$.

It is possible to choose:

two functions $I(\tau)$ and $K(\tau)$ which play a role equivalent to the coefficients $1/E$ and ν/E in the uniaxial linear isotropic law:

- $$\boldsymbol{\varepsilon} = (I+K) \otimes \frac{D\boldsymbol{\sigma}}{D\tau} - K \otimes \frac{D[\mathrm{Tr}(\boldsymbol{\sigma})]}{D\tau}\mathbf{1}$$

or, alternatively, two functions $L(\tau)$ and $M(\tau)$ which play a role equivalent to Lame's coefficients:

- $$\boldsymbol{\sigma} = L \otimes \frac{D[\mathrm{Tr}(\boldsymbol{\varepsilon})]}{D\tau}\mathbf{1} + 2M \otimes \frac{D\boldsymbol{\varepsilon}}{D\tau}.$$

The identification of functions I and K or L and M by simple tests is based upon creep or relaxation tests in tension and in shear.

Identification of functions I and K:

Creep test in simple tension, $\sigma_{11} = \text{constant}$

$\varepsilon_{11} = [I(\tau) + K(\tau)]\sigma_{11} - K(\tau)\sigma_{11}$

$I(\tau) = \varepsilon_{11}(\tau)/\sigma_{11} = J(\tau).$

Creep test in shear, $\sigma_{12} = \text{constant}$

$$\varepsilon_{12} = [I(\tau) + K(\tau)]\sigma_{12}$$
$$K(\tau) = \varepsilon_{12}(\tau)/\sigma_{12} - I(\tau).$$

Identification of functions L and M:

Relaxation test in shear $\varepsilon_{12} = \text{constant}$

$$\sigma_{12} = 2M(\tau)\varepsilon_{12}$$
$$M(\tau) = \sigma_{12}/2\varepsilon_{12}.$$

Relaxation test in simple tension $\varepsilon_{11} = \text{constant}$

$$\sigma_{11} = L(\tau)\varepsilon_{11}(1 - 2\nu) + 2M(\tau)\varepsilon_{11}.$$

The contraction ratio ν can be expressed as a function of L and M by writing the relation between the zero component of stress ($\sigma_{22} = 0$) and the corresponding strain component ($\varepsilon_{22} = -\varepsilon_{11}$):

$$0 = L(\tau)\varepsilon_{11}(1 - 2\nu) - 2M\nu\varepsilon_{11}$$
$$\nu = L/2(L + M).$$

After measuring $\sigma_{11} = R(t)\varepsilon_{11}$, we obtain:

$$L(\tau) = \frac{M(R - 2M)}{3M - R}.$$

It is seen that the identification can be done with the same types of test as those used for the elasticity law, by taking into account the temporal evolution of the properties as given by the creep or relaxation functions.

4.3.4 *Viscoelastic properties of common materials*

Fig. 4.10(*a*) and (*b*) and 4.11(*a*) and (*b*) show graphs of creep and relaxation functions respectively for typical polymers identified as follows.

I An amorphous polymer of low molecular weight (11 000), polyisobutylene at 25 °C: creep and relaxation functions in shear:

$$\varepsilon_{12} = J_c(t)\sigma_{12}^0, \quad \sigma_{12} = R(t)\varepsilon_{12}^0, \quad (A = -3).$$

II An amorphous polymer of high molecular weight (300 000), a polyvinylacetate at 75 °C: creep relaxation functions in shear ($A = -1$).

III An amorphous polymer of still higher molecular weight

(3 620 000), a methacrylate poly-*n*-octyl: creep and relaxation functions in shear ($A = 0$).

IV An amorphous polymer of high molecular weight below its glass transition temperature, a polymethyl methacrylate at $-22\,^{\circ}C$: creep and relaxation functions in shear ($A = -7$).

Fig. 4.10. Creep functions for different, polymers (adapted from *Viscoelastic Properties of Polymers*, F. X. de Charentenay).

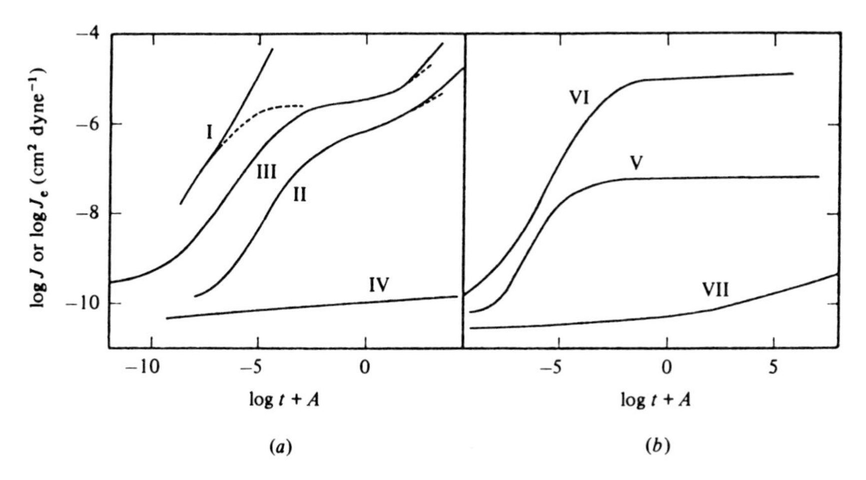

Fig. 4.11. Relaxation functions for different polymers (adapted from *Viscoelastic Properties of Polymers*, F. X. de Charentenay).

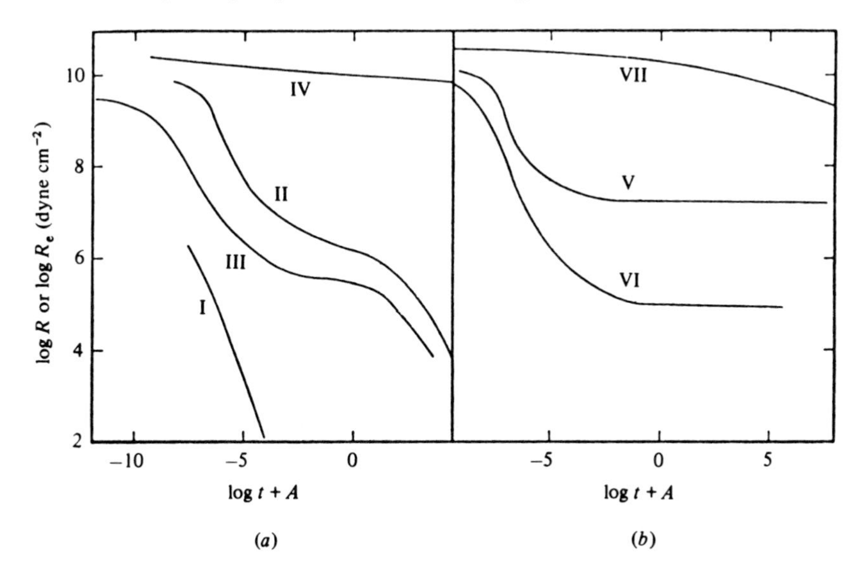

V A weakly cross-linked amorphous polymer, a vulcanized rubber Hévéa at 25 °C: creep and relaxation functions in tension:

$$\varepsilon = J(t)\sigma^0, \quad \sigma = R(t)\varepsilon^0, \quad (A = 0).$$

VI A diluted cross-linked gel at 10% of polyvinyl chloride in a dimethyl thianthrene at 25 °C: creep and relaxation functions in shear ($A = 0$).

VII A highly crystallized polymer, a linear polyethylene of density 0.965 g ml^{-1} at 20 °C: creep and relaxation functions in tension ($A = 2$).

4.3.5 *Elements of viscoelastic analysis of structures*

The calculation of displacements, stresses and strains in linear viscoelastic structures can be reduced to solving classical elasticity problems by means of symbolic calculus: a linear viscoelasticity problem transformed by the Laplace–Carson transform is identical to the corresponding linear elasticity problem.

Laplace–Carson transform

For the purposes of viscoelasticity, it is sufficient to consider functions $f(t)$ of variable t, of real or complex values, piecewise indefinitely differentiable and identical to zero for $t \leqslant 0$.

Applied to any function $f(t)$, the Laplace–Carson transform gives a corresponding function $f^+(p)$ of variable p defined by:

$$f^+(p) = \mathscr{L}(f(t)) = p\int_0^\infty f(t)\exp(-pt)\,dt.$$

The important property of the Laplace–Carson transform is that it transforms a convolution product into an ordinary product. We show this property here by using two functions ε and R:

$$\varepsilon^+ = p\int_0^\infty \varepsilon(\tau)\exp(-p\tau)\,d\tau$$

$$R^+ = p\int_0^\infty R(t)\exp(-pt)\,dt = p\int_\tau^\infty R(t-\tau)\exp(-p(t-\tau))\,dt$$

By using the Heaviside function $H(t-\tau)$ we obtain:

$$R^+\exp(-p\tau) = p\int_0^\infty R(t-\tau)H(t-\tau)\exp(-pt)\,dt.$$

Since $\varepsilon(0)=0$, we can, by using integration by parts, easily show that:

$$\varepsilon^{+}=\int_{0}^{\infty}\frac{\mathrm{d}\varepsilon}{\mathrm{d}\tau}(\tau)\exp(-p\tau)\,\mathrm{d}\tau.$$

The product of the transform can then be written in the following form:

$$\varepsilon^{+}R^{+}=\int_{0}^{\infty}\frac{\mathrm{d}\varepsilon}{\mathrm{d}\tau}R^{+}(p)\exp(-p\tau)\,\mathrm{d}\tau$$

$$\varepsilon^{+}R^{+}=p\int_{0}^{\infty}\mathrm{d}\tau\int_{0}^{\infty}\frac{\mathrm{d}\varepsilon}{\mathrm{d}\tau}R(t-\tau)H(t-\tau)\exp(-pt)\,\mathrm{d}t$$

$$=p\int_{0}^{\infty}\left[\int_{0}^{t}\frac{\mathrm{d}\varepsilon}{\mathrm{d}\tau}(\tau)R(t-\tau)\,\mathrm{d}\tau\right]\exp(-pt)\,\mathrm{d}t=\left(\frac{\mathrm{d}\varepsilon}{\mathrm{d}t}\otimes R\right)^{+}.$$

This proof applies also to piecewise continuous functions by writing the derivatives in the sense of distributions.

In general, we will therefore write:

$$\left(\frac{D\varepsilon}{Dt}(t)\otimes R(t)\right)^{+}=\varepsilon^{+}(p)R^{+}(p)$$

or for the tensor σ_{ij}:

$$\left(\frac{D\sigma_{ij}}{Dt}\otimes A\right)^{+}=\sigma_{ij}^{+}A^{+}.$$

Table 4.1 lists some commonly used transforms.

Equations of viscoelastic problems

Constitutive equation

The property of the Laplace–Carson transform in relation to the convolution product can be used to write the linear viscoelastic constitutive equation in the form:

$$\boldsymbol{\sigma}^{+}=\lambda^{+}\,\mathrm{Tr}\,(\boldsymbol{\varepsilon})^{+}\mathbf{1}+2\mu^{+}\boldsymbol{\varepsilon}^{+}$$

with

$$\lambda^{+}(p)=\mathscr{L}(L(t)),\quad \mu^{+}(p)=\mathscr{L}(M(t)).$$

These expressions are similar to the constitutive law for linear elasticity and this fact suggests an examination of all the equations of the problem transformed by the Laplace–Carson transform.

Knowns and unknowns

We express the knowns and unknowns in the transformed form by

Table 4.1. *Laplace–Carson transforms*

Function	Transform	Function	Transform
$f(t)$	$f^+(p)$	t	$1/p$
$C\cdot f(t)$	$C\cdot f^+(p)$	t^n	$n!/p^n$
$\dfrac{d}{dt}$	p	$\exp(-at)$	$p/(p+a)$
$H(t)$	1	$1-\exp(-at)$	$a/(p+a)$
$H(t-\tau)$	$\exp(-\tau p)$	$\cos\omega t$	$p^2/(p^2+\omega^2)$
$f(t-\tau)$	$\exp(-\tau p)f^+(p)$	$\sin\omega t$	$p\omega/(p^2+\omega^2)$
If $f(t)=0$	for $t\leqslant 0$	$f(t)\exp(-at)$	$\dfrac{p}{p+a}f^+(a+p)$
$f'(t)$	$pf^+(p)$		
$\dfrac{df}{dt}(t)\otimes g(t)$	$f^+(p)g^+(p)$	$(-t)^m f(t)$	$p\dfrac{d^m}{dp^m}\left(\dfrac{f^+(p)}{p}\right)$

applying the Laplace–Carson transform to the case of a mixed boundary value problem. The transforms of the specified data are:

$\vec{f}^+$ in the solid $\mathscr{S}$,
$\vec{u}^+$ on the boundary ∂s_u,
$\vec{F}^+$ on the boundary ∂s_F.

The unknown fields are $\vec{u}^+(x)$, $\sigma^+(x)$, $\varepsilon^+(x)$.

Equations

In the case of quasi-static problems, the equations are:

$$\operatorname{div}\boldsymbol{\sigma}^+ + \vec{f}^+ = 0$$

$$\boldsymbol{\varepsilon}^+ = \tfrac{1}{2}[\operatorname{grad}\vec{u}^+ + (\operatorname{grad}\vec{u}^+)^{\mathrm{T}}]$$

$$\boldsymbol{\sigma}^+ = \lambda^+\operatorname{Tr}(\boldsymbol{\varepsilon}^+)\mathbf{1} + 2\mu^+\boldsymbol{\varepsilon}^+$$

$$\boldsymbol{\sigma}^+\cdot\vec{n} = \vec{F}^+ \quad \text{on} \quad \partial S_F.$$

These equations are the same as those of a linear elastic problem if the boundary conditions do not vary with time (fixed ∂s_F and ∂s_u).

Solution method

The formal analogy between the transformed viscoelasticity problem and the elastic problem leads to the method of solution shown in a schematic way in Fig. 4.12. According to this method, a classical elasticity problem has to be solved first, followed by an inversion or deconvolution. The solutions in the transformed space are fields which depend on p and on coefficients defining the external loads.

The difficulty is focussed to inversion i.e. to the calculation of the original functions of time from the functions of variable p obtained by solving the elasticity problem. The calculations of the inverse are often tricky, particularly if the associated elasticity problem only has an approximate numerical solution. Nevertheless deconvolution codes now exist on computers.

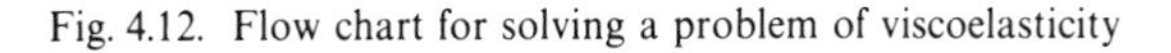

Fig. 4.12. Flow chart for solving a problem of viscoelasticity

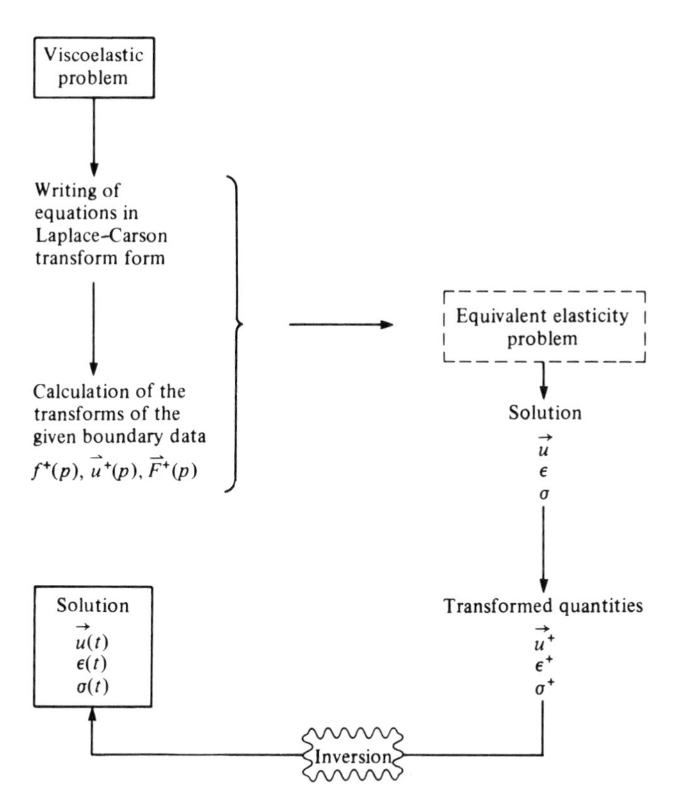

Bibliography

Solomon L. *Elasticité linéaire.* Masson, Paris (1966).

Schreiber E. & Anderson O. *Elastic constants and their measurements.* McGraw Hill, New York (1973).

Tsai S. W. & Thomas Han. *Introduction to composite materials.* Technomic (1980).

Massonet C. & Cescotto S. *Mécanique des matériaux.* Sciences et lettres, Liège (1982).

Courbon J. *Calcul des structures.* Dunod, Paris (1972).

Zienkiewicz O. C. The finite element method in engineering science, McGraw Hill, New York, 1st edn (1967)–3rd edn (1977).

Dhatt G. & Touzot G. *Une présentation de la méthode des éléments finis.* Maloine, Paris (1981).

Salençon J. *Viscoélasticité.* Presses de l'Ecole Nationale des Ponts et Chaussées, Paris (1983).

Findley W. N., Lai J. S. & Onaran K. *Creep and relaxation of non linear visco-elastic materials.* North Holland Publishing Company, Amsterdam (1976).

5

PLASTICITY

Un modèle devient une loi ... ou, sombre dans l'oubli!

This chapter deals with the phenomenological and mathematical modelling of plastic solids according to the schematic classification of Chapter 3: rigid, perfectly plastic solid; elastic, perfectly plastic solid; elastoplastic solid. It can be said that the first scientific work concerning plasticity goes back to Tresca's memoir in 1864 on the maximum shear stress criterion. Although the isotropic flow law was formulated as early as 1871 by St. Venant and Levy, its applications to structures had to wait until about 1950 when limit theorems were discovered. Since 1970 the availability of fast and large computers has led to application of the theory to practical problems. Proportional loading (where principal stresses do not rotate at any point of a structure) provides a large field of application for theories based on isotropic hardening.

Hardening, as we have seen in Chapter 1, is almost always anisotropic. This aspect of plasticity must be considered as soon as the loading is no longer proportional and especially under cyclic loads. Prager, around 1950, gave the first simple formulation of anisotropy, namely kinematic hardening, on which most of the present theories are based.

5.1 Domain of validity and use

The theory of plasticity is the mathematical theory of time-independent irreversible deformations; some comments on its physical nature were given in Chapter 1.

For metals and alloys it involves mainly the movement of dislocations without the influence of viscous phenomena or the presence of decohesion which damages the matter. Its domain of validity is therefore restricted by the following two limitations:

- low temperature usage: a very rough convention is to limit the temperature of usage to one-quarter of the absolute melting temperature of the material under consideration,
- nondamaging loads: for monotonic loads the strains must remain lower than approximately half of those at fracture. For alternating loads the limit depends on the condition of stabilization. Cyclic plasticity is applicable without modification as long as the number of cycles remains below the number which corresponds to stabilization (half life).

For soils the limitation is essentially due to the occurrence of slip surfaces caused by instability.

For polymers and wood, the irreversible deformations are more appropriately accounted for by viscoplasticity.

For concrete, the irreversible deformations are mainly due to microcracks, and hence a model based on coupling between elasticity and damage may be preferable.

These limitations only give indications. In a viscoplasticity problem for example, if it is known *a priori* that the deformation process takes place with a quasi-uniform strain rate field $\dot{\varepsilon}(M)$, it is possible to solve this problem by the theory of time-independent plasticity provided that the constitutive equation is identified by hardening experiments conducted at the considered strain rate $\dot{\varepsilon}$ (and at the considered temperature). In contrast, the relaxation of steel at room temperature (evolving as a function of time) can only be accounted for by an elastoviscoplastic theory.

The theory of plasticity is used to calculate permanent deformations of structures, to predict the plastic collapse of structures, to investigate stability (of a land-fill for example), to calculate forces required in metal forming operations, etc. We can also use the scheme of time-independent plasticity for metals at high temperature to discover asymptotic states which correspond to fixed loads or very high speeds. Another aspect of plasticity consists in its coupling with damage. This is studied in Chapter 7.

5.2 Phenomenological aspects

In this section we have collected the main results directly originating from experiments supposedly conducted on the considered material in the reference state of its use. To interpret these experiments, elementary models are used. Experimental results concerning particular effects (like cyclic hardening or softening, ratchetting, relaxation of mean stress) are mentioned in Chapter 3.

5.2.1 ***Uniaxial behaviour***

The hardening curve in tension–compression completely characterizes the uniaxial behaviour. Two possible types of curves (with discontinuous and continuous tangent moduli) are shown in Fig. 5.1.

Elastic limit

This is the stress above which irreversible deformations appear. Detection of this limit presents an experimental problem since it depends on the precision of the strain measuring device used. In the interest of an objective measure, we adopt a conventional definition of this limit.

The conventional elastic limit is the stress which corresponds to the occurrence of a specified amount of permanent strain.

> For quality control of materials, a conventional value of permanent strain equal to 0.2% is commonly used.
> If the above value of permanent strain is too high (of the same order of magnitude as the accompanying elastic strain) a more refined conventional value of 0.02% permanent strain is sometimes adopted.
> Another method consists in applying the convention to the ratio of the acceptable permanent strain to the accompanying elastic strain. For example
>
> $$\sigma_{Y0,1e} = \sigma_Y(\varepsilon_p/\varepsilon_e = 10\%)$$
>
> corresponds to a permanent strain which is ten times smaller than the elastic strain.

Except for mild steel (with a plateau), the practical determination of the elastic limit is always conventional. Even if the concept of elastic domain is perfectly justifiable at the theoretical level, it is essential to bear in mind the fact that its measure depends on some adopted convention.

Plastic flow, hardening and threshold

The decoupling of elastic and plastic effects can be justified on the basis of the physics and chemistry of solids, thermodynamics, and phenomenological experiments as follows:

> The total strain may be partitioned or separated into the reversible or elastic strain ε_e and the irreversible or inelastic strain ε_p without prejudging the nature of the latter strain which can be further

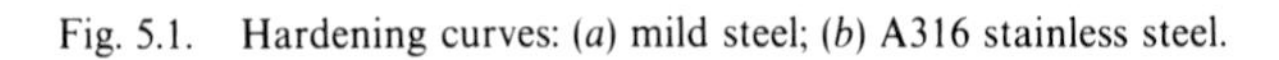

Fig. 5.1. Hardening curves: (*a*) mild steel; (*b*) A316 stainless steel.

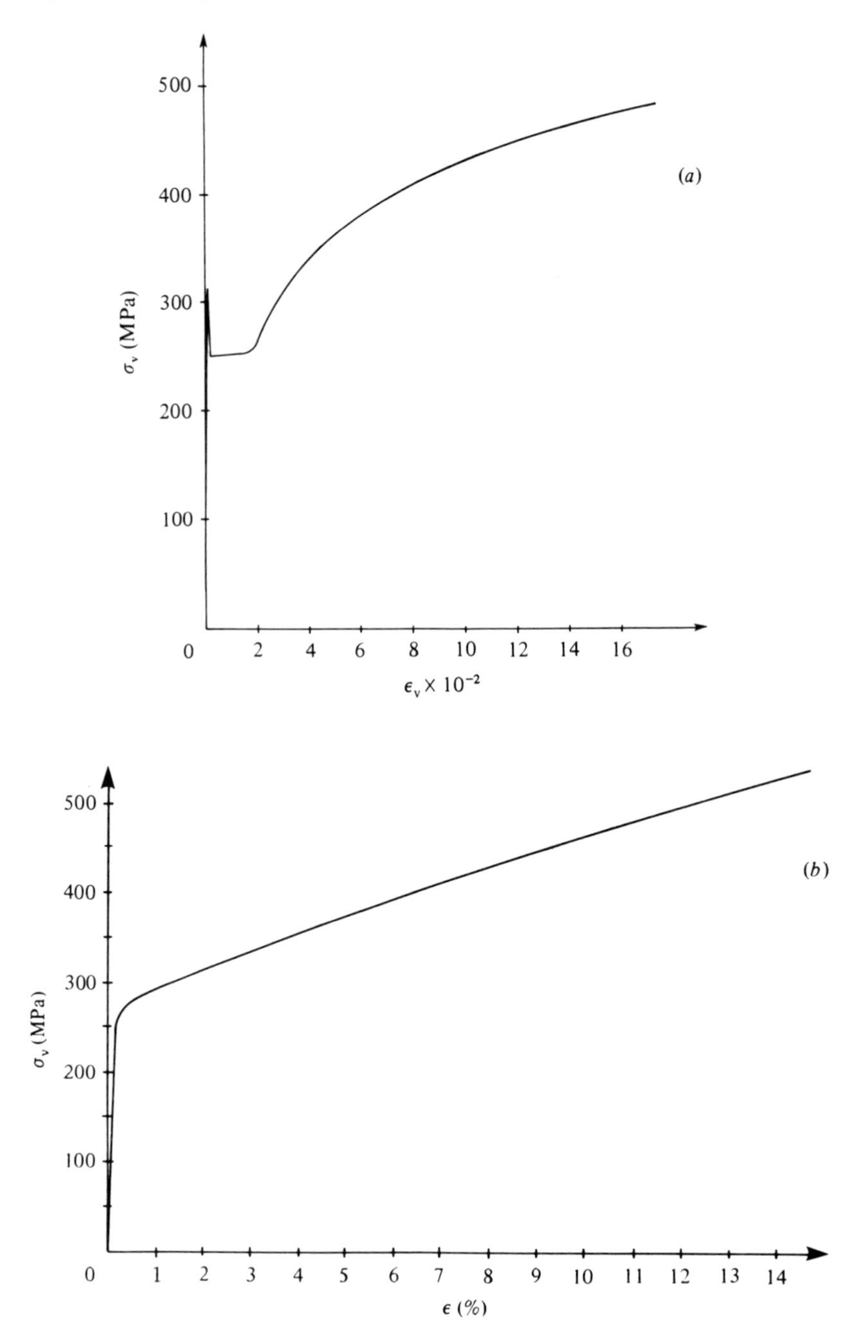

separated into anelastic, plastic, viscoplastic, etc. This partition is justified by the physics of the phenomena (cf. Chapter 3) as elastic deformation schematically corresponds to a variation in the interatomic distances without changes of place, while plastic (or inelastic) deformation implies 'slip' movements with modification of interatomic bonds. Within the framework of elastoplasticity we will therefore use

$$\varepsilon = \varepsilon_e + \varepsilon_p.$$

There exist decoupled constitutive relations for ε_e and ε_p

$$\varepsilon_e = A(\sigma) \qquad \forall \sigma$$
$$\begin{cases} \varepsilon_p = g(\sigma) & \text{for} \quad |\sigma| \geqslant \sigma_Y \\ \varepsilon_p = 0 & \text{if} \quad |\sigma| < \sigma_Y \end{cases}$$

with the latter relation being valid only in the case of continuous plastic flow (without unloading). Fig. 5.2 illustrates these two properties and shows in particular that even within a large range of

Fig. 5.2. Hardening curves with unloading/reloading: 2024 age-hardened aluminium alloy, $T = 20\,°C$.

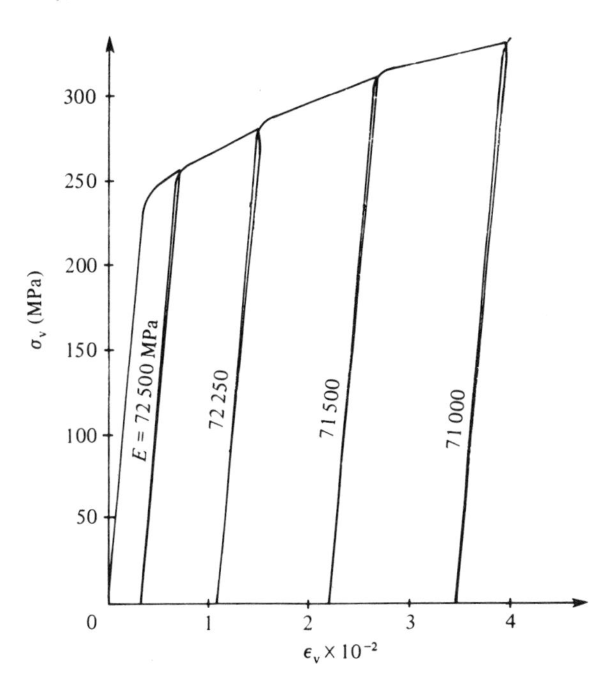

plastic strains the elastic modulus is little affected by the plastic flow.

Perfect plastic flow without hardening corresponds to the case in which the stress remains constant during the flow; this is exhibited by mild steel in the flat (plateau) zone. The hardening effect due to plastic flow manifests itself in two ways:

the flow occurs only if the stress increases, and
the elastic limit increases during the flow; this is verified by unloading and then reloading as in the examples of Fig. 5.2.

On the physical level, hardening is due to an increase in the dislocation density: the dislocations have a tendency to interlock and to block each other. In the first approximation, the increase in the elastic limit follows the increase in the stress, and it is this approximation which constitutes the theoretical basis of classical plasticity. Thus, for monotonic loading, the current limit of elasticity, also called the plasticity threshold or the yield stress, is equal to the highest value of the stress previously attained. For a material with positive hardening, $\mathrm{d}\sigma/\mathrm{d}\varepsilon_\mathrm{p} > 0$, the 'natural' elastic limit σ_Y is the smallest value of the yield stress (which is a function of the history of plastic deformations).

Any point on a monotonic hardening curve can therefore be considered as a representative point of the plasticity threshold, and the characteristic hardening law may be written as

$$\sigma_\mathrm{s} = g^{-1}(\varepsilon_\mathrm{p}).$$

The plastic flow occurs only if $\sigma = \sigma_\mathrm{s}$, i.e.

$$\sigma < \sigma_\mathrm{s} \rightarrow \dot{\varepsilon}_\mathrm{p} = 0$$
$$\sigma = \sigma_\mathrm{s} \rightarrow \exists\, \dot{\varepsilon}_\mathrm{p} \neq 0.$$

A number of analytic expressions have been proposed to model the hardening function g; we will use the one resulting from a calculation which is based on dislocation theory and which shows that the yield stress is proportional to the square root of the density of dislocations ρ_d:

$$\sigma_\mathrm{s} = \kappa b \rho_\mathrm{d}^{1/2}.$$

In reality, the density of dislocations is never zero. Hence, letting ρ_{d_0} be the dislocation density in the initial state corresponding to the elastic limit σ_Y, we may write

$$\sigma_\mathrm{s} = \sigma_\mathrm{Y} + \kappa b(\rho_\mathrm{d} - \rho_{\mathrm{d}_0})^{1/2}.$$

In terms of macroscopic strains, we may, in analogy with the above relation, write

$$\sigma_s = \sigma_Y + K_Y \varepsilon_p^{1/M_Y}.$$

This last relation, generally called the Ramberg–Osgood equation, can be

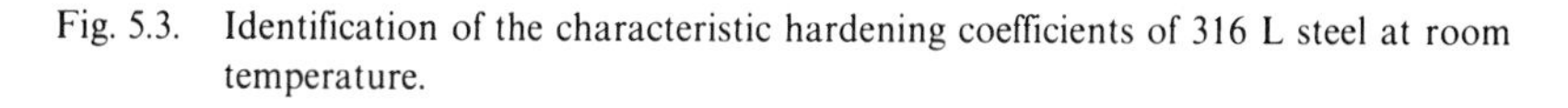

Fig. 5.3. Identification of the characteristic hardening coefficients of 316 L steel at room temperature.

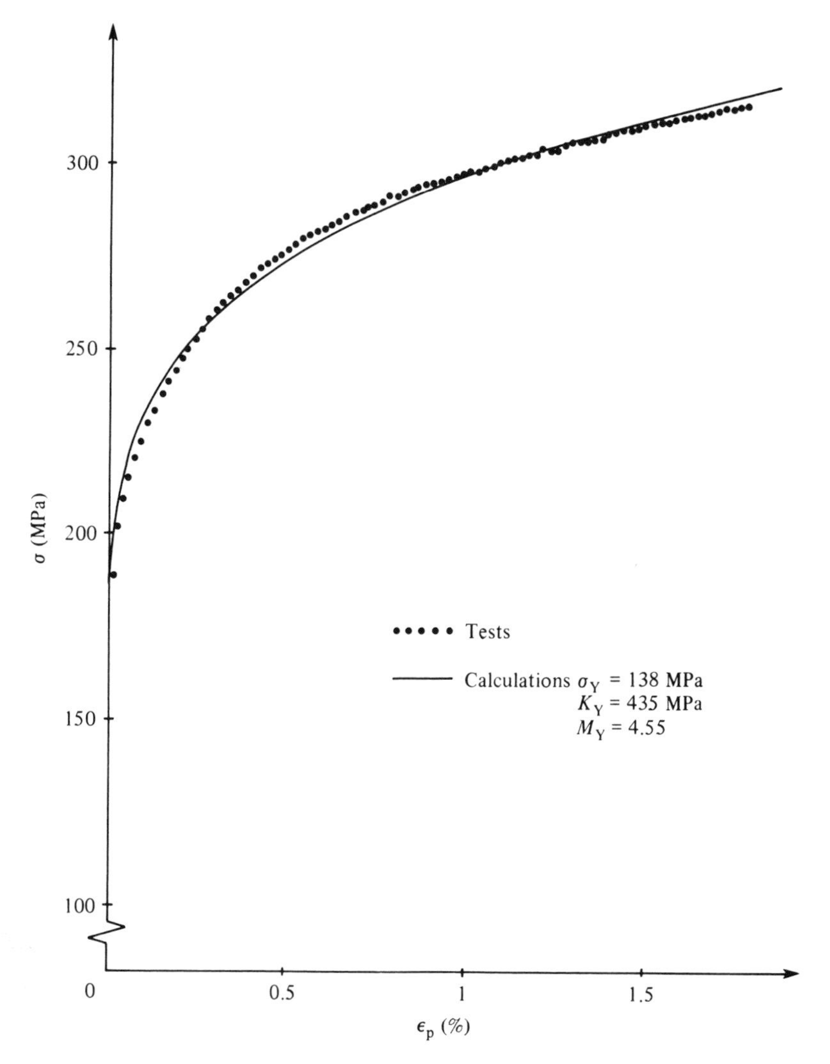

easily inverted to give†

$$\bullet \qquad \varepsilon_p = g(\sigma_s) = \left\langle \frac{\sigma_s - \sigma_Y}{K_Y} \right\rangle^{M_Y}$$

where σ_Y is the elastic limit, K_Y is the coefficient of plastic resistance, M_Y is the hardening exponent and the angular brackets have the following meaning: $\langle x \rangle = x$ if $x > 0$ and $\langle x \rangle = 0$ if $x \leqslant 0$. The identification of this expression, i.e., the determination of the values of the coefficient σ_Y, K_Y and M_Y for a particular material is done through hardening test results. After determining σ_Y from the hardening curve, the straight line closest to the experimental points, plotted as $\ln(\sigma_s - \sigma_Y)$ versus ε_p, gives K_Y and M_Y according to the following straight line equation:

$$\ln(\sigma_s - \sigma_Y) = \ln K_Y + \frac{1}{M_Y} \ln \varepsilon_p.$$

An example is given in Fig. 5.3. Chapter 3 provides an outline of numerical identification methods for non-linear equations such as this one.

The hardening law $\varepsilon_p = g(\sigma_s)$ is written in terms of nominal values of the variables if it is to be used for small strains, but should be written in terms of their true values ($\sigma_v = \sigma(1 + \varepsilon)$, $\varepsilon_v = \ln(1 + \varepsilon)$) if used for 'large strains'.

Constitutive equations

Perfectly plastic solids

In this case the threshold σ_s is constant and is equal in tension and compression. Depending on whether the model has to be used to calculate

Fig. 5.4. Elastic limit and ultimate stress.

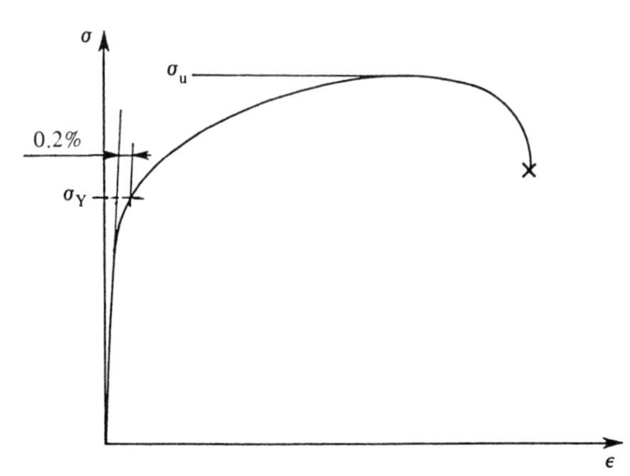

the lower or upper bounds (see Section 6.2), the threshold is chosen as equal to the conventional elastic limit σ_Y or the ultimate stress σ_u – the maximum stress attained in a hardening test (Fig. 5.4). The constitutive model is the following:

- $|\sigma| < \sigma_s \rightarrow \varepsilon = \varepsilon_e = \begin{cases} 0 & \text{(rigid body case)} \\ \sigma/E & \text{(perfect elastic body case)} \end{cases}$
- $|\sigma| = \sigma_s \rightarrow \varepsilon = \varepsilon_e +$ arbitrary ε_p of the same sign as σ.

Hardening plastic solids

The complete model of an initially isotropic elastoplastic solid subjected to monotonic uniaxial loading is the following:

- $\varepsilon = \varepsilon_e + \varepsilon_p,$
- $\varepsilon_e = \sigma/E,$
- $\varepsilon_p = \left\langle \dfrac{|\sigma| - \sigma_Y}{K_Y} \right\rangle^{M_Y} \operatorname{Sgn}(\sigma) \quad \text{with } \sigma = \sigma_s.$

In the case of a rigid-plastic hardening solid, two consistent hypotheses consist in neglecting the elastic strains altogether and taking the elastic limit to be equal to zero. The model is then reduced to

- $\varepsilon = \varepsilon_p = (|\sigma|/K)^M \operatorname{Sgn}(\sigma).$

The above expressions constitute the practical ways to represent the hardening law, but other expressions, or the tension curve itself, defined point by point, can be used. Typical values of the coefficients σ_Y, K_Y, M_Y and K and M for different materials are given in Table 5.1 together with the range of their validity.

Cyclic loadings

It is interesting to relate the (peak to peak) strain range $\Delta\varepsilon$ directly to the (peak to peak) stress range $\Delta\sigma$ of stabilized cycles as defined in Chapter 3 for periodic loads.

Taking the example of a cycle symmetric with respect to the origin, such as that shown in Fig. 5.5, it is easy to express $\Delta\varepsilon_p$ as function of $\Delta\sigma$ according to the monotonic hardening law $\varepsilon_p = g(\sigma)$:

$$\Delta\varepsilon_p/2 = g(\Delta\sigma/2).$$

This corresponds to Masing's rule which assumes a homothetic transformation of ratio 2 between the initial tension curve OA and the alternating half

Table 5.1. *Plasticity characteristics*

Material	T °C	Monotonic loading						Cyclic loading	
		σ_Y MPa	σ_u MPa	M	K	M_Y	K_Y	M_c	K_c
35 NCD 16 steel	20	1200	2000	10.5	2990	3.1	3340	6.8	6230
IN 100 alloy	20	650	875	33	960	5.6	655		
IN 100 alloy	700	600	925	18	1150	6.8	354	3.25	9925
316 L steel	20	133				4.5	435	10.50	811
UDIMET 700 alloy	20	700	1250	17	1374	4.8	900		
TA 6 V alloy	350	300	690	10.2	940	4.3	884	11.00	1583
INCO 718 alloy	550	500	1180	12.2	1657	5.2	1658	11.7	2585
COTAC 744	1000			4.2	2206				
X20 CrV 12.1	350			9.6	853			7.25	951
X20 CrV 12.1	600							7.9	554
IMI 550	20			28.2	1176			11.8	1366
COBALT VO 795	20							2.4	5850
Structural steel (0.14% C–0.35 Si; 1.36 Mn–0.24 Cr)	20							5.7	2360
NIMONIC 90	20							7.3	3940

The values of the coefficients listed in Table 5.1 are with reference to the simple model studied in Section 5.2.1.

BA which corresponds to the positive flow (homothetic of ratio -2 for the alternate AB).

In fact the hardening function in the cyclic behaviour model is different from that in the monotonic model because of the consolidation effects induced by the cyclic loading. It is therefore preferable to identify this relation between the ranges from the cyclic tension–compression tests directly:

$$\Delta\varepsilon_p = g_c(\Delta\sigma)$$

where the function g_c may be modelled by

- $$\Delta\varepsilon_p = (\Delta\sigma/K_c)^{M_c}.$$

This expression defines the 'stabilized cyclic hardening curve', i.e., the peaks of stabilized cycles $(\sigma_{Max}, \varepsilon_{p\,Max})$ on the (σ, ε_p) graph for centred cycles such as those of Fig. 5.6(a).

This cyclic hardening curve can be located 'above' (cyclic hardening) or 'below' (cyclic softening) the monotonic hardening curve. The identification of its expression in terms of stress and strain variables presents a problem linked to the history of deformation. As a matter of fact the curve may not be exactly the same if it were obtained by continuous variation as it would be if it were obtained by imposing finite increments in stress or strain ranges, or if one or several specimens are used.

To identify a simple relation between the stress and strain ranges and to verify the uniqueness of the cyclic hardening curve, several methods can be used:

Fig. 5.5. Stabilized cycle.

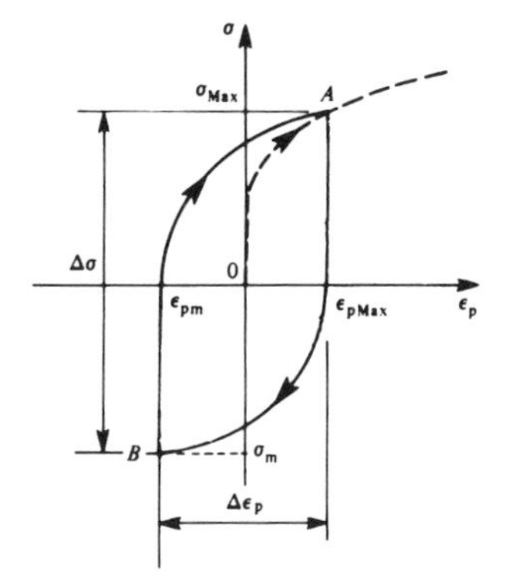

Fig. 5.6. Cyclic hardening curve: 30 NCD 16 nickel–chromium steel (after Lieurade).

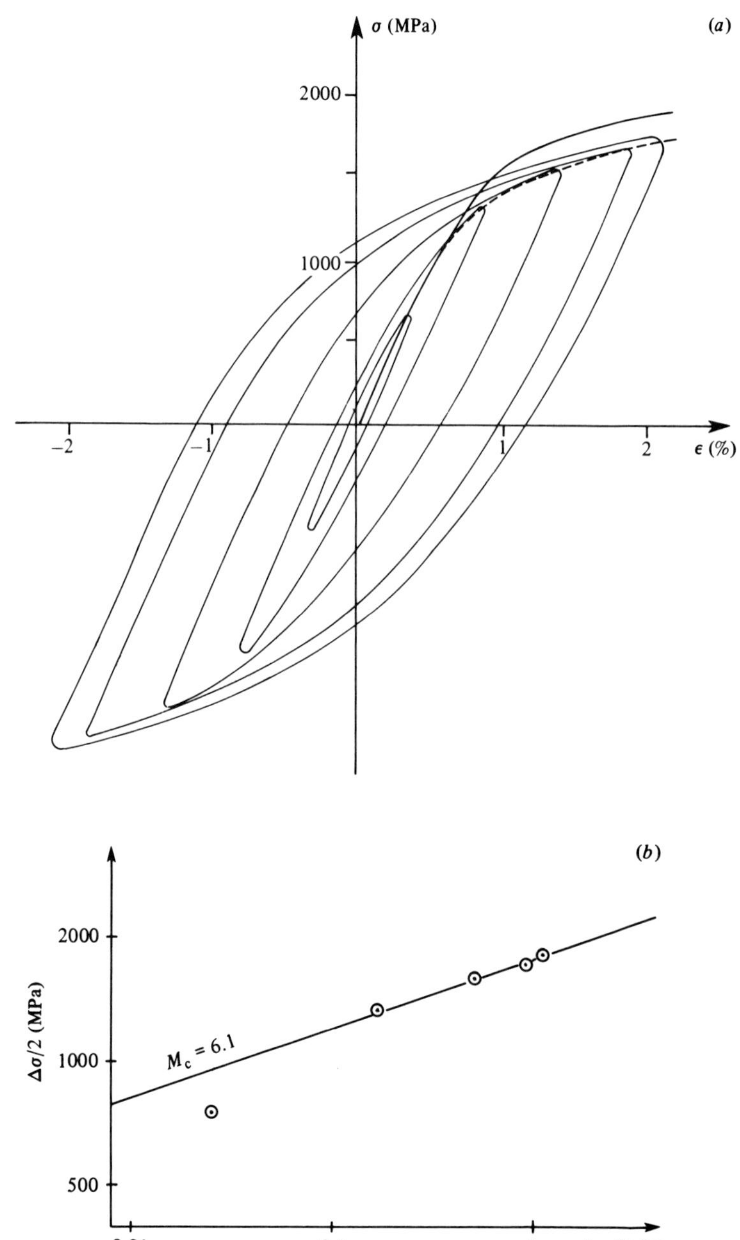

We may compare the curves of amplitudes ($\Delta\sigma$, $\Delta\varepsilon$) with each of the hysterisis loops adjusted to the lower peak ($\sigma - \sigma_m$, $\varepsilon - \varepsilon_m$). If all these curves approximately coincide (Masing's rule), we may conclude that there are no parasitic effects (e.g., Fig. 5.7).
We may compare the cyclic hardening curves obtained by using one specimen per level with those obtained by the incremental test method which consists in subjecting one specimen only to deformation cycles of successively increasing and then decreasing amplitudes until stabilization is achieved (e.g., Fig. 5.8).
We may perform tests in which the strains are varied stepwise, first increasing and then decreasing (e.g., Fig. 5.9 for 316 L steel at 20 °C). After stabilization of hardening at the lowest level, a resumption of hardening at a higher level is an indication of the dependence between the cyclic hardening and the strain amplitude. During decreasing levels the memory effect of the previous load is clearly established by comparison of loops obtained for the same strain level ($\pm 1\%$ and $\pm 1.5\%$ in the example under consideration).

For the identification of the cyclic hardening curve, we recommend this last type of test where the strain is varied in steps; it generally provides sufficiently detailed indications of the uniqueness of this curve with only

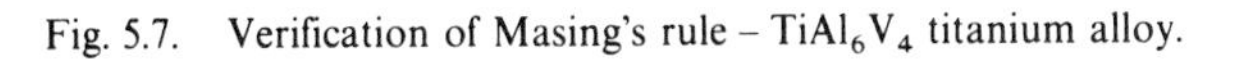

Fig. 5.7. Verification of Masing's rule – $TiAl_6V_4$ titanium alloy.

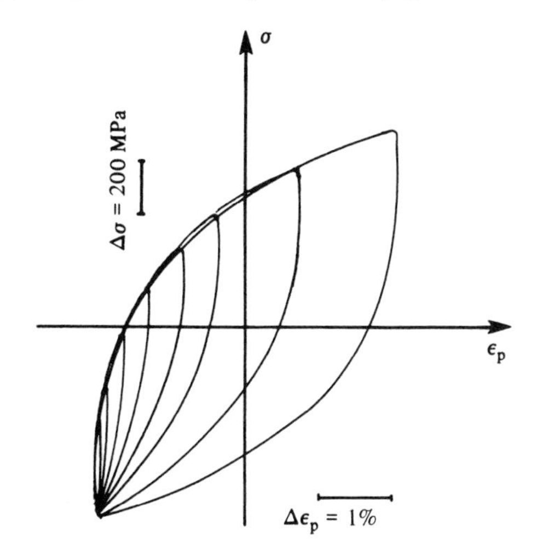

Fig. 5.8. Cyclic hardening curves obtained by different methods: (*a*) evolution of stress and imposed strain; (*b*) NIMONIC 90 alloy; (*c*) 35 CD 4 chromium steel, (*d*) 316 steel (after Lieurade).

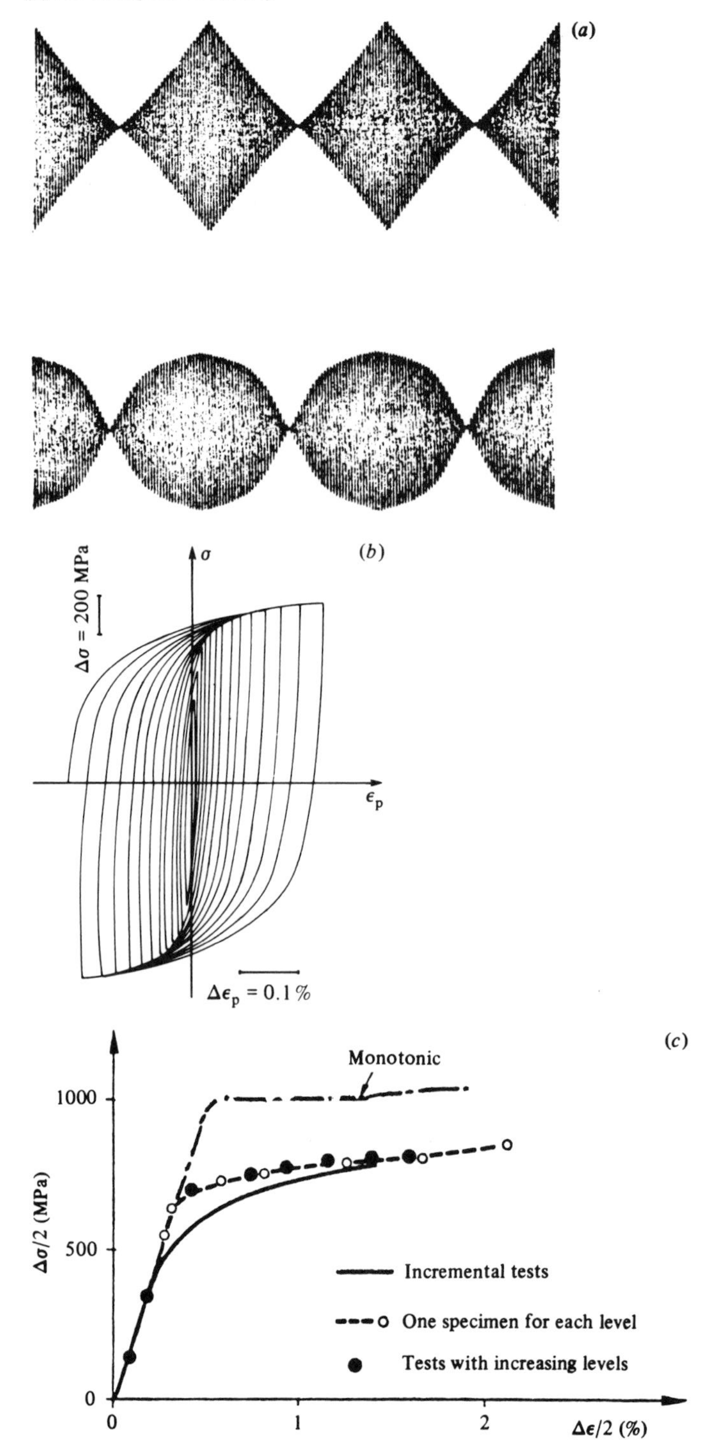

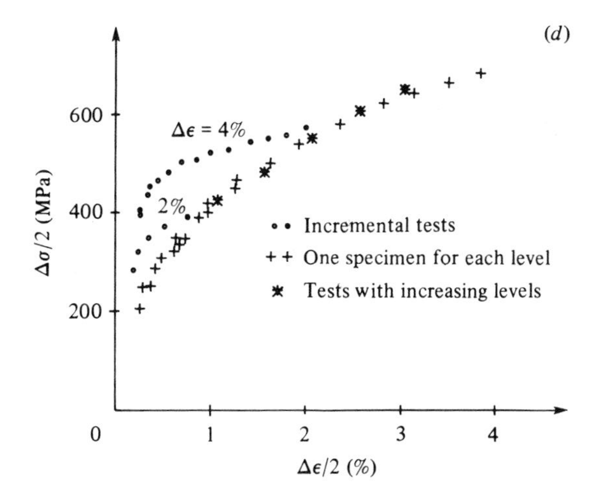

Fig. 5.9. Test with increasing and then decreasing steps, 316 L steel; successive strain ranges (2%, 3%, 4%, 5%, 6%, 2%, 3%).

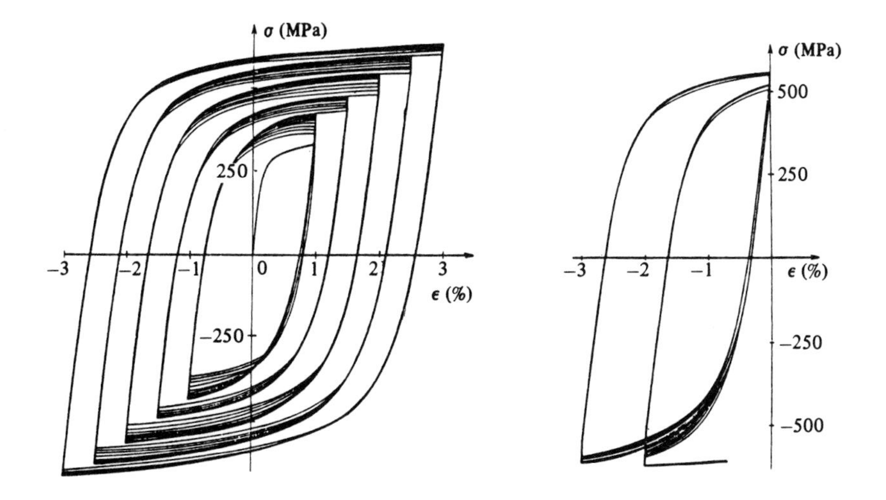

one specimen. On the other hand, when the uniqueness is not evident, the test may be used to define a more complex model which takes into account the influence of the strain path such as the one encountered in Section 5.4.4. Five increasing steps and three decreasing steps in every twenty cycles are usually sufficient.

5.2.2 ***Multiaxial plasticity criteria***

The yield stress of uniaxial plasticity defines the elastic domain in uniaxial stress space (Fig. 5.10). The generalization of this concept to the multiaxial case is the yield criterion. It defines in the stress space of three or six dimensions a domain within which any stress variation generates only variations of elastic strains.

The simplest way to show the existence of this domain in a two-dimensional space is to use the results of tension (or compression)–torsion tests on thin tubes. Successive (proportionally increasing) radial loads, defined by axial force $\vec{F}$ and twisting couple $\vec{C} = \alpha\vec{F}$, are applied to a sample. Each radial loading is identified by a constant value of the scalar α ($-\infty < \alpha < +\infty$). The values of $\vec{F}$ and $\vec{C}$ at which plastic strains become noticeable are recorded; these offset strains lie between 10^{-6} and 10^{-4} depending upon the precision of the apparatus.

Fig. 5.10. One-dimensional elastic domain.

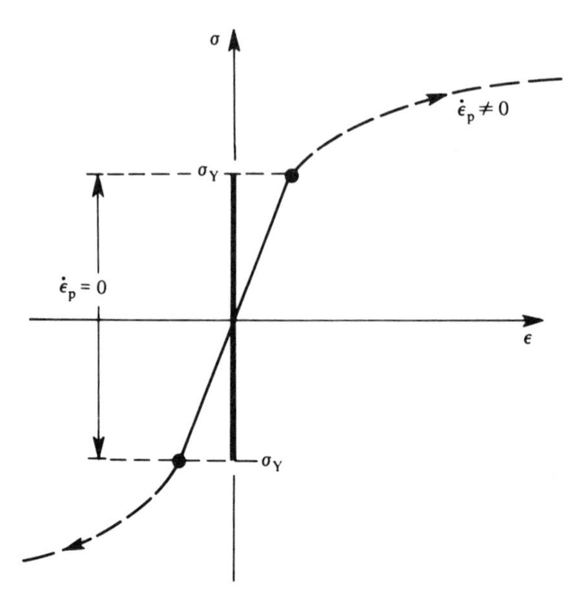

The curve obtained by plotting the normal stress σ_{11} and the corresponding tangential stress σ_{12} at plasticity thresholds then defines the boundary of the elastic domain. Fig. 5.11 shows typical test results for the same material: nonhardened (A), prehardened by a tensile load (B), and prehardened by a tensile load and a shear load (C).

The mathematical modelling of these domains and of their evolution has been the object of a number of propositions. At present we restrict ourselves to the essential concepts most commonly used; however, a more detailed discussion of the three-dimensional behaviour will be taken up in Section 5.4.

A preliminary remark must now be made regarding the terminology used in connection with isotropic or anisotropic yield and hardening criteria. In a first understanding, the terms isotropic or anisotropic yield criteria are applicable to a given state (without modification of hardening). On the other hand, hardening implies the idea of transformation. We therefore speak of isotropic hardening when the transformation is a homothetic transformation (dilatation of the criterion), of kinematic hardening for the translation of the criterion, of combined isotropic–kinematic hardening, and finally of anisotropic hardening when the transformation is arbitrary.

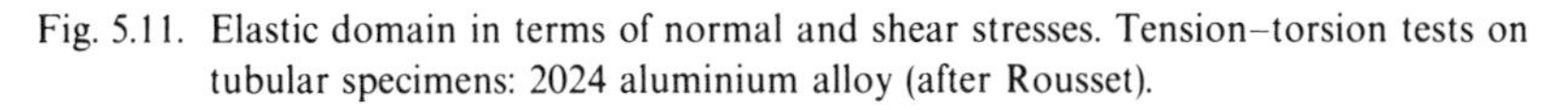

Fig. 5.11. Elastic domain in terms of normal and shear stresses. Tension–torsion tests on tubular specimens: 2024 aluminium alloy (after Rousset).

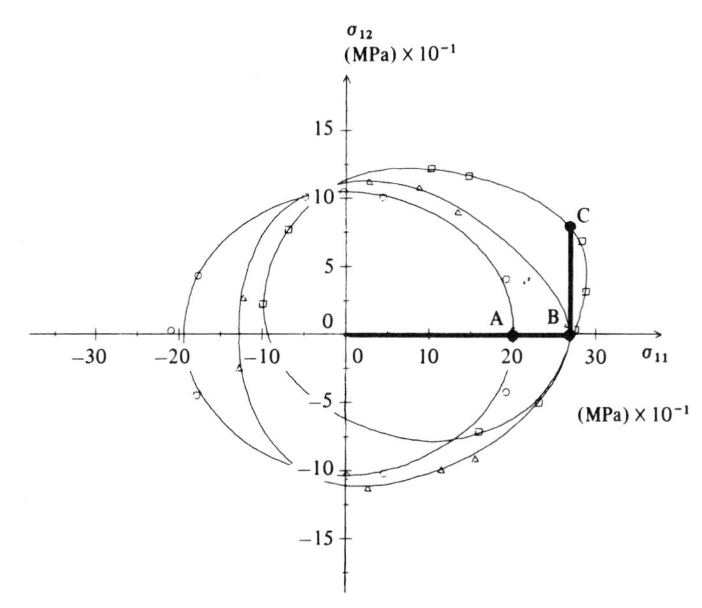

We can thus speak of 'isotropic' hardening of an initially anisotropic material.

Isotropic criteria

These criteria correspond to isotropic-hardening states. The equation of the elastic domain boundary includes, *a priori*, all the components of the stress tensor and a hardening variable which as discussed in Chapter 3 is a scalar variable in the case of isotropic hardening. Choosing this last variable to be the yield stress σ_s in simple tension, the criterion may be expressed as

$$f(\sigma_{ij}, \sigma_s) = 0.$$

Isotropy requires that the boundary of the domain be invariant under a change of axes. Therefore, the function f depends only on the three invariants of the stress tensor (invariants defined in Chapter 2):

$$f(\sigma_{\mathrm{I}}, \sigma_{\mathrm{II}}, \sigma_{\mathrm{III}}, \sigma_s) = 0.$$

Metals generally exhibit plastic incompressibility and yield-independence with respect to hydrostatic stress. It is then sufficient to use the deviatoric stress tensor:

$$\boldsymbol{\sigma}' = \sigma - \tfrac{1}{3}\sigma_{\mathrm{I}}\mathbf{1}.$$

In the isotropic case, we may use the invariants of $\boldsymbol{\sigma}'$, namely $s_{\mathrm{II}} = \frac{1}{2}\mathrm{Tr}\,(\boldsymbol{\sigma}'^2)$ and $s_{\mathrm{III}} = \frac{1}{3}\mathrm{Tr}\,(\boldsymbol{\sigma}'^3)$. Therefore, the relation

$$f(s_{\mathrm{II}}, s_{\mathrm{III}}, \sigma_s) = 0$$

is the general expression of the isotropic criteria of incompressible plasticity. In the space of principal stresses σ_1, σ_2, σ_3, the boundary is represented by a cylinder with its axis equally inclined to the reference axes. Note that instead of the invariants s_{II} and s_{III} of the deviator, the associated homogeneous invariants can also be used.

$$J_2(\boldsymbol{\sigma}) = (3s_{\mathrm{II}})^{1/2} = (\tfrac{2}{3}\sigma'_{ij}\sigma'_{ij})^{1/2}$$
$$J_3(\boldsymbol{\sigma}) = (\tfrac{27}{2}s_{\mathrm{III}})^{1/3} = (\tfrac{9}{2}\sigma'_{ij}\sigma'_{jk}\sigma'_{ki})^{1/3}.$$

The von Mises criterion

Plastic deformation of metals results from slip, i.e., intercrystalline shear governed by the tangential stresses. According to the von Mises criterion, the plasticity threshold is linked to the elastic shear energy. This amounts to neglecting the influence of the third invariant and taking a linear expression for the function f.

The elastic strain energy

$$w_e = \int_0^{\boldsymbol{\varepsilon}^e} \boldsymbol{\sigma}:d\boldsymbol{\varepsilon}^e$$

can be written as the sum of a shear energy w_d and an energy of volumetric deformation

$$w_e = \int_0^{\boldsymbol{\varepsilon}^e} [\boldsymbol{\sigma}' + \tfrac{1}{3}\mathrm{Tr}(\boldsymbol{\sigma})\mathbf{1}]:[d\boldsymbol{\varepsilon}^e + \tfrac{1}{3}\mathrm{Tr}(d\boldsymbol{\varepsilon}^e)\mathbf{1}]$$

$$w_e = \int_0^{\boldsymbol{\varepsilon}^e} \boldsymbol{\sigma}':d\boldsymbol{\varepsilon}^{e'} + \int_0^{\mathrm{Tr}(\boldsymbol{\varepsilon}^e)} \tfrac{1}{3}\mathrm{Tr}(\boldsymbol{\sigma})\,\mathrm{Tr}(d\boldsymbol{\varepsilon}^e).$$

By introducing the linear elastic law

$$d\boldsymbol{\varepsilon}^{e'} = (1/2\mu)\,d\boldsymbol{\sigma}'$$

in the expression for shear energy $w_d = \int_0^{\boldsymbol{\varepsilon}^e} \boldsymbol{\sigma}':d\boldsymbol{\varepsilon}^{e'}$, we obtain

$$w_d = \int_0^{\boldsymbol{\sigma}'} (1/2\mu)\boldsymbol{\sigma}':d\boldsymbol{\sigma}' = (1/4\mu)\boldsymbol{\sigma}':\boldsymbol{\sigma}'.$$

We now equate this energy of a three-dimensional state of stress to that of a one-dimensional state of stress in pure tension at the yield stress $\sigma = \sigma_s$. Since

$$[\sigma] = \begin{bmatrix} \sigma_s & 0 & 0 \\ 0 & 0 & 0 \\ 0 & 0 & 0 \end{bmatrix} \rightarrow \sigma' = \begin{bmatrix} \tfrac{2}{3}\sigma_s & 0 & 0 \\ 0 & -\tfrac{1}{3}\sigma_s & 0 \\ 0 & 0 & -\tfrac{1}{3}\sigma_s \end{bmatrix}$$

we obtain

$$(1/4\mu)\boldsymbol{\sigma}':\boldsymbol{\sigma}' = (1/6\mu)\sigma_s^2$$

so that

$$\tfrac{1}{2}\boldsymbol{\sigma}':\boldsymbol{\sigma}' - \tfrac{1}{3}\sigma_s^2 = 0.$$

The function $f(s_{II}, \sigma_s) = 0$ is therefore expressed by

$$s_{II} - \tfrac{1}{3}\sigma_s^2 = 0$$

or, with the equivalent stress $\sigma_{eq} = J_2(\boldsymbol{\sigma}) = (3s_{II})^{1/2} = (\tfrac{3}{2}\boldsymbol{\sigma}':\boldsymbol{\sigma}')^{1/2}$

- $$f = \sigma_{eq} - \sigma_s = 0.$$

Any three-dimensional state of stress for which $\sigma_{eq} = \sigma_s$ is a state of complex stress, equivalent in the von Mises sense, to a one-dimensional state defined by the yield stress σ_s. In particular, the initial yield criterion is expressed by

$$\sigma_{eq} - \sigma_Y = 0.$$

The expanded expressions of the von Mises yield criterion are:
in the six-dimensional stress space:

$$\tfrac{1}{2}[(\sigma_{11} - \sigma_{22})^2 + (\sigma_{22} - \sigma_{33})^2 + (\sigma_{33} - \sigma_{11})^2 + 6(\sigma_{12}^2 + \sigma_{23}^2 + \sigma_{13}^2)] - \sigma_s^2 = 0$$

in the three-dimensional principal stress space:

- $$\frac{1}{\sqrt{2}}[(\sigma_1 - \sigma_2)^2 + (\sigma_2 - \sigma_3)^2 + (\sigma_3 - \sigma_1)^2]^{1/2} = \sigma_s.$$

This is the equation of a circular cylinder with its axis equally inclined to the three $(\sigma_1, \sigma_2, \sigma_3)$ reference axes and which has a radius $R = \sqrt{2/3}\ \sigma_s$ (Fig. 5.12).

The deviatoric plane is the plane which passes through the origin with its normal equally inclined to the three reference principal stress axes. The orthogonal projection of a point with coordinate $(\sigma_s, 0, 0)$ on this plane gives a point in Fig. 5.12 with coordinates $\sigma'_1 = \sqrt{\tfrac{2}{3}}\sigma_s$, $\sigma'_2 = \sigma'_3 = -\tfrac{1}{2}\sqrt{\tfrac{2}{3}}\sigma_s$. Conversely, the coordinates of this representative point of the uniaxial deviatoric stress tensor, are $\tfrac{2}{3}\sigma_s$, $-\tfrac{1}{3}\sigma_s$ and $-\tfrac{1}{3}\sigma_s$ with respect to the principal stress axes.

The Tresca criterion

According to this criterion, the plasticity threshold is not linked to the energy but to the shear stress: the maximum shear stress. It is expressed by:

$$\tfrac{1}{2}\operatorname*{Sup}_{i \neq j}(|\sigma_i - \sigma_j|).$$

By equating it to its value in the one-dimensional state corresponding to yielding at stress σ_s, we obtain the following expression for the criterion

$$\tfrac{1}{2}\operatorname*{Sup}_{i \neq j}(|\sigma_i - \sigma_j|) = \tfrac{1}{2}\sigma_s.$$

or

- $$f = \operatorname*{Sup}_{i \neq j}(|\sigma_i - \sigma_j|) - \sigma_s = 0.$$

This function can be expressed in terms of the invariants s_{II} and s_{III} of the deviatoric stress tensor. The equation below defines the set of the six planes of the Tresca criterion.

$$f(s_{\text{II}}, s_{\text{III}}) = 4s_{\text{II}}^2 - 27s_{\text{III}}^2 - 9\sigma_s^2 s_{\text{II}}^2 + 6\sigma_s^4 s_{\text{II}} - \sigma_s^6 = 0.$$

The initial yield according to the Tresca criterion is expressed by

$$\operatorname*{Sup}_{i \neq j}(|\sigma_i - \sigma_j|) - \sigma_Y = 0.$$

Fig. 5.12. Geometric representation of the von Mises criterion on the deviatoric plane.

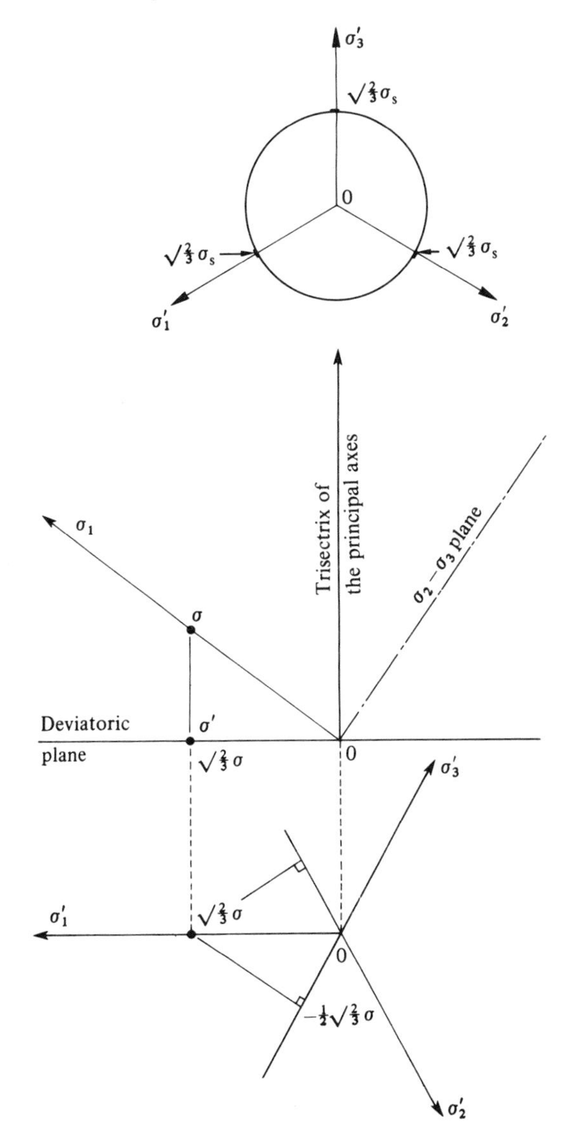

In the principal stress space, the Tresca criterion is represented by an orthogonal prism of hexagonal base with its axis equally inclined to the σ_1, σ_2, σ_3 axes (Fig. 5.13). It is inscribed within the von Mises cylinder.

Experimental validity

It is usually in the vicinity of the initial yield surface that metals are isotropic. For isotropic metals, the experimental points are situated between the von Mises and Tresca criteria, with very ductile metals closer to the Tresca criterion. An example of the yield locus determined by tension (or compression)–torsion tests is given in Fig. 5.14.

A criterion intermediary between those of von Mises and Tresca allows us to match these experimental results better. Proposed by Edelman and Drucker, it introduces a combination of the second and third invariants, J_2 and J_3, of the deviatoric stress tensor and a coefficient C dependent on the material. This criterion describes the elastic domain limit (or the yield surface) by means of the following expression:

$$f = 4s_{\text{II}}^3 - 27Cs_{\text{III}}^2 - C\sigma_s^2(3s_{\text{II}} - \sigma_s^2)^2 - \tfrac{4}{27}(1 - C)\sigma_s^6 = 0.$$

As a function of the J_2 and J_3 invariants, the above criterion is expressed by:

$$J_2^6 - \sigma_s^6 - C[J_3^6 - \sigma_s^6 + \tfrac{27}{4}\sigma_s^2(J_2^2 - \sigma_s^2)^2] = 0.$$

We recover the von Mises criterion for $C = 0$ and the Tresca criterion for $C = 1$. An example of comparison with experiments is given in Fig. 5.14 where $C = 0.384$.

In applications, however, the von Mises criterion is the one most commonly used because of its easy implementation in numerical calculations. The Tresca criterion is more difficult to use because of the

Fig. 5.13. Geometric representation of the Tresca criterion on the deviatoric plane.

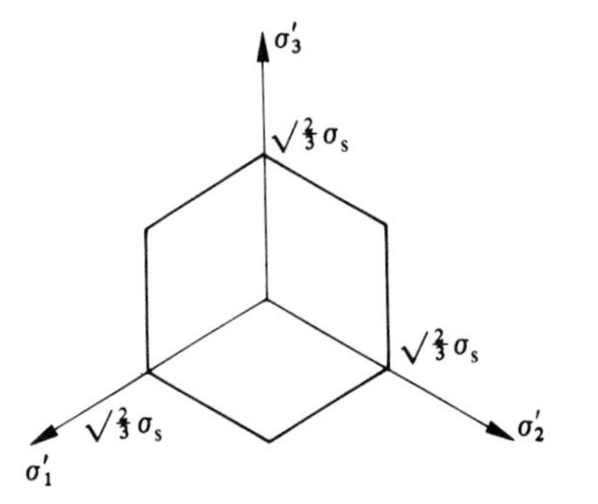

discontinuity of the normals at the corners of the yield surface (see Section 5.3.3).

Anisotropic yield criteria

These criteria cannot be expressed as functions of the stress invariants; preferred directions of anisotropy also appear. For incompressible plastic materials, the classical anisotropic criteria involve a rotation in the stress space; thus, the generalization of the von Mises criterion is of the form

$$(\mathbf{C}:\boldsymbol{\sigma}'):\boldsymbol{\sigma}' = 1$$

where $\mathbf{C}$ is a fourth order tensor dependent on the material properties, and which possesses the symmetries

$$C_{ijkl} = C_{klij} = C_{jikl} = C_{ijkl}.$$

The same criterion in terms of the components of the stress tensor $\boldsymbol{\sigma}$ is

$$(\mathbf{C}':\boldsymbol{\sigma}):\boldsymbol{\sigma} = 1$$

with $C'_{ijkk} = C'_{iikl} = 0$.

Fig. 5.14. Boundary of the elastic limit (or yield locus).

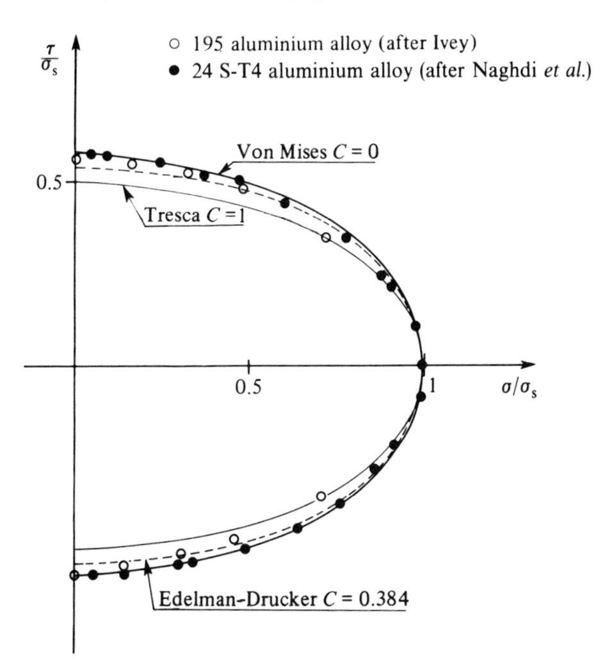

The Hill criterion

The Hill criterion corresponds to a particular kind of anisotropy in which three planes of symmetry are conserved during hardening of the material. The intersection of these three planes are the principal axes of anisotropy. The criterion is formulated with respect to these axes as the reference axes, say $(0, x_1, x_2, x_3)$. This criterion can be deduced from the general expression above with

$$\begin{array}{lll} C'_{1111} = F + H & C'_{2222} = F + G & C'_{3333} = G + H \\ C'_{1122} = -F & C'_{2233} = -G & C'_{3311} = -H \\ C'_{1212} = \tfrac{1}{2}L & C'_{2323} = \tfrac{1}{2}M & C'_{3131} = \tfrac{1}{2}N \end{array}$$

and the values obtained by virtue of the symmetry of C_{ijkl}, which then give

$$\begin{aligned} \bullet \quad & F(\sigma_{11} - \sigma_{22})^2 + G(\sigma_{22} - \sigma_{33})^2 + H(\sigma_{33} - \sigma_{11})^2 \\ & + 2L\sigma_{12}^2 + 2M\sigma_{23}^2 + 2N\sigma_{13}^2 = 1. \end{aligned}$$

F, G, H, L, M, N are the six scalar parameters which characterize the state of anisotropic hardening. We may determine them with the help of three experiments in simple tension and three in simple shear.

tensile yield stress σ_{s_1} in the direction $x_1 \rightarrow F + H = 1/\sigma_{s_1}^2$
tensile yield stress σ_{s_2} in the direction $x_2 \rightarrow F + G = 1/\sigma_{s_2}^2$
tensile yield stress σ_{s_3} in the direction $x_3 \rightarrow G + H = 1/\sigma_{s_3}^2$
shear yield stress $\sigma_{s_{12}}$ in the plane $(0, x_1, x_2) \rightarrow L = \frac{1}{2}\sigma_{s_{12}}^2$
shear yield stress $\sigma_{s_{23}}$ in the plane $(0, x_2, x_3) \rightarrow M = \frac{1}{2}\sigma_{s_{23}}^2$
shear yield stress $\sigma_{s_{31}}$ in the plane $(0, x_3, x_1) \rightarrow N = \frac{1}{2}\sigma_{s_{31}}^2$

Although anisotropic, this criterion does not take into account the Bauschinger effect; the yield stresses are of identical magnitude in tension and compression. However, it does take into account approximately the initial hardening of rolled sheets. An example is given in Fig. 5.15.

The Tsai criterion

The Tsai criterion allows for the eventual differences in hardening states in tension and in compression. In its most general form, it is expressed as the sum of a linear form and a quadratic form of the six stress components. Denoting the stress components by σ_α ($\sigma_1 = \sigma_{11}$, $\sigma_2 = \sigma_{22}$, $\sigma_3 = \sigma_{33}$, $\sigma_4 = \sigma_{23}$, $\sigma_5 = \sigma_{31}$, $\sigma_6 = \sigma_{12}$), the criterion can be expressed as

$$F_\alpha \sigma_\alpha + F_{\alpha\beta} \sigma_\alpha \sigma_\beta = 1.$$

The hypothesis of orthotropy and plastic incompressibility (with no hydrostatic stress influence) leads to a reduction in the number of material coefficients F_α and $F_{\alpha\beta}$ from 27 to 8. The criterion can then be written in a form similar to that of the Hill criterion.

$$\begin{aligned}&F'(\sigma_{11}-\sigma_{22})^2+G'(\sigma_{22}-\sigma_{33})^2+H'(\sigma_{33}-\sigma_{11})^2\\&\quad+2L'\sigma_{12}^2+2M'\sigma_{23}^2+2N'\sigma_{31}^2\\&\quad+P\sigma_{11}+Q\sigma_{22}-(P+Q)\sigma_{33}=1.\end{aligned}$$

The experimental determination of the eight coefficients for a particular material is rather tricky requiring tests in tension, compression and in shear in different directions.

This criterion is quite representative of the yielding of composite fibre–resin materials and wood. As an example Fig. 5.16 gives the graph of the

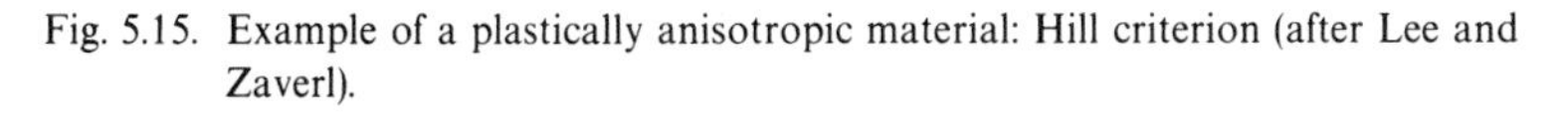

Fig. 5.15. Example of a plastically anisotropic material: Hill criterion (after Lee and Zaverl).

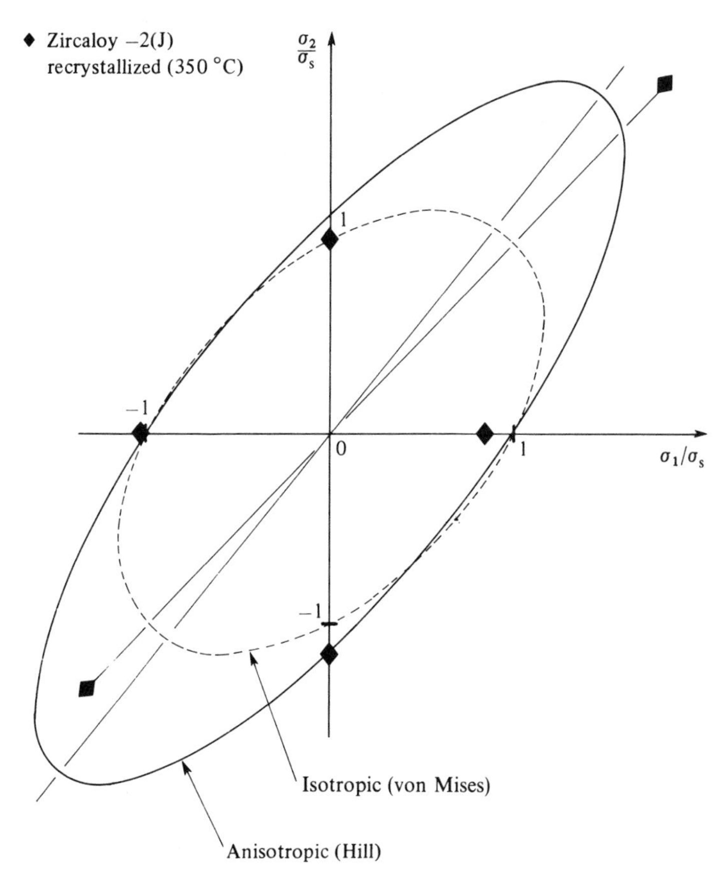

criterion in tension–compression in the L (parallel to the fibre direction) T (tangent to the growth rings) plane for the case of the massive tropical foliaceous tree 'Wana-Kouali'.

5.3 Formulation of general constitutive laws

5.3.1 *Partition hypothesis*

The partition of strains into elastic and plastic parts, already used in the preceding section, is justified by the nature of the physical phenomena. This hypothesis, which will be applied to all the theories of this chapter, is not strictly necessary for the formulation of plasticity laws; the hereditary theories based on the thermodynamics of memory processes do not use it. But, the adoption of this hypothesis greatly simplifies the problems of experimental identification and numerical calculations.

Within the framework of a general theory where large deformations are included, the hypothesis of strain partition is related to the existence of a relaxed intermediate configuration as defined in Chapter 2. The developments of the present chapter are limited to the case of small strains and linear elastic behaviour. Their generalization to the case of large deform-

Fig. 5.16. Tsai criterion for "Wana-Kouali" wood (after Gautherin).

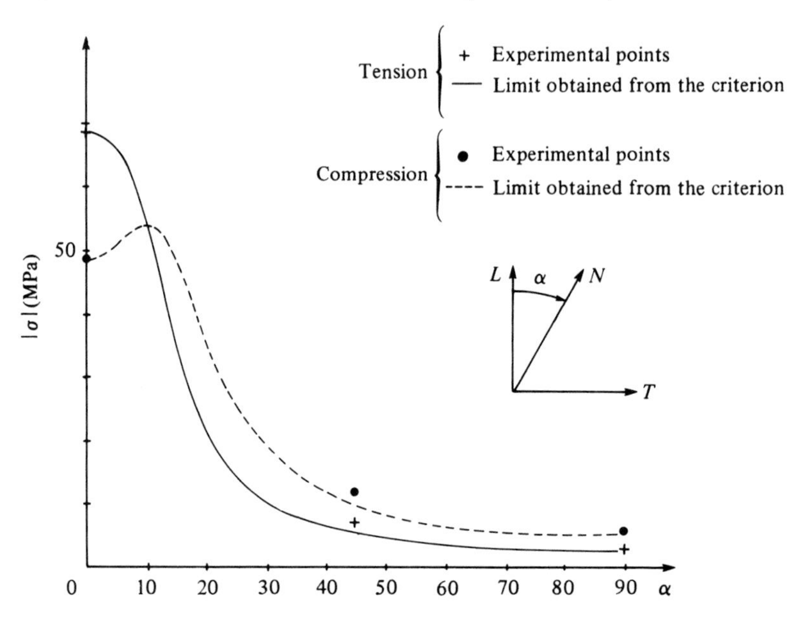

ations does not present any major problems when the elastic strains remain small. Moreover, we assume here decoupling between the elastic and plastic behaviours (Young's modulus and Poisson's ratio, for example, are supposed to be independent of hardening).

The plastic strain is defined by the difference

$$\boldsymbol{\varepsilon}^{\mathrm{p}} = \boldsymbol{\varepsilon} - \boldsymbol{\varepsilon}^{\mathrm{e}},$$

where $\boldsymbol{\varepsilon}$ is the total strain and where $\boldsymbol{\varepsilon}^{\mathrm{e}}$ is the elastic strain which is related linearly to the stress $\boldsymbol{\sigma}$ by the elasticity law. We recall here that, within the framework of a small deformation theory, the total strain is the symmetric part of the displacement gradient:

$$\boldsymbol{\varepsilon} = \tfrac{1}{2}[\operatorname{grad} \vec{u} + (\operatorname{grad} \vec{u})^{\mathrm{T}}].$$

5.3.2 *Choice of thermodynamic variables*

As mentioned in Chapter 2, the state or independent variables are the observable variables: the total strain $\boldsymbol{\varepsilon}$ and temperature T, and the internal variables $\boldsymbol{\varepsilon}^{\mathrm{p}}$, V_k. The partition hypothesis and the application of the Clausius–Duhem inequality to the case of time-independent, elastoplastic deformations allow us to write the stress tensor and the specific entropy, the thermodynamic variables associated to $\boldsymbol{\varepsilon}^{\mathrm{e}}$ and T respectively, in the form (see Chapter 2):

$$\boldsymbol{\sigma} = \rho(\partial\Psi/\partial\boldsymbol{\varepsilon}^{\mathrm{e}}), \qquad s = -\,\partial\Psi/\partial T$$

where ρ is the mass density and Ψ is the specific free energy, dependent on observable as well as internal variables:

$$\Psi = \Psi(\boldsymbol{\varepsilon} - \boldsymbol{\varepsilon}^{\mathrm{p}}, T, V_{\mathrm{k}}) = \Psi(\boldsymbol{\varepsilon}^{\mathrm{e}}, T, V_{\mathrm{k}}).$$

The Clausius–Duhem inequality is then reduced to

$$\boldsymbol{\sigma}{:}\dot{\boldsymbol{\varepsilon}}^{\mathrm{p}} - \rho(\partial\Psi/\partial V_{\mathrm{k}})\dot{V}_k - (1/T)\vec{q}\cdot\overrightarrow{\operatorname{grad}}\ T \geqslant 0.$$

The internal variables V_{k}, of scalar or tensorial nature, represent the current state of the material, i.e., the state of hardening; classically, a scalar variable (an isotropic hardening variable) of one of the following types is used:

the accumulated plastic strain, expressed by:

$$p = \int_0^{\mathrm{t}} [\tfrac{2}{3}\dot{\boldsymbol{\varepsilon}}^{\mathrm{p}}(\tau){:}\dot{\boldsymbol{\varepsilon}}^{\mathrm{p}}(\tau)]^{1/2}\,\mathrm{d}\tau;$$

the dissipated plastic work expressed by:

$$w_{\mathrm{p}} = \int_0^t \boldsymbol{\sigma}(\tau):\dot{\boldsymbol{\varepsilon}}^{\mathrm{p}}(\tau)\,\mathrm{d}\tau;$$

plus one or several tensorial variables or kinematic hardening variables. In what follows, unless stated otherwise, only one kinematic variable denoted by $\boldsymbol{\alpha}$ will be used. The definitions of variables p and w_{p} are illustrated in Fig. 5.17(*a*) for the tension–compression case.

Very schematically, the scalar variables are associated with the density of dislocations in the current state while the kinematic variables correspond to the incompatibilities of plastic deformations within the polycrystal. Generally, unless a recovery of crystalline structure occurs which is essentially governed by time, the variables p and w_{p} are increasing; they do not therefore allow representation of the alternating hardening effects observed

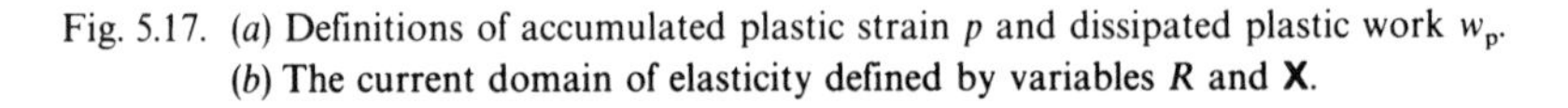

Fig. 5.17. (*a*) Definitions of accumulated plastic strain p and dissipated plastic work w_{p}. (*b*) The current domain of elasticity defined by variables R and $\mathbf{X}$.

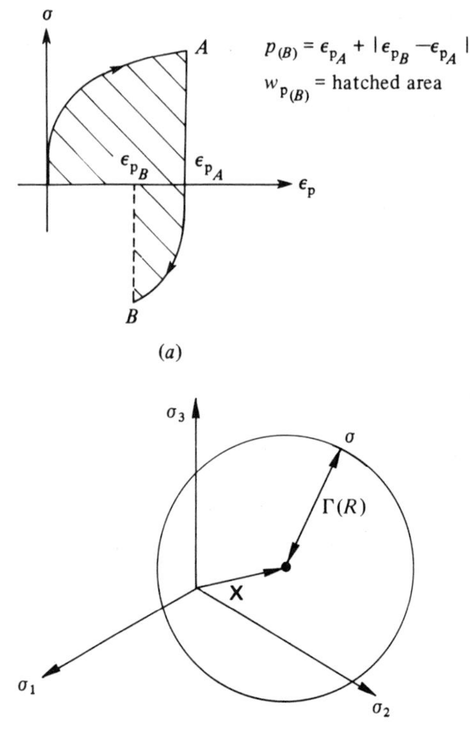

under cyclic loading. On the other hand, the kinematic variables are more directly related to the current state of deformation. In the case of cyclic hardening, they can represent real behaviour because of their evolution during each cycle even in a stabilized regime. A kinematic variable which is often used is the plastic strain itself ($\boldsymbol{\alpha} = \boldsymbol{\varepsilon}^{\mathrm{p}}$) as for example, in Prager's hardening model (linear kinematic hardening).

The decoupling between elastic behaviour and hardening entails writing the free energy in the form

$$\Psi = \Psi_{\mathrm{e}}(\boldsymbol{\varepsilon}^{\mathrm{e}}, T) + \Psi_{\mathrm{p}}(p, \boldsymbol{\alpha}, T).$$

The associated thermodynamic force variables can be deduced from it as

$$R = \rho(\partial\Psi/\partial p), \qquad \mathbf{X} = \rho(\partial\Psi/\partial\boldsymbol{\alpha}).$$

Figure 5.17(*b*) schematically shows the roles of these two variables in the description of the hardening state by the growth of the elastic domain; the size of the domain is a function of the isotropic variable R while its centre is identified by the kinematic variable $\mathbf{X}$.

Within the context of decoupling between intrinsic and thermal dissipations (which does not imply decoupling of the effects), and with the above variables, the Clausius–Duhem inequality expresses the positive character of these two dissipations:

$$\boldsymbol{\sigma}:\dot{\boldsymbol{\varepsilon}}_{\mathrm{p}} - R\dot{p} - \mathbf{X}:\dot{\boldsymbol{\alpha}} \geqslant 0, \qquad -(1/T)\vec{q}\cdot\overrightarrow{\mathrm{grad}}\ T \geqslant 0.$$

The intrinsic dissipation represents the difference between the energy dissipated by virtue of plastic deformations and the energy blocked within the volume element by dislocation networks for example.

5.3.3 *Loading surface and dissipation potential*

Loading surface, loading–unloading criterion

Time-independent plastic behaviour must be considered as a particular case of the more general schematic representation of viscoplasticity studied in Chapter 6. Very schematically, we may imagine, embedded in the stress space, a family of equipotential surfaces or, equivalently of surfaces of equal dissipation (Fig. 5.18). For a fixed hardening state, represented for example by the centre and the dimension of the interior surface $\Omega = 0$, the rate of flow is higher the farther the stress point is from the centre, $\mathbf{X}$. On the surface closest to the centre, $\Omega = 0$, the rate of flow is zero, whereas on the surface farthest from the centre it is infinite. The domain of viscoplasticity is situated between the two; the domain of elasticity being represented, of course, by $\Omega < 0$.

Thus the scheme of time-independent plasticity is applicable

> either to infinitely slow loads or to the attainment of asymptotic states under fixed loads;
> or to extremely rapid loads.

The responses for these two cases differ to varying extents: for metals loaded at temperatures lower than one quarter of their absolute melting temperature, they are very close which for them justifies the use of the time-independent plasticity analysis.

In the second case of an infinitely large rate, viscous deformations can be considered negligible (as they have no time to occur!). Thus, regardless of the case under consideration, a surface, defined in the stress space exists from which plastic flow can occur. For stress states contained within this surface, the material behaviour is entirely elastic. This $f = 0$ surface is called the loading surface or the flow surface. Initially, it is identical to the initial yield surface (see Section 5.2.2).

In order to describe the possibility of plastic flow completely, it is necessary to introduce still one more criterion, that of loading–unloading.

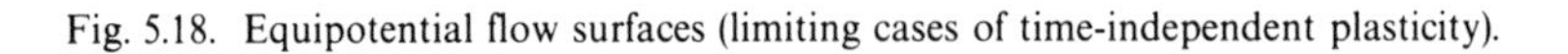

Fig. 5.18. Equipotential flow surfaces (limiting cases of time-independent plasticity).

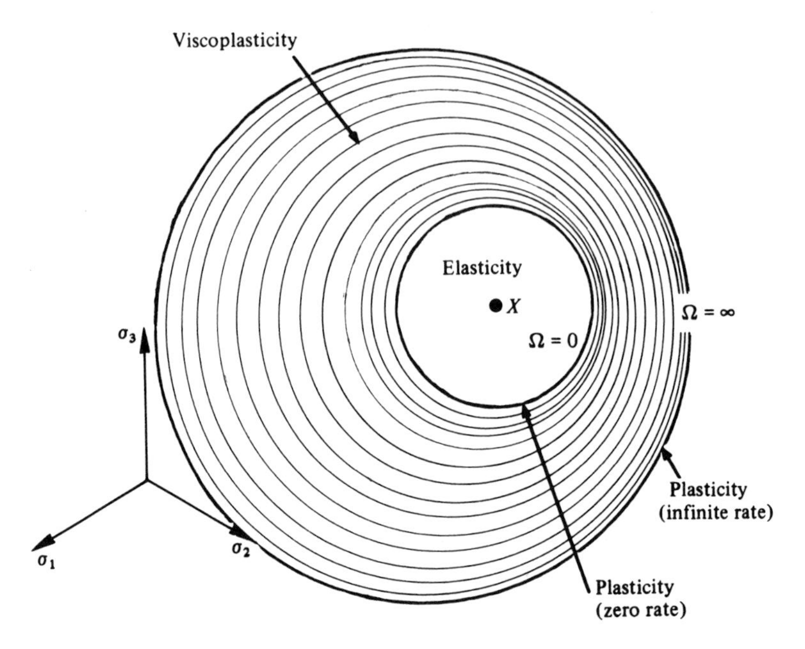

Plastic flow occurs only if the following two conditions are simultaneously satisfied (see Fig. 5.19).

(1) The representative point of the stress state is situated on the loading surface

$$f(\boldsymbol{\sigma}^*, V_k) = 0.$$

(2) The framework of classical plasticity requires that the representative point of the stress state does not leave the loading surface ($f > 0$ is impossible). During the continuous flow, the consistency condition,

$$\mathrm{d}f(\boldsymbol{\sigma}^*) = \frac{\partial f}{\partial \boldsymbol{\sigma}} : \mathrm{d}\boldsymbol{\sigma}^* + \frac{\partial f}{\partial V_k} \mathrm{d}V_k = 0$$

must be satisfied, which implies that the point representative of the stress $(\boldsymbol{\sigma}^* + \mathrm{d}\boldsymbol{\sigma}^*)$ remains on the loading surface.

On the other hand, stress variation is allowed to displace the stress point towards the interior of the loading surface. This is the case of unloading characterized by

$$\mathrm{d}f(\boldsymbol{\sigma}^*) < 0.$$

According to this scheme no plastic flow can occur during unloading; the behaviour becomes elastic as soon as the unloading starts. In summary:

- $f < 0$: ⟶ elastic behaviour

Fig. 5.19. Loading–unloading criterion for a positive hardening.

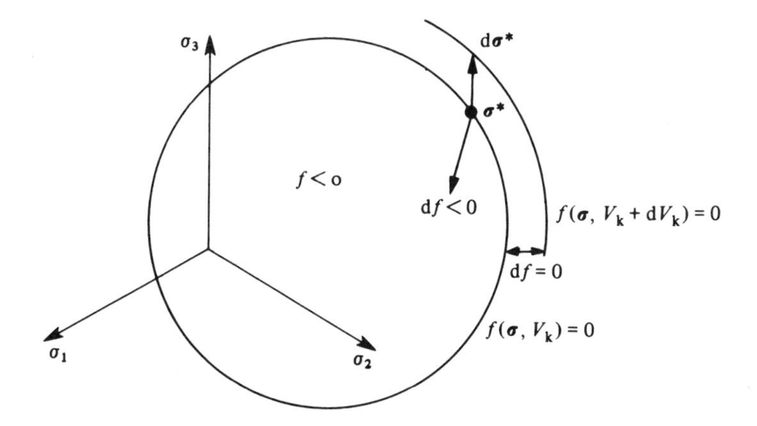

- $f = 0$ and $df = 0$: → plastic flow
- $f = 0$ and $df < 0$: → elastic unloading.

For materials with positive hardening, the loading surface is expressible as a function of the components of the stress tensor and depends on the hardening state through the internal variables. These variables are generally the thermodynamic forces $\mathbf{X}$ and R as defined previously. Eventually it also depends on the temperature:

$$f = f(\boldsymbol{\sigma}, R, \mathbf{X}, T).$$

For materials with negative hardening, it may be of more interest to use a more general formulation, in terms of strain, e.g.,

$$f' = f'(\boldsymbol{\varepsilon}, p, \boldsymbol{\alpha}, T).$$

Fig. 5.20 shows that in the case of simple tension, when the hardening is negative ($d\sigma/d\varepsilon_p < 0$), the loading condition $\dot{\sigma} > 0$ cannot be used.

Nonassociated plasticity

In the general framework of thermodynamics, the existence of a dissipation potential was postulated; its knowledge provides the laws of evolution of plastic deformation and internal variables. For plasticity, the use of this potential leads to the schematic representation of associated plasticity as will be seen in Section 5.3.3.4. This is a representation in which the yield surface (or the loading or flow surface) is assimilated with an 'equipotential surface'. A more general case is that of 'nonassociated plasticity' which is useful for describing certain materials or phenomena, and is practically

Fig. 5.20. Loading–unloading criterion in tension for positive or negative hardening.

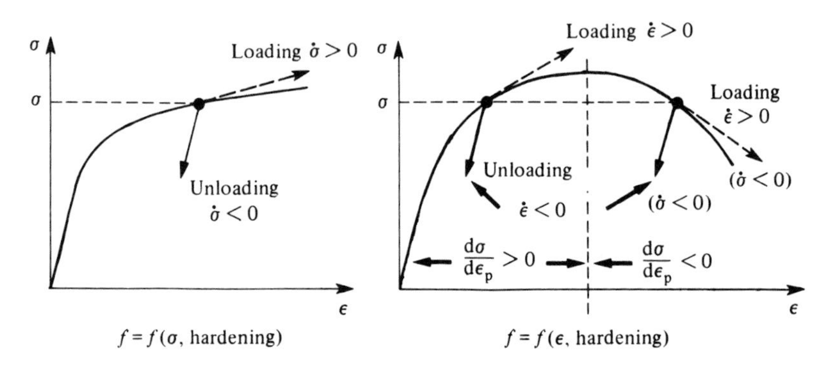

indispensible in soil mechanics. This schematic representation is not covered directly by the formalism of 'generalized standard materials' introduced in Chapter 1.

The schematic representation of nonassociated plasticity requires the use of three potentials: the free energy Ψ; the yield surface, $f = 0$; and a potential surface, $F =$ constant, which determines the direction of plastic flow in the space of generalized force variables (it ultimately depends on the state variables, especially on the temperature);

$$F = F(\boldsymbol{\sigma}, R, \mathbf{X}, T).$$

The generalized normality hypothesis associated with the instantaneous dissipative phenomena described in Section 2.4.3 allows us to write:

$$\begin{aligned} \dot{\boldsymbol{\varepsilon}}_{\mathrm{p}} &= \dot{\lambda}\partial F/\partial \boldsymbol{\sigma} \\ -\dot{p} &= \dot{\lambda}\partial F/\partial R \\ -\dot{\boldsymbol{\alpha}} &= \dot{\lambda}\partial F/\partial \mathbf{X} \end{aligned}$$

where $\dot{\lambda}$ is the multiplier of time-independent plasticity which is determined later. $\dot{\lambda}$ is equal to zero when there is no flow, i.e. when

$$f < 0 \qquad \text{or} \qquad f = 0 \qquad \text{and} \qquad \frac{\partial f}{\partial \boldsymbol{\sigma}}:\dot{\boldsymbol{\sigma}} \leqslant 0.$$

We may note that this concept, although it does not exactly satisfy the hypotheses of generalized standard materials, allows us to verify *a priori* the second principle of thermodynamics. The Clausius–Duhem inequality becomes:

$$\Phi = \left[\boldsymbol{\sigma}:\frac{\partial F}{\partial \boldsymbol{\sigma}} + R\frac{\partial F}{\partial R} + \mathbf{X}:\frac{\partial F}{\partial \mathbf{X}}\right]\dot{\lambda} = 0.$$

To ensure this condition, it is sufficient that F be convex, positive as soon as there is plastic flow, and include the origin ($F \geqslant f = 0$, $F(0,0,0;T) = 0 \forall T$). We can then easily show that

$$\Phi \geqslant F\dot{\lambda} \geqslant 0.$$

Consistency condition – expression for plasticity multiplier

We have seen that the loading–unloading criterion requires the imposition of $f = 0$ and $\dot{f} = 0$ during plastic flow. This last condition implies that

$$\frac{\partial f}{\partial \boldsymbol{\sigma}}:\dot{\boldsymbol{\sigma}} + \frac{\partial f}{\partial R}\dot{R} + \frac{\partial f}{\partial \mathbf{X}}:\dot{\mathbf{X}} + \frac{\partial f}{\partial T}\dot{T} = 0.$$

By using the relations between the internal variables and the associated thermodynamic forces, we have

$$\dot{R} = L_{pp}\dot{p} + \mathbf{L}_{\alpha p}:\dot{\boldsymbol{\alpha}} + L_{pT}\dot{T}$$
$$\dot{\mathbf{X}} = \mathbf{L}_{\alpha p}\dot{p} + \mathbf{L}_{\alpha\alpha}:\dot{\boldsymbol{\alpha}} + \mathbf{L}_{\alpha T}\dot{T}$$

where the operator $\mathbf{L}$ is derived from the free energy Ψ by:

$$L_{pp} = \frac{\partial^2 \Psi}{\partial p^2} \qquad \mathbf{L}_{\alpha T} = \frac{\partial^2 \Psi}{\partial \boldsymbol{\alpha} \partial T} \qquad L_{pT} = \frac{\partial^2 \Psi}{\partial p \partial T}$$

$$\mathbf{L}_{\alpha p} = \frac{\partial^2 \Psi}{\partial \boldsymbol{\alpha} \partial p} \qquad \mathbf{L}_{\alpha\alpha} = \frac{\partial^2 \Psi}{\partial \boldsymbol{\alpha} \partial \boldsymbol{\alpha}}.$$

Substitution in $\dot{f} = 0$ and the use of normality relations leads to

$$\frac{\partial f}{\partial \boldsymbol{\sigma}}:\dot{\boldsymbol{\sigma}} + \left(\frac{\partial f}{\partial T} + \frac{\partial f}{\partial \mathbf{X}}:\mathbf{L}_{\alpha T} + \frac{\partial f}{\partial R}L_{pT}\right)\dot{T}$$
$$-\dot{\lambda}\left[\frac{\partial f}{\partial \mathbf{X}}:\mathbf{L}_{\alpha\alpha}:\frac{\partial F}{\partial \mathbf{X}} + \left(\frac{\partial f}{\partial \mathbf{X}}\frac{\partial F}{\partial R} + \frac{\partial f}{\partial R}\frac{\partial F}{\partial \mathbf{X}}\right):\mathbf{L}_{\alpha p} + \frac{\partial f}{\partial R}\frac{\partial F}{\partial R}L_{pp}\right] = 0$$

from which $\dot{\lambda}$ can be determined as a function of the different potentials, stress rate, and temperature.

Only materials with positive hardening can be covered within the context of the present formalism. For negative hardening, a flow theory dependent upon a loading function expressed in terms of strain should be developed. To say that the hardening is positive amounts to considering the hardening modulus as positive:

$$h = \frac{\partial f}{\partial \mathbf{X}}:\mathbf{L}_{\alpha\alpha}:\frac{\partial F}{\partial \mathbf{X}} + \left(\frac{\partial f}{\partial \mathbf{X}}\frac{\partial F}{\partial R} + \frac{\partial f}{\partial R}\frac{\partial F}{\partial \mathbf{X}}\right):\mathbf{L}_{\alpha p} + \frac{\partial f}{\partial R}\frac{\partial F}{\partial R}L_{pp} \geqslant 0.$$

In the present framework, if h is positive, the expression for the plasticity multiplier becomes

$$\dot{\lambda} = \frac{H(f)}{h}\left\langle \frac{\partial f}{\partial \boldsymbol{\sigma}}:\dot{\boldsymbol{\sigma}} + \left(\frac{\partial f}{\partial T} + \frac{\partial f}{\partial \mathbf{X}}:\mathbf{L}_{\alpha T} + \frac{\partial f}{\partial R}L_{pT}\right)\dot{T}\right\rangle$$

where H denotes the Heaviside step function: $H(f) = 0$ if $f < 0$, $H(f) = 1$ if $f \geqslant 0$ and the symbol $\langle \quad \rangle$ means that $\langle u \rangle = 0$ if $u < 0$, $\langle u \rangle = u$ if $u \geqslant 0$. The introduction of these two symbols enables us directly to ensure $\dot{\lambda} = 0$ when $f < 0$ or when there is unloading ($\dot{f} < 0$).

Associated plasticity. Generalized standard materials

The theory of associated plasticity is the theory developed by identifying the function F with the loading function f. The potential φ^* then plays the role of the indicator function of the convex set defined by $f=0$.

$$\varphi^* = \varphi^*(\boldsymbol{\sigma}, R, \mathbf{X}; T)$$

is obtained by application of the Legendre–Fenchel transformation to the dissipation potential:

$$\varphi = \varphi(\dot{\boldsymbol{\varepsilon}}^{\mathrm{p}}, \dot{p}, \dot{\boldsymbol{\alpha}}; T).$$

In time-independent plasticity, the potential φ^* is equal to zero inside the elastic domain $f < 0$, and infinite outside. It is then easy to appreciate how plasticity can be considered as a particular case of viscoplasticity (see the discussion about the loading–unloading criterion, subsection 5.3.3, and Fig. 5.18).

Many classical laws follow from this specialization of the generalized standard materials. The expression for the hardening modulus becomes

$$h = \left(\frac{\partial f}{\partial R}\right)^2 L_{pp} + \frac{\partial f}{\partial \mathbf{X}} : \mathbf{L}_{\alpha\alpha} : \frac{\partial f}{\partial \mathbf{X}} + 2\frac{\partial f}{\partial R}\frac{\partial f}{\partial \mathbf{X}} : \mathbf{L}_{\alpha p}$$

and the flow law is expressed by:

$$\dot{\boldsymbol{\varepsilon}}^{\mathrm{p}} = \dot{\lambda}(\partial f/\partial \boldsymbol{\sigma})$$
$$-\dot{p} = \dot{\lambda}(\partial f/\partial R)$$
$$-\dot{\boldsymbol{\alpha}} = \dot{\lambda}(\partial f/\partial \mathbf{X}).$$

The expression for the plasticity multiplier is the same as above except f replaces F.

5.4 Particular flow laws

5.4.1 *Different types of criteria and flow laws*

Before dealing with particular models, it is advisable to give definitive meanings to the terms isotropy, kinematic and anisotropy. Two aspects are distinguished: one related to flow criteria and the other to hardening, i.e., the evolution of the criterion. The words isotropic and anisotropic describe both aspects, while the term kinematic is applied only to the idea of evolution. Schematically, there are six levels of theory as illustrated below, where we have restricted ourselves to the use of the second invariant of stress (see Table 5.2).

Table 5.2. *Different plasticity criteria and flow rules*

Case	Initial criterion	Hardening variables	Transformed criterion due to hardening	
1	$f(J(\boldsymbol{\sigma}))$ isotropic	p	$f(J(\boldsymbol{\sigma})) - R(p)$	isotropic
2	$f(J(\boldsymbol{\sigma}))$ isotropic	$\mathbf{X}$	$f(J(\boldsymbol{\sigma} - \mathbf{X}))$	kinematic
3	$f(\mathbf{C}:\boldsymbol{\sigma}':\boldsymbol{\sigma}')$ anisotropic	p	$f(\mathbf{C}:\boldsymbol{\sigma}':\boldsymbol{\sigma}') - R(p)$	isotropic
4	$f(\mathbf{C}:\boldsymbol{\sigma}':\boldsymbol{\sigma}')$ anisotropic	$\mathbf{X}$	$f(\mathbf{C}:(\boldsymbol{\sigma}' - \mathbf{X}'):(\boldsymbol{\sigma}' - \mathbf{X}'))$	kinematic
5	$f(J(\boldsymbol{\sigma}))$ isotropic	$p, \mathbf{X}, \boldsymbol{\varepsilon}^{p}$	$f(\mathbf{C}^{*}:(\boldsymbol{\sigma}' - \mathbf{X}'):(\boldsymbol{\sigma}' - \mathbf{X}')) - R(p)$	isotropic + kinematic + anisotropic
6	$f(\mathbf{C}:\boldsymbol{\sigma}':\boldsymbol{\sigma}')$ anisotropic	$p, \mathbf{X}, \boldsymbol{\varepsilon}^{p}$	$f(\mathbf{C}^{*}:(\boldsymbol{\sigma}' - \mathbf{X}'):(\boldsymbol{\sigma}' - \mathbf{X}')) - R(p)$	isotropic + kinematic + anisotropic

(1) Isotropic criterion – isotropic hardening: the invariant $J(\boldsymbol{\sigma})$ and a scalar variable p are sufficient (J is, for example, the second invariant, $J = J_2$).

(2) Isotropic criterion – kinematic hardening: the translation of the criterion by a tensor $\mathbf{X}$ is used.

(3) Anisotropic criterion – isotropic hardening: the criterion is expressed in a more complex form, for example with $(\mathbf{C}:\boldsymbol{\sigma}'):\boldsymbol{\sigma}'$ where $\mathbf{C}$ depends on the material. However, this form of the criterion is retained during the hardening process by simple dilatation, and therefore, involves only a scalar variable p.

(4) Anisotropic criterion – kinematic hardening: the form of the criterion is retained during hardening but the surface is subjected to a translation represented by the tensor $\mathbf{X}$.

(5) Isotropic criterion – anisotropic hardening (nonkinematic): the invariant $J_2(\boldsymbol{\sigma})$ is transformed into a more complex expression, eventually leading to a kinematic effect in which $\mathbf{C}^*(\boldsymbol{\varepsilon}^p)$ expresses an anisotropy depending on the present state of plastic strain.

(6) Anisotropic criterion – anisotropic hardening. This is the generalization of case (5); $\mathbf{C}$ describes the initial anisotropy in a state of zero plastic strain.

Note that the distinction between the 'initial' state and 'hardened' state is subtle as any hardened state can possibly be considered as an initial state.

The anisotropy introduced by the operator $\mathbf{C}$ is not the most general one;

Fig. 5.21. Anisotropic hardening (rotation of the surface) superposed on kinematic hardening (translation) and on isotropic softening (reduction) for 1Cr–$\frac{1}{2}$Mo–$\frac{1}{4}$V steel (after Moreton *et al.*).

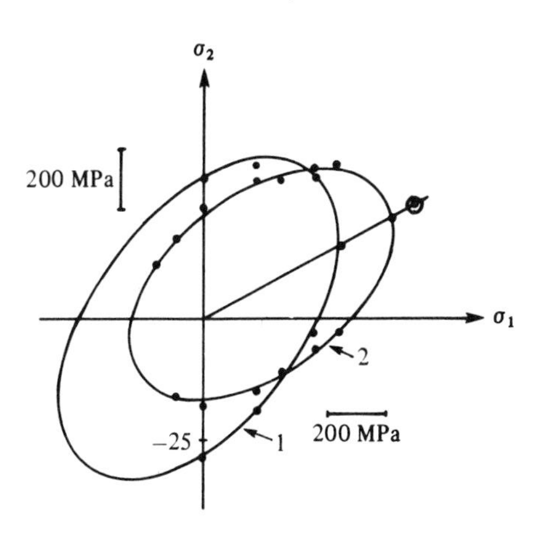

it only implies rotation of the criterion. For example, in the (σ_1, σ_2) plane of principal stresses it results in a change of eccentricity of the von Mises ellipse or in a rotation of the axes of this ellipse (see, e.g., Figs. 5.15 and 5.21).

Experiments show that in addition to rotation and uniform expansion or contraction of the surface, deformation of the surface can occur by the appearance of 'corners', but modelling of this kind of anisotropy is very complex. In what follows, we will deal only with cases of isotropic and kinematic hardening (cases (1)–(4)).

5.4.2 *Isotropic hardening rules*

In these rules, the evolution of the loading surface is governed only by one scalar variable, either the dissipated plastic work, or the accumulated plastic strain p, or any associated variable such as the thermodynamic force R. These rules are easily written in the general context of associated plasticity, and therefore we limit ourselves to the identity between the dissipation potential and the loading function. To simplify the presentation, the rules are developed assuming the temperature to be constant, or at least using criteria which are temperature independent. Generalization to the non-isothermal case does not present any particular problems in following the formalism given in Section 5.3.3. Thus we have

$$f = f(\boldsymbol{\sigma}, R).$$

Fig. 5.22. Isotropic hardening: representation in the stress space and in tension–compression.

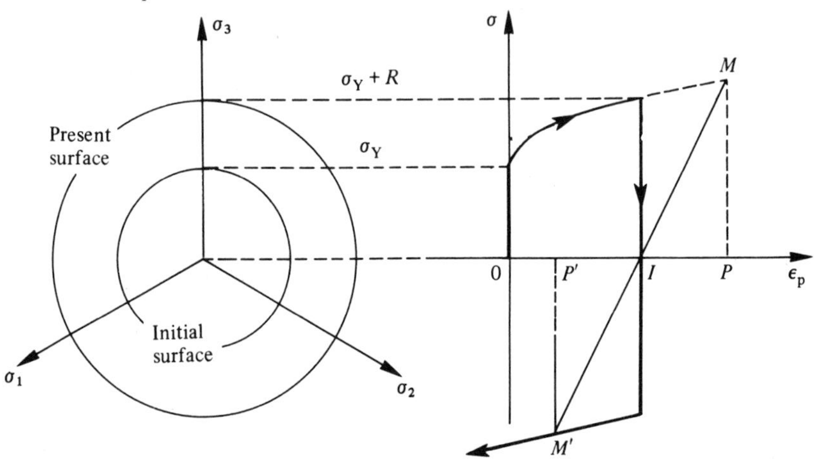

The hypothesis of isotropic hardening greatly facilitates the statement of flow equations. Whether accumulated plastic strain or accumulated plastic work is used, it is easy to identify the hardening model with any expression of the monotonic uniaxial stress–strain curve. The only differences will be in either the criterion chosen for the loading function or the dissipation expression.

The loading function is expressed in the form

$$f = f_{\mathrm{Y}}(\boldsymbol{\sigma}) - \Gamma(R)$$

where the function f_{Y} indicates the form of the yield criterion, and the function Γ introduces hardening through the relation between the thermodynamic force R and the hardening variable chosen (p or w_{p}). For an isotropic material (with an initially isotropic yield surface) f_{Y} is a function of the stress tensor invariants.

Isotropic hardening corresponds to a uniform expansion of the initial criterion. Fig. 5.22 schematically shows the evolution of the criterion in the stress space, and the stress–plastic strain curve in tension and compression. It also shows why the accumulated plastic strain can be used as a variable of isotropic hardening: the points M and M' have the same state and the same accumulated plastic strain $0I + IP = 0I + IP'$.

Prandtl–Reuss equation

Formulation

The Prandtl–Reuss equation is a flow law in an elastoplastic regime with isotropic hardening. It is based on the following assumptions.

(1) Hypothesis of plastic incompressibility: plastic strain occurs at constant volume and flow does not depend on the hydrostatic stress $\sigma_{\mathrm{H}} = \frac{1}{3}\mathrm{Tr}\,(\boldsymbol{\sigma})$. The loading function depends only on the deviatoric stress and the internal variables, i.e.,

$$\partial f/\partial\sigma_{\mathrm{H}} = 0.$$

(2) Hypothesis of initial isotropy and isotropic hardening: the loading function depends only on the invariants J_2 and J_3 of the deviatoric stress tensor:

$$J_2(\boldsymbol{\sigma}) = \sigma_{\mathrm{eq}} = (\tfrac{3}{2}\boldsymbol{\sigma}':\boldsymbol{\sigma}')^{1/2}$$
$$J_3(\boldsymbol{\sigma}) = (\tfrac{9}{2}\boldsymbol{\sigma}'\cdot\boldsymbol{\sigma}':\boldsymbol{\sigma}')^{1/3}.$$

(3) Associated plasticity and normality hypotheses:

$$d\boldsymbol{\varepsilon}^p = d\lambda(\partial f/\partial \boldsymbol{\sigma})$$
$$dp = -d\lambda(\partial f/\partial R).$$

(4) Choice of the von Mises loading function, independent of the third invariant, in the form:

$$f = \sigma_{eq} - R - \sigma_Y = 0$$

where σ_Y is the initial yield stress in tension.

The constitutive equation, in fact, the hardening curve, is expressed by the relation

$$R = k(p) = \rho(\partial \Psi/\partial p)$$

with $R(0) = k(0) = 0$.

The general procedure described in Section 5.3.3 is easily applicable. First we write

$$d\boldsymbol{\varepsilon}^p = d\lambda(\partial f/\partial \boldsymbol{\sigma}) = \tfrac{3}{2} d\lambda(\boldsymbol{\sigma}'/\sigma_{eq})$$
$$dp = -d\lambda(\partial f/\partial R) = d\lambda = (\tfrac{2}{3} d\boldsymbol{\varepsilon}^p : d\boldsymbol{\varepsilon}^p)^{1/2}$$

where we get back the definition of the accumulated strain increment. The consistency condition in the presence of flow ($f = 0$ and $df = 0$) gives

$$df = d\sigma_{eq} - k'(p)\, dp = 0$$

which can be used to express the plasticity multiplier as

$$d\lambda = dp = H(f)(d\sigma_{eq}/k'(p)).$$

For a material with positive hardening, i.e., with $k'(p) > 0$, there is no flow except when $d\sigma_{eq}$ is positive (i.e., when the equivalent stress increases). For a material with negative hardening, $d\sigma_{eq} < 0$, the plasticity multiplier must be zero, and the symbol $\langle x \rangle$ is used:

$$d\lambda = dp = H(f)(\langle d\sigma_{eq} \rangle / k'(p)).$$

The flow equation is therefore written as:

$$d\boldsymbol{\varepsilon}^p = \tfrac{3}{2} H(f) \frac{\langle d\sigma_{eq} \rangle}{k'(p)} \frac{\boldsymbol{\sigma}'}{\sigma_{eq}}.$$

It can also be expressed solely in terms of the stresses by introducing

$$p = k^{-1}(R) = k^{-1}(\sigma_{eq} - \sigma_Y)$$

and

$$g'(\sigma_{eq}) = \{k'[k^{-1}(\sigma_{eq} - \sigma_Y)]\}^{-1}.$$

Taking into account the partition of strains and the linear isotropic elasticity law, the Prandtl–Reuss equation is expressed by

- $$d\boldsymbol{\varepsilon} = d\boldsymbol{\varepsilon}^e + d\boldsymbol{\varepsilon}^p$$
- $$d\boldsymbol{\varepsilon}^e = \frac{1+\nu}{E} d\boldsymbol{\sigma} - \frac{\nu}{E} d(\mathrm{Tr}(\boldsymbol{\sigma}))\mathbf{1}$$
- $$d\boldsymbol{\varepsilon}^p = \tfrac{3}{2} H(f) g'(\sigma_{eq}) \frac{\langle d\sigma_{eq} \rangle}{\sigma_{eq}} \boldsymbol{\sigma}'$$

or

$$d\varepsilon^p_{ij} = \tfrac{3}{2} H(f) g'(\sigma_{eq}) \frac{\langle d\sigma_{eq} \rangle}{\sigma_{eq}} \sigma'_{ij}$$

while remembering that the last relation holds only when $f = 0$, i.e., when $\sigma_{eq} - R - \sigma_Y = 0$. If we neglect the elastic deformations we obtain

$$d\boldsymbol{\varepsilon} = \tfrac{3}{2} H(f) g'(\sigma_{eq}) \frac{\langle d\sigma_{eq} \rangle}{\sigma_{eq}} \boldsymbol{\sigma}'$$

which is known as the Levy–Mises relation.

Identification

A simple tension test gives meaning to the variables R and p and allows the identification of the function $g(\sigma_{eq})$ of the Prandtl–Reuss equation. In this case the sole nonzero component of stress is $\sigma_1 = \sigma$, whereas the nonzero components of the deviatoric stress tensor are $\sigma'_{11} = \frac{2}{3}\sigma$, $\sigma'_{22} = \sigma'_{33} = -\frac{1}{3}\sigma$, and we find that $p = \varepsilon_p$ and $J_2 = \sigma_{eq} = \sigma$. The flow law therefore becomes:

$$d\varepsilon_{p11} = d\varepsilon_p = g'(\sigma)\, d\sigma.$$

The function $[g'(\sigma)]^{-1}$ plays the role of the tangent modulus of the hardening curve:

$$\frac{1}{g'(\sigma)} = \frac{d\sigma}{d\varepsilon_p}$$

and indeed the variable R has the meaning:

$$R = k(p) = \int_0^{\varepsilon} k'(\varepsilon_p)\, d\varepsilon_p = \int_{\sigma_Y}^{\sigma} \frac{1}{g'(\sigma)} d\varepsilon_p(\sigma) = \int_{\sigma_Y}^{\sigma} d\sigma = \sigma - \sigma_Y$$

which is consistent with the expression of the loading function for simple tension:

$$f = |\sigma| - R - \sigma_Y = 0.$$

Fig. 5.23 illustrates this identification, which is very simple in this case.

In the particular case where the power-law expression (see p. 169) is chosen for the hardening curve, the variable R can be written as:

$$R = \sigma_{eq} - \sigma_Y = K_Y p^{1/M_Y} = k(p)$$

with the derivative:

$$k'(p) = \frac{K_Y}{M_Y} p^{(1-M_Y)/M_Y} = \frac{K_Y}{M_Y}\left(\frac{\sigma_{eq} - \sigma_Y}{K_Y}\right)^{1-M_Y}.$$

The Prandtl–Reuss equation can then be written in one of the following forms:

$$\mathrm{d}\boldsymbol{\varepsilon}^p = \frac{3}{2}\frac{M_Y}{K_Y} p^{(M_Y-1)/M_Y} \frac{\langle \mathrm{d}\sigma_{eq} \rangle}{\sigma_{eq}} \boldsymbol{\sigma}'$$

- $$\mathrm{d}\boldsymbol{\varepsilon}^p = \frac{3}{2}\frac{M_Y}{K_Y}\left\langle\frac{\sigma_{eq} - \sigma_Y}{K_Y}\right\rangle^{M_Y-1} \frac{\langle \mathrm{d}\sigma_{eq} \rangle}{\sigma_{eq}} \boldsymbol{\sigma}'$$

or

$$\mathrm{d}\varepsilon^p_{ij} = \frac{3}{2}\frac{M_Y}{K_Y}\left\langle\frac{\sigma_{eq} - \sigma_Y}{K_Y}\right\rangle^{M_Y-1} \frac{\langle \mathrm{d}\sigma_{eq} \rangle}{\sigma_{eq}} \sigma'_{ij}.$$

Fig. 5.23. Schematic representation of the tension curve and interpretation by isotropic hardening.

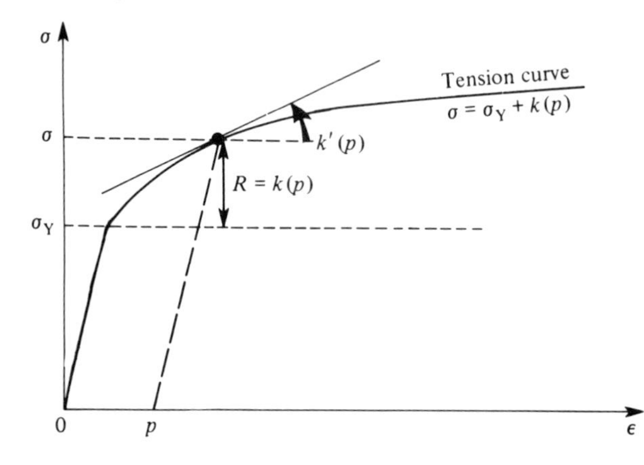

In the case of simple tension, the second expression gives:

$$d\varepsilon_p = \frac{M_Y}{K_Y}\left\langle \frac{\sigma - \sigma_Y}{K_Y} \right\rangle^{M_Y - 1} d\sigma$$

or, when integrated for $\sigma > \sigma_Y$

$$\varepsilon_p = \left\langle \frac{\sigma - \sigma_Y}{K_Y} \right\rangle^{M_Y}.$$

Case of perfect plasticity

This is the case in which there is no hardening; the variable R is zero and

$$f = \sigma_{eq} - \sigma_Y = 0.$$

The consistency condition $df = d\sigma_{eq} = 0$ cannot be used to obtain the plastic multiplier $d\lambda$. The plastic strains are now indeterminate:

- $$d\boldsymbol{\varepsilon}^p = \tfrac{3}{2} d\lambda \frac{\boldsymbol{\sigma}'}{\sigma_{eq}}$$

where $d\lambda$ is arbitrary but positive.

Other isotropic hardening laws

By remaining within the framework of 'associated plasticity' and using the generalized normality hypothesis, it is possible to formulate more general isotropic hardening laws than the Prandtl–Reuss equation. This may be done by using the loading function which corresponds to Tresca criterion or to one of the anisotropic criteria (see Section 5.2.2).

Isotropic plasticity law using the Tresca criterion

Let us recall the normality law

$$d\boldsymbol{\varepsilon}^p = d\lambda(\partial f/\partial \boldsymbol{\sigma})$$

and for simplicity, let us study the components of $d\boldsymbol{\varepsilon}^p$ with respect to the principal stress directions with

$$\sigma_3 < \sigma_2 < \sigma_1.$$

This corresponds to the 'face' regime in the sense that the stress point is situated on one of the faces of the hexagonal prism of the Tresca criterion, which is expressed by

$$f = \mathrm{Sup}(|\sigma_i - \sigma_j|) - \sigma_s = 0$$
$$f = \sigma_1 - \sigma_3 - \sigma_s = 0.$$

We then have

$$\partial f/\partial \sigma_1 = 1, \qquad \partial f/\partial \sigma_2 = 0, \qquad \partial f/\partial \sigma_3 = -1$$

from which

$$\mathrm{d}\varepsilon_1^{\mathrm{p}} = \mathrm{d}\lambda \qquad \mathrm{d}\varepsilon_2^{\mathrm{p}} = 0 \qquad \mathrm{d}\varepsilon_3^{\mathrm{p}} = -\mathrm{d}\lambda.$$

In the case where two principal components are equal (e.g., $\sigma_1 = \sigma_2$) the load point is on an edge. This regime requires special handling because of the nonuniqueness of the normal to the loading surface.

The multiplier $\mathrm{d}\lambda$ is identified from the hardening curve in simple tension as was done for the Prandtl–Reuss law:

$$\mathrm{d}\lambda = \mathrm{d}\varepsilon_1^{\mathrm{p}} = g'(\sigma_{\mathrm{s}})\,\mathrm{d}\sigma_{\mathrm{s}}$$

from which

$$\left.\begin{array}{l} \bullet \quad \mathrm{d}\varepsilon_1^{\mathrm{p}} = g'(\sigma_1 - \sigma_3)\,\mathrm{d}(\sigma_1 - \sigma_3) \\ \bullet \quad \mathrm{d}\varepsilon_2^{\mathrm{p}} = 0 \\ \bullet \quad \mathrm{d}\varepsilon_3^{\mathrm{p}} = -g'(\sigma_1 - \sigma_3)\,\mathrm{d}(\sigma_1 - \sigma_3) \end{array}\right\} \begin{array}{l} \text{if } \sigma_1 - \sigma_3 - \sigma_{\mathrm{s}} = 0 \\ (\text{with } \sigma_3 < \sigma_2 < \sigma_1). \end{array}$$

If the hardening curve is expressed by $\varepsilon_{\mathrm{p}} = (\sigma/K)^M$, then

$$g'(\sigma_1 - \sigma_3) = \frac{M}{K}\left(\frac{\sigma_1 - \sigma_3}{K}\right)^{M-1}.$$

Isotropic flow rule using anisotropic criteria

If the criterion is expressed in the form:

$$f = (\mathbf{C}:\boldsymbol{\sigma}'):\boldsymbol{\sigma}' - R - \sigma_{\mathrm{Y}} = 0$$

we obtain the flow equation:

$$\mathrm{d}\boldsymbol{\varepsilon}^{\mathrm{p}} = \frac{H(f)}{k'(p)}\langle \mathbf{C}:\boldsymbol{\sigma}':\mathrm{d}\boldsymbol{\sigma}' \rangle \frac{\mathbf{C}:\boldsymbol{\sigma}'}{R + \sigma_{\mathrm{Y}}}$$

or, in index notation:

$$\mathrm{d}\varepsilon_{ij}^{\mathrm{p}} = \frac{H(f)}{k'(p)}\langle C_{pqrs}\sigma'_{pq}\,\mathrm{d}\sigma'_{rs} \rangle \frac{C_{ijkl}\sigma'_{kl}}{R + \sigma_{\mathrm{Y}}}.$$

This type of hardening rule can be used to describe materials which are strongly anisotropic in their initial state: anisotropy due to metal working, forming texture, composite materials, etc.

As was seen in Section 5.2.2, it is possible to specialize the criterion by using only six constants which describe the anisotropic state of the material. This remains fixed during the plastic flow; the principal directions of anisotropy remain unchanged.

5.4.3 *Linear kinematic hardening rules*

Prager's kinematic hardening rule

General form

Kinematic hardening corresponds to translation of the loading surface. The hardening variable $\mathbf{X}$ is of a tensorial nature; it indicates the present position of the loading surface:

$$f = f_Y(\boldsymbol{\sigma} - \mathbf{X}) - k.$$

Fig. 5.24 shows in a schematic way the movement of this surface in the stress space and the corresponding modelling in tension–compression on the stress–strain diagram. The value of the yield stress k is generally different from σ_Y used in the previous sections.

Generally, isotropy is considered to exist in the unstressed state represented by $\mathbf{X}$ and plastic deformation is supposed to take place at constant volume. Therefore, in the function f_Y only the invariants $J_2(\boldsymbol{\sigma} - \mathbf{X})$ and $J_3(\boldsymbol{\sigma} - \mathbf{X})$ of the deviatoric parts of $\boldsymbol{\sigma}$ and $\mathbf{X}$ appear. These assumptions

Fig. 5.24. Kinematic hardening: representation in stress space and in tension–compression. The scales coincide on the axes σ, σ_1, σ_2, σ_3 (but not on the axes σ'_1, σ'_2, σ'_3 of the deviatoric plane).

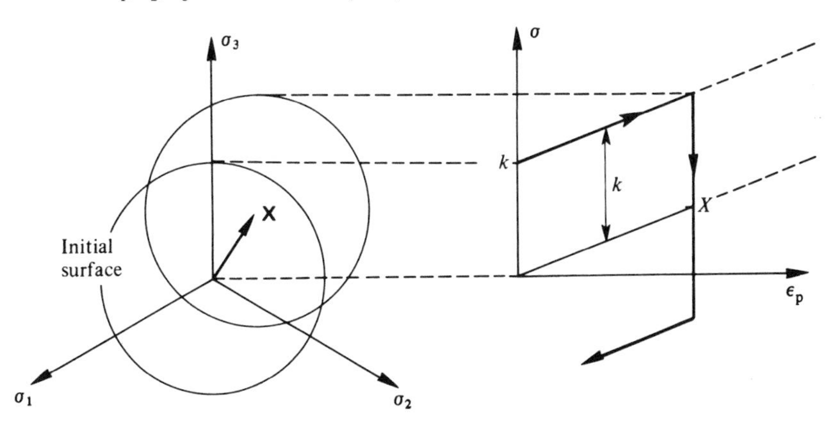

have been verified for a number of materials. Fig. 5.25 shows the example of the $2\frac{1}{4}$Cr–1 Mo steel.

The generalized normality hypothesis implies linear kinematic hardening (the hardening variable $\boldsymbol{\alpha}$ is identical to the present plastic strain):

$$d\varepsilon^p = d\lambda(\partial f/\partial \boldsymbol{\sigma})$$

$$d\boldsymbol{\alpha} = -\,d\lambda(\partial f/\partial \mathbf{X}) = d\lambda(\partial f/\partial \boldsymbol{\sigma}) = d\varepsilon^p.$$

It is difficult, even impossible, to introduce a nonlinear relation between $\boldsymbol{\alpha}$ and $\mathbf{X}$; it would lead to a one-to-one nonlinearity that would not correspond to the experimental observations under cyclic loads. (Fig. 5.26(*a*) schematically represents the nonlinearity effect that would

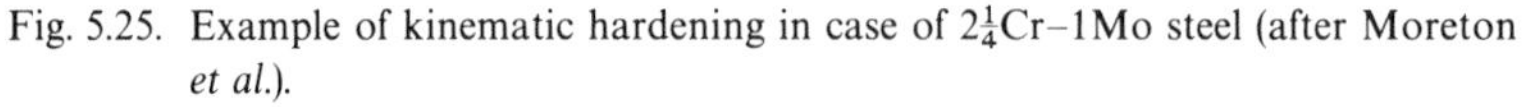
Fig. 5.25. Example of kinematic hardening in case of $2\frac{1}{4}$Cr–1Mo steel (after Moreton *et al.*).

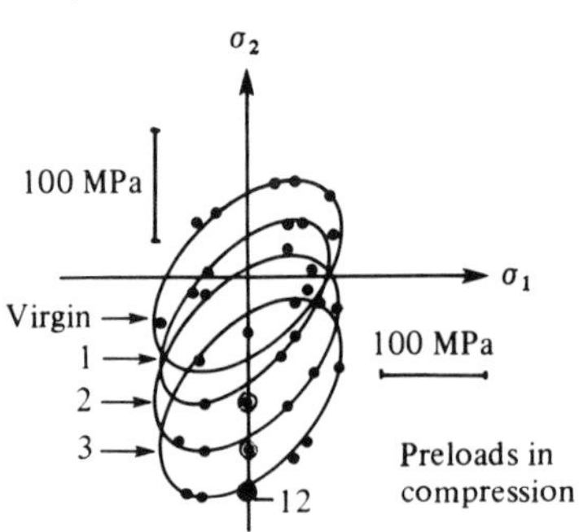

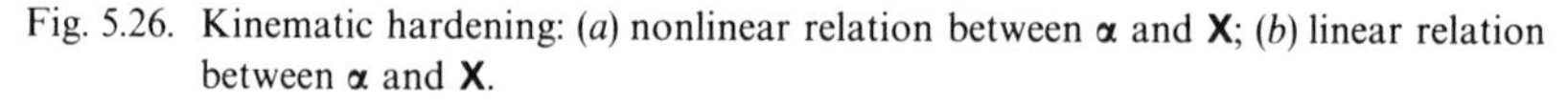
Fig. 5.26. Kinematic hardening: (*a*) nonlinear relation between $\boldsymbol{\alpha}$ and $\mathbf{X}$; (*b*) linear relation between $\boldsymbol{\alpha}$ and $\mathbf{X}$.

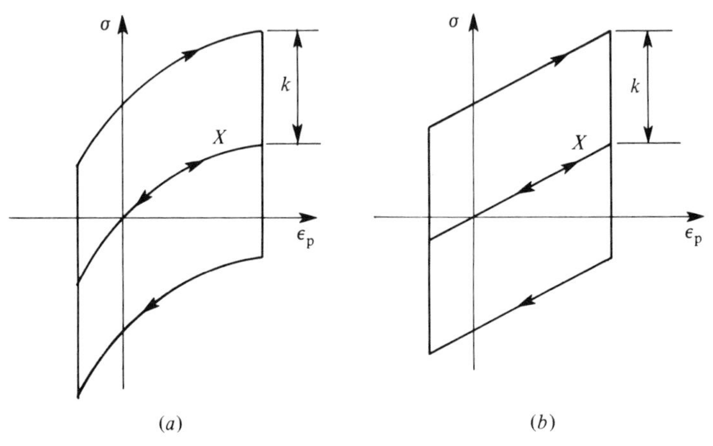

occur, for example, in a tension–compression cycle). We are led therefore to use the specific free energy in a form quadratic in $\boldsymbol{\alpha}$, which gives:

$$\mathbf{X} = \rho(\partial\Psi/\partial\boldsymbol{\alpha}) = C_0\boldsymbol{\alpha}.$$

Prager's kinematic hardening law is based on this schematic representation:

- $$\mathrm{d}\boldsymbol{\varepsilon}^{\mathrm{p}} = \mathrm{d}\lambda\frac{\partial f}{\partial\boldsymbol{\sigma}},$$
- $$\mathrm{d}\mathbf{X} = C_0\mathrm{d}\boldsymbol{\varepsilon}^{\mathrm{p}}.$$

The consistency condition when plastic flow occurs allows us to write:

$$\mathrm{d}f = \frac{\partial f}{\partial\boldsymbol{\sigma}}:\mathrm{d}\boldsymbol{\sigma} + \frac{\partial f}{\partial\mathbf{X}}:\mathrm{d}\mathbf{X} = \frac{\partial f}{\partial\boldsymbol{\sigma}}:\mathrm{d}\boldsymbol{\sigma} - C_0\frac{\partial f}{\partial\boldsymbol{\sigma}}:\mathrm{d}\boldsymbol{\varepsilon}^{\mathrm{p}}$$

$$= \frac{\partial f}{\partial\boldsymbol{\sigma}}:\mathrm{d}\boldsymbol{\sigma} - C_0\frac{\partial f}{\partial\boldsymbol{\sigma}}:\frac{\partial f}{\partial\boldsymbol{\sigma}}\mathrm{d}\lambda = 0$$

which leads to the following expression for the plastic multiplier

$$\mathrm{d}\lambda = \frac{H(f)}{C_0}\frac{\left\langle\dfrac{\partial f}{\partial\boldsymbol{\sigma}}:\mathrm{d}\boldsymbol{\sigma}\right\rangle}{\dfrac{\partial f}{\partial\boldsymbol{\sigma}}:\dfrac{\partial f}{\partial\boldsymbol{\sigma}}}.$$

The plastic strain increment is:

$$\mathrm{d}\boldsymbol{\varepsilon}^{\mathrm{p}} = \frac{H(f)}{C_0}\frac{\left\langle\dfrac{\partial f}{\partial\boldsymbol{\sigma}}:\mathrm{d}\boldsymbol{\sigma}\right\rangle}{\dfrac{\partial f}{\partial\boldsymbol{\sigma}}:\dfrac{\partial f}{\partial\boldsymbol{\sigma}}}\frac{\partial f}{\partial\boldsymbol{\sigma}}.$$

We note that $\mathrm{d}\boldsymbol{\varepsilon}^{\mathrm{p}}$ is equal to the projection of the increment $\mathrm{d}\boldsymbol{\sigma}/C_0$ on the normal to the loading surface. In fact:

$$\mathrm{d}\boldsymbol{\varepsilon}^{\mathrm{p}}:\frac{\partial f}{\partial\boldsymbol{\sigma}} = \frac{1}{C_0}\left\langle\frac{\partial f}{\partial\boldsymbol{\sigma}}:\mathrm{d}\boldsymbol{\sigma}\right\rangle.$$

Case of the von Mises criterion

In particular, if the loading surface is described by the von Mises criterion, the function f_{Y} depends only on the second invariant:

$$J_2(\boldsymbol{\sigma} - \mathbf{X}) = [\tfrac{3}{2}(\boldsymbol{\sigma}' - \mathbf{X}'):(\boldsymbol{\sigma}' - \mathbf{X}')]^{1/2}$$

and then it makes no difference to take f_{Y} as identical to the above function.

We then have

$$\frac{\partial f}{\partial \boldsymbol{\sigma}} = \frac{3}{2} \frac{\boldsymbol{\sigma}' - \mathbf{X}'}{J_2(\boldsymbol{\sigma} - \mathbf{X})}$$

with its modulus (the usual norm) equal to $\sqrt{\frac{3}{2}}$. The exterior unit normal is then:

$$\mathbf{n} = \sqrt{\frac{3}{2}} \frac{\boldsymbol{\sigma}' - \mathbf{X}'}{J_2(\boldsymbol{\sigma} - \mathbf{X})}.$$

In this particular case, we can express the kinematic hardening law in the form

$$\mathrm{d}\mathbf{X} = \tfrac{2}{3} C \, \mathrm{d}\boldsymbol{\varepsilon}^{\mathrm{p}}.$$

The plastic multiplier becomes

$$\mathrm{d}\lambda = \frac{3H(f)}{2C} \frac{\langle (\boldsymbol{\sigma}' - \mathbf{X}') : \mathrm{d}\boldsymbol{\sigma} \rangle}{J_2(\boldsymbol{\sigma} - \mathbf{X})}$$

and the increment of plastic strain

$$\mathrm{d}\boldsymbol{\varepsilon}^{\mathrm{p}} = \frac{9H(f)}{4C} \left\langle \frac{(\boldsymbol{\sigma}' - \mathbf{X}') : \mathrm{d}\boldsymbol{\sigma}}{J_2(\boldsymbol{\sigma} - \mathbf{X})} \right\rangle \frac{\boldsymbol{\sigma}' - \mathbf{X}'}{J_2(\boldsymbol{\sigma} - \mathbf{X})}.$$

We may replace $J_2(\boldsymbol{\sigma} - \mathbf{X})$ by k in view of the fact that $f = 0$. The complete law can then be written as:

- $f = J_2(\boldsymbol{\sigma} - \mathbf{X}) - k \leqslant 0$
- $\mathrm{d}\boldsymbol{\varepsilon}^{\mathrm{p}} = \dfrac{9H(f)}{4Ck^2} \langle (\sigma' - \mathbf{X}') : \mathrm{d}\boldsymbol{\sigma} \rangle (\boldsymbol{\sigma}' - \mathbf{X}')$
- $\mathrm{d}\mathbf{X} = \tfrac{2}{3} C \, \mathrm{d}\boldsymbol{\varepsilon}^{\mathrm{p}}.$

In index notation, we may write:

$$\mathrm{d}\varepsilon_{ij}^{\mathrm{p}} = \frac{9H(f)}{4Ck^2} \langle (\sigma'_{kl} - X'_{kl}) \, \mathrm{d}\sigma_{kl} \rangle (\sigma'_{ij} - X'_{ij}).$$

Identification

Identification is done from simple tension tests. The matrix of plastic strain is written in the usual way while, analogous to the deviatoric stress tensor, the matrix of the deviator of internal stress is written as:

$$[X] = \begin{bmatrix} \frac{2}{3}X & 0 & 0 \\ 0 & -\frac{1}{3}X & 0 \\ 0 & 0 & -\frac{1}{3}X \end{bmatrix}.$$

The criterion and the plastic flow equation can be written in the form:

$$f = |\sigma - X| - k = 0$$
$$\mathrm{d}\varepsilon_{11}^{\mathrm{p}} = \mathrm{d}\varepsilon_{\mathrm{p}} = (1/C)\langle \mathrm{d}\sigma \rangle$$

which allows us to verify that the hardening modulus is constant and equal to C, as could be shown directly from the relation $\mathrm{d}X = C\,\mathrm{d}\varepsilon_{\mathrm{p}}$ which is illustrated in Fig. 5.26(*b*). At each instant, it is evidently possible to write:

$$\sigma = X \pm k = C\varepsilon_{\mathrm{p}} \pm k$$

where the use of + or − depends upon the direction of the plastic flow.

In principle, the determination of the kinematic hardening law of a given material is particularly simple. In practice, the choice of the straight line which best describes the tension curve depends on the field of envisaged applications and therefore on the range of strain to be encountered in the applications (Fig. 5.27). A simple rule, which allows the definition of a law from generally known characteristics of a given material, consists in using the 0.2% yield stress, the fracture stress, and the elongation at fracture:

$$C \approx \frac{\sigma_{\mathrm{u}} - \sigma_{\mathrm{Y}}}{\varepsilon_{\mathrm{u}}}.$$

The linearity of the hardening rule offers the advantage of constructing more stable and less expensive algorithms for the numerical analysis of structures. On the other hand, if in the case of cyclic loads the Bauschinger

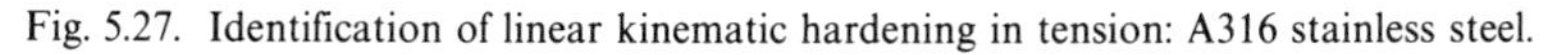
Fig. 5.27. Identification of linear kinematic hardening in tension: A316 stainless steel.

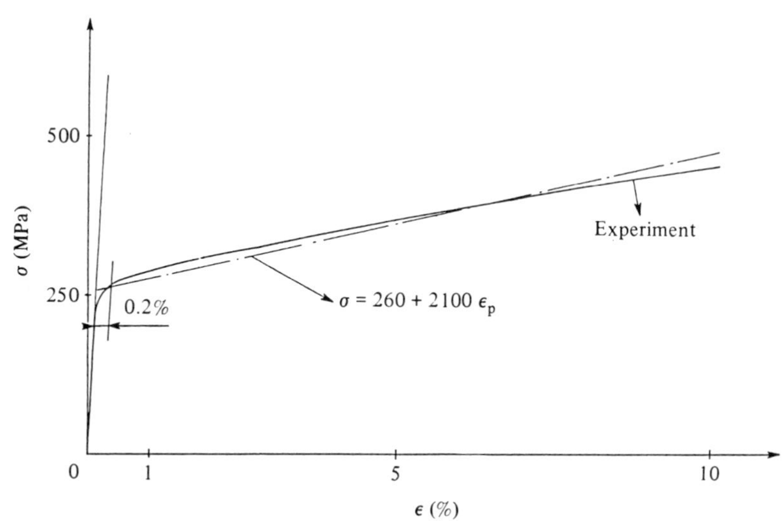

effect is qualitatively represented, ratchetting effects are not described. For any cyclic load there exists a stabilized cycle without progressive deformation as shown schematically in Fig. 5.28 for tension–compression and in Fig. 5.29 for cyclic torsion in the presence of a constant tension. In this latter case as long as the plastic strain ε_{p11} increases, the loading surface undergoes displacement in the σ_{11} direction. The stabilization occurs when $X_{11} = \sigma_{11}$, i.e., when the normal to the surface is directed along σ_{12}.

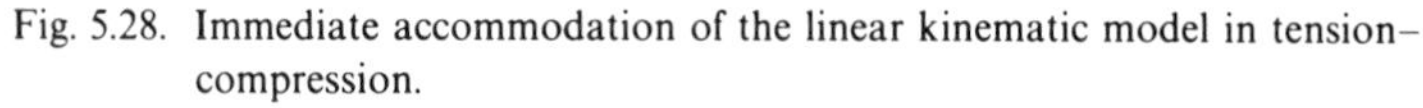

Fig. 5.28. Immediate accommodation of the linear kinematic model in tension–compression.

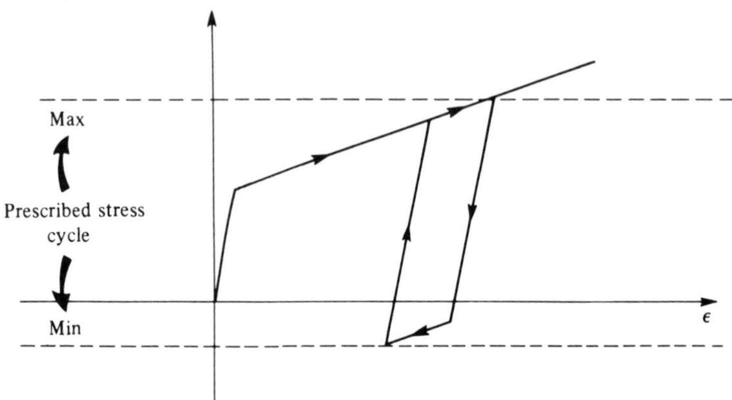

Fig. 5.29. Progressive accommodation of the linear kinematic model in cyclic torsion with constant tension.

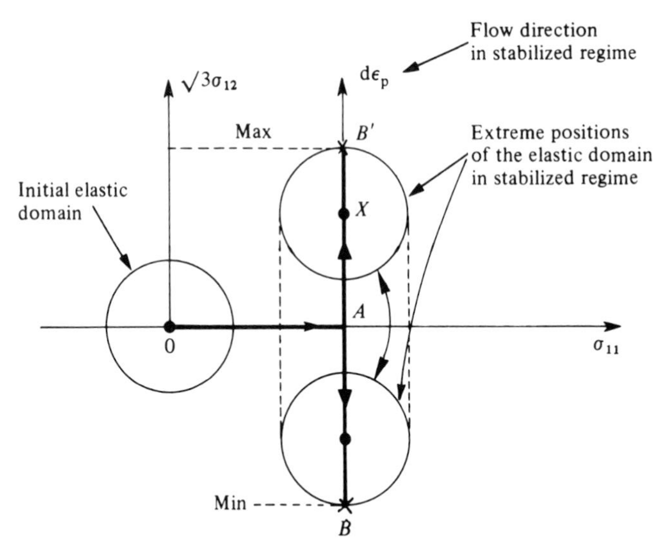

Prager–Ziegler equation and more general formulations

Ziegler's modification of Prager's rule does not come within the framework of the generalized normality hypothesis. It consists in replacing the hardening law by:

$$\mathrm{d}\mathbf{X} = (\boldsymbol{\sigma} - \mathbf{X})\,\mathrm{d}\mu.$$

The use of the tensor $(\boldsymbol{\sigma} - \mathbf{X})$ instead of its deviator allows us to account for the influence of hydrostatic stress on the evolution of the kinematic internal variable. Under a complex loading, therefore, we obtain flow laws different from that of Prager, but the two formulations are identical in the case of incompressible plasticity. The multiplying factor $\mathrm{d}\mu$, determined from the consistency condition $\mathrm{d}f = 0$, is

$$\mathrm{d}\mu = H(f)\frac{\langle(\partial f/\partial\boldsymbol{\sigma}):\mathrm{d}\sigma\rangle}{(\boldsymbol{\sigma} - \mathbf{X}):(\partial f/\partial\boldsymbol{\sigma})}$$

and the plastic multiplier $\mathrm{d}\lambda$ can take any value depending, for example, on the invariants of $\boldsymbol{\sigma}$ or $\mathbf{X}$. This allows the introduction of a nonlinear kinematic hardening by writing for example:

$$\mathrm{d}\lambda = \frac{H(f)}{C(\boldsymbol{\sigma})}\frac{\langle(\partial f/\partial\boldsymbol{\sigma}):\mathrm{d}\boldsymbol{\sigma}\rangle}{(\partial f/\partial\boldsymbol{\sigma}):(\partial f/\partial\boldsymbol{\sigma})}$$

Fig. 5.30. Different types of kinematic hardening.

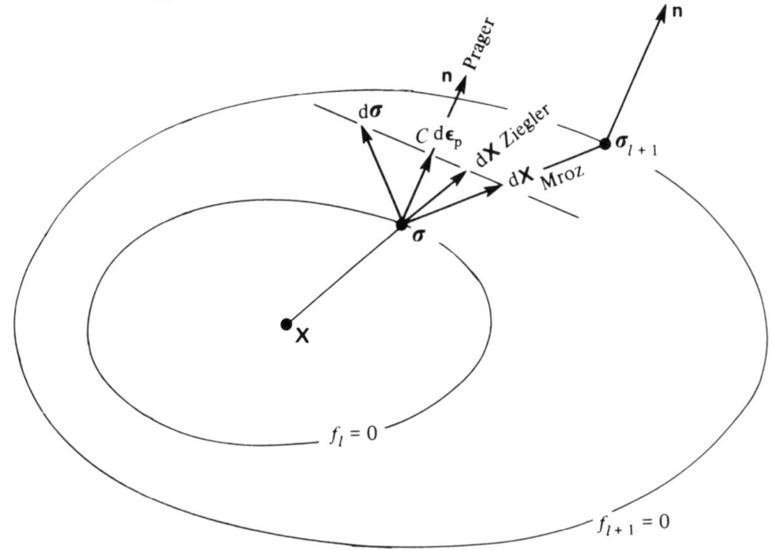

which again expresses the fact that the quantity $C(\boldsymbol{\sigma})\,\mathrm{d}\varepsilon^{\mathrm{p}}$ represents the projection of $\mathbf{d}\boldsymbol{\sigma}$ on the normal to the loading surface. It is also the projection of $\mathrm{d}\mathbf{X}$ on the normal (see Fig. 5.30). The choice of the function $C(\boldsymbol{\sigma})$ which specifies the evolution of the hardening modulus and therefore permits a nonlinear flow, presents some problems and leads to the prediction of abnormal behaviour in certain cyclic loading cases.

Kinematic hardening can also occur in a material which is already anisotropic in the elastic domain. Suppose that this domain is expressed in the form:

$$f = f_{\mathrm{Y}}(\mathbf{C}:\boldsymbol{\sigma}':\boldsymbol{\sigma}') - k = 0.$$

A kinematic hardening in which the principal directions of anisotropy remain preserved, may be defined simply by using:

$$f = f_{\mathrm{Y}}(\mathbf{C}:(\boldsymbol{\sigma}' - \mathbf{X}'):(\boldsymbol{\sigma}' - \mathbf{X}')) - k = 0.$$

The general treatment which we have already employed several times, can be used to determine explicitly the plastic multiplier $\mathrm{d}\lambda$ of the flow law expressed by:

$$\mathrm{d}\boldsymbol{\varepsilon}^{\mathrm{p}} = f'_{\mathrm{Y}}\mathbf{C}:(\boldsymbol{\sigma}' - \mathbf{X}')\,\mathrm{d}\lambda$$
$$\mathrm{d}\mathbf{X} = C f'_{\mathrm{Y}}\mathbf{C}:(\boldsymbol{\sigma}' - \mathbf{X}')\,\mathrm{d}\lambda.$$

Such a theory, which leads to more complex considerations, does not, however, allow us to vary the rate of anisotropy development (or the principal axes of anisotropy) as a function of the hardening.

5.4.4 *Flow rules under cyclic or arbitrary loadings*

The combination of isotropic hardening and linear kinematic hardening is quite sufficient for all cases where the load is nearly monotonic and includes the case where stresses vary almost proportionally (quasi-radial loads). However, these simple theories prove inadequate in cases of cyclic loading.

We have seen in Section 5.4.3 that linear kinematic hardening cannot be used to describe ratchetting effects correctly. It also cannot describe the relaxation effect on the average stress. Furthermore, it provides only a basic description of the Bauschinger effect and of hysteresis loops. With regard to isotropic hardening, the only stabilized cycles under prescribed stress to which this can lead are completely adapted without any plastic flow (line CC' on Fig. 5.31). Under prescribed strain, stabilization occurs according to a 'perfectly plastic' cycle (parallelogram $ABA'B'$ in Fig. 5.31).

The combination of linear kinematic hardening and isotropic hardening can be used to represent some cyclic hardening, but it does not eliminate the shortcomings mentioned above. In particular, in stabilized regimes the influence of isotropic hardening must disappear and we must again find a hysteresis cycle with the linear kinematic hardening parallelograms of Figs. 5.26 and 5.28).

The aim of this section is to show different means of improving this combination, by introducing a nonlinearity due to kinematic hardening.

Modelling with updating of characteristic coefficients

These models result directly from uniaxial models generalized by isotropic hardening, i.e., without the introduction of the tensorial variables related to the kinematic hardening. The general idea is to use the points of maximum stress (points of loading–unloading) as particular points of loading: at each point the characteristic coefficients of an isotropic hardening law are updated, e.g., the parameters of the Ramberg–Osgood relation which were discussed in Section 5.2.1:

$$\sigma = \sigma_{\mathrm{Y}} + K_{\mathrm{Y}} \varepsilon_{\mathrm{p}}^{1/M_{\mathrm{Y}}}.$$

Fig. 5.31. Cyclic behaviour with an isotropic hardening model.

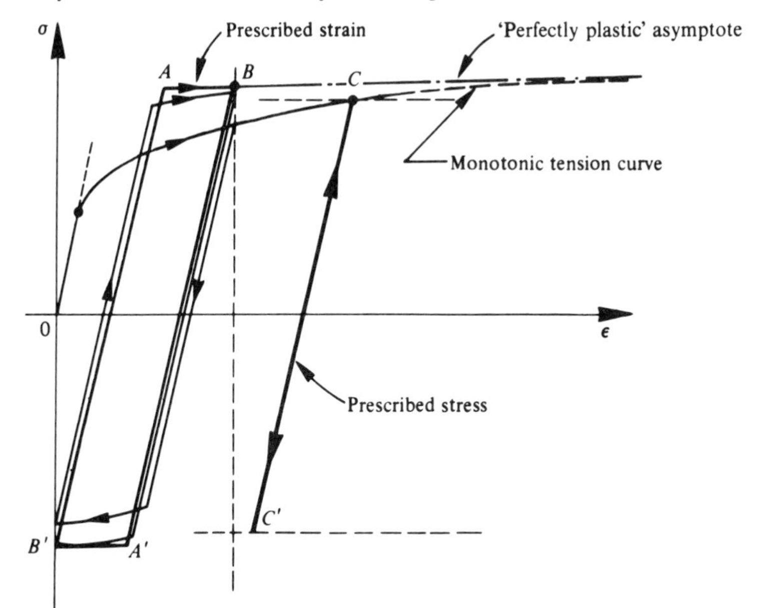

Under tension-compression, for example, with the notation of Fig. 5.32, we write that at the nth cycle (n can take the values $\frac{1}{2}$, 1, $\frac{3}{2}, \ldots, k$, $(2k+1)/2, \ldots$):

$$\sigma = \sigma_n + (-1)^{2n+1}(2\sigma_{Yn} + K_n|\varepsilon_p - \varepsilon_{pn}|^{1/M_n})$$

where the characteristic coefficients of the nth cycle (σ_{Yn}, K_n, M_n) depend on the loading history, for example the accumulated plastic strain, and on the state at the beginning of the cycle, i.e., σ_n and ε_{pn}:

$$\sigma_{Yn} = F_Y(p, \sigma_n, \varepsilon_{pn})$$
$$K_n = F_K(p, \sigma_n, \varepsilon_{pn})$$
$$M_n = F_M(p, \sigma_n, \varepsilon_{pn}).$$

We see that this type of model amounts to taking, besides the accumulated plastic strain, particular values of stress or strain states as internal variables. In fact, in this way, discrete memory variables of each stress peak are introduced, and they can actually be incorporated in the thermodynamic formalism with internal variables.

More difficulties occur when arbitrary loads are considered, even in the case of tension–compression. It is thus necessary to keep track of many previous states and to define a rule which governs the closure of each hysteresis loop and ensures that it merges with the preceding loop on return, as shown schematically in Fig. 5.33.

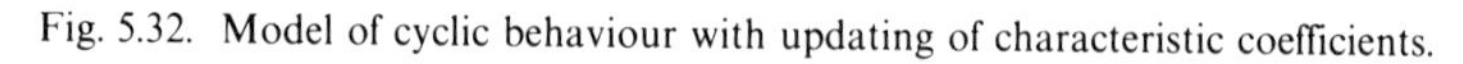
Fig. 5.32. Model of cyclic behaviour with updating of characteristic coefficients.

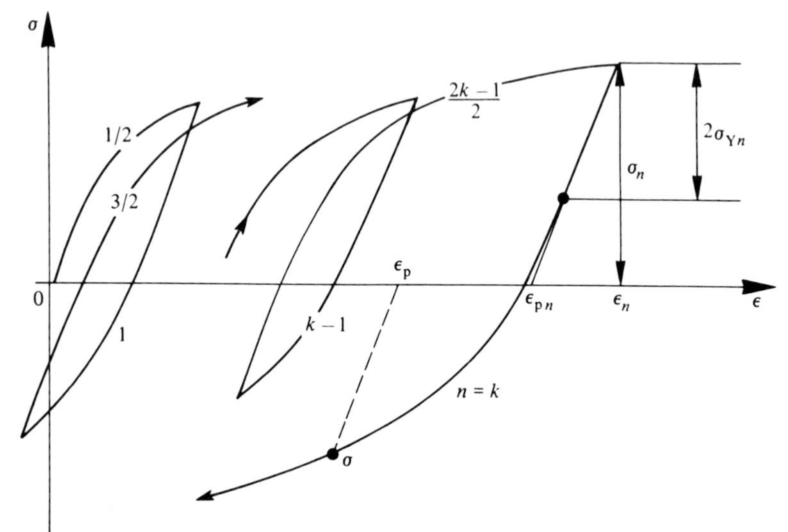

In tension–compression (or in the case of a simple load) this algorithm is quite suitable for computer calculations, but cases of nonproportional loading require much more involved considerations. Moreover, in such a theory, the points of loading–unloading play a crucial role: what would happen if cycles occurred without unloading like the 'spiraling' load shown in Fig. 5.34?

Fig. 5.33. Problem of the 'closure' of the cycle in updating rules.

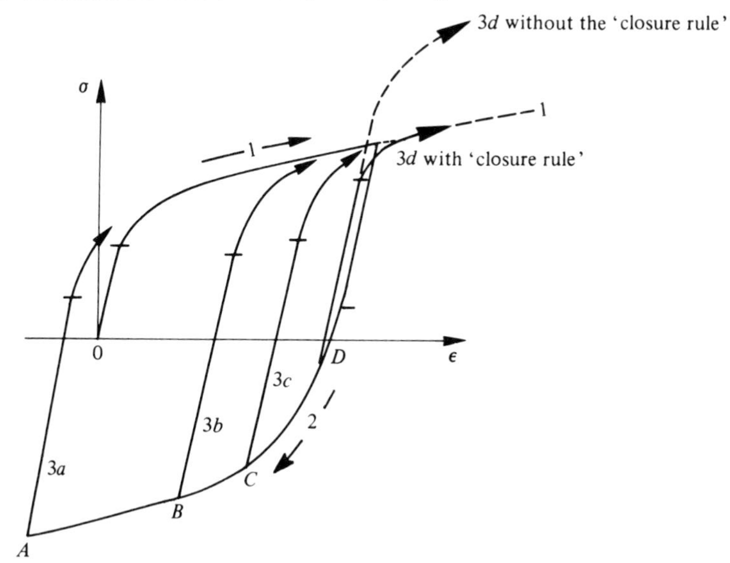

Fig. 5.34. Problem of the definition of a cycle in updating rules.

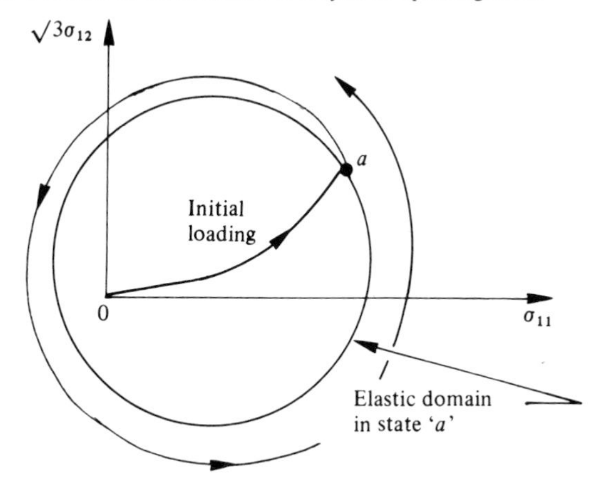

It is, however, to be noted that these concepts have been developed by several authors, including some in the field of viscoplasticity, and that the corresponding models and algorithms have been incorporated in several structural analysis computer codes.

Multilayer models and the Mroz formulation

Kinematic hardening generally corresponds to the internal stresses generated in the polycrystal by the incompatibilities of local plastic deformations from one grain to the next. It is then quite natural to improve Prager's linear hardening model by superposing several elementary models corresponding to several grains: this is the multilayer model. Work has been done on this model for a long time, and has been expanded in different ways. We have seen an example of this approach in relation to generalized St Venant model in Chapter 3: an assembly of perfectly elastoplastic elements with different moduli and yield stresses can be used to represent nonlinear kinematic hardening. Moreover (cf. Fig. 3.30) the concavity of tension and compression curves is correctly represented, as is Masing's approximation corresponding to a homothetic transformation of factor 2 between each hysteresis loop and the monotonic loading curve. Such a model can evidently be generalized by introducing isotropic and (or) linear kinematic hardening for each of the elements.

The Mroz formulation is a generalization of the multilayer model to the multiaxial case. It consists of a certain number of surfaces in the stress space, 'inserted' one into another with (linear) kinematic and (or) isotropic transformations. Denoting by $\mathbf{X}_l$ and R_l the current centre and size of the surface l (see Fig. 5.35), and assuming that it obeys the von Mises criterion, we may write,

$$f_l = J_2(\boldsymbol{\sigma} - \mathbf{X}_l) - R_l = 0.$$

Starting from a stress state located on the elastic domain (the smallest surface) successive flow will occur with each surface. Each surface is associated with a different hardening modulus which allows us to simulate the multilayer model. The last surface encountered (in loading) is the active surface and the flow direction is assumed to be normal to it:

$$\mathrm{d}\boldsymbol{\varepsilon}^{\mathrm{p}} = \mathrm{d}\lambda(\partial f_l / \partial \boldsymbol{\sigma}).$$

There is no flow ($\mathrm{d}\lambda = 0$) when the stress state is within the smallest surface: $f_l < \cdots < f_1 < f_0 < 0$. In what follows we assume that there is flow with $f_l = 0$. The coherence of the approach requires that the surfaces $f_j = 0$ for $j < l$ 'follow' the applied stress. Thus implicitly:

$$f_0 = f_1 = \cdots = f_{l-1} = f_l = 0.$$
$$d\boldsymbol{\varepsilon}^{\mathrm{p}} = d\lambda_0(\partial f_0/\partial \boldsymbol{\sigma}) = \cdots = d\lambda_l(\partial f_l/\partial \boldsymbol{\sigma}).$$

Fig. 5.35. Behaviour of the Mroz model: (*a*) in tension–compression; (*b*) in a biaxial stress state

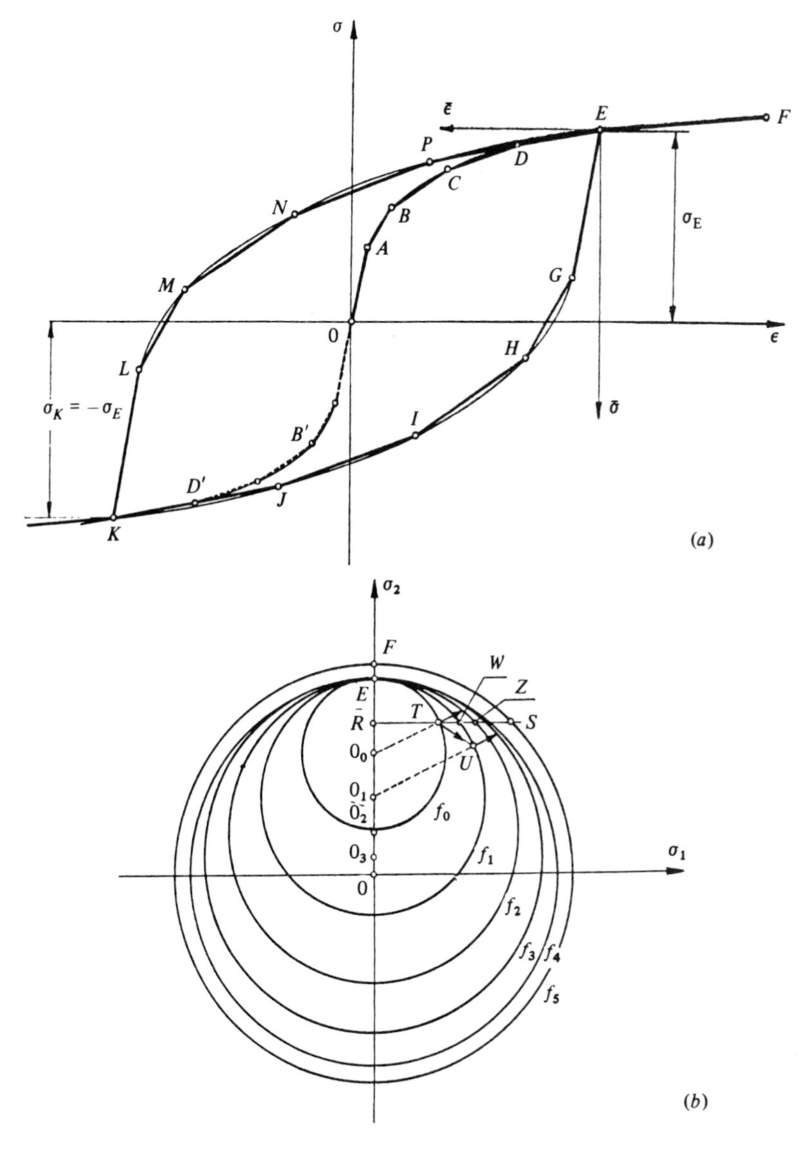

This requires the normals to the different surfaces at the point of applied stress to be identical and the plastic multipliers to be equal: $d\lambda_0 = \cdots = d\lambda_l = d\lambda$. From this it also follows (consistency conditions $df_0 = \cdots = df_l = 0$) that the movements of the centres are identical:

$$d\mathbf{X}_0 = \cdots = d\mathbf{X}_{l-1} = d\mathbf{X}_l.$$

On the other hand, the flow, and therefore the hardening between the current stress state (on surface l) and the next surface $f_{l+1} = 0$ must occur in such a way that the surfaces meet only at the point of applied stress, with identical normals. With this aim, Mroz defined the following rule: kinematic hardening occurs in the direction $\boldsymbol{\sigma}_{l+1} - \boldsymbol{\sigma}$ where $\boldsymbol{\sigma}_{l+1}$ is the stress on the surface $l+1$ such that the outward normal at this point is identical to the outward normal to the surface l at the current stress point $\boldsymbol{\sigma}$. Fig. 5.30 illustrates this hardening hypothesis which can be written in one of the following forms

$$d\mathbf{X}_l = [(R_{l+1} - R_l)\boldsymbol{\sigma} - (R_{l+1}\mathbf{X}_l - R_l\mathbf{X}_{l+1})]\frac{d\mu}{R_l}$$

or

● $$d\mathbf{X}_l = (\boldsymbol{\sigma}_{l+1} - \boldsymbol{\sigma})\, d\mu.$$

Here $d\mu$ is a multiplier whose value is determined by the consistency condition $df_l = 0$:

$$d\mu = H(f_l)\frac{\langle(\partial f_l/\partial\boldsymbol{\sigma}):d\boldsymbol{\sigma}\rangle}{\partial f_l/\partial\boldsymbol{\sigma}:(\boldsymbol{\sigma}_{l+1} - \boldsymbol{\sigma})}.$$

The multiplier $d\lambda$ of the flow relation can be an arbitrary function of $\boldsymbol{\sigma}$ or $\mathbf{X}_l$ obeying the relation

$$d\lambda = \frac{H(f_l)}{C_l(\boldsymbol{\sigma})}\frac{\langle(\partial f_l/\partial\boldsymbol{\sigma}):d\boldsymbol{\sigma}\rangle}{(\partial f_l/\partial\boldsymbol{\sigma}):(\partial f_l/\partial\boldsymbol{\sigma})}$$

which shows that the quantity $C_l\, d\boldsymbol{\varepsilon}^p$ is the projection of $d\boldsymbol{\sigma}$ on the normal to the surface $f_l = 0$. As in Ziegler's formulation, the hardening modulus may depend on $\boldsymbol{\sigma}$ or, what amounts to the same, on $\mathbf{X}_l$. However, because of the inconveniences already mentioned, it is preferable to use a constant value.

Nonlinear kinematic hardening rules

In this subsection we will limit ourselves to the use of the von Mises criterion and to one linear function f_Y of the invariant J_2. The generalization to another kind of material does not present any problem from the criterion or the plastic incompressibility point of view.

Formulation

The plasticity criterion is always expressed in the following form:

$$f = J_2(\boldsymbol{\sigma} - \mathbf{X}) - k.$$

The inconvenience of Prager's rule (proportionality between $d\boldsymbol{\varepsilon}^p$ and $\mathbf{X}$) is eliminated by a recall term which introduces a fading memory effect of the strain path

- $$d\mathbf{X} = \tfrac{2}{3} C \, d\boldsymbol{\varepsilon}^p - \gamma \mathbf{X} \, dp$$

where dp is the increment of the accumulative plastic strain, and C and γ are the characteristic coefficients of the material. We generally assume the tensor $\mathbf{X}$ to be zero in the initial state. The other elements of plasticity theory are not modified. The normality hypothesis and the consistency condition $df = 0$ lead to the expression

- $$d\boldsymbol{\varepsilon}^p = d\lambda \frac{\partial f}{\partial \boldsymbol{\sigma}} = \frac{H(f)}{h} \left\langle \frac{\partial f}{\partial \boldsymbol{\sigma}} : d\boldsymbol{\sigma} \right\rangle \frac{\partial f}{\partial \boldsymbol{\sigma}}$$

where the hardening modulus h now depends on the kinematic stress:

$$h = \tfrac{2}{3} C \frac{\partial f}{\partial \boldsymbol{\sigma}} : \frac{\partial f}{\partial \boldsymbol{\sigma}} - \gamma \mathbf{X} : \frac{\partial f}{\partial \boldsymbol{\sigma}} \left(\frac{2}{3} \frac{\partial f}{\partial \boldsymbol{\sigma}} : \frac{\partial f}{\partial \boldsymbol{\sigma}} \right)^{1/2}.$$

With the von Mises criterion, we have $d\lambda = dp$ and the hardening modulus becomes

- $$h = C - \frac{3}{2} \gamma \mathbf{X} : \frac{\boldsymbol{\sigma}' - \mathbf{X}'}{k}.$$

It is clear that when kinematic hardening (represented by $\mathbf{X}$) increases the hardening modulus decreases. This is even more apparent in tension–compression where the hardening modulus is expressed (see the following section) by

$$h = C - \gamma X \, \mathrm{Sgn}\,(\sigma - X).$$

In pure tension, for example, this decrease in the hardening modulus induces a concavity of the stress–strain curve directed downwards. In compression the concavity is directed upwards. Thus two drawbacks of Prager's rule are eliminated: the linearity and the one-to-one correspondence between the hardening and the plastic strain (Fig. 5.36). In order

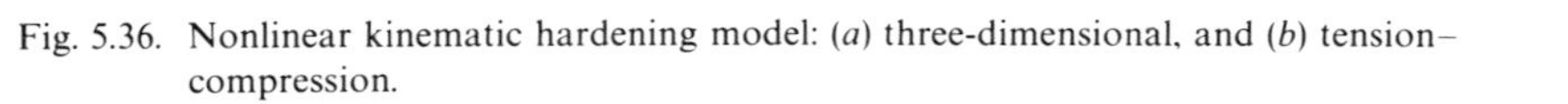

Fig. 5.36. Nonlinear kinematic hardening model: (*a*) three-dimensional, and (*b*) tension–compression.

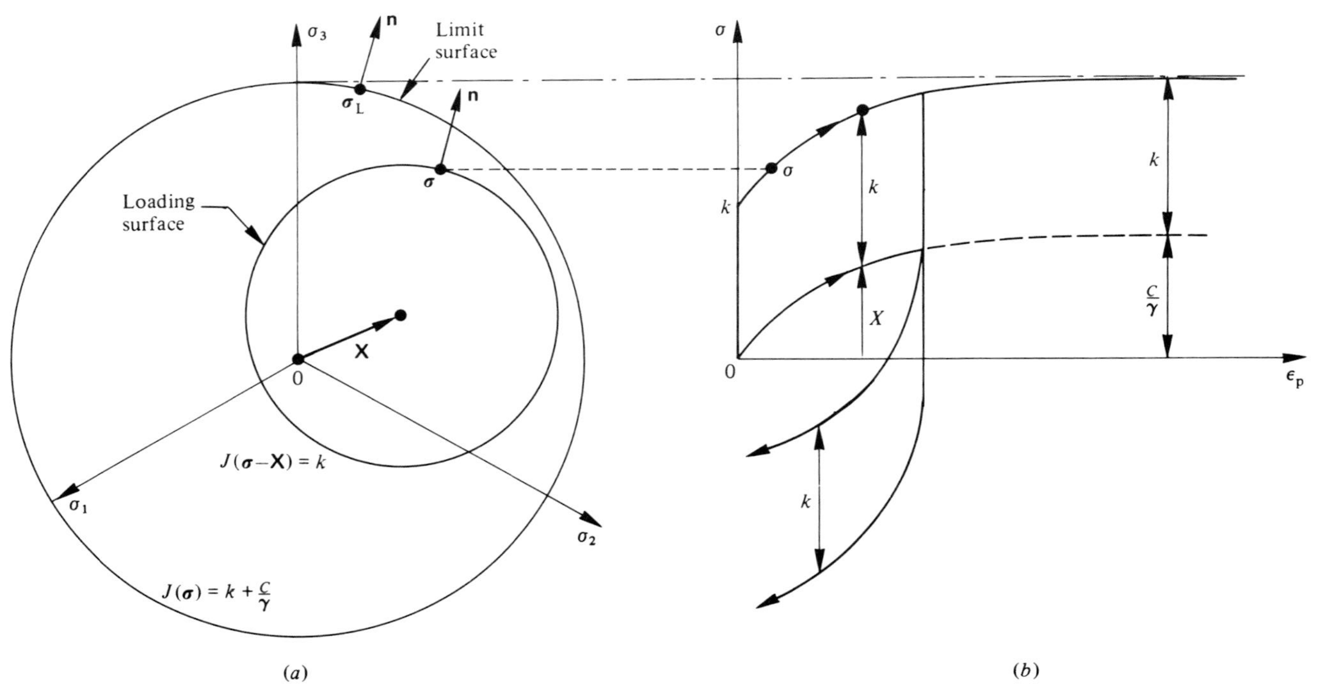

to incorporate this nonlinear kinematic modelling in the general framework of Section 5.3.3 (generalized normality) it is necessary to abandon the associated plasticity and choose a flow potential F different from the loading surface f. Here we take

$$F = J_2(\boldsymbol{\sigma} - \mathbf{X}) + \frac{3\gamma}{4C}\mathbf{X}:\mathbf{X} - k = f + \frac{3\gamma}{4C}\mathbf{X}:\mathbf{X}.$$

The hypothesis of normal dissipativity then yields (a particular case of a von Mises type material):

$$\mathrm{d}\boldsymbol{\varepsilon}^{\mathrm{p}} = \frac{\partial F}{\partial \boldsymbol{\sigma}}\mathrm{d}\lambda = \frac{3}{2}\frac{\boldsymbol{\sigma}' - \mathbf{X}'}{J_2}\mathrm{d}p$$

$$\mathrm{d}\boldsymbol{\alpha} = -\frac{\partial F}{\partial \boldsymbol{\sigma}}\mathrm{d}\lambda = \mathrm{d}\boldsymbol{\varepsilon}^{\mathrm{p}} - \frac{3}{2}\frac{\gamma}{C}\mathbf{X}\,\mathrm{d}p.$$

By using the same thermodynamic potential as for Prager's rule, we again find:

$$\mathbf{X} = \rho\frac{\partial \Psi}{\partial \boldsymbol{\alpha}} = \tfrac{2}{3}C\boldsymbol{\alpha}$$

$$\mathrm{d}\mathbf{X} = \tfrac{2}{3}C\,\mathrm{d}\boldsymbol{\alpha} = \tfrac{2}{3}C\,\mathrm{d}\boldsymbol{\varepsilon}^{\mathrm{p}} - \gamma\mathbf{X}\,\mathrm{d}p.$$

When there is plastic flow ($f = 0$), the dissipation potential is not zero except when $\mathbf{X} = 0$:

$$F = \frac{3\gamma}{4C}\mathbf{X}:\mathbf{X}.$$

This potential always lies between 0 and $C/2\gamma$ as can easily be verified by calculating the maximum value of the modulus $J_2(\mathbf{X})$ of $\mathbf{X}$ when $\mathrm{d}\mathbf{X} = 0$ with:

$$J_2(\mathbf{X}) = (\tfrac{3}{2}\mathbf{X}':\mathbf{X}')^{1/2} \leqslant C/\gamma.$$

We note that under the initial condition $\mathbf{X}(0) = 0$, the evolution equation of $\mathbf{X}$ implies, because of the plastic incompressibility, that $\mathbf{X}$ is deviatoric: $\mathbf{X}' = \mathbf{X}$.

Tension–compression case

In the case of tension–compression, the tensors are expressed as in the case of linear kinematic hardening (Section 5.4.3). The criterion and the equations of flow and hardening can be expressed in the form:

- $f = |\sigma - X| - k = 0,$
- $d\varepsilon_p = \frac{1}{h}\left\langle \frac{\sigma - X}{k} d\sigma \right\rangle \frac{\sigma - X}{k} = \frac{d\sigma}{h} \quad \text{if } (\sigma - X)\,d\sigma > 0,$
- $dX = C\,d\varepsilon_p - \gamma X\,|d\varepsilon_p|,$
- $h = C - \gamma X\,\mathrm{Sgn}\,(\sigma - X).$

In tension–compression, and more generally, in proportional loading, the evolution equation of hardening can be integrated analytically to give:

$$X = \nu\frac{C}{\gamma} + \left(X_0 - \nu\frac{C}{\gamma}\right)\exp\left[-\nu\gamma(\varepsilon_p - \varepsilon_{p0})\right]$$

where $\nu = \pm 1$ according to the direction of flow, and ε_{p0} and X_0 denote the initial values, for example at the beginning of each plastic flow.

It may be remarked that here it is not necessary to use a process of updating the variables: the state (ε_{p0}, X_0) results from the previous flow, with the flow always expressed by the same evolutionary equation. At each moment the stress is given by:

$$\sigma = X + \nu k$$

as illustrated in Fig. 5.36(*b*).

Properties
One of the first characteristics to be noted is that the nonlinear kinematic model corresponds to a formulation of classical plasticity with two surfaces:

the yield surface coincident with the loading surface;
a limiting surface induced by virtue of the property mentioned above regarding the limit value of the modulus $J_2(\mathbf{X})$ of the kinematic variable $\mathbf{X}$. It is easily verified that this condition implies:

$$J_2(\boldsymbol{\sigma}) = J_2(\boldsymbol{\sigma} - \mathbf{X} + \mathbf{X}) \leqslant J_2(\boldsymbol{\sigma} - \mathbf{X}) + J_2(\mathbf{X}) = k + J_2(\mathbf{X})$$

from which

$$J_2(\boldsymbol{\sigma}) \leqslant k + C/\gamma.$$

This property is illustrated in Fig. 5.36, where the multiaxial situation is represented on the deviatoric stress plane, and the corresponding modelling on the stress–strain diagram. Unlike the theories which postulate *a priori* the existence of this limit surface, the present theory results directly from the choice of a kinematic evolution equation.

A second interesting property of the nonlinear kinematic hardening rule is that it agrees with the basic assumptions of the Mroz formulation. It is a particular two-surface Mroz model, but with a continuously evolving hardening modulus.

Let $\boldsymbol{\sigma}_L$ be the stress state on the limit surface with the same outward normal $\mathbf{n}$ as the direction of plastic flow at the point $\boldsymbol{\sigma}$. In the nonlinear kinematic model, the limit surface does not move. We, therefore, have

$$\frac{\boldsymbol{\sigma}'_L}{k + C/\gamma} = \frac{\boldsymbol{\sigma}' - \mathbf{X}'}{k}.$$

Replacing $d\boldsymbol{\varepsilon}^p$ in the evolution law of $\mathbf{X}$ be obtain:

$$d\mathbf{X} = \left(C \frac{\boldsymbol{\sigma}' - \mathbf{X}'}{k} - \gamma \mathbf{X}' \right) dp.$$

Or, combining the two expressions:

$$d\mathbf{X} = \gamma(\boldsymbol{\sigma}'_L - \boldsymbol{\sigma}')\, dp.$$

This last relation is identical to the Mroz hypothesis (Section 5.4.4) for a criterion of the von Mises type. The essential difference comes from the plastic multiplier which is chosen to be a constant in the Mroz theory. The advantage of the present formulation comes therefore from the fact that, with only two surfaces, it is possible to describe the nonlinearity of the hardening, and to do it in a continuous way.

The existence of a limiting surface can be used to describe the ratchetting effect either in tension–compression (nonsymmetric stress) or in tension–torsion (constant tension and cyclic torsion). Fig. 5.37 shows how these effects manifest themselves. It will be noted that the curve $X(\varepsilon_p)$ is defined to within a translation (ε_p) starting with the value X_0 (the model is insensitive to mean strains). Several consequences follow from this:

with a prescribed stress, the response of the model is stabilized after a single cycle;

the cyclic stabilization is possible only for a symmetrical stress cycle (otherwise ratchetting occurs) (Fig. 5.37(*a*));

the relation between the stress amplitude and the plastic strain amplitude for the stabilized cycle is easily obtained by integrating on two half cycles of tension and compression

$$\frac{\Delta\sigma}{2} = X_{\text{Max}} + k = \frac{C}{\gamma} \tan h\left(\gamma \frac{\Delta\varepsilon_p}{2} \right) + k;$$

Fig. 5.37. Ratchetting in (*a*) tension–compression and (*b*) tension–torsion.

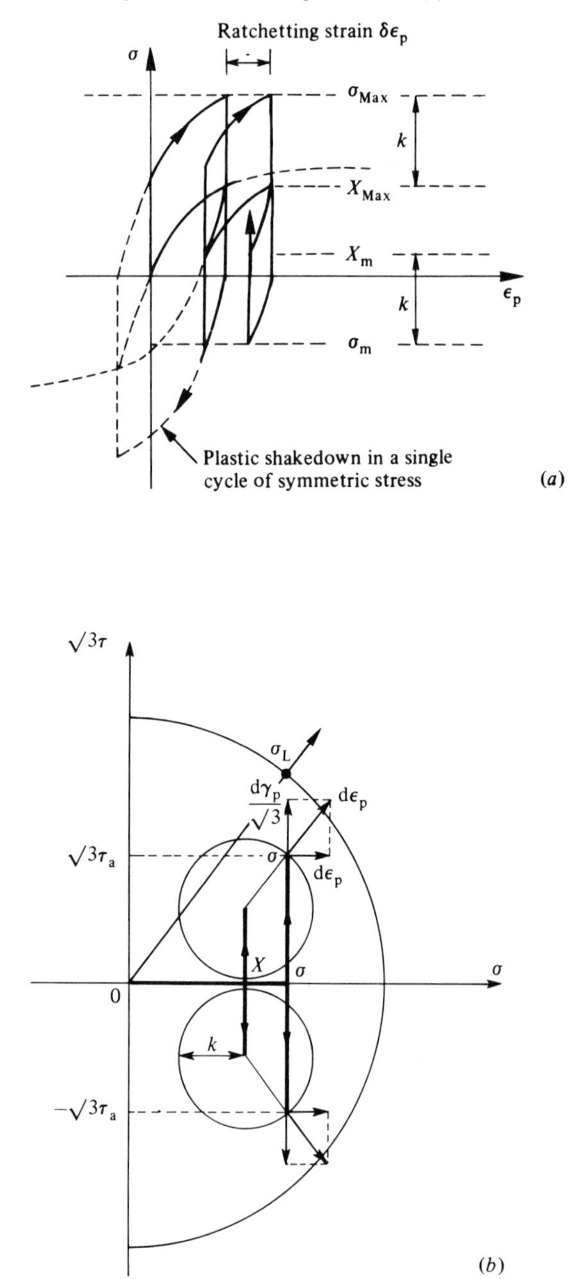

with controlled strain, when the mean strain is not zero, the initial asymmetry of the stresses progressively disappears: this is the mean stress relaxation effect observed in a number of materials (Fig. 5.38). In this case, however, the stabilization is slower than in the case of prescribed stress;

progressive ratchetting strain in tension–compression under prescribed stress occurs as soon as the mean stress is nonzero (and the stress range $\Delta\sigma = \sigma_{Max} - \sigma_{min}$ is more than $2k$). It is easy to show that this progressive strain per cycle is expressed by:

$$\delta\varepsilon_p = \frac{1}{\gamma}\log\left[\frac{\left(\frac{C}{\gamma}\right)^2 - (\sigma_{min} + k)^2}{\left(\frac{C}{\gamma}\right)^2 - (\sigma_{Max} - k)^2}\right]$$

at the maximum prescribed stress; the ratchetting is maximum when $\sigma_{min} = -k$ (see Fig. 5.37(*a*)).

We now reconsider the ratchetting effect in tension–torsion (Fig. 5.37(*b*)) under a constant tension and an alternating torsion of stress amplitude τ_a. Knowing that stabilization implies $dX_{11} = 0$, it can be shown that:

depending on the relative values of σ and τ_a, we will have elastic

Fig. 5.38. Mean stress relaxation

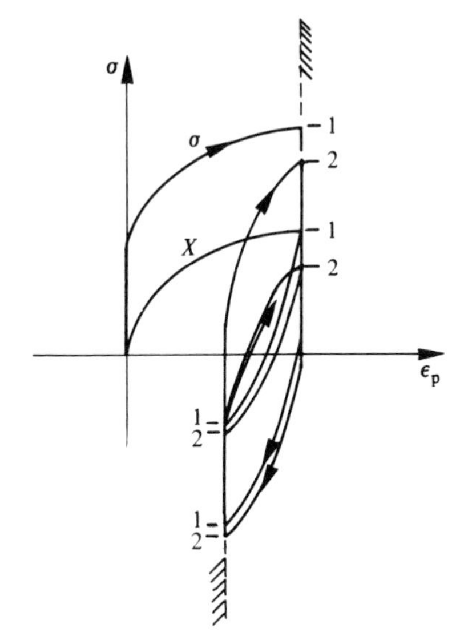

shakedown (no more flow) or ratchetting in the direction of tension (progressive strain) if:

$$\sqrt{3}\tau_a > \frac{k}{k + C/\gamma}\left[\left(k + \frac{C}{\gamma}\right)^2 - \sigma^2\right]^{1/2};$$

the ratchetting strain per cycle is proportional to the amplitude of the shear strain, with the factor of proportionality depending on the tensile stress and the dimension of the limit surface:

$$\delta\varepsilon_p = \frac{4}{\sqrt{3}}\frac{\sigma}{[(k + C/\gamma)^2 - \sigma^2]^{1/2}}\Delta\varepsilon_{12}^p.$$

Identification

The identification of characteristic coefficients k, C, γ of the material is done in tension–compression from the stabilized hysteresis loops which correspond to different strain amplitudes. For this purpose either numerical methods (see Chapter 3) or graphical methods may be used:

(1) determine k approximately from the elastic domain. This yield stress is higher than σ_Y because the initial hardening modulus C is not infinite;

(2) determine C/γ as an asymptotic value of the measure $(\Delta\sigma/2) - k$ as $\Delta\varepsilon$ increases;

(3) determine the coefficient C by fitting the results obtained by using the relation established in the previous section:

$$\frac{\Delta\sigma}{2} - k = \frac{C}{\gamma}\tanh\left(\gamma\frac{\Delta\varepsilon_p}{2}\right).$$

Figures 5.39 and 5.40 show an example of identification performed by means of three stabilized cycles on the INCO 718 alloy at 550 °C. The material is considered as plastic with the loading consisting of a prescribed plastic strain amplitude. Fig. 5.39 can be used to determine k from the experimentally obtained loops. Fig. 5.40 illustrates the determination of coefficients C and γ. The loops obtained from the calculated coefficients ($k = 520$ MPa, $C = 140\,600$ MPa, $\gamma = 380$) are shown in Fig. 5.39.

Cyclic plasticity properties

We list in Table 5.3 the coefficients of the nonlinear kinematic model which describes the material behaviour in stabilized conditions for some materials. The calculated and observed cyclic curves for some materials in Table 5.3 are shown in Fig. 5.41.

Fig. 5.39. Identification of the coefficient k from the experimental loops and verification of the model: INCO 718 alloy, $T = 550\,^{\circ}\mathrm{C}$.

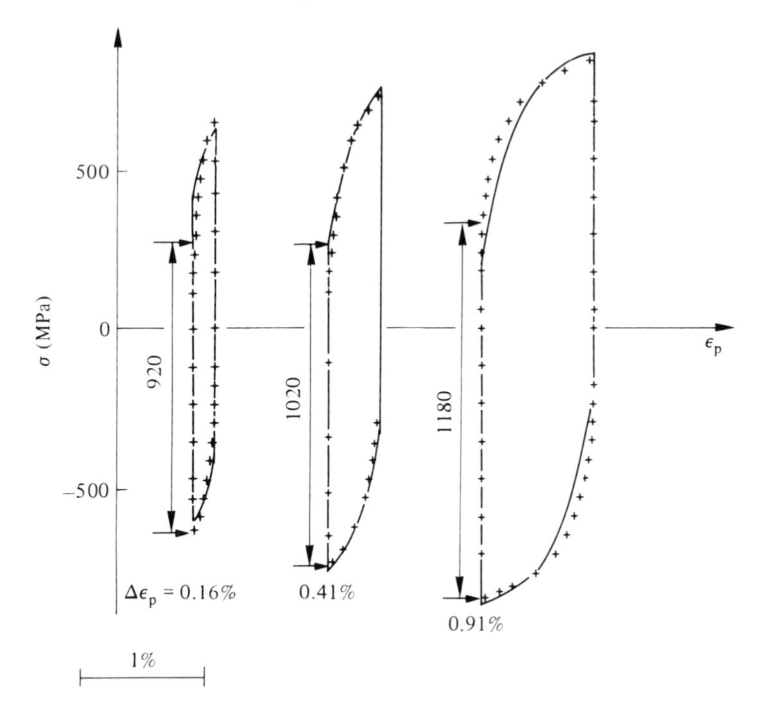

Fig. 5.40. Identification of the coefficients C and γ: INCO 718 alloy, $T = 550\,^{\circ}\mathrm{C}$.

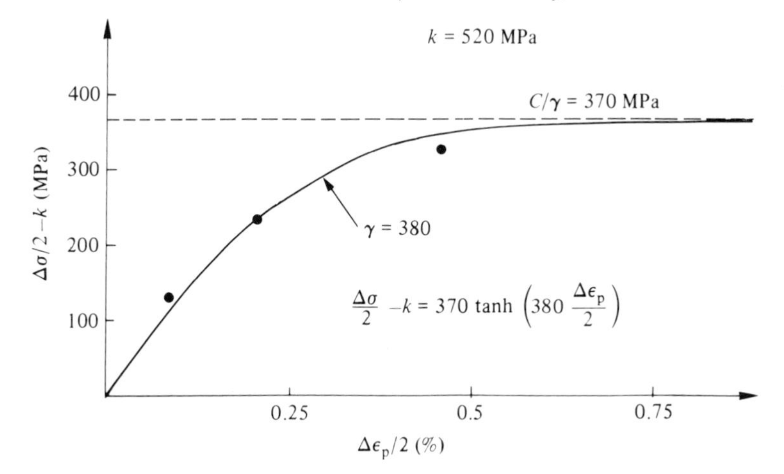

Table 5.3. *Cyclic plasticity properties*

Material	Temperature (°C)	k (MPa)	C (MPa)	γ
316L steel	20	300	30 000	60
INCO 718 alloy	550	520	140 600	380
TA6V titanium alloy	350	310	131 000	570
VO 795 cobalt	20	85	142 000	400
Structural steel	20	225	280 000	1300
NIMONIC 90 alloy	20	505	666 000	1800
TiAl 6 V4 titanium alloy	20	260	560 000	1000
35 NCD 16 nickel–chromium steel	20	980	228 000	300

Combined hardening laws

Superposition of isotropic hardening

Superposition of isotropic hardening on a nonlinear kinematic hardening results in a modification of the elastic domain by translation and uniform expansion. The state variables used in describing the isotropic hardening are the accumulated plastic strain p and the associated thermodynamic

Fig. 5.41. Cyclic curves of some materials and their interpretation by the nonlinear kinematic model.

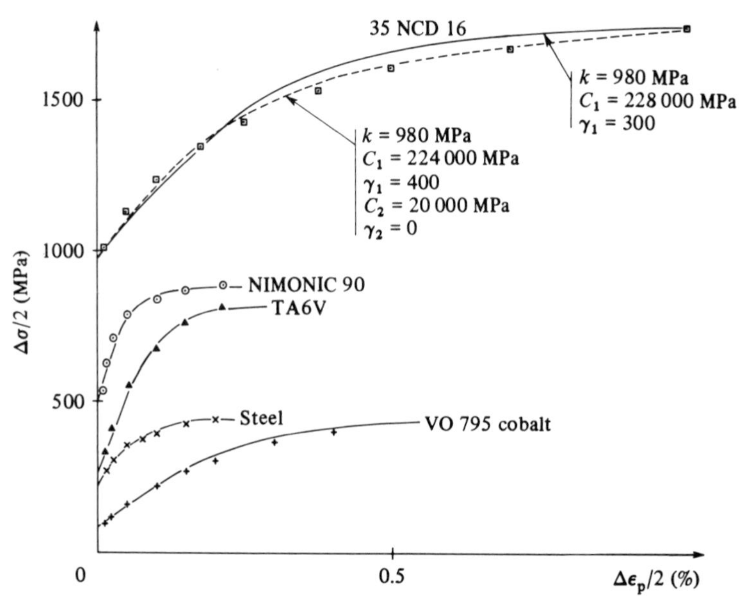

force R which represents the change in the size of the elastic domain (cf. Section 5.4.2). This domain is expressed by:

$$f = J_2(\boldsymbol{\sigma} - \mathbf{X}) - R - k \leqslant 0$$

where k represents the initial elastic limit in tension.

The evolution of R as a function of p represents the progression of hardening: for cyclic effects this evolution is slow and can occur in either an increasing fashion (cyclic hardening) or a decreasing fashion (cyclic softening). In the latter case, the formulation is valid only in the range of strain where the superposition of nonlinear kinematic hardening and isotropic softening results globally in positive hardening during each plastic flow. (Otherwise the formulation must be replaced by that in terms of strains as mentioned in Section 5.4.3).

It may prove advantageous to specialize the evolution of R by means of an equation similar to that employed for kinematic hardening:

$$\mathrm{d}R = b(Q - R)\,\mathrm{d}p$$

where b and Q are two constants. Q is the asymptotic value which corresponds to a regime of stabilized cycles, and b indicates the speed of the stabilization. The initial value of R may be taken as zero or nonzero: $R(0) = R_0$. This evolution law is quite a good representation of the cyclic hardening effects, for example in controlled strain situations. The integration of this relation and the application of the given criterion to each uniaxial cycle gives:

$$\sigma_{\mathrm{Max}} = X_{\mathrm{Max}} + k + Q[1 - \exp(-bp)].$$

Assuming X_{Max} to be a constant and the plastic strain range $\Delta\varepsilon_{\mathrm{p}}$ to be approximately constant, we obtain for the Nth cycle:

$$\frac{\sigma_{\mathrm{Max}} - \sigma_{\mathrm{Max}\,0}}{\sigma_{\mathrm{Max}\,s} - \sigma_{\mathrm{Max}\,0}} = 1 - \exp(-2b\Delta\varepsilon_{\mathrm{p}}N)$$

where $\sigma_{\mathrm{Max}0}$ and $\sigma_{\mathrm{Max}s}$ are respectively the maximum stress in the first cycle and that in the stabilized cycle. Fig. 5.42 effectively shows a very good agreement between this last relation and the experimental results on the 316 steel. (For this material the value of the coefficient Q must, however, be adjusted at each test, as mentioned later.)

Isotropic hardening, considered as a cyclic hardening effect, can also be applied to the kinematic variable by introducing functions of the ac-

Fig. 5.42. Verification of the evolution relation of the isotropic hardening. 316 steel.

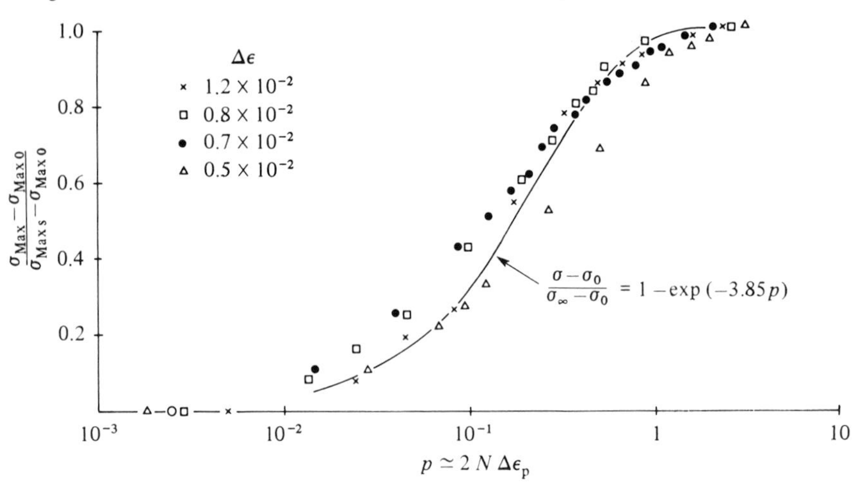

Fig. 5.43. Superposition of isotropic hardening (1) on R, (2) on C, (3) on γ.

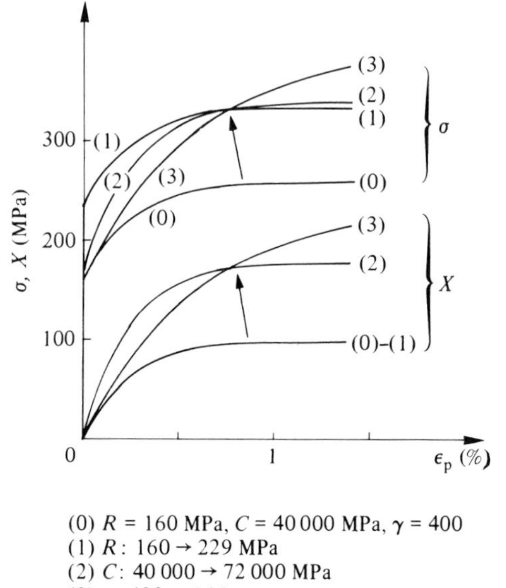

(0) R = 160 MPa, C = 40 000 MPa, γ = 400
(1) R: 160 → 229 MPa
(2) C: 40 000 → 72 000 MPa
(3) γ; 400 → 166

Fig. 5.44. Identification of the first and the stabilized cycles in a strain-controlled test on 316L steel at 20 °C (after Marquis).

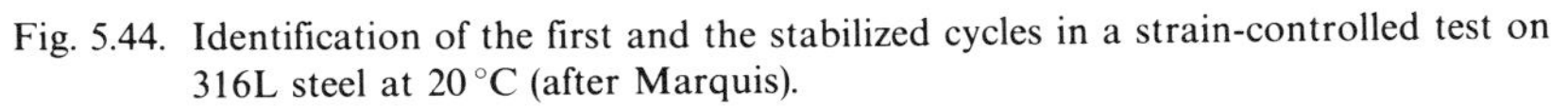

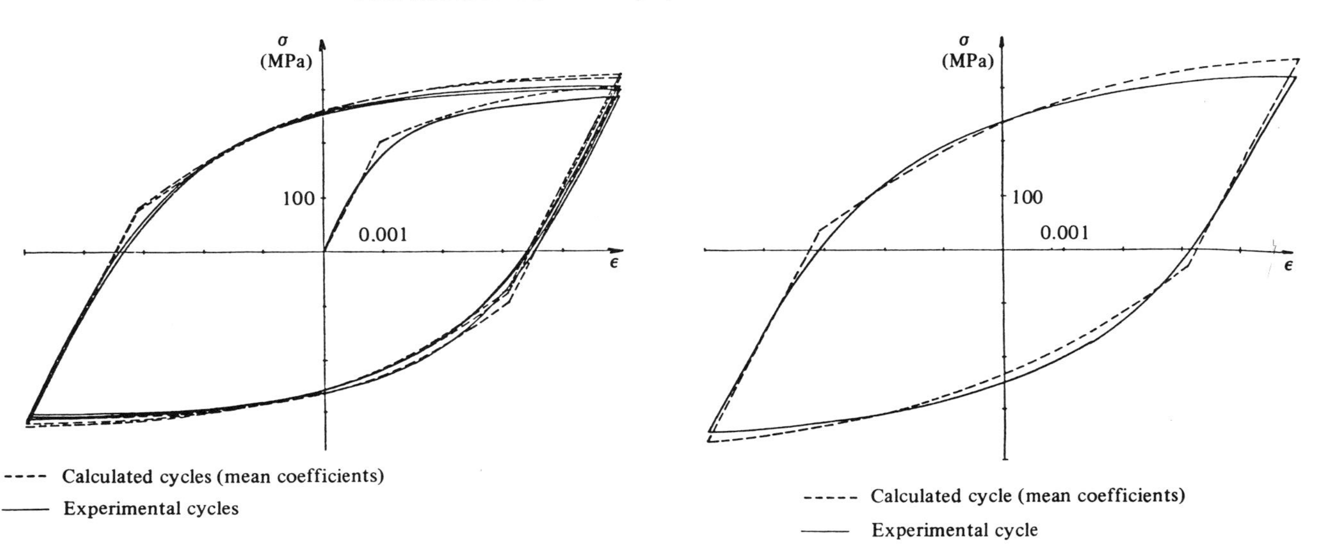

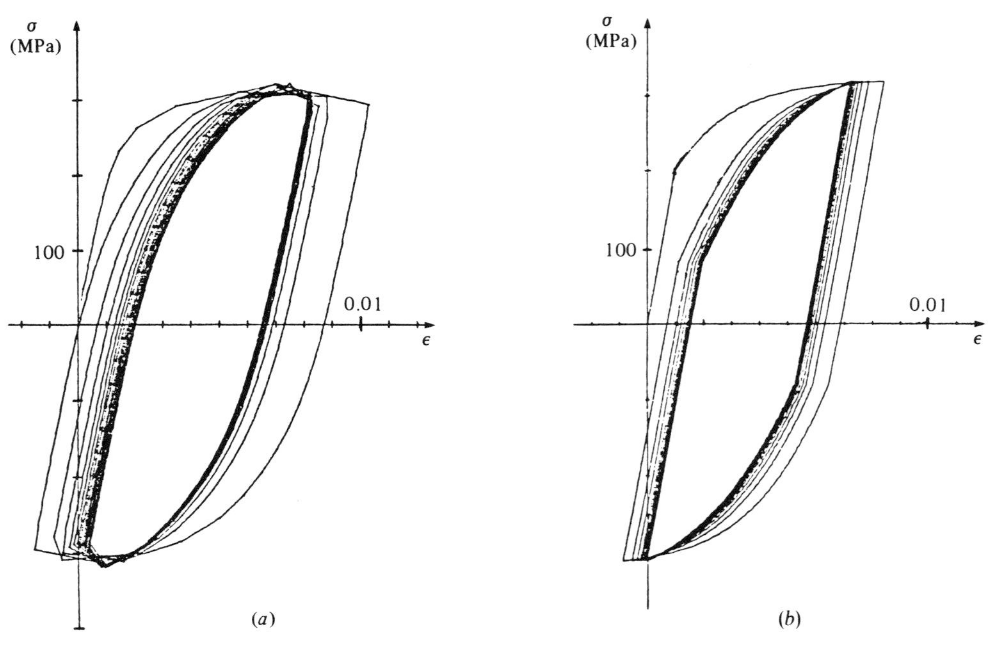

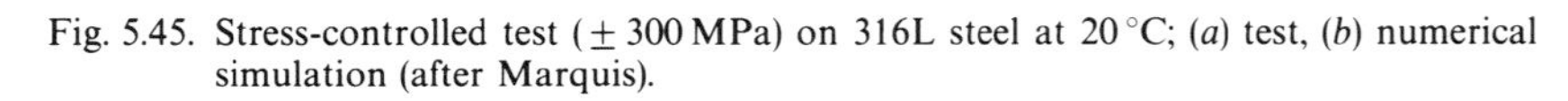

Fig. 5.45. Stress-controlled test (± 300 MPa) on 316L steel at 20 °C; (*a*) test, (*b*) numerical simulation (after Marquis).

cumulated plastic strain, for example

$$\mathrm{d}\mathbf{X} = \tfrac{2}{3} C(p)\, \mathrm{d}\boldsymbol{\varepsilon}^{\mathrm{p}} - \gamma(p)\mathbf{X}\, \mathrm{d}p.$$

Cyclic hardening can be represented by either an increasing function $C(p)$ or a decreasing function $\gamma(p)$. These two variations as well as the case when R is variable, are illustrated in Fig. 5.43 which can be used to judge the modelling differences by the shape of the stress-strain curve. An example of experimental simulation is given in Fig. 5.44 for the 316L stainless steel at room temperature, in the case where C is constant and γ decreases exponentially.

By comparing the first cycle and the stabilized cycle (in a strain controlled test) the effect of cyclic hardening can be observed on both the stress amplitude and the hardening modulus (i.e., the value of $\mathrm{d}\sigma/\mathrm{d}\varepsilon_{\mathrm{p}}$) at the maximum of the hysteresis loop.

Figure 5.45 shows the comparison for a reversed stress-controlled test. There is no ratchetting effect after the stabilization of the cyclic hardening effect. The same model, with the same values of the coefficients, correctly represents the first cycle, the effect of progressive hardening, and the stabilized cycle.

In summary, the set of equations for a model with combined isotropic and kinematic hardening can be written as follows for a material obeying the von Mises criterion:

- $\mathrm{d}\boldsymbol{\varepsilon}_{\mathrm{p}} = \tfrac{3}{2}\mathrm{d}\lambda \dfrac{\boldsymbol{\sigma}' - \mathbf{X}'}{J_2(\boldsymbol{\sigma} - \mathbf{X})},$
- $\mathrm{d}\mathbf{X} = \tfrac{2}{3} C(p)\, \mathrm{d}\boldsymbol{\varepsilon}^{\mathrm{p}} - \gamma(p)\mathbf{X}\, \mathrm{d}p,$
- $\mathrm{d}R = b(Q - R)\, \mathrm{d}p,$
- $f = J_2(\boldsymbol{\sigma} - \mathbf{X}) - R - k,$
- $\mathrm{d}\lambda = \dfrac{1}{h} H(f) \langle \tfrac{3}{2}(\boldsymbol{\sigma}' - \mathbf{X}') : \mathrm{d}\boldsymbol{\sigma} / J_2(\boldsymbol{\sigma} - \mathbf{X}) \rangle,$
- $h = C - [\tfrac{3}{2}\gamma\mathbf{X} : (\boldsymbol{\sigma}' - \mathbf{X}')/J_2(\boldsymbol{\sigma} - \mathbf{X})] + b(Q - R).$

Superposition of several kinematic models

Despite the definite improvements provided by the nonlinear kinematic hardening model over the linear one, it furnishes an inadequate description when the range of changes in strain is significant: for very low strains we recover the linear case with a poor representation of the elastoplastic transition; whereas for high strains the limiting value is reached very rapidly. It is easy to remedy such a deficiency by superposing several analogous models. The criterion is always expressed by $f = J_2(\boldsymbol{\sigma} - \mathbf{X}) - R - k = 0$ with:

$$\mathbf{X} = \sum_l \mathbf{X}_l.$$

Each of the kinematic variables $\mathbf{X}_l$ is independent, and obeys an evolution rule such that:

$$\mathrm{d}\mathbf{X}_l = \tfrac{2}{3} C_l \, \mathrm{d}\boldsymbol{\varepsilon}^{\mathrm{p}} - \gamma_l \mathbf{X}_l \, \mathrm{d}p.$$

The possibility of analytical integration under proportional loading, and the principal properties remain unchanged. Additional degrees of freedom allow us to extend the domain of validity in the manner shown in Fig. 5.46. Three kinematic variables are amply sufficient to remove the deficiencies mentioned above (each one covers a range of strain). If we wish to retain a linear kinematic hardening for large deformations, it is always possible to do so (Fig. 5.46, $\gamma_3 = 0$). One of the examples of Fig. 5.41 has been recalculated with such a model.

Another advantage of this superposition is in relation to describing the ratchetting effect: the basic model leads to a ratchetting effect which is often too significant in comparison with the experimental observations. The superposition of a linear model eliminates this problem at least after a

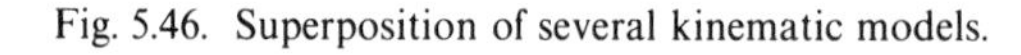

Fig. 5.46. Superposition of several kinematic models.

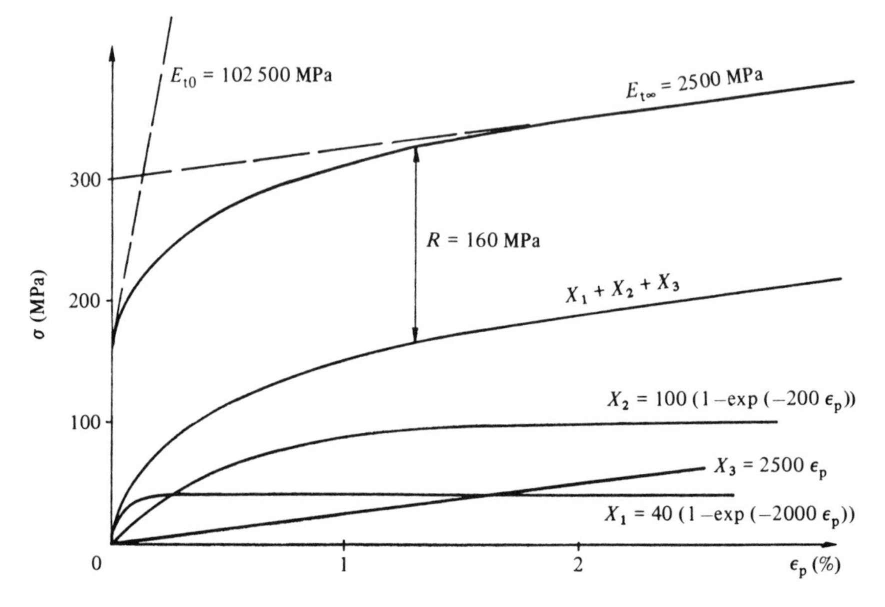

certain number of cycles. The superposition of an almost linear model ($\gamma_l \ll C_l$) leads to a less pronounced ratchetting effect.

Introduction of a hardening memory

The nonlinear kinematic model describes a hardening with fading memory (as a function of accumulated strain). Regardless of the previous history, the stabilized cycle is unique for a given cyclic load. This property is sometimes lacking, for example in some steels. To account for it, it is necessary to introduce an additional state variable (which cannot be the accumulated plastic strain as its influence becomes saturated at the stabilized cycle).

The observation of cycles under sequential loads (e.g., 316L stainless steel in Fig. 5.9) shows that in tension–compression this memory effect may be stored with the maximum plastic strain range. The comparison between the monotonic curve and the cyclic curve (Fig. 5.49) shows that cyclic hardening becomes more and more important as the amplitude of the prescribed strain increases. In other words, the asymptotic value Q of the isotropic variable R which represents the cyclic hardening (see above) is no longer a constant but depends on the strain amplitude.

The additional variable, having a memory of the maximum plastic strains can be introduced with the help of an index surface in the space of plastic strains (Fig. 5.47):

$$F = \tfrac{2}{3}J_2(\boldsymbol{\varepsilon}^{\mathrm{p}} - \boldsymbol{\zeta}) - q = 0.$$

The movement of this enveloping surface of previous strains only occurs when the current state of strain is on the surface ($F = 0$) and that the flow

Fig. 5.47. Index surface defining memory in the space of plastic strains.

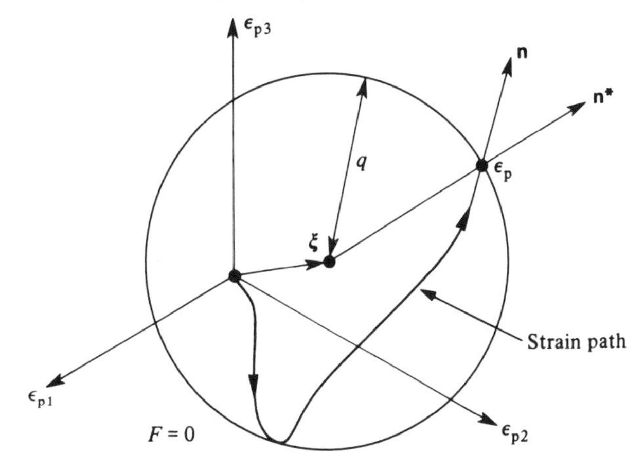

Fig. 5.48. Simulation of the test at (*a*) increasing levels, and then (*b*) decreasing levels, on 316L steel at 20 °C.

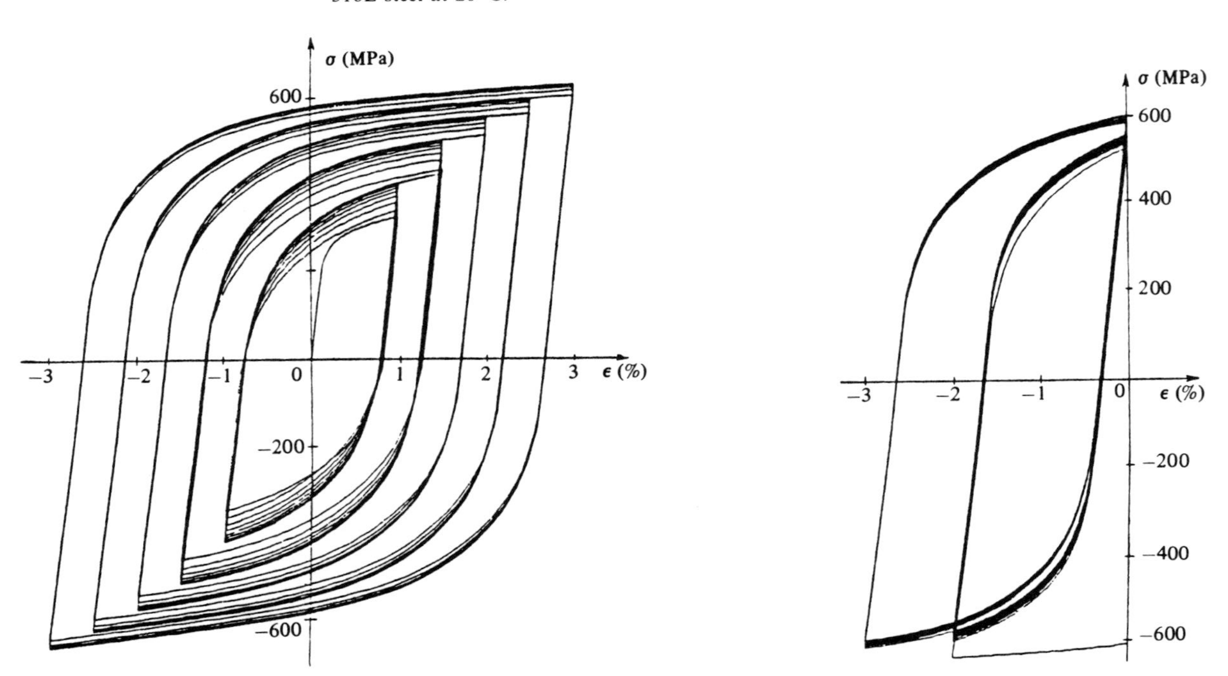

occurs in the direction external to the surface. The loading–unloading criterion used is of the same type as for classical plasticity, and the evolution equations are:

$$d\zeta = \tfrac{1}{2}(\mathbf{n}^*:d\boldsymbol{\varepsilon}^p)\mathbf{n}^* H(F)$$

$$dq = \eta\, dp\, H(F)$$

where η is determined by the consistency condition $dF = 0$:

$$\eta = \tfrac{1}{2}\langle \mathbf{n}:\mathbf{n}^* \rangle$$

where $\mathbf{n}$ and $\mathbf{n}^*$ are the unit outward normals, in $\boldsymbol{\sigma}$ and $\boldsymbol{\varepsilon}^p$ respectively, to the loading surface $f = 0$ and the index surface $F = 0$.

The size q of this enveloping surface gives the memory of deformation: in tension–compression, for example, we can easily verify that $q = \tfrac{1}{2}\Delta\varepsilon_{pMax}$. The choice of the dependence between the asymptotic value Q of isotropic hardening and the variable q enables us to complete the modelling process. For example, it is possible to take:

$$Q = Q_M - (Q_M - Q_0)\exp(-2\mu q)$$

where Q_0, Q_M and μ are three constants.

The superposition of this memory effect may be used to expand the modelling possibilities considerably. In the case of 316L steel, for example, (with a total of 13 coefficients) we can describe correctly:

- the progressive hardening (some ten cycles) for each level of the test with increasing strain levels (compare Figs. 5.9 and 5.48);
- the residual difference observed for a cycle at a lower level ($\pm 1\%$ and $\pm 1.5\%$) after one at a high level ($\pm 3\%$);
- the pronounced difference between the monotonic curve and the cyclic curve (one specimen per level), the difference becoming more pronounced as the prescribed strain increases (Fig. 5.49);
- the incremental cyclic tests, which for this material differ greatly depending on the strain amplitudes (Fig. 5.50).

The memory effect introduced in this way is perfect, that is permanent! However, a certain progressive fading can be observed. Thus, we may say that the reality is between the two extreme cases: permanent memory and completely faded memory. A more precise description is possible only at the expense of great complexity, and is of practical interest only for a detailed follow-up of the transient cyclic effects. In structural analysis under cyclic loads, we must often limit ourselves to analysis of the stabilized state. It is

Fig. 5.49. Simulation of monotonic and cyclic curves: 316L steel at 20 °C.

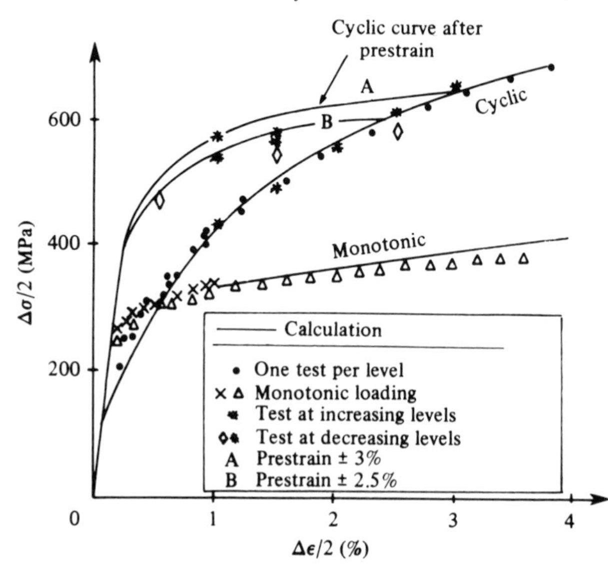

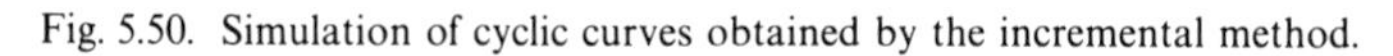

Fig. 5.50. Simulation of cyclic curves obtained by the incremental method.

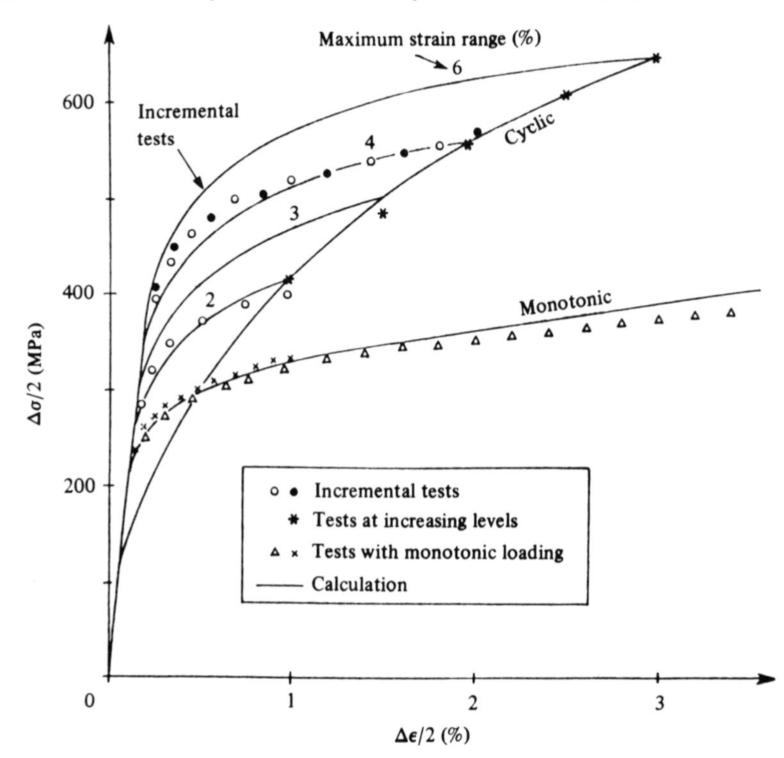

Table 5.4. *Validity of the different models.*

Models	Monotonic hardening	Bauschinger effect	Cyclic hardening or softening	Ratchetting effect	Memory effect
Prandtl–Reuss	x		x		
linear kinematic	x	x			
Mroz	x	x	x		
nonlinear kinematic	x	x		x	
kinematic + isotropic	x	x	x	x	
kinematic + isotropic + memory	x	x	x	x	x

then possible to accelerate the passage from the first cycle to the stabilized cycle by proceeding in two steps:

analysis of the first cycle to obtain an approximation (which is generally sufficient) of the value of the memory q;
saturation of the coefficients at their stabilized value $R = Q = Q(q)$ and analysis of one or more cycles under these conditions (to take into account the effects of stress redistribution).

5.4.5 *Classification of different models*

In order to define roughly the domain of validity of the different models studied, they are listed in Table 5.4 together with the phenomena they take into account.

5.5 Proportional loading

5.5.1 *Definition*

In plasticity, the current deformation of solids depends on the history of the load path. This is represented by the incremental character of the flow law which requires integration for each particular load case. An important exception is the case of proportional (or radial, or simple), loading for which the integration depends only on a scalar and can be done once and for all. These are the Hencky–Mises equations.

A loading is said to be proportional, if at any point M of the solid the stress tensor $\boldsymbol{\sigma}(M, t)$ is proportional to a tensor $\mathbf{S}(M)$, which is independent of time, the proportionality factor being a monotonic function of time $\alpha(t)$ such that $\alpha(0) = 0$:

$$\boldsymbol{\sigma}(M, t) = \alpha(t)\mathbf{S}(M).$$

During the loading the principal directions of stress remain constant.

It is easy to deduce from this that the principal directions of the stress deviator also remain constant.

$$\begin{aligned}\boldsymbol{\sigma}'(M, t) &= \boldsymbol{\sigma} - \tfrac{1}{3}\operatorname{Tr}(\boldsymbol{\sigma})\mathbf{1}\\ &= \alpha(t)[\mathbf{S} - \tfrac{1}{3}\operatorname{Tr}(\mathbf{S})\mathbf{1}]\\ \boldsymbol{\sigma}'(M, t) &= \alpha(t)\mathbf{S}'(M).\end{aligned}$$

The equivalent von Mises stress is written:

$$\sigma_{\text{eq}} = J_2(\boldsymbol{\sigma}) = (\tfrac{3}{2}\boldsymbol{\sigma}':\boldsymbol{\sigma}')^{1/2}.$$

5.5.2 *Integrated Hencky–Mises law. Equivalent stress and strain*

The condition of proportional loading allows formal integration of the Prandtl–Reuss plasticity law by assuming the material to be nonhardened in the initial state: $\boldsymbol{\varepsilon}^{\mathrm{p}}(t=0)=0$

$$\mathrm{d}\boldsymbol{\varepsilon}^{\mathrm{p}} = \tfrac{3}{2}g'(\sigma_{\mathrm{eq}})\frac{\mathrm{d}\sigma_{\mathrm{eq}}}{\sigma_{\mathrm{eq}}}\boldsymbol{\sigma}'$$

$$\boldsymbol{\varepsilon}^{\mathrm{p}} = \int_0^t \tfrac{3}{2}g'(\sigma_{\mathrm{eq}})\alpha(t)\mathbf{S}'\frac{\mathrm{d}\sigma_{\mathrm{eq}}}{\sigma_{\mathrm{eq}}}.$$

To integrate with respect to the variable σ_{eq}, we need to express α as a function of σ_{eq}:

$$\alpha = \frac{\sigma_{\mathrm{eq}}}{(\tfrac{3}{2}\mathbf{S}':\mathbf{S}')^{1/2}} = \frac{\sigma_{\mathrm{eq}}}{J_2(\mathbf{S})}$$

$$\boldsymbol{\varepsilon}^{\mathrm{p}} = \frac{\mathbf{S}'}{J_2(\mathbf{S})}\int_0^t \tfrac{3}{2}g'(\sigma_{\mathrm{eq}})\,\mathrm{d}\sigma_{\mathrm{eq}}.$$

After integration and multiplication of numerator and denominator by α we obtain the Hencky–Mises relation:

- $$\boldsymbol{\varepsilon}^{\mathrm{p}} = \tfrac{3}{2}g(\sigma_{\mathrm{eq}})\frac{\boldsymbol{\sigma}'}{\sigma_{\mathrm{eq}}}.$$

This relation leads to the concept of equivalent stress and strain, equivalence with respect to the uniaxial case of tension–compression.

The expression for the von Mises criterion:

$$\sigma_{\mathrm{eq}} - \sigma_{\mathrm{s}} = 0$$

already shows that any multiaxial state represented by σ_{eq} can be compared to the uniaxial case represented by σ_{s}. This is why σ_{eq} is called the equivalent stress in the von Mises sense. A similar concept can be defined for strain by expression the Hencky–Mises equation in a scalar form, by calculating the second invariant of the strain tensor:

$$(\tfrac{2}{3}\boldsymbol{\varepsilon}^{\mathrm{p}}:\boldsymbol{\varepsilon}^{\mathrm{p}})^{1/2} = \frac{3}{2}\frac{g(\sigma_{\mathrm{eq}})}{\sigma_{\mathrm{eq}}}(\tfrac{2}{3}\boldsymbol{\sigma}':\boldsymbol{\sigma}')^{1/2} = g(\sigma_{\mathrm{eq}}).$$

The quantity $\varepsilon^{\mathrm{p}}_{\mathrm{eq}} = (\tfrac{2}{3}\boldsymbol{\varepsilon}^{\mathrm{p}}:\boldsymbol{\varepsilon}^{\mathrm{p}})^{1/2}$ can be used to express any multiaxial evolution (under proportional loading) in a manner equivalent to the uniaxial hardening law, and is called the equivalent strain in the sense of von Mises:

- $\varepsilon_{\text{peq}} = g(\sigma_{\text{eq}})$

with

- $\varepsilon_{\text{peq}} = (\frac{2}{3}\boldsymbol{\varepsilon}^{\text{p}}:\boldsymbol{\varepsilon}^{\text{p}})^{1/2}$ and $\sigma_{\text{eq}} = (\frac{3}{2}\boldsymbol{\sigma}':\boldsymbol{\sigma}')^{1/2}$.

All multiaxial states under proportional loading are represented by a unique graph in the space of equivalent stress and equivalent strain.

5.5.3 *Existence theorem for proportional loading*

Sufficient conditions for the existence of a proportional loading, for a given structure and load, are as follows:

The external forces increase proportionally with a single parameter.
The initial state is an undeformed, nonhardened state.
The material obeys the Prandtl–Reuss law.
The hardening law is expressed in the form of a power function.
Elastic deformations are negligible.

Let a solid $\mathscr{S}$ be loaded by

body forces of density $\vec{f}(M,t)$ per unit volume in $\mathscr{S}$,
surface forces of intensity $\vec{F}(M,t)$ per unit area on the outer surface of $\mathscr{S}$.

We seek conditions under which at any instant the solution of the problem in terms of stress, i.e., $\boldsymbol{\sigma}(M,t)$, can be expressed as a scalar function of the solution $\boldsymbol{\sigma}^*(M,t^*)$ at a particular instant t^*. Thus, supposing $\boldsymbol{\sigma}^*(M,t^*)$ to be known

$$\boldsymbol{\sigma}(M,\, t) = (\alpha(t)/\alpha^*)\boldsymbol{\sigma}^*(M,\, t^*)$$

where $\alpha(t)$ is a scalar function of time such that:

$$\alpha(0) = 0, \qquad \dot{\alpha} \geqslant 0 \qquad \text{and} \qquad \alpha(t^*) = \alpha^*.$$

The equations of static equilibrium and the boundary conditions at the instant t^*

$$\operatorname{div}\boldsymbol{\sigma} + \vec{f} = 0$$

$$\boldsymbol{\sigma}\cdot\vec{n} = \vec{F}$$

will be satisfied, independent of time, if

$$\vec{f} = (\alpha/\alpha^*)\vec{f}^*(t^*) \qquad \vec{F} = (\alpha/\alpha^*)\vec{F}^*(t^*)$$
$$(\alpha/\alpha^*)\,\mathrm{div}\,\boldsymbol{\sigma} + (\alpha/\alpha^*)\vec{f}^* = 0$$
$$(\alpha/\alpha^*)\boldsymbol{\sigma}\cdot\vec{n} = (\alpha/\alpha^*)\vec{F}^*.$$

Choosing the Prandtl–Reuss equation, the plasticity relations can be integrated to give the Hencky–Mises relation if:

$$\boldsymbol{\varepsilon}^{\mathrm{p}}(t=0) = 0.$$

Under this condition

$$\boldsymbol{\varepsilon}^{\mathrm{p}} = \frac{3}{2} g\left(\frac{\alpha}{\alpha^*}\sigma^*_{\mathrm{eq}}\right)\frac{\boldsymbol{\sigma}^*}{\sigma^*_{\mathrm{eq}}}.$$

If g is of the form $g(\sigma_{\mathrm{eq}}) = [\sigma_{\mathrm{eq}}/K]^M$, then

$$\boldsymbol{\varepsilon}^{\mathrm{p}} = (\alpha/\alpha^*)^M \boldsymbol{\varepsilon}^{\mathrm{p}^*}(t^*).$$

The elasticity law shows that

$$\boldsymbol{\varepsilon}^{\mathrm{e}} = (\alpha/\alpha^*)\boldsymbol{\varepsilon}^{\mathrm{e}^*}(t^*).$$

The strain–displacement relations have still to be satisfied, but

$$\boldsymbol{\varepsilon}^* = \tfrac{1}{2}[\mathrm{grad}\,\vec{u}^* + (\mathrm{grad}\,\vec{u}^*)^{\mathrm{T}}]$$

formally leads to

$$\boldsymbol{\varepsilon} = \boldsymbol{\varepsilon}^{\mathrm{e}} + \boldsymbol{\varepsilon}^{\mathrm{p}} = \tfrac{1}{2}[\mathrm{grad}\,\vec{u} + (\mathrm{grad}\,\vec{u})^{\mathrm{T}}]$$

for arbitrary t only if $\boldsymbol{\varepsilon}^{\mathrm{e}} = 0$.

From these equations we obtain the set of conditions stated before the proof. These conditions are rather restrictive, but, in practice, we very often meet conditions of almost proportional (i.e., almost radial) loading.

5.6 Elements of computational methods in plasticity

The aim of this Section is to give only a few indications of how to use the models and plasticity properties in the traditional computational methods described in some of the works cited as references.

5.6.1 Structural analysis

Analysis of stress concentration by Neuber's method

This is an approximate technique that can be used to take into account the redistribution of stress caused by plastic flow in a zone of stress concentration, a case often met in practice. The conditions of validity of this method are that the plastic zone must remain sufficiently contained (surrounded by an elastic zone) and that the loading must be radial. The elastic solution of the problem under the action of the external forces is assumed to be known: let $\boldsymbol{\sigma}_E$ be the elastically computed stress at the most stressed point.

Plane stress

For plane structures, of small thickness, we may consider the stress state at the root of a notch to be uniaxial. Neuber's method assumes that the product of the stress and the local strain is independent of the plastic flow. In other words, we write:

- $\sigma\varepsilon = \sigma_E \varepsilon_E = \sigma_E^2 / E$

where σ_E and ε_E are the local stress and strain calculated on the basis of

Fig. 5.51. Application of Neuber's method for a monotonic loading.

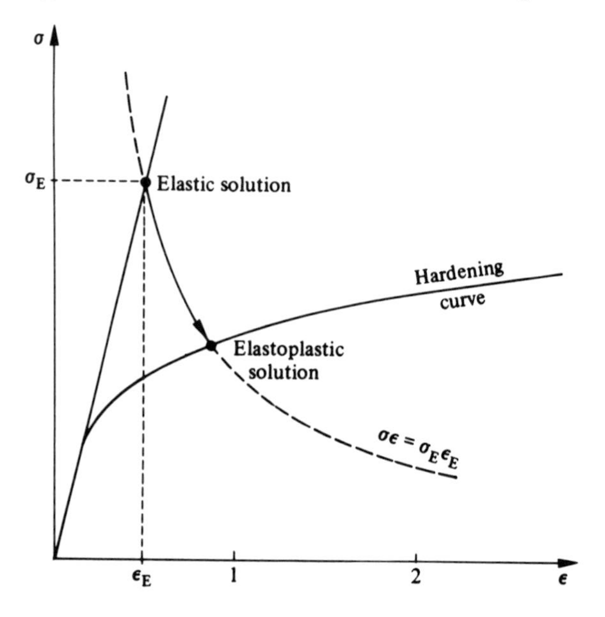

elasticity (E is Young's modulus). We then have a graphical construction (Fig. 5.51) using the hardening curve of the material. The (approximate) local elastoplastic solution is given by the intersection of this curve with the hyperbola corresponding to the elastic solution (σ_E, ε_E).

For cyclic loading, the method can be generalized branch by branch. Since the unloading is elastic, the same construction is repeated from the last loading point (e.g., Fig. 5.52) by writing:

$$\sigma'\varepsilon' = (\sigma - \sigma_0)(\varepsilon - \varepsilon_0) = (\sigma_E - \sigma_{E0})(\varepsilon_E - \varepsilon_{E0}) = (\sigma_E - \sigma_{E0})^2/E$$

and by using a change of axes. This requires knowledge of the constitutive equation for each reversal. We may also use the cyclic curve of the material

Fig. 5.52. Application of Neuber's method to a cyclic loading case.

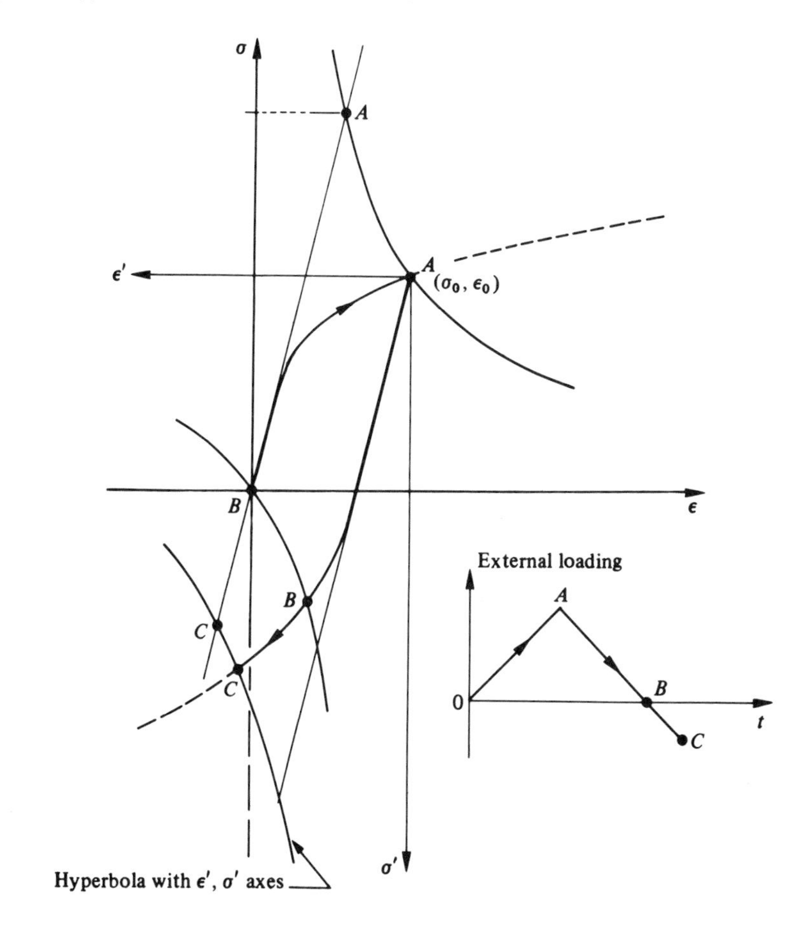

directly and think in terms of amplitudes:

$$\Delta\sigma\Delta\varepsilon = \Delta\sigma_E\Delta\varepsilon_E = (\Delta\sigma_E)^2/E.$$

More general case

When the local stress is not uniaxial, several hypotheses can be proposed to generalize the method. Elementary invariants such as the von Mises equivalent stress and equivalent strain, or the maximum principal stress and maximum principal strain can be used. The most direct generalization, which is satisfactory because it is based on an energy concept, can be written:

$$(\boldsymbol{\sigma} - \boldsymbol{\sigma}_0):(\boldsymbol{\varepsilon} - \boldsymbol{\varepsilon}_0) = (\boldsymbol{\sigma}_E - \boldsymbol{\sigma}_{E0}):(\boldsymbol{\varepsilon}_E - \boldsymbol{\varepsilon}_{E0}).$$

Few results are available that could be used to validate or invalidate one or the other of these generalizations, bearing in mind the fact that these are approximate methods.

Finite element methods in elastoplasticity

In Chapter 4, we outlined the application of the finite element method within the framework of linear elasticity. We will limit ourselves here to showing briefly how this method can be generalized to the case of plasticity by solving in a step-by-step fashion a succession of elasticity problems.

For an elastic material without initial stress, the formulation leads to the linear problem[†]

$$Kq = Q$$

where K is the stiffness matrix of the structure, obtained by assembling the stiffness matrices of each element, q is the column of nodal unknowns, i.e., nodal displacements, and Q is the column of nodal forces equivalent to the external forces.

The initial stress σ_0, assumed known, can be taken into account in the constitutive law, by writing, in the notation of Chapter 4:

$$\sigma = a\varepsilon + \sigma_0.$$

The above system is then transformed into:

$$Kq = Q + Q_0$$

[†] The brackets of matrix symbols have been omitted in this section.

where Q_0 is the column of nodal forces equivalent to the initial stress:

$$Q_0 = -\int_{\mathscr{V}} B^{\mathrm{T}} \sigma_0 \, \mathrm{d}V.$$

The simplest method of solving an elastoplastic problem incrementally is the method of 'initial strain' in which the existing plastic strain is treated as the initial strain:

$$\sigma = a\varepsilon_{\mathrm{e}} = a\varepsilon - \underbrace{a\varepsilon_{\mathrm{p}}}_{\sigma_0}.$$

The solution for the load F being known, we write the preceding system in incremental form

$$K\Delta q = \Delta Q + \Delta Q_0$$

Fig. 5.53. Iterative integration schemes: (*a*) initial strain method; (*b*) variable stiffness method; (*c*) initial stress method; (*d*) mixed method (variable stiffness + initial stress).

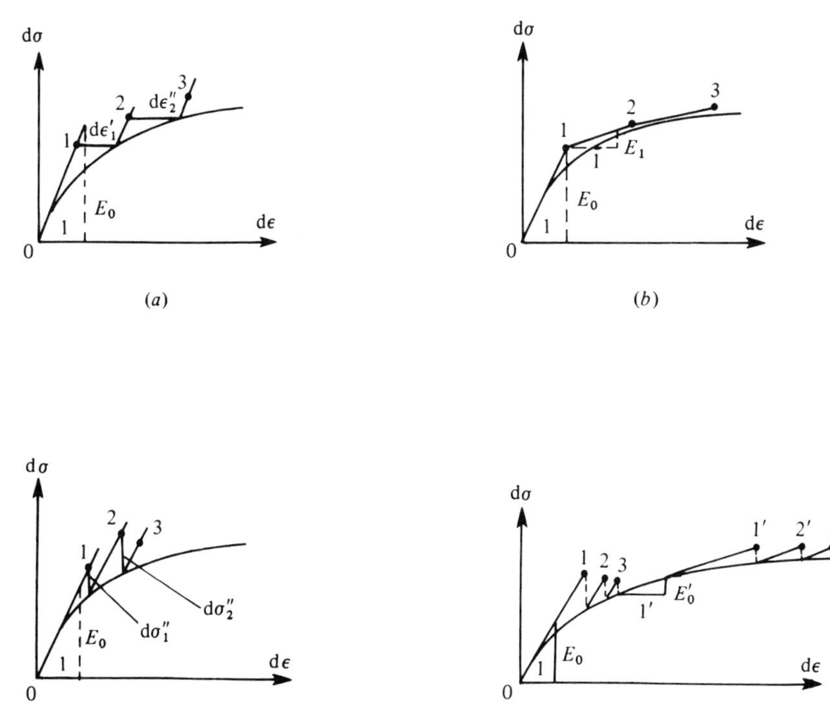

with

$$\Delta Q_0 = \int_{\mathscr{S}} B^{\mathrm{T}} a \Delta\varepsilon_{\mathrm{p}} \, \mathrm{d}V.$$

The iterative scheme for progressing from load level F to the following level $F + \Delta F$ is now apparent, and is illustrated in Fig. 5.53(*a*).

(i) Start with $\Delta\varepsilon_{\mathrm{p}} = 0$.
(ii) Solve the linear system to find Δq, $\Delta\sigma$.
(iii) Use the constitutive equation to obtain the new increments $\Delta\varepsilon_{\mathrm{p}}$ in plastic strain. This step may require one or more internal iterations to satisfy the hardening law correctly (requiring intermediate steps).
(iv) Repeat the procedure starting from (ii).

This method converges but sometimes with difficulty, particularly when the hardening moduli are low. We are then led to modify it by varying the stiffness of the structure. The plasticity law can be written in the form:

$$\mathrm{d}\varepsilon_{\mathrm{p}} = a_{\mathrm{p}}^{-1} \, \mathrm{d}\sigma$$

where a_{p} denotes the hardening moduli matrix which depends on the current state (and therefore on the unknown qs).

The constitutive equation may be written as

$$\mathrm{d}\sigma = a \, \mathrm{d}\varepsilon - a \, \mathrm{d}\varepsilon_{\mathrm{p}} = a \, \mathrm{d}\varepsilon - a \cdot a_{\mathrm{p}}^{-1} \, \mathrm{d}\sigma$$

from which it is easy to obtain

$$\mathrm{d}\sigma = a_{\mathrm{ep}} \, \mathrm{d}\varepsilon$$

with

$$a_{\mathrm{ep}} = (a^{-1} + a_{\mathrm{p}}^{-1})^{-1}.$$

In fact, the stiffness is not varied at each iteration (see Fig. 5.53(*b*)). We then write

$$\mathrm{d}\sigma = a_{\mathrm{ep0}} \, \mathrm{d}\varepsilon - a_{\mathrm{ep0}} \, \mathrm{d}\varepsilon^*$$

with

$$\mathrm{d}\varepsilon^* = (a_{\mathrm{p}}^{-1} - a_{\mathrm{p0}}^{-1}) \, \mathrm{d}\sigma^*$$

where $\mathrm{d}\sigma^*$ is the value of $\mathrm{d}\sigma$ in the preceding iteration.

The 'variable stiffness' method consists in using the following linear system

$$K(q)\Delta q = \Delta Q + \Delta Q^*$$

where $K(q)$ is the tangent stiffness matrix for the load F:

$$K(q) = \int_{\mathscr{S}} B^{\mathrm{T}} a_{\mathrm{ep}} B \,\mathrm{d}V$$

and ΔQ^* corresponds to the initial stresses induced by the strain increments $\mathrm{d}\varepsilon^*$. The stiffness matrix $K(q)$ is constant during iterations in going from F to $F + \Delta F$. The iteration procedure is the same as before. This method performs much better although at the cost of requiring extra computations in the reevaluation and reduction of the stiffness matrix. A number of variants can be developed.

5.6.2 ***Limit analysis***

Two limit theorems are given here without proof. They enable us to bound, in the energy sense, the solution of a given plasticity problem by choosing in a class of particular stress and strain fields those that are closest to the solution.

Admissible fields

The particular fields used, called admissible fields, must satisfy a number of conditions.

Admissible stress field

A stress field is an admissible field for a given problem if it is

statically admissible:

the functions $\sigma_{ij}(x_1, x_2, x_3)$ are continuous and piecewise continuously differentiable in the domain under consideration;

they satisfy the static equilibrium equation

$$\operatorname{div} \boldsymbol{\sigma} + \vec{f} = 0;$$

they satisfy the given boundary conditions

$$\boldsymbol{\sigma} \cdot \vec{n} = \vec{F}^{\mathrm{d}};$$

plastically admissible: the plasticity criterion is satisfied at all points of the domain:

$$f(\boldsymbol{\sigma}) \leqslant 0.$$

Admissible rate of displacement field

A velocity field is an admissible field for a given problem if it is

kinematically admissible:
the functions $v_i(x_1, x_2, x_3)$ are continuous and piecewise continuously differentiable in the domain under consideration;

they satisfy the boundary conditions

$$\vec{v} = \vec{v}^{\mathrm{d}};$$

plastically admissible: the functions v_i satisfy the condition of incompressibility of total strain:

$$\dot{\boldsymbol{\varepsilon}} = \tfrac{1}{2}[\operatorname{grad}\vec{v} + (\operatorname{grad}\vec{v})^{\mathrm{T}}], \quad \operatorname{Tr}(\dot{\boldsymbol{\varepsilon}}) = 0$$

which implies that the elastic deformations are negligible.

Limit theorems

The principle of virtual work, the convexity of the plasticity criterion, and the normality law can be used to prove the following two theorems.

The lower bound theorem

In a class of admissible stress fields for a given problem, the field closest to the solution in the sense of total energy is that which maximizes the functional of the power of external forces.

If $\vec{F}^*$ and $\vec{F}$ are respectively the surface intensities of forces of the admissible field and the real field, where the velocities $\vec{v}^{\mathrm{d}}$ are the given velocities, then

- $$\int_{\partial\mathscr{S}_v} \vec{F}^* \cdot \vec{v}^{\mathrm{d}} \, \mathrm{d}S \leqslant \int_{\partial\mathscr{S}_v} \vec{F} \cdot \vec{v}^{\mathrm{d}} \, \mathrm{d}S.$$

The upper bound theorem

In a class of admissible velocity fields for a given problem, the field closest to the solution in the sense of total energy is the field that minimizes the functional difference between the power of the internal forces and that part of the power of the external forces which corresponds only to the given forces.

With asterisks denoting the admissible field:

- $$\int_{\mathscr{S}} \boldsymbol{\sigma} : \dot{\boldsymbol{\varepsilon}} \, \mathrm{d}V - \int_{\partial\mathscr{S}_F} \vec{F}^{\mathrm{d}} \cdot \vec{v} \, \mathrm{d}S \leqslant \int_{\mathscr{S}} \boldsymbol{\sigma}^* \cdot \dot{\boldsymbol{\varepsilon}}^* \, \mathrm{d}V - \int_{\partial\mathscr{S}_F} \vec{F}^{\mathrm{d}} \cdot \vec{v}^* \, \mathrm{d}S$$

where $\boldsymbol{\sigma}^*$ is related to $\dot{\boldsymbol{\varepsilon}}^*$ by the constitutive equation.

These theorems, which are usually used with the assumption of rigid perfectly plastic solids, require the choice of the yield stress value to be made in a manner consistent with the approximation of the theorems:

$\sigma_s = \sigma_Y$ for the lower bound theorem ($\sigma_Y =$ elastic limit)

$\sigma_s = \sigma_u$ for the upper bound theorem ($\sigma_u =$ ultimate strength).

5.6.3 *Approximate global method of uniform energy*

In the problems of metal forming (rolling, spinning, forging, die stamping, embossing, etc.) the unknowns are the external forces necessary to transform a given initial geometry into a specified final geometry. These problems may be solved by the methods of limit analysis. From a number of approximate methods (the method of slices, the method of characteristics, etc.) we mention the uniform energy method as the one which is especially easy to use.

Its basis lies in the gross assumption that the energy density or the power of deformation is uniform in the zone of the solid which is undergoing plastic deformations. Under radial load, neglecting elastic strains, and using the concept of equivalent stress and strain, this power density is expressed by:

$$\dot{w} = \boldsymbol{\sigma}:\dot{\boldsymbol{\varepsilon}} = \sigma_{eq}\dot{\varepsilon}_{eq}.$$

By considering the material to be rigid-perfectly plastic with a yield stress σ_s in the deformed zone of volume V_p, the total power of the internal stress is

$$\dot{W} = \sigma_s \dot{\varepsilon}_{eq} V_p.$$

We calculate the 'mean' $\dot{\varepsilon}_{eq}$ according to the final and the initial geometries. This deformation is assumed to have occurred at a constant rate during a time Δt which is expressed as a function of the geometry and of the velocity v at the point of application of the unknown force F. By equating the power of the internal forces to that of the external forces i.e.,

● $$\sigma_s \dot{\varepsilon}_{eq} V_p = Fv$$

we can easily determine F.

This method gives results which are either too high or too low, depending on the particular case, but usually too low: the material being characterized by σ_s, it is advisable to take

$$\sigma_s = \sigma_u.$$

Bibliography

Hill R. *The mathematical theory of plasticity.* The Clarendon Press, Oxford (1950–71).

Kachanov L. M. *Foundations of the theory of plasticity.* North-Holland Publication Company, Amsterdam (1971).

Argon A. S. *Constitutive equations in plasticity.* M.I.T. Press, Cambridge, Mass. (1975).

Mandel J. *Plasticité et viscoplasticité.* Cours C.I.S.M. Udine. Springer-Verlag, Berlin (1971).

Prager W. *Problèmes de plasticité théorique.* Dunod, Paris (1958).

Zyckowski M. *Combined loading in the theory of plasticity.* Polish Scientific Publishers, Varsovie (1981).

Salençon *Calcul à la rupture et analyse limite.* Presses de l'ENPC, Paris (1983).

Hult J. & Lemaitre J. *Physical non-linearities in structural analysis.* I.U.T.A.M. Symposium Springer-Verlag, Berlin (1981).

Sawczuk A. & Bianchi G. *Plasticity today.* (Conf. C.I.S.M. Udine). Elsevier Applied Sciences Publisher (1985).

Baque P., Felder E., Hyafil J. & Descatha. *Mise en forme des métaux.* Dunod, Paris (1973).

CNRS. *Mise en forme des métaux et alliages.* Edition du C.N.R.S., Paris (1976).

Szczepinski W. *Introduction to the mechanics of plastic forming of metals.* Sijthoff and Noordhoff, Groningen (1979).

6

VISCOPLASTICITY

Tout solide est un fluide qui s'ignore.

With this Chapter, we come to problems related to the temporal growth of permanent deformations, i.e., to the study of the perfectly viscoplastic, perfectly elastic viscoplastic, and hardening elastoviscoplastic solids, described schematically in Chapter 3.

The important dates in the development of the mathematical modelling of viscoplasticity are:

> 1910, with the representation of primary creep by Andrade's law;
> 1929, with Norton's law which links the rate of secondary creep to the stress;
> 1934, with Odqvist's generalization of Norton's law to the multiaxial case.

The first IUTAM Symposium 'Creep in Structures' organized by Hoff took place in 1960. It became the starting point of a great development in viscoplasticity with the works of Hoff, Rabotnov, Perzyna, Hult and Lemaitre for the isotropic hardening laws, and those of Kratochvil, Malinin and Khadjinsky, Ponter and Leckie, and Chaboche for the kinematic hardening laws.

These works are now realizing their full potential in view of the possibilities offered by modern computers for analysing structures subjected to high temperature creep. Some of the basic principles of these methods are given at the end of this Chapter, which is essentially devoted to the formulation and identification of constitutive laws.

The formalism used is that of the thermodynamics of irreversible processes, guided by the phenomenological aspects described at the beginning of the chapter. Particular multiaxial constitutive laws are presented in detail in terms of modelling, identification and particular material characteristics: perfect viscoplasticity, viscoplasticity with isotropic hardening, and viscoplasticity with kinematic hardening. These laws are studied and developed here within the framework of the small strain assumption.

6.1 Domain of validity and use

The theory of viscoplasticity describes the flow of matter by creep, which in contrast to plasticity, depends on time. For metals and alloys, it corresponds to mechanisms linked to the movement of dislocations in grains – climb, deviation, polygonization – with superposed effects of intercrystalline gliding. These mechanisms begin to arise as soon as the temperature is greater than approximately one third of the absolute melting temperature. This is a schematic and indicative limit. In fact, certain alloys exhibit viscoplasticity at room temperature (300 K) even though their melting point is greater than 1400 K. For a large temperature range, the choice between the theories of plasticity and viscoplasticity depends on the type of application envisaged. Another aspect introduced in this Chapter is the effects of time, which are produced independently of any macroscopic deformation, such as recovery of crystalline structure, ageing, etc. Sometimes these mechanisms occur simultaneously with deformation and in such cases the modelling is even more complicated. For polymers, wood, bitumen, etc. the theory of viscoplasticity must be used as soon as the load has passed the limit of elasticity or viscoelasticity.

As was the case with plasticity, we will be concerned here only with nondamaging loads, which further limit the strains to half of their values at fracture, and the number of possible cycles to that for stabilization in cyclic softening or hardening. The coupling between viscoplasticity and damage is studied in Chapter 7.

6.2 Phenomenological aspects

Depending on the method adopted, it is from the qualitative analysis of the results of characteristic tests that a model is obtained. In the domain considered here, these characteristic tests are hardening tests in tension at constant strain rate or constant stress rate, creep tests at constant force or constant stress, relaxation tests at constant strain, and cylic tests.

6.2.1 *Results derived from hardening tests*

The hardening curves of a viscoplastic material are not significantly

Fig. 6.1. Hardening tests at different strain rates: (*a*) wrought Udimet 700, 927 °C (after Laflen and Stouffer); (*b*) hardened AU4G, 200 °C.

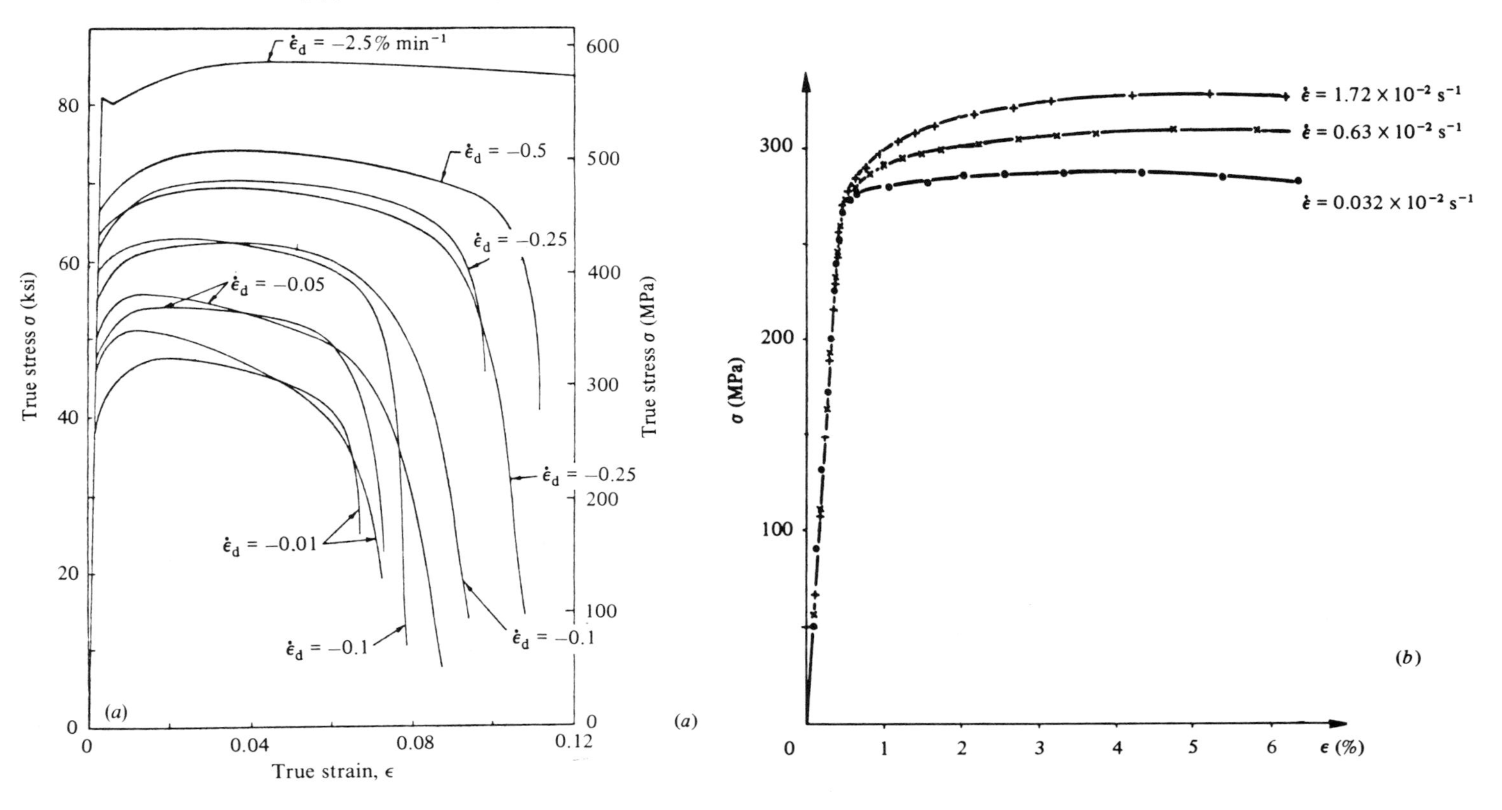

different from those of plastic materials (curves with plateaus are, however, never observed). Nevertheless, three essential differences are apparent:

- the influence of the rate of strain (or stress): at the same strain, the higher the rate of strain (or stress) is, the higher will be the stress. Two examples are shown in Figs. 6.1(*a*) and (*b*);
- a change in the rate of strain during the test results in an immediate change in the stress–strain curve which tends to make it rejoin the monotonic stress–strain curve which corresponds to the new rate (see e.g., Fig. 6.2);
- the concept of a plastic yield limit is no longer strictly applicable; plastic flow can occur at a stress lower than any applied previously. The elastic limit, or the initial yield stress, is also difficult to define; at high temperatures it is often convenient to regard it as zero and schematize the viscoplastic behaviour by 'very' viscous fluid models.

If the unloading is performed at high enough a rate ($\dot{\varepsilon} > 10^{-3}\,\mathrm{s}^{-1}$) the elastic behaviour is observed to remain unaffected by the viscoplastic deformations. This means that the hypothesis of partitioning the strains by decoupling is still applicable in most cases (where the strains are small):

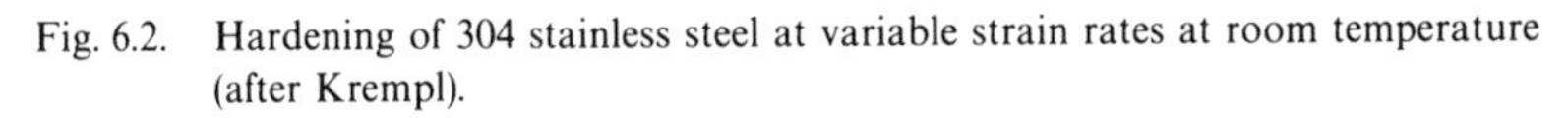

$$\varepsilon = \varepsilon_e + \varepsilon_p$$

where ε_e is the linear elastic strain and ε_p is the viscoplastic strain. In some

Fig. 6.2. Hardening of 304 stainless steel at variable strain rates at room temperature (after Krempl).

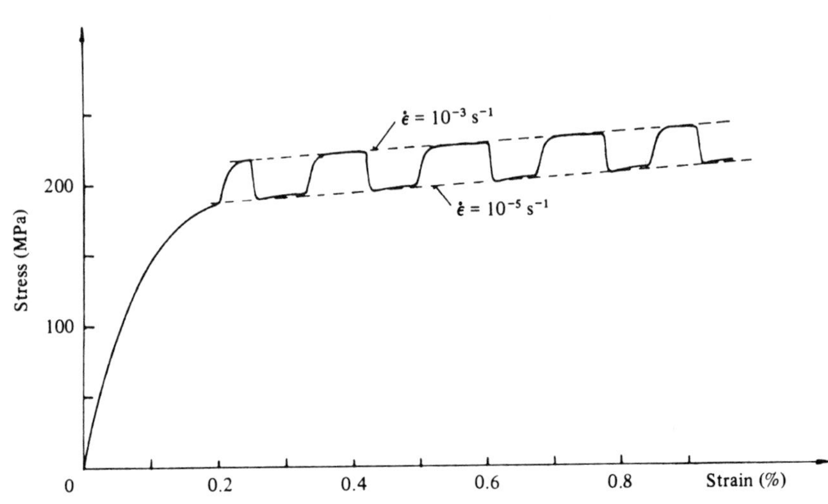

rare cases, it may become necessary to add an instantaneous plastic strain, but this is not explicitly considered in this chapter: in fact, we will be using a combination of hardening due to deformation (strain-hardening) and hardening due to time (time-hardening) to account for the superposition of the so-called instantaneous plastic strain and viscoplastic strain.

Finally, we may note that the hardening curves for different rates can roughly be deduced from one another through an affinity with the stresses. Hardening is therefore governed essentially by the amount of strain, and can be expressed analytically by expressions similar to those used in plasticity.

6.2.2 *Results derived from creep tests*

The classical creep curve represents the evolution of strain as a function of time in a material subjected to uniaxial stress at a constant temperature (Fig. 6.3). This curve generally shows three phases or periods of behaviour:

a 'primary' creep phase ($0 \leqslant \varepsilon < \varepsilon_1$) during which hardening of the material leads to a decrease in the rate of flow which is initially very high;
a 'secondary' creep phase ($\varepsilon_1 \leqslant \varepsilon < \varepsilon_2$) during which the rate of flow is almost constant;
a 'tertiary' creep phase ($\varepsilon_2 \leqslant \varepsilon < \varepsilon_R$) in which the usual increase in the strain rate up to the fracture strain is due to two factors: the reduction in the cross-sectional area of the specimens in tests conducted at constant force, and the occurrence of the phenomenon of damage which progressively reduces the resistance of the material.

The primary creep of a number of metals and alloys is rather well represented by Andrade's law

$$\sigma = At^{1/q}$$

where A and q are coefficients which depend on the material and the temperature; the values of q are generally close to 3.

Different creep tests performed at different stress levels, but at the same temperature, can be used to show the relation which exists between $\dot{\varepsilon}_p^*$, the rate of secondary creep, and the stress; this relation is Norton's law:

● $$\dot{\varepsilon}_p^* = (\sigma/\lambda^*)^{N^*},$$

where λ^* and N^* are other coefficients, characteristic of each material and temperature. An example is given in Fig. 6.4.

The value of the exponent N^* (and consequently that of λ^*) often depends on the range of rate considered: for high rates the geometric effect of the reduction in the cross-sectional area becomes dominant and the values obtained for N^* are generally higher. On the other hand for low

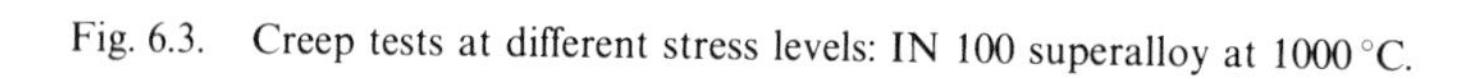

Fig. 6.3. Creep tests at different stress levels: IN 100 superalloy at 1000 °C.

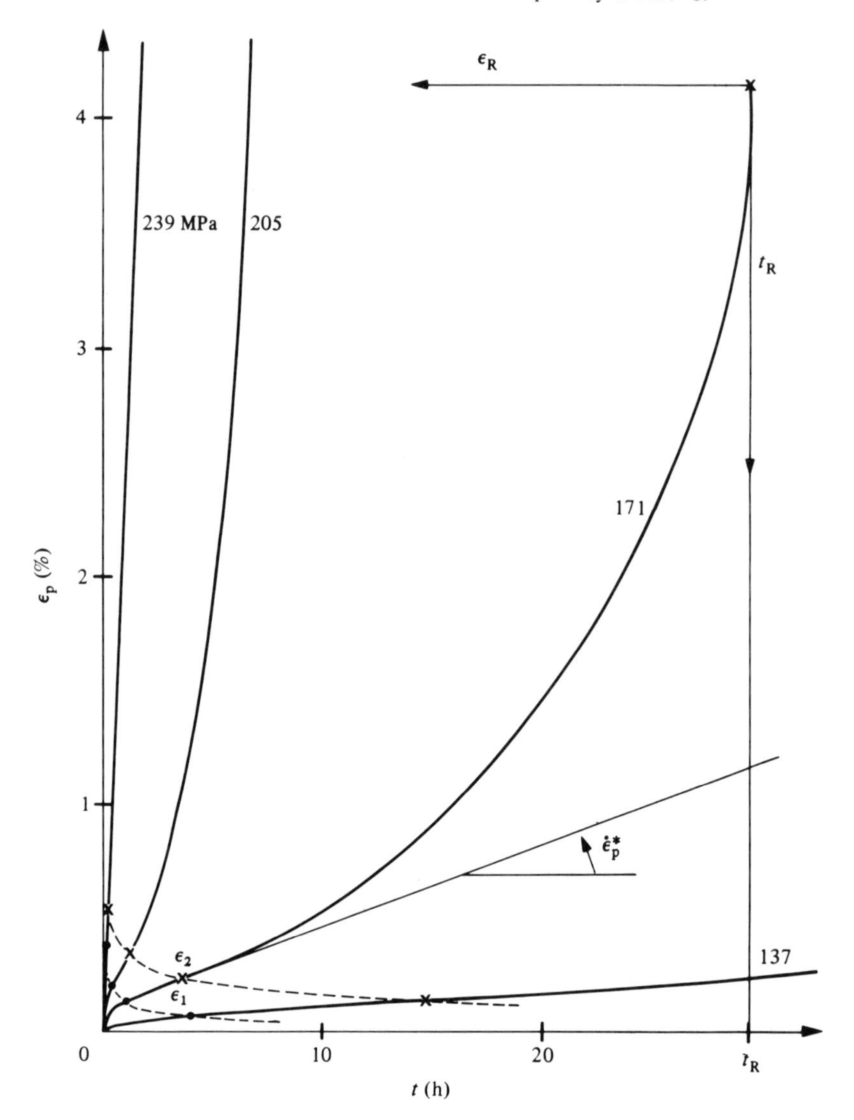

rates, the exponent has a tendency to decrease, as shown in Fig. 6.5. Such variations are often observed even after correction for the area reduction effect. Kinematic-hardening and time-recovery effects (Section 6.4.4) enable us to take these variations into consideration. Another way is to modify the power function, for example by a secondary creep law such as:

$$\dot{\varepsilon}_p^* = (\sigma/\lambda_0)^{N_0} \exp(\alpha \sigma^{N_0+1}).$$

This corresponds to a power law in which the exponent of the stress is equal

Fig. 6.4. Norton's diagram: IN 100, $T = 1000\,°C$ ($N^* = 9.7$); AU2GN T6 aluminium alloy, $T = 180\,°C$ ($N^* = 26$).

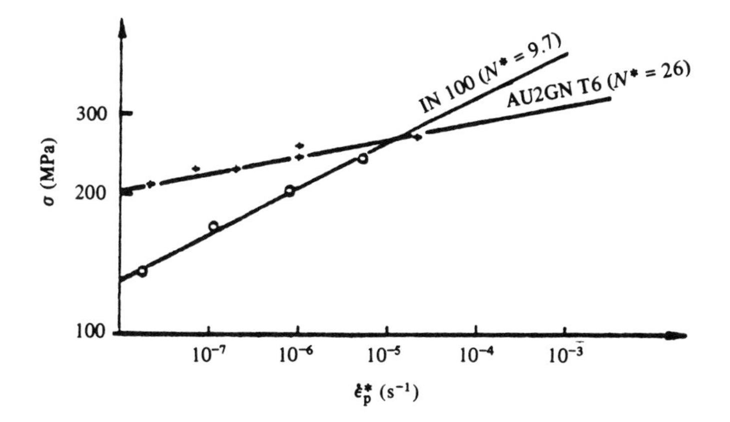

Fig. 6.5. Evolution of Norton's exponent as function of stress: a magnesium alloy at 260 °C (after Störakers).

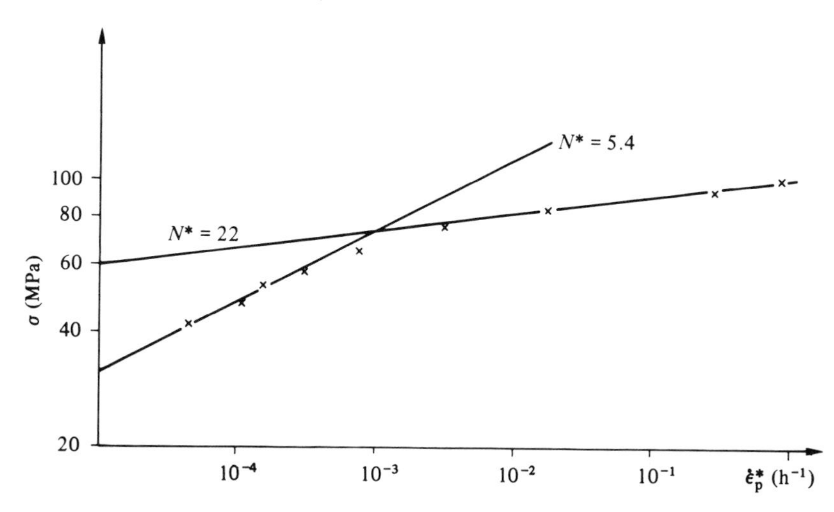

to:

$$N_0 + \alpha(N_0 + 1)\sigma^{N_0+1}$$

and thus the exponent varies between N_0 for small σ and infinity for large σ. This agrees rather well with the experimental observations, as shown in Fig. 6.5.

In addition, the exponent of Norton's law generally decreases as a function of temperature: we thus come back to the conclusion that the material becomes more and more viscous (a large value of N^* corresponds to a weak rate effect).

Fig. 6.6 and Table 6.1 give some examples of the values of Norton's exponent. Usually the values of the exponent lie between 3 and 20. The values given were determined in a region where N^* is approximately constant (for a range of strain rate between 10^{-8} and $10^{-3}\,s^{-1}$).

Tertiary creep will be studied in Chapter 7 in the framework of coupling between viscoplasticity and damage. For certain materials (e.g., TA6V titanium alloy) loaded at low strain rates ($\dot{\varepsilon} \approx 10^{-3}$–$10^{-4}\,s^{-1}$), under certain temperature conditions ($T \approx 850$–$950\,°C$), tertiary creep is delayed until very large strains ($\varepsilon \approx 1000$–2000%) are reached; this is the regime of super(visco)plasticity.

Fig. 6.6. Evolution of Norton's exponent as a function of temperature for some materials.

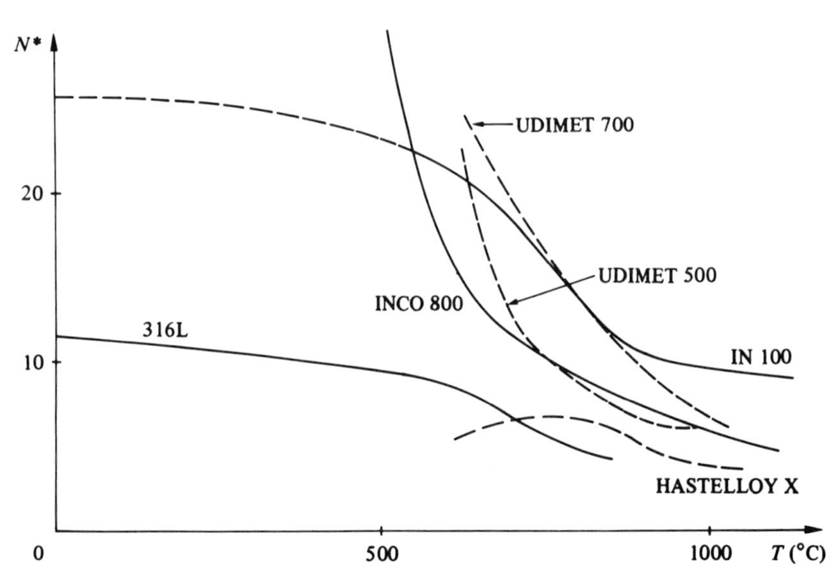

Table 6.1. *Norton's law coefficients*

Material	State	T (°C)	N^*	λ^* (MPa h)
IN 100	Coated	1000	9.7	860
UDIMET 700	Wrought	900	9.5	1340
HASTELLOY X		730	6.6	1909
INCO 800		650	13.4	715
INCO 800		980	6.1	196
316L	Quenched	600	8.2	765
316L		760	5.3	1380
UDIMET 500		900	6.5	1980
L 605		980	7.6	446
B 1900		980	5.7	3240
MAR M 509		980	6.6	912
AU2GN		180	26	412
Steel, 13% Cr	Wrought	400	7.7	2028
Steel, ASTM 321	Laminated	500	5.6	1780
Steel, ASTM 316	Laminated	732	6.4	138
Steel, ASTM 310	Laminated	700	5	553
Nimonic 75	Wrought	650	2.7	25650

Other elementary facts may be derived from special creep tests, i.e., tests at several levels of stress. Fig. 6.7(a) shows schematically how such tests can be used to make a clear choice between two commonly accepted hardening hypotheses (from among the simplest ones) to represent primary creep:

> strain-hardening for which the creep curve at the second level (σ_2) is deducible from the reference curve (AC) at stress σ_2 by shifting it parallel to the time axis (same strain);
> time-hardening for which the curve BC is shifted parallel to the strain axis (same time).

In the primary creep phase, these two hypotheses lead to very different results, especially when $\sigma_2 > \sigma_1$. Usually, experiments performed on different materials show either an agreement with the hypothesis of strain-hardening, or that this hypothesis slightly underestimates the rates (in the case where $\sigma_2 > \sigma_1$). Fig. 6.7(b) shows an example of such comparisons.

6.2.3 *Results derived from relaxation tests*

Relaxation tests demonstrate the decrease in the stress which results from maintaining a volume element in uniaxial loading at constant strain

(Fig. 6.8). In fact these tests characterize the viscosity and can be used to determine the relation which exists between the stress and the rate of viscoplastic strain. In terms of rates, the strain partitioning may be written as:

$$\dot{\varepsilon} = \dot{\varepsilon}_e + \dot{\varepsilon}_p$$

where, with the linear elasticity law $\dot{\varepsilon}_e = \dot{\sigma}/E$, and for the case of relaxation where $\dot{\varepsilon} = 0$,

$$\dot{\varepsilon}_p = -\dot{\sigma}/E.$$

Thus, each point on the relaxation curve $\sigma(t)$ gives the stress and rate of viscoplastic strain $\dot{\varepsilon}_p = -\dot{\sigma}/E$.

Fig. 6.7. Creep tests at two levels: (*a*) illustration of strain-hardening (SH) and time-hardening (TH); (*b*) D16T aluminium alloy test results compared with strain-hardening theory (after Rabotnov).

(*a*)

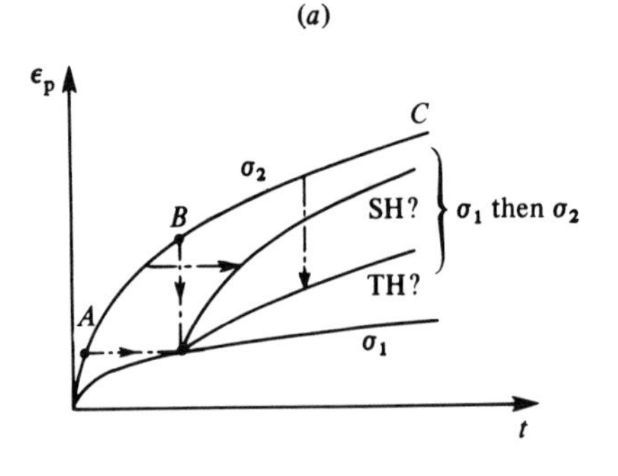

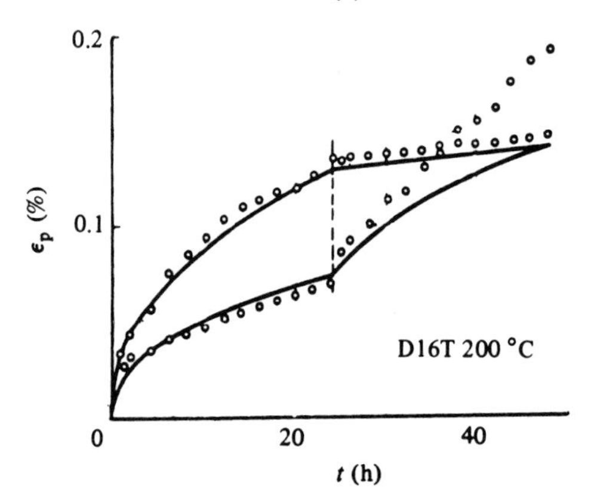

Fig. 6.8. Three relaxation tests: AU4G age-hardened alloy, 200 °C.

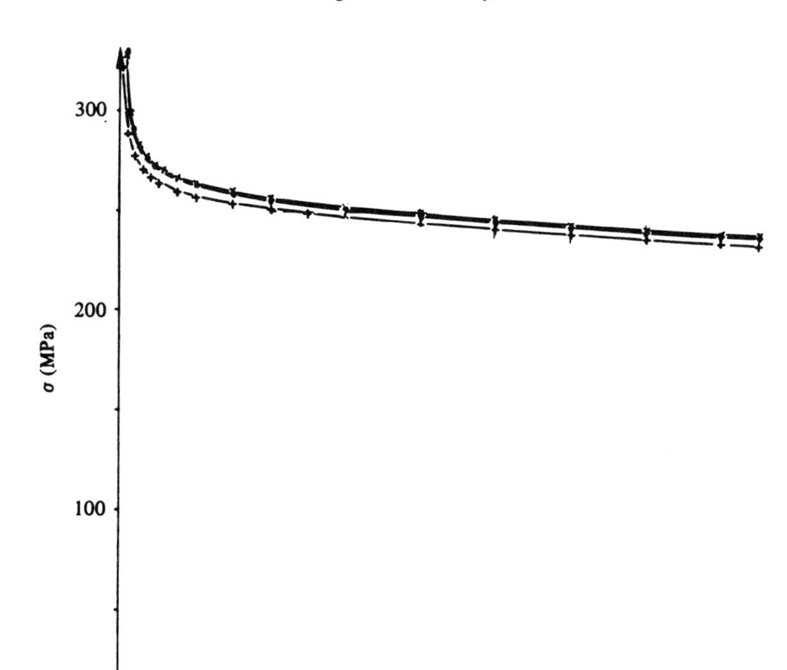

Fig. 6.9. Representation of the three basic tests in the space of the variables ε_p and $\dot{\varepsilon}_p$.

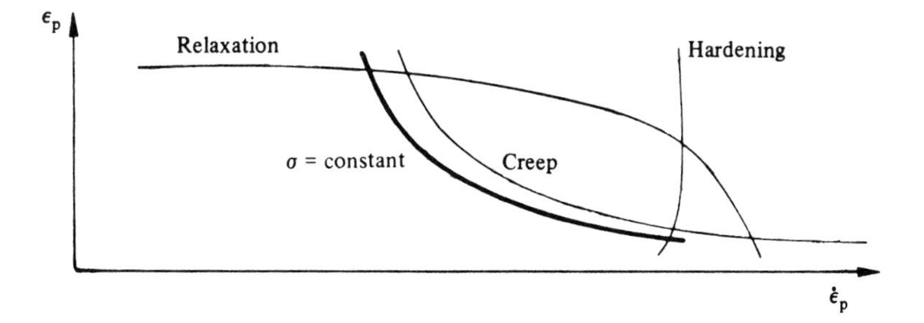

Experiments show that in this way very low viscoplastic strain rates ($\dot{\varepsilon}_p \approx 10^{-10}$–$10^{-12}\,s^{-1}$) can be reached in a relatively short time. This property is used to advantage in the 'creep-relaxation' method described in Section 6.4.2.

6.2.4 *Viscosity-hardening law*

The different results which have just been described can be put together in a uniaxial constitutive law with three parameters which represents the viscoplastic phenomena under monotonically increasing strain ($\dot{\varepsilon} \geqslant 0$) rather well.

For each of the three kinds of tests, the evolution of ε_p and $\dot{\varepsilon}_p$ can be calculated so as to represent them in a (ε_p, $\dot{\varepsilon}_p$) graph with stress as the parameter (Fig. 6.9). For many materials, a set of hardening, creep, and relaxation tests, at a fixed temperature, can be used to draw isostress curves in the ($\varepsilon_p, \dot{\varepsilon}_p$) plane within a margin of error which is not more than the scatter in the measurements. This proves experimentally that a mechanical law of state:

$$\sigma = f(\varepsilon_p, \dot{\varepsilon}_p)$$

can be defined in the domain of variation of ε_p and $\dot{\varepsilon}_p$. There we find again, in a more general context, the results derived from the creep tests at two stress levels, and we may therefore say that the plastic strain allows us to describe the hardening state of a material with reasonable accuracy.

Taking Norton's law and the affinity of hardening curves into account, a reasonable specification of the function f consists in taking it as the product of power functions. By including the linear elasticity law, the viscosity-hardening law can be written as:

- $\varepsilon = \varepsilon_e + \varepsilon_p,$
- $\sigma = E\varepsilon_e,$
- $\sigma = K\varepsilon_p^{1/M}\dot{\varepsilon}_p^{1/N}.$

where N, M and K are three parameters, which are functions of temperature and depend on the material: N is the viscosity exponent with a value of the order of 2 for very viscous materials and of 100 for those slightly viscous materials which warrant a plasticity law; M is the hardening exponent which varies approximately between 2 and 50; K is the coefficient of resistance with an order of magnitude for metals which varies from 100 to 10 000 MPa s. Specific values of N, M, K for some materials are listed in Table 6.2.

Table 6.2. *Coefficients of the viscosity-hardening law*

Material	T (°C)	N	M	K (MPa s)
Steel 0, 35% C	450	15	6	762
MARM 509	900	11	16	650
Steel Z10 CNT 18	800	13	6	400
HASTELLOY X	650	5.6	8.4	8843
AISI 304	20	65	5	752
IN 100	800	17	13.2	2140
	900	12.2	10.5	2110
	1000	10.8	9.8	1450
	1100	9.8	9.5	875
UDIMET 700	800	13.5	57	1620
	900	9.5	63.3	1480
	1000	7	70	1080
TA6V	350	120	11	970
INCO 718	550	140	13	1695
	600	88	18	1538
316L	20	65	14	458
	550	120	6	494
MG alloy	260	24.6	18.9	165
2024 annealed	208	12.8	15.3	262
2024 age-hardened	130	50	21	502
	200	27	33.4	386

Fig. 6.10 shows an example of the representation of the viscosity-hardening law and its comparison with the results of hardening, creep and relaxation tests. We note that this law, written as a product of two power functions, includes as particular cases:

the simplest hardening law already used in the preceding chapter in the framework of plasticity (with $\sigma_Y = 0$): it is sufficient to take large enough N;

Andrade's law for primary creep: by integrating the equation $\sigma = K\varepsilon_p^{1/M}\dot{\varepsilon}_p^{1/N}$ for a constant stress (with the initial condition $\varepsilon_p(0) = 0$), we obtain in effect:

$$\varepsilon_p = \left[\frac{N+M}{M}\left(\frac{\sigma}{K}\right)^N\right]^{M/(N+M)} t^{M/(N+M)}.$$

For the usual values of N and M, we obtain the exponent $1/q = M/(N+M)$ with q lying between 1.5 and 4;

Norton's law for secondary creep is not found directly but can be

Fig. 6.10. Graphs of (*a*) hardening, and (*b*) relaxation functions: 2024 alloy, annealed, $T = 208\,°\mathrm{C}$.

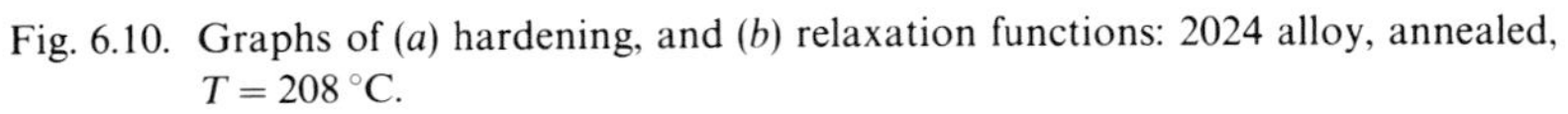

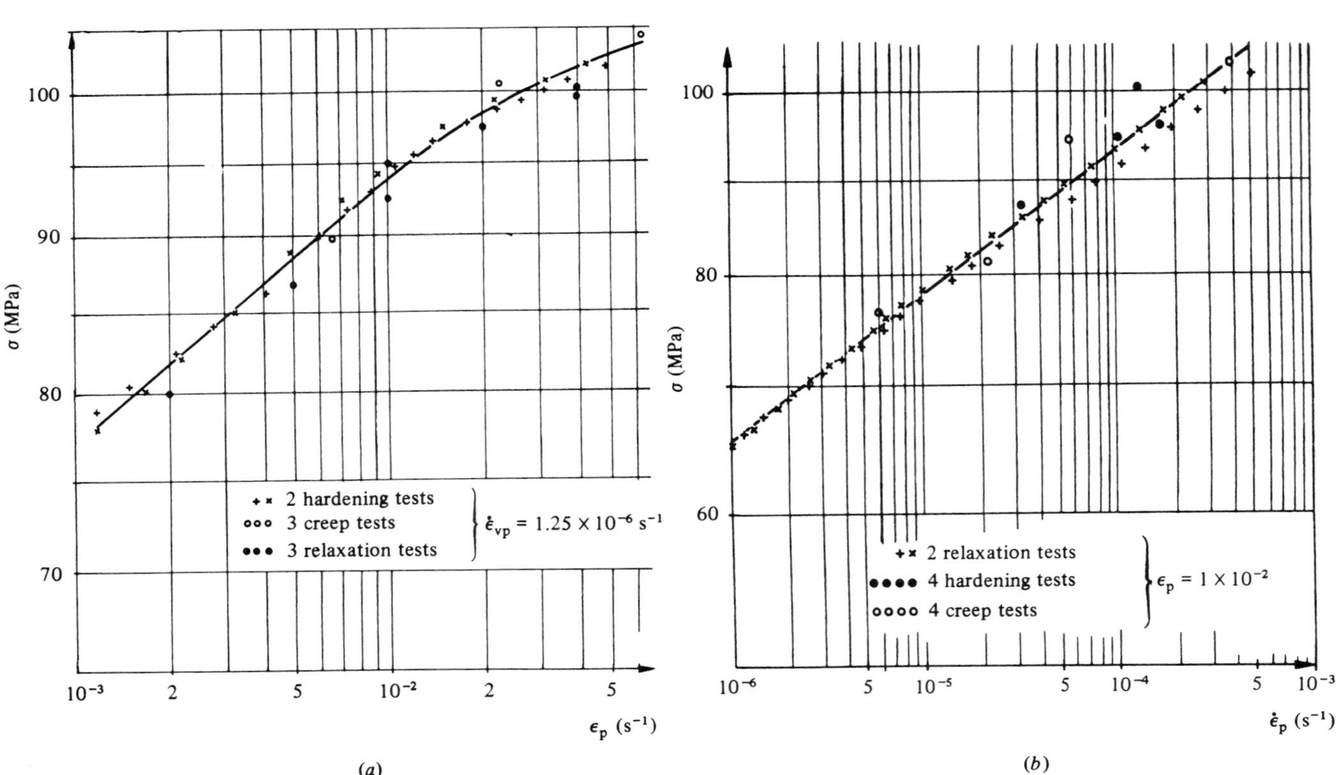

obtained as the limiting case by writing (Fig. 6.11)

$$\dot{\varepsilon}_p = \dot{\varepsilon}_p^* = \text{constant for } t \geqslant t_1 \quad \text{or } \varepsilon_p \geqslant \varepsilon_{p_1}$$

where t_1 depends on the applied stress. For example, we take the following power relation between t_1 and the strain at the start of the secondary creep, ε_{p_1}:

$$\varepsilon_{p_1} = A^* t_1^{-\alpha}.$$

Such a decrease is generally observed. With the help of the previous relations we can identify the strain rate $\dot{\varepsilon}_p^* = \dot{\varepsilon}_p(\varepsilon_{p_1})$ with that given by Norton's law. Then α and A^* are:

$$\alpha = \frac{1 - N^*/N}{N^*(N+M)/NM - 1},$$

$$A^* = \left[\left(\frac{N+M}{M}\right)^{1-N^*/N}\left(\frac{K}{\lambda^*}\right)^{N^*}\right]^{1/(1-N^*(N+M)/NM)}.$$

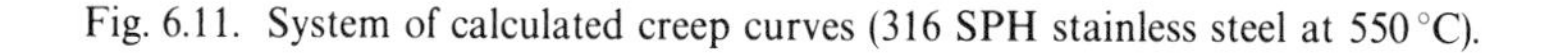

Fig. 6.11. System of calculated creep curves (316 SPH stainless steel at 550 °C).

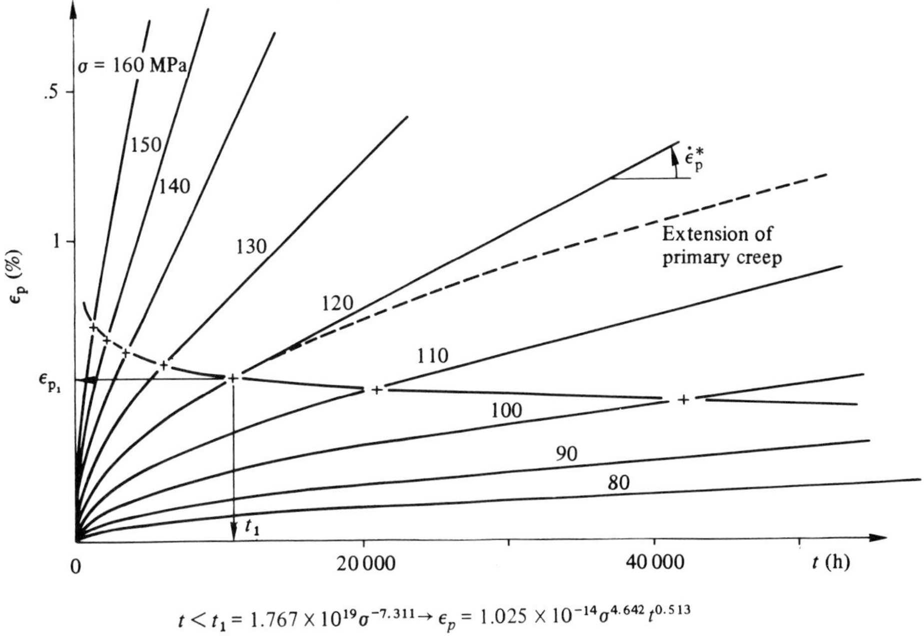

$t < t_1 = 1.767 \times 10^{19}\sigma^{-7.311} \rightarrow \epsilon_p = 1.025 \times 10^{-14}\sigma^{4.642}t^{0.513}$

$t > t_1 \rightarrow \epsilon_p = 2.268 \times 10^{-24}\sigma^{8.2}(t - t_1) + 7.666 \times 10^{-5}\sigma^{-2.669}$

$N^* = 8.2 \quad \lambda^* = 764.6 \quad N = 9.04 \quad M = 9.54 \quad K = 1111.3$ MPa h

The fact that the strain at the start of the secondary creep decreases with the stress shows that the Norton exponent N^* satisfies

$$NM/(N+M) < N^* < N.$$

6.2.5 *Influence of temperature*

The viscosity phenomena which occur in the grains and at the grain boundaries of metals, or at the intermolecular links in polymers, are highly influenced by temperature. A system of curves, such as that shown in Fig. 6.12, shows how the stress which generates a given viscoplastic strain in a given time diminishes as a function of temperature. Each experimental point in this system is the point ($\varepsilon_p = 0.2\%$, t) of a creep curve defined by the stress σ and the temperature T.

One way of accounting for this temperature influence is to consider creep as a thermally activated phenomenon. The creep rate is considered to be proportional to the number of active sources, which is assumed to be given by the Maxwell probability law. If h is the energy associated with a source and if ΔH is the activation energy of each source, then:

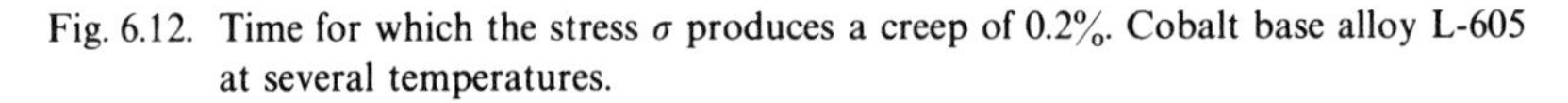

$$\dot{\varepsilon}_p \sim \int_{\Delta H}^{\infty} \frac{1}{kT} \exp(-h/kT)\,\mathrm{d}h = \exp(-\Delta H/kT).$$

Fig. 6.12. Time for which the stress σ produces a creep of 0.2%. Cobalt base alloy L-605 at several temperatures.

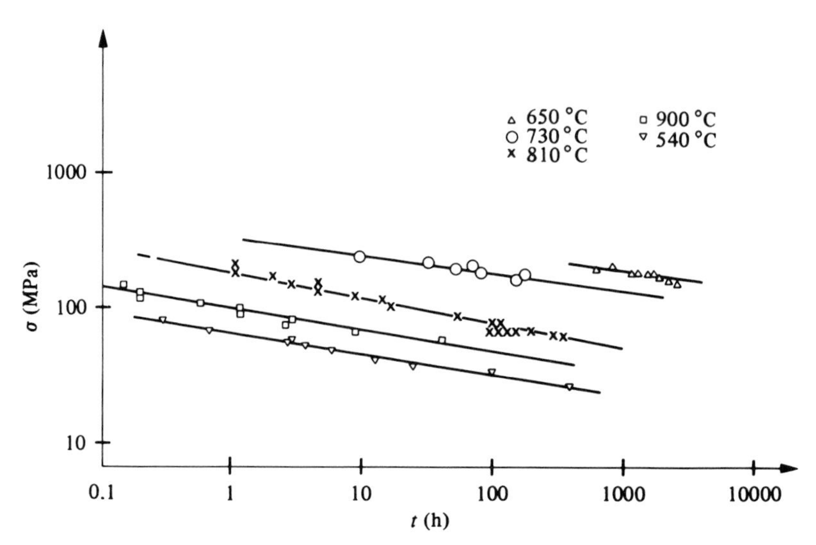

We thus define an 'equivalence' between the creep rate and the temperature represented by the Dorn parameter:

$$\dot{\varepsilon}_p \exp(-\Delta H/kT).$$

The coefficient $\Delta H/k$ is of the order of 15 000–35 000 K for common metals at temperatures between 500 and 1000 K.

The viscosity-hardening law, taking the temperature into account, can therefore be expressed as:

- $$\sigma = K'\varepsilon_p^{1/M}[\dot{\varepsilon}_p \exp(\Delta H/kT)]^{1/N}.$$

Unfortunately, the activation energy ΔH varies with the stress and the temperature, so that this formulation is no simpler than taking K, M and N as functions of temperature, except, however, for small variations in temperature:

> if $T_0 < T < T_1$ so that $|(T_1 - T_0)/T_1| \leqslant 5\%$, the above formula can be used to define the viscosity-hardening law at any temperature between T_0 and T_1 from the values of the coefficients K_0, M_0, N_0 determined at the temperature T_0 and the value of K_1

Fig. 6.13. Evolution of the coefficients N, M, K as functions of temperature: IN 100 uncoated refractory alloy.

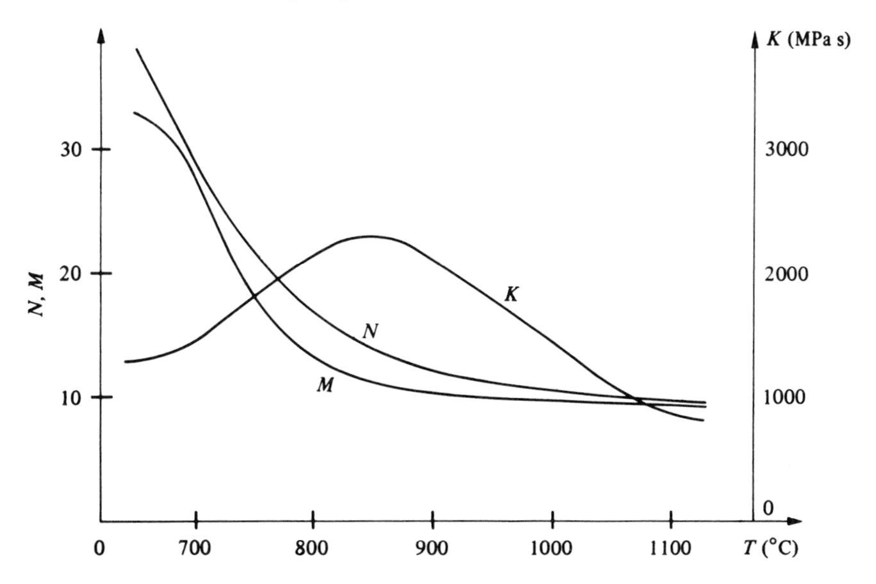

Fig. 6.14. (*a*) Cyclic softening in stress controlled tests on INCO 718 alloy at 550 °C.
(*b*) Cyclic hardening in strain controlled tests on 316 stainless steel at 550 °C.
(Here t_T is the hold time in tension and t_C is that in compression.)

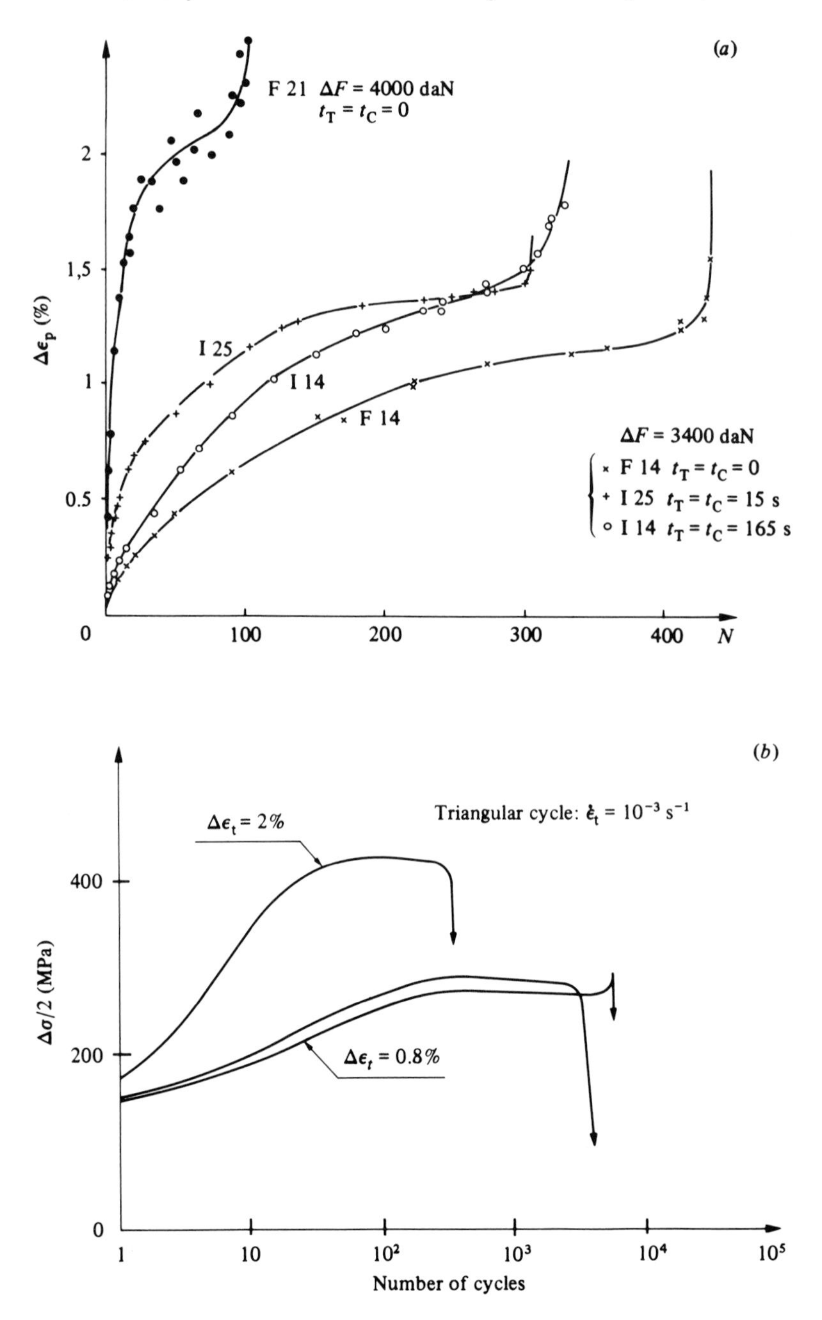

determined at temperature T_1:

$$K = \frac{K_0^{T_0/T_1}}{K_1^{(T_1/T)(T-T_0)/(T_0-T_1)}};$$

if $|(T - T_0)/T| > 5\%$, it is necessary to introduce three functions of temperature $K(T)$, $M(T)$ and $N(T)$. Fig. 6.13 shows the form of their variation for a particular example.

6.2.6 *Results derived from cyclic tests*

The results of cyclic tests within controlled stress or strain ranges on metals at temperatures which cause noticeable creep, show the same phenomena of cyclic hardening and softening as observed in pure plasticity at lower temperatures. Fig. 6.14(*a*) shows an example of the evolution of the plastic strain range as a function of the number of cycles in stress controlled tests of a material which softens cyclically. A material which exhibits cyclic hardening at room temperature shows similar effects at a higher temperature (for example 316 steel in strain controlled tests, Fig. 6.14(*b*).

Two kinds of time effects may be observed in cyclic tests:

(a) the influence of the material's viscosity is detected by cyclic creep tests (with equal holding times in tension and compression in stress-controlled tests) (Fig. 6.15(*a*) shows an example at the stabilized cycle), cyclic relaxation tests (keeping the strain constant, generally in tension as indicated in Fig. 6.16), and tests with different frequencies (as shown in Fig. 6.15(*b*)).

In each case the plastic strain, for the same stress, has more time to develop. Consequently, the cyclic curve (stress amplitude–strain amplitude) is no longer unique as it is in plasticity; it depends on the frequency and on the hold time (e.g., 316 steel in Fig. 6.17).

For a material which exhibits strong cyclic hardening, we may observe an increase in the maximum stress corresponding to an apparent increase of the elastic domain (Fig. 6.16(*a*)) during the cycling and an increase in the relaxed stress during the hold time;

(b) the time-recovery effects which can be observed indirectly either by unloading in creep, or in the cyclic tests themselves. In the former, after complete unloading, we may measure a partial recovery of the viscoplastic strain, after a sufficiently long time (this is the recovery test illustrated schematically in Fig. 6.18(*a*)). For a partial unloading in creep by $\delta\sigma$ we may observe the phenomenon of creep delay

Fig. 6.15. (*a*) A stabilized cycle in cyclic creep tests on IN 100 alloy at 800 °C. (*b*) The effect of the strain rate on stabilized cycles of Hastelloy-X alloy at 871 °C (after Walker).

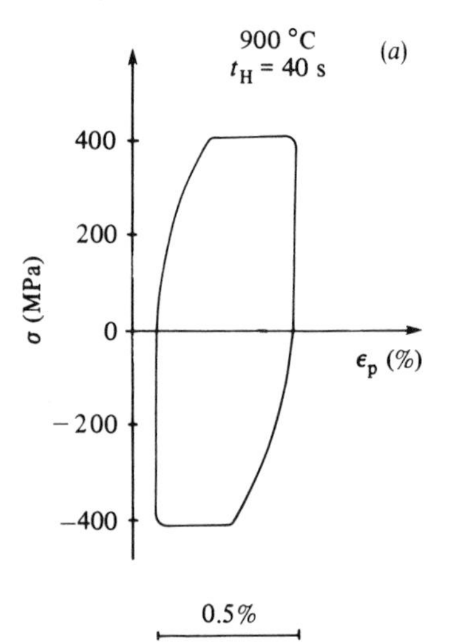

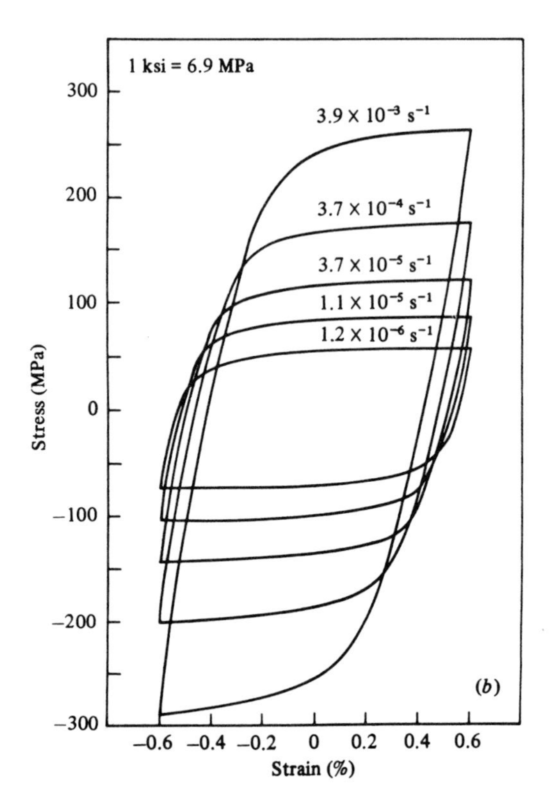

shown schematically in Fig. 6.18(*b*): the strain rate, initially zero during an incubation time increases progressively until it reaches the nominal rate (under the stress $\sigma - \delta\sigma$). Note that these tests may be used to prove the existence of an internal stress which is such that the strain rate is proportional to $\sigma - \sigma_i$:

$$\dot{\varepsilon}_p = f(\sigma, \ldots)(\sigma - \sigma_i)$$

and to quantify the time-recovery effect of this internal stress, which corresponds to a slow restoration of the crystalline structure. For cyclic tests with controlled strain, the influence of the recovery effect diminishes the stress amplitude at the stabilized cycle when the hold time is increased (see the example of 316 stainless steel in Fig. 6.16(*b*)). The recovery effect, which becomes more important as the duration of the test increases, partly compensates the cyclic strain-hardening.

The different observations presented in this section are only examples, as many complex effects can interact and sometimes lead to contradictory results: in the high-temperature regime, metals are sometimes imperfectly stable, which complicates the study of their mechanical behaviour, because then time, which characterizes the age, becomes a state variable (see Section 6.4.4).

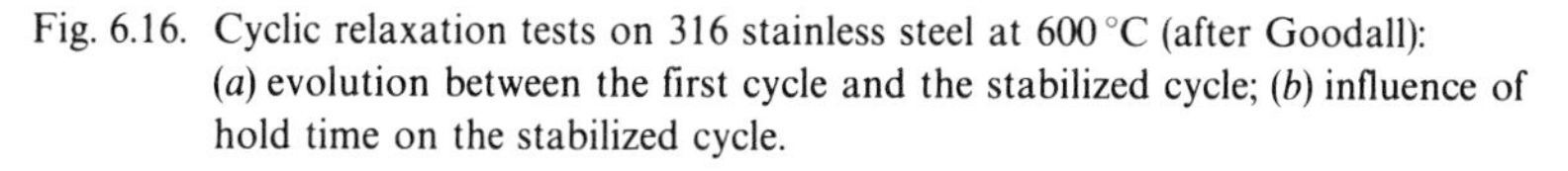

Fig. 6.16. Cyclic relaxation tests on 316 stainless steel at 600 °C (after Goodall): (*a*) evolution between the first cycle and the stabilized cycle; (*b*) influence of hold time on the stabilized cycle.

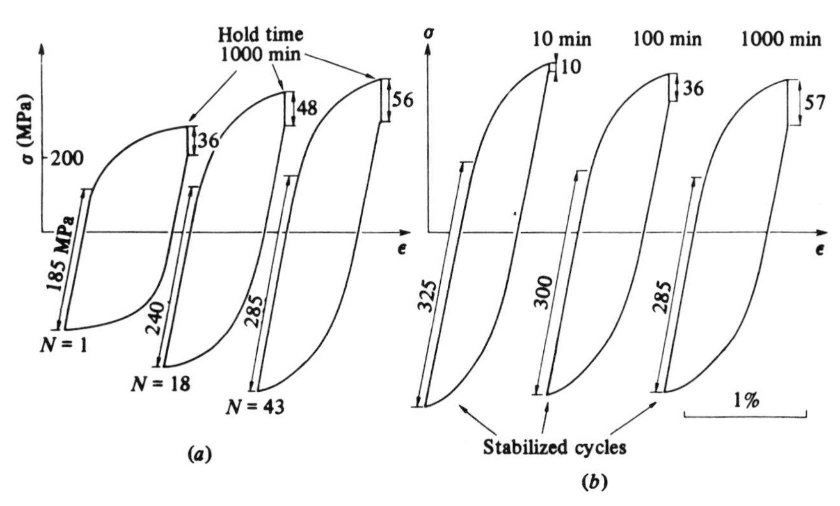

6.2.7 *Results derived from multiaxial tests*

Unfortunately, because of the difficulties of complex loadings and the problems associated with the use of average and high temperatures only a few multiaxial test results exist. The few results from tension–torsion–internal pressure tests performed on tubular samples, or biaxial tension

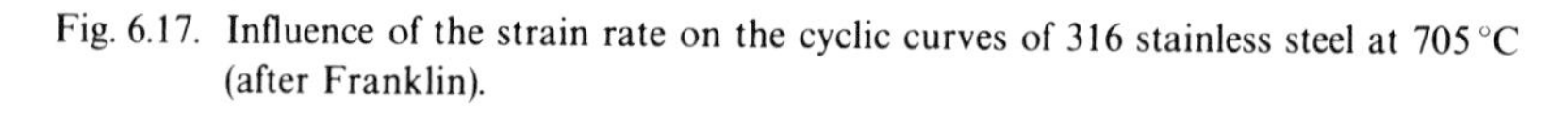

Fig. 6.17. Influence of the strain rate on the cyclic curves of 316 stainless steel at 705 °C (after Franklin).

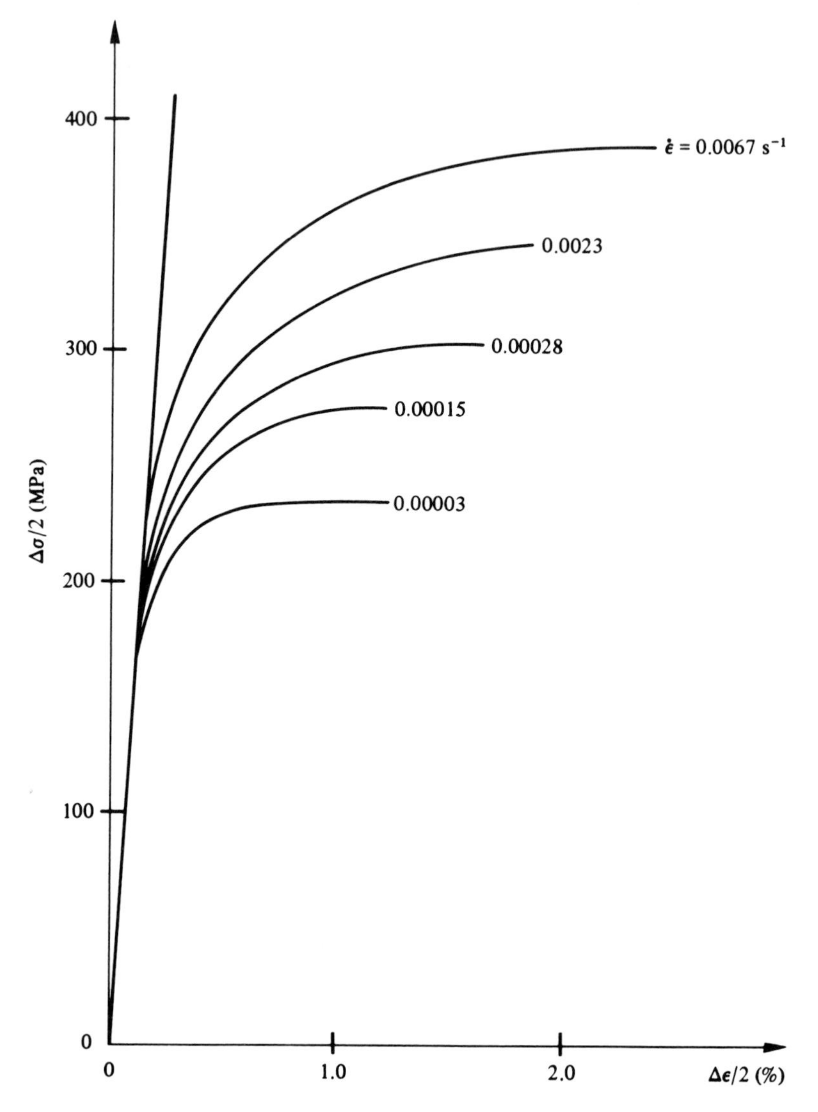

tests on cruciform specimens show that in radial loading the concepts of equivalent stress and equivalent strain in the sense of the von Mises criterion can be used to define in the stress space, surfaces of equal strain rates for a given strain.

The assumption of viscoplastic incompressibility is generally well confirmed by experiments on metals and alloys. Similarly, the normality rule is well satisfied and some experiments have demonstrated the major influence of the second invariant of the deviatoric stress tensor (the third invariant plays only a minor role).

The initial equipotential surfaces generally closely satisfy the von Mises criterion. Moreover, creep tests for different states of stress $(\sigma_{11}, \sigma_{12})$, after an initial creep, show the essentially kinematic-hardening character of the

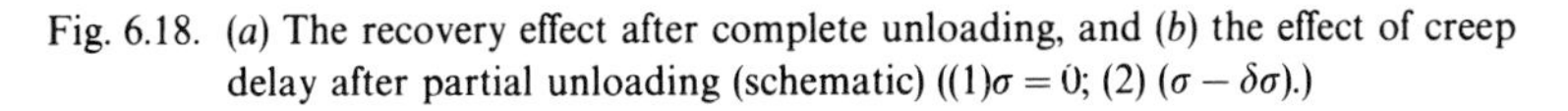

Fig. 6.18. (*a*) The recovery effect after complete unloading, and (*b*) the effect of creep delay after partial unloading (schematic) ((1)$\sigma = 0$; (2) $(\sigma - \delta\sigma)$.)

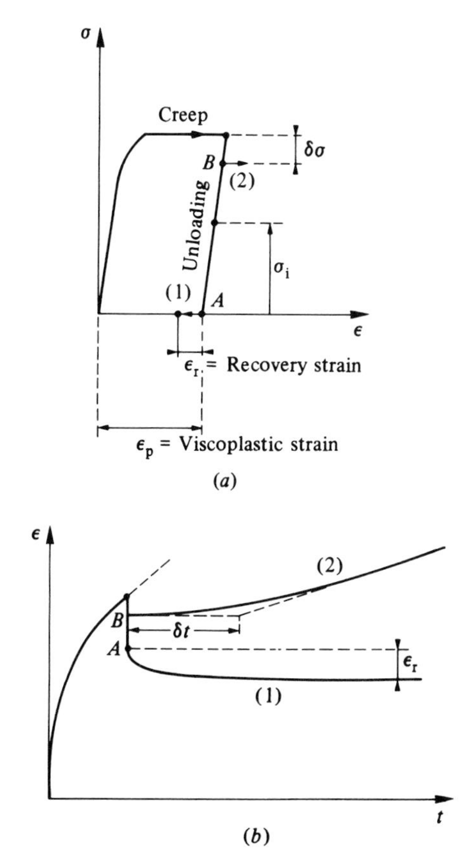

material, even at high temperatures. Fig. 6.19 enables us to observe this tendency from the measurement of the direction of the flow ($\dot{\varepsilon}_{p11}, \dot{\varepsilon}_{p12}$). The magnitude of the strain rate is, of course, larger for stress states farther from the centre of the surfaces (initial stress or kinematic stress). Thus, these multiaxial tests confirm the observations already made regarding the existence of an internal stress for creep under uniaxial tension.

6.3 General formulation of the constitutive equations

The formalism used for viscoplasticity is the same as that used for plasticity in Chapter 5. We will therefore restrict ourselves here to a summary of what was presented in the Chapter 5, emphasizing only the new aspects related to the effect of time.

6.3.1 *Partition of strains*

Here again, we use the hypothesis of the partition of total strain into an elastic (reversible) strain and an inelastic strain. For small strains:

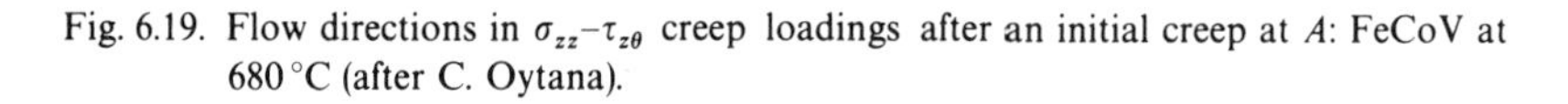

$$\boldsymbol{\varepsilon} = \boldsymbol{\varepsilon}^{e} + \boldsymbol{\varepsilon}^{in}.$$

Inelastic (non-instantaneously reversible) strain corresponds to the following set of physical phenomena:

Fig. 6.19. Flow directions in σ_{zz}–$\tau_{z\theta}$ creep loadings after an initial creep at A: FeCoV at 680 °C (after C. Oytana).

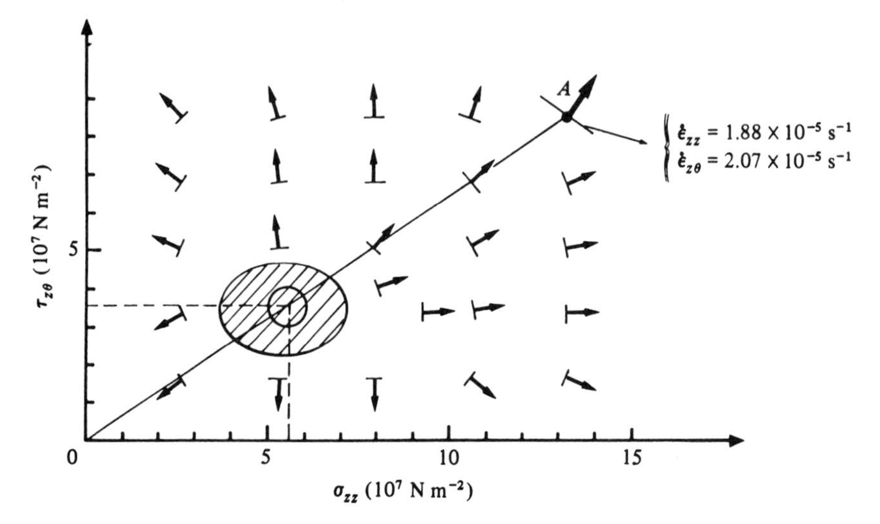

an instantaneous plastic strain, independent of time,
a viscoplastic strain (sometimes qualified as creep strain),
an anelastic strain, or that due to delayed elasticity.

The first type of inelastic strain, already studied in Chapter 5 will be assumed to be nonexistent in the framework of the present treatment. How it is taken into account will, however, be referred to in Section 6.4.4, although for most materials and for a broad range of temperature, the theory of viscoplasticity alone will be sufficient to describe the set of experimental observations.

Anelastic strain can be treated in the framework of viscoelasticity (Chapter 4). It is difficult to detect in metals and alloys, and therefore it will not be studied here. On the other hand, the effects of partial recovery of plastic strain (which is what has sometimes justified the introduction of anelasticity) may be described by a proper choice of internal variables.

In accordance with the above remarks, therefore only one irreversible strain is considered, and the hypothesis of strain partitioning may be written as:

$$\boldsymbol{\varepsilon} = \boldsymbol{\varepsilon}^{e} + \boldsymbol{\varepsilon}^{p}$$

where $\boldsymbol{\varepsilon}^{p}$ is the viscoplastic strain.

6.3.2 *Choice of thermodynamic variables*

The thermodynamic or independent variables are the observable variables, the total strain $\boldsymbol{\varepsilon}$ and the temperature T, and the internal variables which describe the current state of the material: $\boldsymbol{\varepsilon}^{p}$, V_{k}. The general treatment presented in Chapter 2 shows that the free energy depends on the elastic strain, temperature, and the internal variables V_{k}:

$$\Psi = \Psi(\boldsymbol{\varepsilon}^{e}, T, V_{k}).$$

Stress and entropy are the thermodynamic variables associated to $\boldsymbol{\varepsilon}^{e}$ and T (ρ is the mass density):

$$\boldsymbol{\sigma} = \rho(\partial\Psi/\partial\boldsymbol{\varepsilon}^{e}), \qquad s = -\partial\Psi/\partial T.$$

As for time-independent plasticity, an isotropic hardening variable, scalar in nature and denoted by r, will be used as an internal variable. As a particular case we may take this variable to be the accumulated plastic strain p:

$$p = \int_{0}^{t} (\tfrac{2}{3}\dot{\boldsymbol{\varepsilon}}^{p}(\tau):\dot{\boldsymbol{\varepsilon}}^{p}(\tau))^{1/2}\,\mathrm{d}\tau.$$

We also use hardening variables which are tensorial in nature, called kinematic variables, such as the variable α introduced earlier.

From a physical point of view, these hardening variables are associated with, on one hand, the dislocation density (necessarily isotropic) and, on the other, the incompatibility of various viscoplastic strains within the polycrystal. The isotropic hardening variable r will generally be increasing, unless a possible recovery takes place (then r is no longer exactly the accumulated plastic strain p). For cyclic loading, the kinematic variable $\boldsymbol{\alpha}$ will, on the contrary, not have a continuous evolution.

To these hardening variables, other internal variables may be added to describe the current state of the microstructure. These variables can be made to appear in the macroscopic description by introducing the difference between the normal, metastable state of the material and the state which exists during microstructural instability. Of course, for physico-chemically stable materials, such variables do not appear.

Here again, we assume decoupling between the elastic behaviour and hardening, with the specific free energy being decomposed as:

$$\Psi = \Psi_{\mathrm{e}}(\boldsymbol{\varepsilon}^{\mathrm{e}}, T) + \Psi_{\mathrm{p}}(\boldsymbol{\alpha}, r, m_{\mathrm{k}}, T)$$

and the thermodynamic variables associated with the hardening variables are

$$R = \rho(\partial\Psi/\partial r), \qquad \mathbf{X} = \rho(\partial\Psi/\partial\boldsymbol{\alpha}).$$

These represent, on the one hand, the size of the existing elasticity domain (or the size of the equipotentials of dissipation), and, on the other the centre of this domain, as indicated schematically in Fig. 5.17. Moreover, we note:

$$M_{\mathrm{k}} = \rho(\partial\Psi/\partial m_{\mathrm{k}}).$$

As in plasticity, we adopt the hypothesis of decoupling between thermal and intrinsic dissipations. The second principle of thermodynamics then reduces to:

$$\Phi_1 = \boldsymbol{\sigma}:\dot{\boldsymbol{\varepsilon}}^{\mathrm{p}} - \mathbf{X}:\dot{\boldsymbol{\alpha}} - R\dot{r} - M_{\mathrm{k}}\dot{m}_{\mathrm{k}} \geqslant 0.$$

where the symbol : denotes the tensorial product contracted on two indices.

6.3.3 ***Dissipation potential***

Normality laws

The concept of a loading surface and the loading–unloading criterion which was used in plasticity is no longer necessary. It is replaced by a family of equipotential surfaces. These are the surfaces in stress space at each point of which the magnitude of the strain rate is the same (i.e., the dissipation is the same). The surface of zero potential is the surface delimiting the elastic domain. This leads us again to Fig. 5.18, in which the surface of infinite potential is pushed to infinity: instantaneous plastic strains are considered to be negligible.

The generalization of the concept of equipotential surfaces to all the variables leads to the definition of the dissipation potential which, expressed in the dual form, is:

$$\varphi^* = \varphi^*(\boldsymbol{\sigma}, \mathbf{X}, R, M_k; T, \boldsymbol{\alpha}, r, m_k)$$

where the state variables T, $\boldsymbol{\alpha}$, r and m_k are only parameters. As in Chapter 2, we deduce from this the normality laws of generalized standard materials:

$$\dot{\boldsymbol{\varepsilon}}^p = \partial\varphi^*/\partial\boldsymbol{\sigma}$$
$$\dot{\boldsymbol{\alpha}} = -\,\partial\varphi^*/\partial\mathbf{X}$$
$$\dot{r} = -\,\partial\varphi^*/\partial R$$
$$\dot{m}_k = -\,\partial\varphi^*/\partial M_k.$$

The intrinsic dissipation can now be expressed as:

$$\Phi_1 = \boldsymbol{\sigma}:\frac{\partial\varphi^*}{\partial\boldsymbol{\sigma}} + \mathbf{X}:\frac{\partial\varphi^*}{\partial\mathbf{X}} + R\frac{\partial\varphi^*}{\partial R} + M_k\frac{\partial\varphi^*}{\partial M_k}.$$

We assume that the dissipation potential satisfies the following:

> φ^* is a convex function of the components of each of the dependent variables $\boldsymbol{\sigma}$, $\mathbf{X}$, R, M_k;
> φ^* is always positive and contains the origin, where $\sigma_{ij} = X_{ij} = R = M_k = 0$ and it assumes the zero value: $\varphi^*(0, 0, 0, 0; T, \boldsymbol{\alpha}, r, m_k) = 0$ regardless of T, $\boldsymbol{\alpha}$, r, m_k.

Under these two conditions, the mechanical dissipation is such that (cf. Chapter 2):

$$\Phi_1 \geqslant \varphi^* \geqslant 0$$

which allows us to satisfy the second principle of thermodynamics.

Specialization of the potential

In practice, the viscoplasticity laws, that can eventually include recovery effects, use a dissipation potential that can be decomposed as:

$$\varphi^* = \Omega_p + \Omega_r$$
$$\Omega_p = \Omega_p(J(\boldsymbol{\sigma} - \mathbf{X}) - R - k, \mathbf{X}, R, M_k; T, \boldsymbol{\alpha}, r, m_k)$$
$$\Omega_r = \Omega_r(\mathbf{X}, R, M_k; T, \boldsymbol{\alpha}, r, m_k).$$

where Ω_p denotes the equipotential surfaces in the stress space. These may possibly depend on $\mathbf{X}$, R and the variables M_k; Ω_r is the recovery potential. It is present even when $\Omega_p = 0$.

$J(.)$ denotes a norm in the stress space with $J(\boldsymbol{\sigma} - \mathbf{X}) - R - k = 0$ corresponding, as in plasticity, to the extent of the existing elastic domain. The convexity of Ω_p is then necessary only *vis-a-vis* the quantities $J(\boldsymbol{\sigma} - \mathbf{X}) - R - k$, X_{ij}, R, M_k, and that of Ω_r *vis-a-vis* X_{ij}, R, M_k.

It should be noted that, in the above decomposition, the normality rule is necessarily implied:

$$\dot{\boldsymbol{\varepsilon}}^p = \frac{\partial \Omega_p}{\partial J} \frac{\partial J}{\partial \boldsymbol{\sigma}}$$

$$\dot{r} = -\frac{\partial \Omega_p}{\partial J} - \frac{\partial \Omega_r}{\partial R}.$$

In the particular case of a von Mises material and when the isotropic variable R appears only in the expression

$$J - R - k = J(\boldsymbol{\sigma} - \mathbf{X}) - R - k = [\tfrac{3}{2}(\boldsymbol{\sigma}' - \mathbf{X}'):(\boldsymbol{\sigma}' - \mathbf{X}')]^{1/2} - R - k$$

where $\boldsymbol{\sigma}'$ and $\mathbf{X}'$ are the deviators of $\boldsymbol{\sigma}$ and $\mathbf{X}$, we find:

$$\dot{\boldsymbol{\varepsilon}}^p = \frac{3}{2} \frac{\partial \Omega_p}{\partial J} \frac{\boldsymbol{\sigma}' - \mathbf{X}'}{J} = \frac{3}{2} \dot{p} \frac{\boldsymbol{\sigma}' - \mathbf{X}'}{J}$$

$$\dot{r} = \dot{p} - \frac{\partial \Omega_r}{\partial R}.$$

Here p is, of course, the accumulated plastic strain since:

$$\dot{p} = (\tfrac{2}{3} \dot{\boldsymbol{\varepsilon}}^p : \dot{\boldsymbol{\varepsilon}}^p)^{1/2}$$

and the internal variable r is identifiable with p when recovery effects are negligible ($\Omega_r = 0$).

6.4 Particular constitutive equations

A number of laws are presented in the following sections in order of increasing complexity, which corresponds to increasing possibilities of modelling. For each type of equation, we present a general formulation, the particular case of uniaxial loading, and the process of identification from experiments. We also give some characteristic results concerning different metallic materials. These equations are introduced in the framework of the general formulation. The potential φ^* is usually derived from the visco-plastic potential $\Omega = \Omega_p$. Note that this potential could be employed in the stress space, without invoking the generalized normality hypothesis.

6.4.1 *Laws of perfect viscoplasticity*

Multiaxial formulation

General form

The law of perfect viscoplasticity is the multiaxial generalization of the relation between the viscoplastic strain rate and the stress established in tension for secondary creep (Section 6.2.2). It is presented here in the framework of isotropic materials in which viscoplastic strains occur at constant volume.

Perfect viscoplasticity corresponds to the case of a material which has no hardening effect (no primary creep): no internal variable is therefore used. Following the general formalism of the preceding section we define a dissipation potential:

$$\varphi^* = \Omega = \Omega(\boldsymbol{\sigma}; T)$$

which represents a family of equipotential surfaces in the stress space. The assumption of incompressibility and isotropy implies dependence only on J_2 and J_3, so that:

$$\Omega(J_2(\boldsymbol{\sigma}), J_3(\boldsymbol{\sigma}); T).$$

In general, we neglect the influence of the third invariant. This is partly justified by experiments and results in an additional simplification. We are thus left with a viscoplastic material which is an analogue of a plastic material which obeys the von Mises criterion (the equipotentials are cylinders with axes equally inclined to the σ_1, σ_2, σ_3 axes in the space of principal stresses):

$$\Omega = \Omega(\sigma_{eq}, T) \quad \text{with} \quad \sigma_{eq} = J_2(\boldsymbol{\sigma}) = (\tfrac{3}{2}\boldsymbol{\sigma}':\boldsymbol{\sigma}')^{1/2}$$

where $\boldsymbol{\sigma}'$ is, as may be recalled, the deviator of $\boldsymbol{\sigma}$.

The law of viscoplasticity is completely defined by the knowledge of this potential since the normality rule gives:

$$\dot{\boldsymbol{\varepsilon}}^{\mathrm{p}} = \frac{\partial \Omega}{\partial \boldsymbol{\sigma}} = \frac{3}{2}\frac{\partial \Omega}{\partial \sigma_{\mathrm{eq}}}\frac{\boldsymbol{\sigma}'}{\sigma_{\mathrm{eq}}}.$$

If we wish to take into account purely elastic behaviour, the potential to be used is of the form:

$$\Omega = \Omega^*(\langle \sigma_{\mathrm{eq}} - k(T) \rangle)$$

where Ω^* assumes a zero value when $\langle \sigma_{\mathrm{eq}} - k(T) \rangle$ is 0, keeping in mind the convention that $\langle x \rangle = x$ if $x > 0$ and $\langle x \rangle = 0$ if $x \leqslant 0$. The elastic limit depends on the temperature. The viscoplastic potential Ω is completely defined by knowledge of the uniaxial secondary creep equation (see p. 283).

Odqvist's law

Odqvist's law is a direct generalization of Norton's law by considering the elastic domain to be negligible, reduced to a point at the origin.

$$\Omega = \frac{\lambda^*}{N^* + 1}\left(\frac{\sigma_{\mathrm{eq}}}{\lambda^*}\right)^{N^* + 1},$$

$$\dot{\boldsymbol{\varepsilon}}^{\mathrm{p}} = \frac{3}{2}\left(\frac{\sigma_{\mathrm{eq}}}{\lambda^*}\right)^{N^*}\frac{\boldsymbol{\sigma}'}{\sigma_{\mathrm{eq}}}.$$

This law of perfect viscoplasticity is very often used for the following reasons:

its simplicity of use in numerical calculations by virtue of the absence of hardening and the elastic domain;

its immediate incorporation in the framework of generalized standard materials which allows proofs of existence and uniqueness theorems, and helps to obtain simple analytical solutions (elastic analogy, upper and lower bounds, limit analysis, viscoplastic buckling etc.). In this sense, for viscoplasticity, it is the equivalent of the law of perfect plasticity.

Equation with elastic domain

If a power function approximation is satisfactory, we may, for example, express the potential in the form:

$$\Omega = \frac{\lambda_{\mathrm{Y}}}{N_{\mathrm{Y}} + 1}\left\langle \frac{\sigma_{\mathrm{eq}} - k}{\lambda_{\mathrm{Y}}} \right\rangle^{N_{\mathrm{Y}} + 1}$$

where N_{Y} and λ_{Y} are coefficients which depend on the temperature. The viscoplastic strain rate can then be written as:

$$\bullet \qquad \dot{\boldsymbol{\varepsilon}}^{\mathrm{p}} = \frac{3}{2}\left\langle \frac{\sigma_{\mathrm{eq}} - k}{\lambda_{\mathrm{Y}}} \right\rangle^{N_{\mathrm{Y}}} \frac{\boldsymbol{\sigma}'}{\sigma_{\mathrm{eq}}}.$$

Note that this relation is of interest only when the specific exponent of Norton's law has a tendency to increase with low stresses (see below).

Equation with an exponential term

This equation enlarges the domain of validity of the law of perfect viscoplasticity when the exponent of Norton's law decreases with low stresses (see Fig. 6.5). The dissipation potential is:

$$\Omega = \frac{\exp(\alpha_0 \sigma_{\mathrm{eq}}^{N_0 + 1})}{\alpha_0 (N_0 + 1) \lambda_0^{N_0}}$$

and this yields the plastic strain rate:

$$\bullet \qquad \dot{\boldsymbol{\varepsilon}}^{\mathrm{p}} = \frac{3}{2}\left(\frac{\sigma_{\mathrm{eq}}}{\lambda_0} \right)^{N_0} \exp(\alpha_0 \sigma_{\mathrm{eq}}^{N_0 + 1}) \frac{\boldsymbol{\sigma}'}{\sigma_{\mathrm{eq}}}$$

in which the material is represented by the three coefficients λ_0, N_0 and α_0.

Identification from tension tests.

In a uniaxial state of stress, under a tensile stress σ, it is obvious that $\sigma_{\mathrm{eq}} = \sigma$ and the rate of strain $\dot{\varepsilon}_{\mathrm{p}11} = \dot{\varepsilon}_{\mathrm{p}}$ can be obtained immediately from the multiaxial law. For example, for the three preceding laws, we obtain:

$$\dot{\varepsilon}_{\mathrm{p}} = (\sigma/\lambda^*)^{N^*}$$

$$\dot{\varepsilon}_{\mathrm{p}} = \left\langle \frac{\sigma - k}{\lambda_{\mathrm{Y}}} \right\rangle^{N_{\mathrm{Y}}}$$

$$\dot{\varepsilon}_{\mathrm{p}} = (\sigma/\lambda_0)^{N_0} \exp(\alpha_0 \sigma^{N_0 + 1}).$$

In these three cases the law is completely identified by the results of secondary creep: two interpretations of secondary creep are possible as shown schematically in Fig. 6.20:

The secondary creep rate is defined by the minimum value of the rate during the creep test (the tangent at the inflexion point of the curve). In this case the strain due to primary creep, represented by the ordinate at the origin $\varepsilon_{\mathrm{po}}$, is neglected.

The creep rate is given by the straight line passing through the origin ($\varepsilon_{\mathrm{po}} = 0$) and the point of inflexion of the curve. This second approximation takes the primary creep into account, but only roughly.

For the first equation (Norton's law) the two coefficients λ^* and N^* can be obtained from the graph ($\ln \sigma$ versus $\ln \dot{\varepsilon}_{\mathrm{p}}$) by hand fitting (as in the examples of Fig. 6.4) or by numerical means (by the linear least squares method described in Chapter 3). The values obtained for the exponent N^* with the above two definitions of the secondary creep rate are practically identical; only the coefficient λ^* differs.

For the second equation, the coefficients N_{Y}, λ_{Y}, k are obtained by a process of successive approximations for the choice of the elastic limit k, by including, for example, the values of the pairs ($\ln(\sigma - k)$, $\ln \dot{\varepsilon}_{\mathrm{p}}$). A suitable criterion consists in obtaining N_{Y} as constant as possible (see Sections 6.2.2 and 6.4.2). Note that the elastic limit k always has a low value, much smaller than that which is observed in hardening tests; this is due to two factors:

the large nonlinearity of the viscoplastic phenomenon;
the possibility of viscoplastic flows (which is slow but not necessarily negligible) under low stresses, as, for example, during a relaxation test.

Fig. 6.20. Two schematic representations of secondary creep.

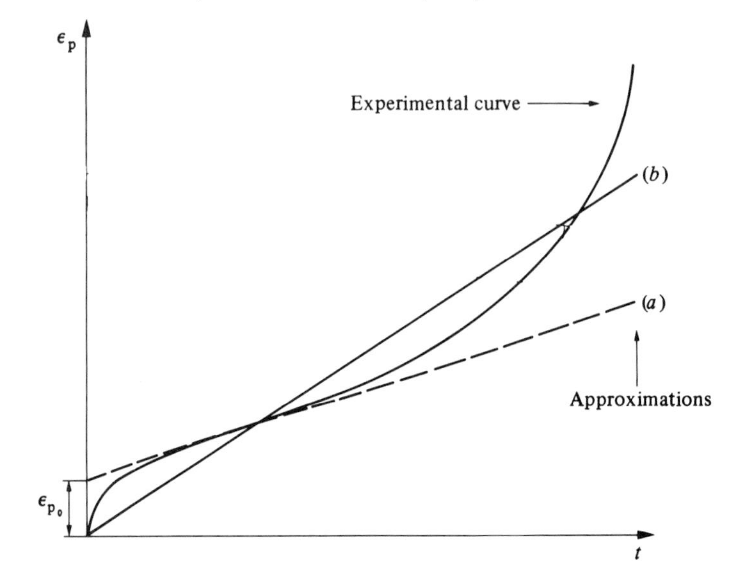

Fig. 6.21. Identification of the equations with an exponential term for a magnesium alloy at 260 °C: (*a*) secondary creep points; (*b*) apparent exponent of the secondary creep.

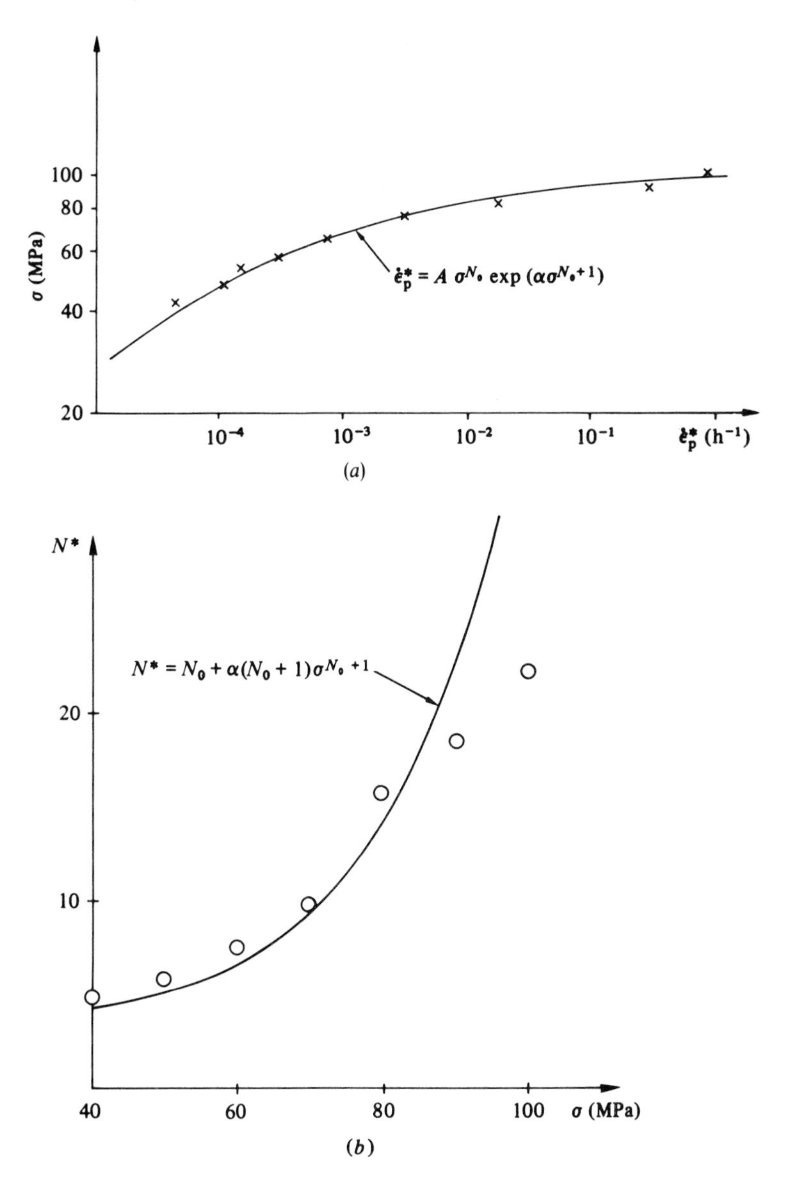

We also note that the introduction of an elastic domain leads to a decrease in the value of the exponent $N (N < N^*)$.

Identification of the third equation can be performed by numerical means starting from measured values of σ and $\dot{\varepsilon}_p$ (using the nonlinear least squares method described in Chapter 3). Another procedure consists in the identification of the variation of Norton's exponent as a function of the stress. This exponent is determined by the tangent to the curve of Fig. 6.21(*a*). Its evolution for this example is shown in Fig. 6.21(*b*). The variation of the equivalent exponent can be established by using the third equation to give:

$$N^*(\sigma) = \mathrm{d}\ln\dot{\varepsilon}_p/\mathrm{d}\ln\sigma = N_0 + \alpha(N_0 + 1)\sigma^{N_0+1}.$$

The value of N_0 is easily determined by successive approximations; for example, the curve of Fig. 6.21(*b*) is obtained with $N_0 = 4$ and $\alpha = 65 \times 10^{-9}$. The improvement in the relation between σ and $\dot{\varepsilon}_p$ is noticeable when these values are used (together with $\lambda_0 = 86.3$).

Domain of validity

We have already presented in the previous sections a number of results concerning Norton's law and its generalization. Some values characteristic of Norton's law are shown on Table 6.1. Fig. 6.6 shows the evolution of Norton's exponent as a function of temperature for some materials.

The domain of validity of Norton's law is limited by:

the possibility of approximating the creep curves by a straight line: the law is clearly a better approximation for materials in which the strain due to primary creep is small in comparison to the strain during secondary creep;

the relation obtained experimentally between σ and $\dot{\varepsilon}_p^*$: depending on the material (and the temperature), the power function approximation is valid in a quite wide range of strain rate;

the temperature range itself: for low temperatures, two problems generally arise:

a low viscosity which leads to very high values of N^*;

a creep with a somewhat logarithmic character, that is without real secondary creep.

The latter limitation, which is undoubtedly the most important, is related to the loading history: Norton's law is derived essentially through a creep test (under constant stress); since it does not take hardening effects into account, it gives only a rough approximation to the tension curves. The example using IN 100 in Fig. 6.22 illustrates this fact well by comparing experimental curves with calculated ones, using the coefficients N^* and λ^* determined by the second interpretation of secondary creep. Moreover, Norton's law does not take into account either cyclic tests or effects related to recovery.

The chief advantage of Norton's law lies in the convenience with which analytical integration may be performed. For structural analysis, this amounts to the possibility of carrying out simplified analyses. This convenience can be illustrated by the following example of relaxation in tension:

$$\dot{\varepsilon} = \dot{\varepsilon}_e + \dot{\varepsilon}_p = \dot{\sigma}/E + (\sigma/\lambda^*)^{N^*} = 0.$$

The integration of the differential equation, without the second term, is easily performed to give:

$$\sigma = \sigma_0 \left[1 + (N^* - 1) \frac{E}{\lambda^*} \left(\frac{\sigma_0}{\lambda^*} \right)^{N^* - 1} t \right]^{-1/(N^* - 1)}$$

where σ_0 is the stress at the start of relaxation.

Fig. 6.22. Hardening followed by relaxation of the refractory alloy IN 100: calculation by Norton's law.

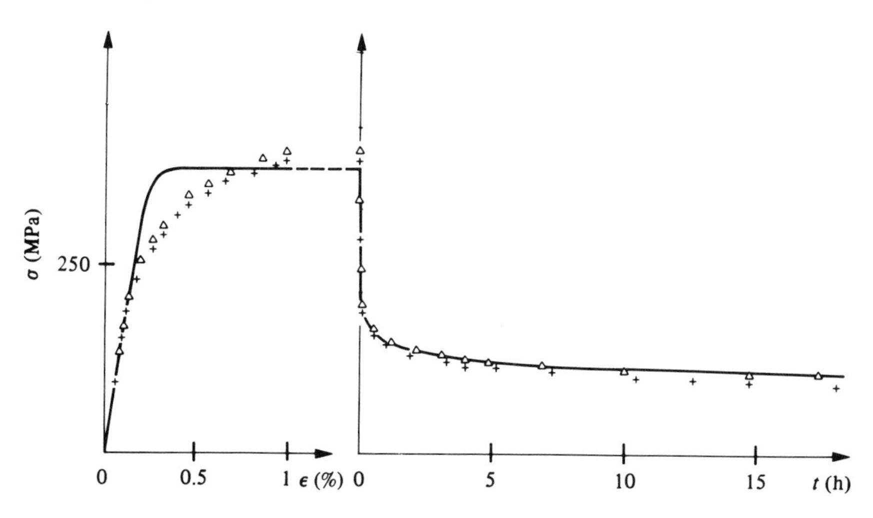

6.4.2 *Viscoplasticity laws with isotropic hardening*

Multiaxial formulation

General form

For isotropic hardening, the only internal variable is the scalar variable r: it represents the current hardening state of the material and is equal to the accumulated plastic strain p in the absence of recovery processes. The specific free energy depends therefore on $\boldsymbol{\varepsilon}^e$, T and r. The thermodynamic force associated with r is

$$R = \rho(\partial\Psi/\partial r)$$

and the dissipation potential depends on the stress tensor, the internal stress R, and ultimately on the state variable r which acts as a parameter:

$$\varphi^* = \Omega = \Omega(\boldsymbol{\sigma}, R; T, r).$$

For metals, the assumption of incompressibility and isotropy means that the stress appears through the invariants J_2 and J_3 of its deviator:

$$\Omega = \Omega(J_2(\boldsymbol{\sigma}), J_3(\boldsymbol{\sigma}), R; T, r)$$

and since we usually neglect the influence of the third invariant (von Mises material):

$$\Omega = \Omega(\sigma_{eq}, R; T, r).$$

The variable R is associated with the size of the current elastic domain. We may then express the dissipation potential as a function of $\sigma_{eq} - R - k$ where k is the initial elastic limit. Moreover, we can independently introduce a flow potential (with hardening) and a recovery potential:

$$\Omega = \Omega_p(\sigma_{eq} - R - k; T, r) + \Omega_r(R; T).$$

The function Ω_p takes the value zero when $\sigma_{eq} - R - k$ does, and Ω_r becomes zero with R. The generalized normality law then yields the following equations:

$$\dot{\boldsymbol{\varepsilon}}^p = \frac{\partial\Omega}{\partial\boldsymbol{\sigma}} = \frac{3}{2}\frac{\partial\Omega_p}{\partial\sigma_{eq}}\frac{\boldsymbol{\sigma}'}{\sigma_{eq}} = \frac{3}{2}\dot{p}\frac{\boldsymbol{\sigma}'}{\sigma_{eq}}$$

$$\dot{r} = -\frac{\partial\Omega}{\partial R} = \frac{\partial\Omega_p}{\partial\sigma_{eq}} - \frac{\partial\Omega_r}{\partial R} = \dot{p} - \frac{\partial\Omega_r}{\partial R}$$

where $\dot{p}$ is the invariant of the plastic strain rate:

$$\dot{p} = \left(\tfrac{2}{3}\dot{\boldsymbol{\varepsilon}}^p:\dot{\boldsymbol{\varepsilon}}^p\right)^{1/2}.$$

Thus the hardening and eventual recovery effects are clearly apparent in the evolution of the variable r.

Knowledge of the three potentials Ψ, Ω_p and Ω_r completely defines the viscoplasticity law with hardening and recovery. In the following subsections we study two classical forms of the law which represent hardening, while neglecting the recovery effect ($\Omega_r \equiv 0$).

Additive viscosity-hardening law

Hardening takes place with a decrease in the effective stress and the variable r is absent from the potential Ω_p. The potential is often expressed in the form of a power function:

$$\Omega = \Omega_p = \frac{K_a}{N_a + 1}\left\langle \frac{\sigma_{eq} - R - k}{K_a} \right\rangle^{N_a + 1}$$

which yields a strain rate defined by:

- $$\dot{\boldsymbol{\varepsilon}}^p = \frac{3}{2}\dot{p}\frac{\boldsymbol{\sigma}'}{\sigma_{eq}}$$

where

- $$\dot{p} = \dot{r} = \left\langle \frac{\sigma_{eq} - R - k}{K_a} \right\rangle^{N_a}.$$

The relation between R and p must be chosen so as to approximate the hardening and the primary creep curves correctly. A simple relation uses a linear term and an exponential term:

$$R = Q_1 p + Q_2[1 - \exp(-bp)]$$

or, in the differential form:

- $$\dot{R} = b(Q_1 p + Q_2 - R)\dot{p} + Q_1\dot{p}.$$

This choice corresponds to an expression of the free energy in the form:

- $$\rho\Psi_p = \tfrac{1}{2}Q_1 p^2 + Q_2\left[p + \frac{1}{b}\exp(-bp) - \frac{1}{b}\right].$$

The coefficients N_a, K_a, k, Q_1, Q_2 and b may depend on temperature. This type of law, although it directly belongs to the formalism of standard materials, is rarely used. Instead, the product form, which is easier to use and identify, is preferred.

Multiplicative viscosity-hardening law

Here, we no longer have a strictly standard law since it is necessary to use the variable r as a parameter (in the case under consideration we have $r = p$). We assume that the dissipation potential can be expressed in the form:

$$\Omega = \Omega_p(\sigma_{eq} - R - k + h'(p))\zeta(p)$$

with the free energy written as:

$$\rho\Psi = \rho\Psi_e + h(p).$$

Usually power functions are chosen to represent Ω_p and ζ:

$$\Omega_p = \frac{K}{N+1}\left\langle \frac{\sigma_{eq} - R - k + h'(p)}{K} \right\rangle^{N+1} p^{\gamma}$$

where the coefficients N, K, k and γ, as well as the function h, may depend on temperature.

The strain rate is expressed by:

$$\dot{\boldsymbol{\varepsilon}}^p = \frac{\partial \Omega}{\partial \boldsymbol{\sigma}} = \frac{3}{2}\dot{p}\frac{\boldsymbol{\sigma}'}{\sigma_{eq}}$$

$$\dot{p} = -\frac{\partial \Omega}{\partial R} = \left\langle \frac{\sigma_{eq} - R - k + h'(p)}{K} \right\rangle^{N} p^{\gamma}.$$

Taking into account the fact that

$$R = \rho(\partial \Psi / \partial p) = h'(p)$$

we may eliminate R from the flow law and obtain:

$$\dot{p} = \left\langle \frac{\sigma_{eq} - k}{K} \right\rangle^{N} p^{\gamma}.$$

The hardening represented by the variable p, has the effect of decreasing the strain rate: the exponent γ is therefore negative. By writing $\gamma = -N/M$ and assuming that k is zero, we obtain the multiaxial viscous-hardening law studied in Section 6.2.4:

- $$\dot{\boldsymbol{\varepsilon}}^p = \frac{3}{2}\dot{p}\frac{\boldsymbol{\sigma}'}{\sigma_{eq}} = \frac{3}{2}\left(\frac{\sigma_{eq}}{Kp^{1/M}}\right)^{N}\frac{\boldsymbol{\sigma}'}{\sigma_{eq}}.$$

Note that this law can easily be integrated to give a relation between strain and stress, with stress as a given function of time:

$$\dot{p} = (\tfrac{2}{3}\dot{\boldsymbol{\varepsilon}}^p : \dot{\boldsymbol{\varepsilon}}^p)^{1/2} = \left(\frac{\sigma_{eq}}{Kp^{1/M}}\right)^{N} \frac{(\tfrac{3}{2}\boldsymbol{\sigma}' : \boldsymbol{\sigma}')^{1/2}}{\sigma_{eq}} = \left(\frac{\sigma_{eq}}{Kp^{1/M}}\right)^{N}$$

which, with the initial condition $p = 0$ at $t = 0$, gives

$$p(t) = \left[\frac{N+M}{M} \int_0^t \left(\frac{\sigma_{eq}(t)}{K} \right)^N \mathrm{d}t \right]^{M/(N+M)}$$

and

$$\boldsymbol{\varepsilon}^{\mathrm{p}}(t) = \int_0^t \frac{3}{2} \left(\frac{\sigma_{eq}(t)}{K p(t)^{1/M}} \right)^N \frac{\boldsymbol{\sigma}'(t)}{\sigma_{eq}(t)} \mathrm{d}t.$$

Study of the dissipation

The intrinsic dissipation is expressed by:

$$\Phi_1 = \boldsymbol{\sigma} : \dot{\boldsymbol{\varepsilon}}^{\mathrm{p}} - R\dot{r}.$$

We can immediately verify that it is always positive:

$$\Phi_1 = \frac{3}{2} \dot{p} \frac{\boldsymbol{\sigma}' : \boldsymbol{\sigma}'}{\sigma_{eq}} - R\dot{p} + R \frac{\partial \Omega_{\mathrm{r}}}{\partial R} = \dot{p}(\sigma_{eq} - R) + R \frac{\partial \Omega_{\mathrm{r}}}{\partial R}$$

$$\Phi_1 = \dot{p}(\sigma_{eq} - R - k) + R(\partial \Omega_{\mathrm{r}} / \partial R) + k\dot{p} \geqslant 0.$$

Indeed, if $\sigma_{eq} - R - k < 0$, we have $\dot{p} = 0$, and if Ω_{r} is a convex function such that $\Omega_{\mathrm{r}}(0) = 0$, the product $R(\partial \Omega_{\mathrm{r}} / \partial R)$ is always positive.

For the additive hardening law without recovery effects ($\Omega_{\mathrm{r}} = 0$), the different dissipation terms can be visualized as shown in Fig. 6.23(*a*) for the

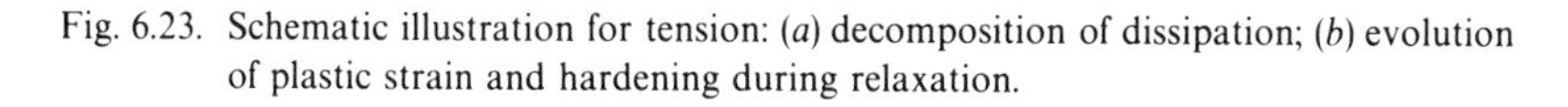

Fig. 6.23. Schematic illustration for tension: (*a*) decomposition of dissipation; (*b*) evolution of plastic strain and hardening during relaxation.

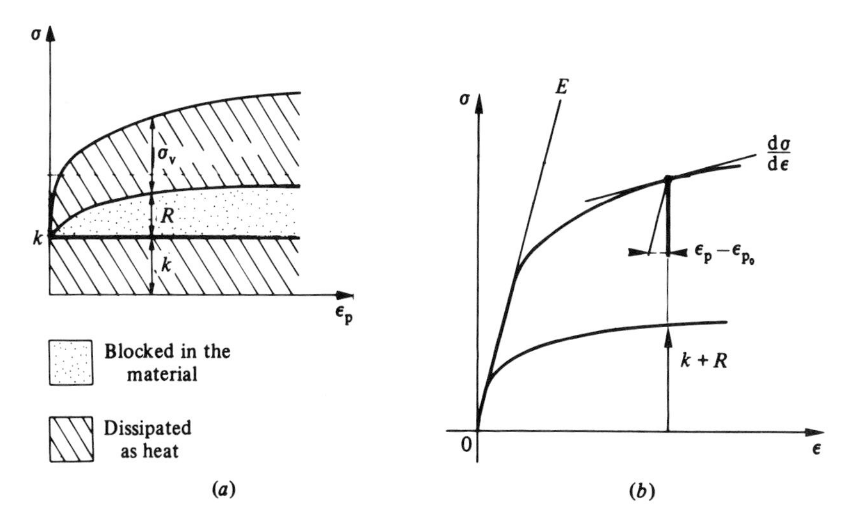

case of tensile loading. The dissipated plastic power $\boldsymbol{\sigma}:\dot{\boldsymbol{\varepsilon}}^{\mathrm{p}}$ may be divided in two parts:

> a part dissipated as heat, represented by the intrinsic dissipation $\Phi_1 = (\sigma_{\mathrm{eq}} - R)\dot{p}$ which corresponds to the viscous dissipation (σ_{v}) and the perfectly plastic dissipation (k);
> a part blocked in the system (in the system of dislocations or internal stresses), represented by the term $R\dot{p}$.

For the product law the dissipation is positive provided that the free energy is assumed to be independent of r (or p), the function h is constant, and $R = 0$. The mechanical dissipation is then reduced to $\boldsymbol{\sigma}:\dot{\boldsymbol{\varepsilon}}^{\mathrm{p}}$ which means that all plastic power is dissipated as heat.

Identification according to the tension tests

Additive viscosity-hardening law

For uniaxial tension, the stress invariant $J_2(\boldsymbol{\sigma})$ reduces to $\sigma_{\mathrm{eq}} = J_2(\boldsymbol{\sigma}) = \sigma$, with the first component of the deviatoric stress tensor equal to $\frac{2}{3}\sigma$. The first of the above-mentioned laws is reduced to:

$$\dot{\varepsilon}_{\mathrm{p}} = \left\langle \frac{\sigma - R(\varepsilon_{\mathrm{p}}) - k}{K_{\mathrm{a}}} \right\rangle^{N_{\mathrm{a}}}$$

which can be inverted to give:

$$\sigma = k + R(\varepsilon_{\mathrm{p}}) + K_{\mathrm{a}}\dot{\varepsilon}_{\mathrm{p}}^{1/N_{\mathrm{a}}}.$$

It is still necessary to specify the form of the hardening function $R(\varepsilon_{\mathrm{p}})$. It may, for example, be deduced directly from the relation presented in the preceding section:

$$R(\varepsilon_{\mathrm{p}}) = Q_1\varepsilon_{\mathrm{p}} + Q_2[1 - \exp(-b\varepsilon_{\mathrm{p}})].$$

The constitutive relation clearly shows the decomposition of the stress as the sum of a hardening term which represents the current size of the elastic domain, and a viscosity term, often called the viscous stress:

$$\sigma_{\mathrm{v}}(\dot{\varepsilon}_{\mathrm{p}}) = K_{\mathrm{a}}\dot{\varepsilon}_{\mathrm{p}}^{1/N_{\mathrm{a}}}.$$

As for the multiplicative law, it is preferable to perform hardening and relaxation tests because they allow separation of the influence of ε_{p} from that of $\dot{\varepsilon}_{\mathrm{p}}$. Fig. 6.24(*b*) shows schematically that $\dot{\varepsilon}_{\mathrm{p}}$ is approximately constant towards the end of a hardening test conducted at constant strain

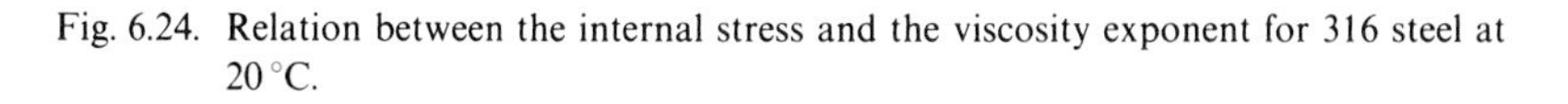

Fig. 6.24. Relation between the internal stress and the viscosity exponent for 316 steel at 20 °C.

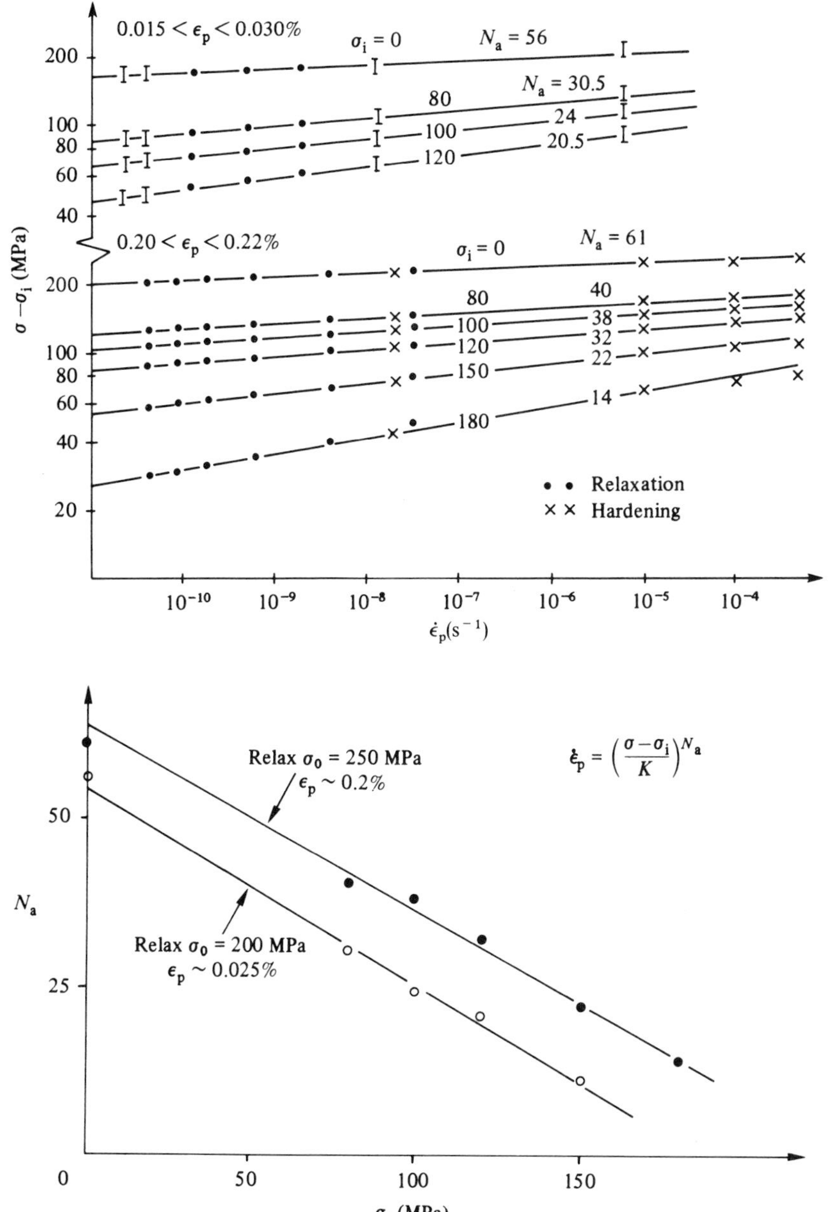

rate:

$$\dot{\varepsilon}_{\mathrm{p}} = \dot{\varepsilon} - \frac{\dot{\sigma}}{E} = \dot{\varepsilon}\left(1 - \frac{1}{E}\frac{\mathrm{d}\sigma}{\mathrm{d}\varepsilon}\right) \approx \dot{\varepsilon}.$$

On the other hand, in a relaxation test, it is the plastic strain that varies little and the hardening state remains almost unchanged.

From the hardening test we can therefore derive an approximation of the function $R(\varepsilon_{\mathrm{p}})$ for large ε_{p} (within an additive constant); this identifies the coefficient Q_1. From the relaxation test we can, therefore, derive a relation between the exponent N_{a} and the value of the nonviscous stress $\sigma_{\mathrm{i}} = k + R(\varepsilon_{\mathrm{p}})$. This relation is obtained by drawing a log–log graph of $\sigma - \sigma_{\mathrm{i}}$, as a function of $\dot{\varepsilon}_{\mathrm{p}}$ for different choices of σ_{i} (see Fig. 6.24).

By choosing a value of k which is dictated by the initial size of the apparent elastic domain, only four coefficients of the constitutive equation have still to be determined. This number is reduced to three when the experimental relation between N_{a} and σ_{i} is obtained. We may also use a numerical process for the identification of the coefficients, which is either entirely automatic (see Chapter 3), or which resorts to an interactive system.

As can be seen from the case of the 316 steel at room temperature (a difficult case since the viscosity is low), this law furnishes an excellent

Fig. 6.25. Description of relaxation tests by the additive law (for 316 stainless steel at 20 °C).

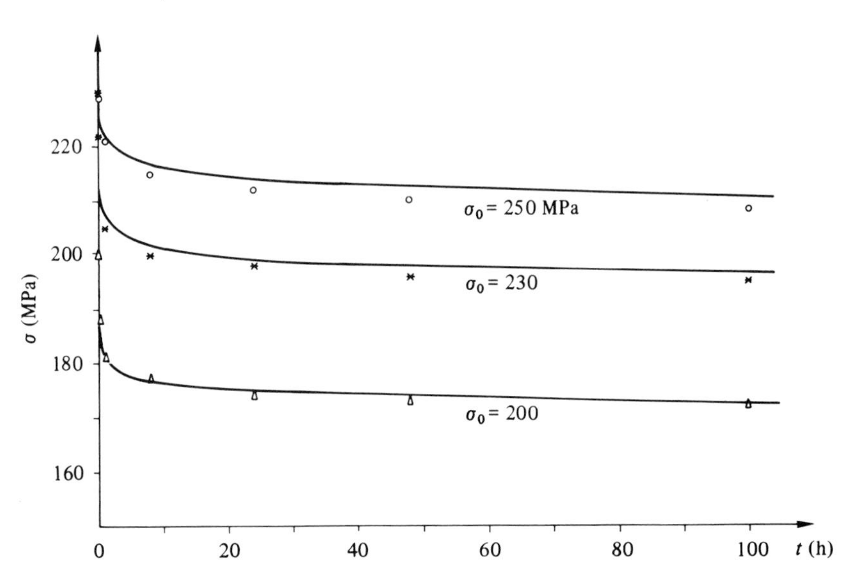

modelling of relaxation (Fig. 6.25) and hardening tests at different strain rates (Fig. 6.26).

Multiplicative viscosity-hardening law
The uniaxial expression of this equation (in simple tension):

$$\dot{\varepsilon}_p = (\sigma / K \varepsilon_p^{1/M})^N \quad \text{or} \quad \sigma = K \varepsilon_p^{1/M} \dot{\varepsilon}_p^{1/N}$$

is that already studied in Section 6.2.4 and which takes into account creep, hardening, and relaxation tests, simultaneously.

Fig. 6.26. Description of hardening tests at different strain rates by the additive law (for 316 steel at 20 °C).

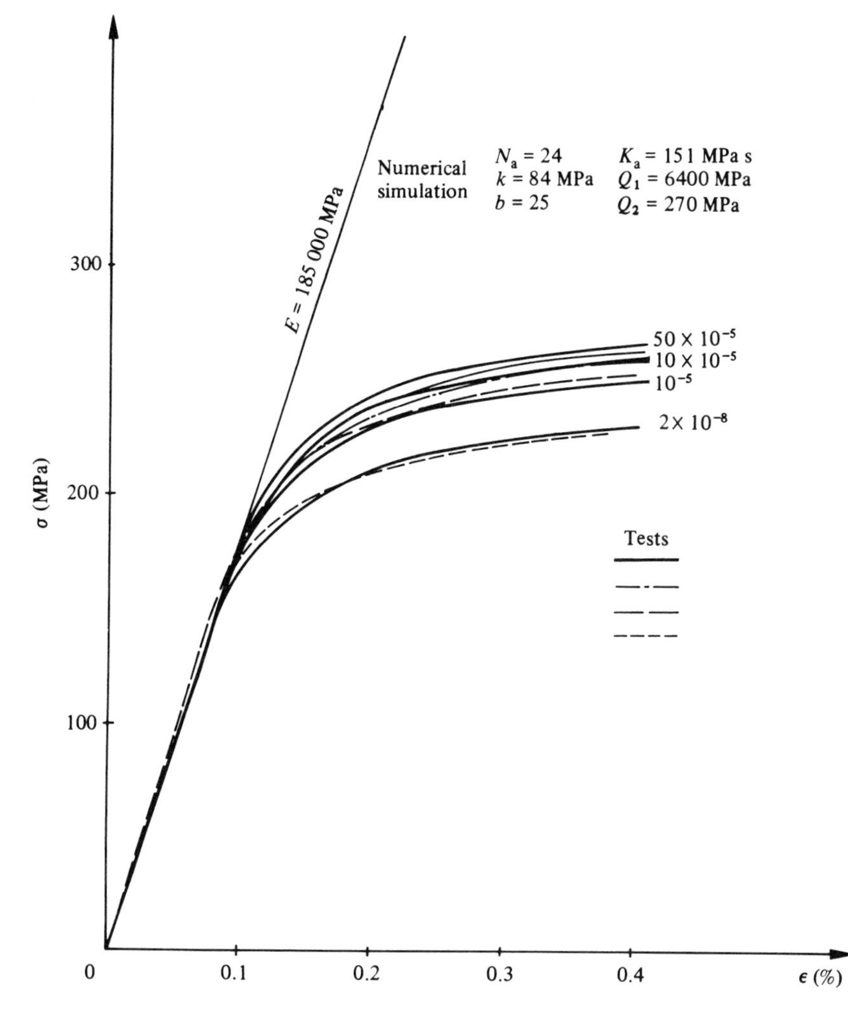

The identification of this law is particularly easy, and this is the reason why it is used so often. Three types of method may be used to identify the characteristic coefficients K, M and N.

(i) If a wide enough range of experimental results (generally from hardening, relaxation and primary creep), i.e. a range which shows a sufficiently significant variation of the experimental values of σ, ε_p, $\dot{\varepsilon}_p$ (at the same instant) is available, we may use the linear least squares method. The law can be written for each experimental point j:

$$\frac{1}{N}\ln \dot{\varepsilon}_{pj} + \frac{1}{M}\ln \varepsilon_{pj} + \ln K = \ln \sigma_j$$

as a linear relation between $1/N$, $1/M$ and $\ln K$. By finding the plane which passes closest to the experimental points in the space of variables $\ln \dot{\varepsilon}_p$, $\ln \varepsilon_p$, and $\ln \sigma$, we may obtain the values of N, M and K.

(ii) Hardening–relaxation tests furnish a quick method of identification. As we have seen above (Fig. 6.23(b)), a hardening test is performed at almost a constant plastic strain rate, while a relaxation test corresponds to

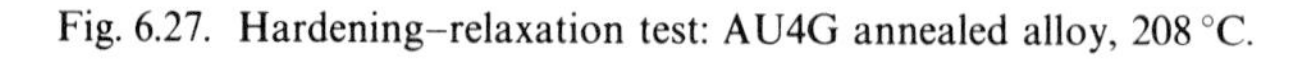

Fig. 6.27. Hardening–relaxation test: AU4G annealed alloy, 208 °C.

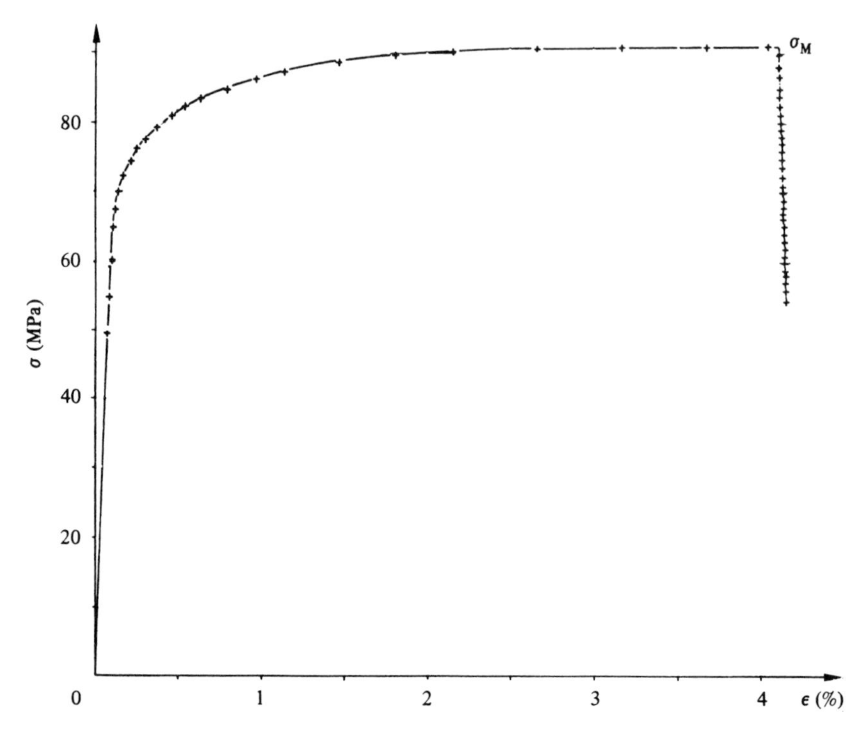

an approximately constant plastic strain. By plotting the points (σ, ε_p) and $(\sigma, \dot{\varepsilon}_p)$ as two log–log plots we immediately obtain (by drawing the best straight line) the values of $1/M$ and $1/N$. Figs. 6.27 and 6.28 are examples of this sort of procedure.

For relaxations at low strain, the variation of plastic strain can be quite important. Similarly the start of a hardening test involves a variable strain rate $\dot{\varepsilon}_p$. We are thus led to make a correction which constitutes the first step of an iteration in recording the following points on a log–log plot:

values of $\sigma/\varepsilon_p^{1/M}$ as a function of $\dot{\varepsilon}_p$
values of $\sigma/\dot{\varepsilon}_p^{1/N}$ as a function of ε_p,

where M and N are the values determined from the first graph. Experimental points from the hardening or relaxation (or even creep) tests can be used together. The two lines which pass closest to the two sets of points and which intersect at a point on the line $\ln \varepsilon_p = \ln \dot{\varepsilon}_p = 0$ furnish the final values of $1/N$ and $1/M$. The ordinate at the origin gives $\ln K$. Such a procedure is illustrated in Fig. 6.29.

When using this procedure and eventually the preceding one, it is preferable to perform a verification of the obtained coefficients by recalculating the creep tests in the primary domain (see later).

(iii) When mostly creep test data are available, the product law offers the advantage of being explicitly integrable (for an imposed stress) because of the separability of the variables. In creep, we find again an Andrade type law for the relation between plastic strain and time:

$$\varepsilon_p = \left[\frac{M+N}{M}\left(\frac{\sigma}{K}\right)^N t\right]^{M/(N+M)}.$$

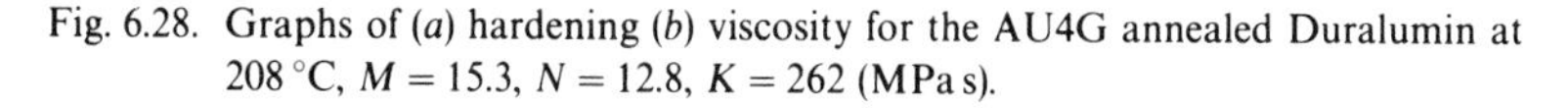
Fig. 6.28. Graphs of (*a*) hardening (*b*) viscosity for the AU4G annealed Duralumin at 208 °C, $M = 15.3$, $N = 12.8$, $K = 262$ (MPa s).

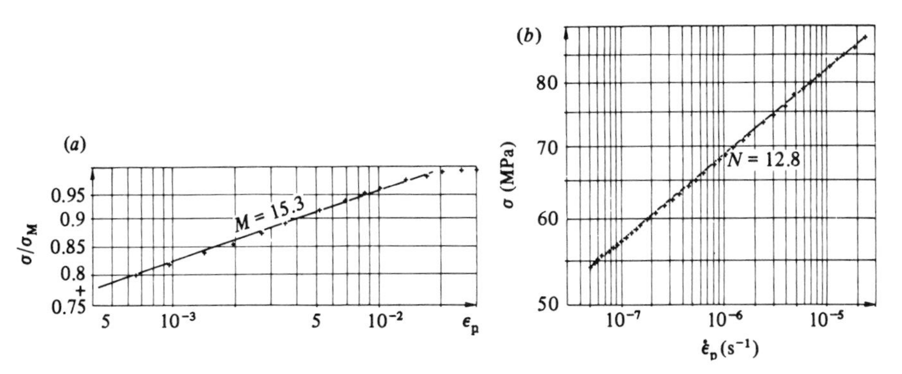

By plotting the primary creep (ε_p, t) points on a log–log plot, we can determine the mean value of the exponent $M/(M+N)$. Then, we may plot the values of $\varepsilon_p/t^{M/(M+N)}$ as a function of σ for different stresses. From the value obtained for the exponent $MN/(M+N)$, it is easy to derive the coefficients M and N and then a mean value of K from the set of tests. Fig. 6.30 shows an example of such a determination. When employing this procedure, it is useful to check the values of the coefficients by recalculating a hardening–relaxation test.

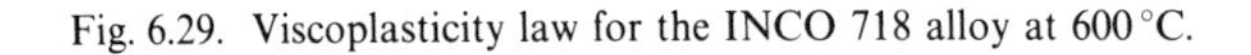

Fig. 6.29. Viscoplasticity law for the INCO 718 alloy at 600 °C.

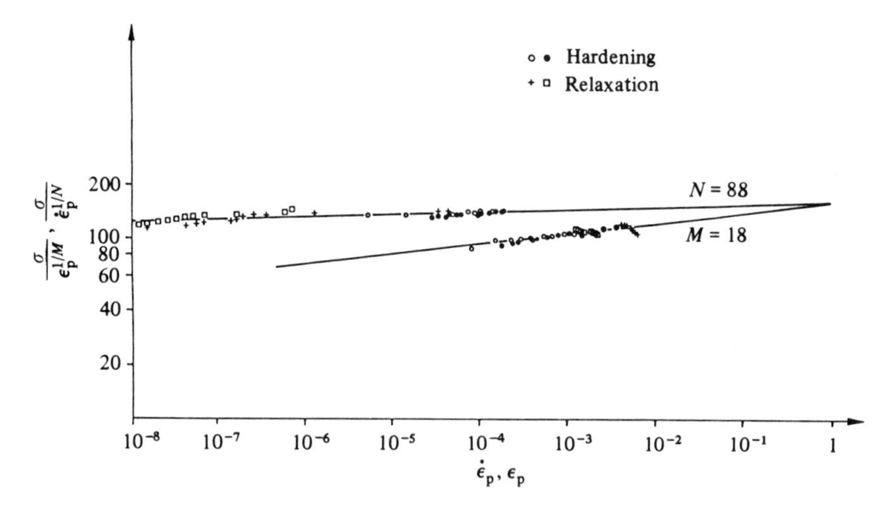

Fig. 6.30. Determination of the multiplicative law from the creep tests (Hastelloy X, 650 °C).

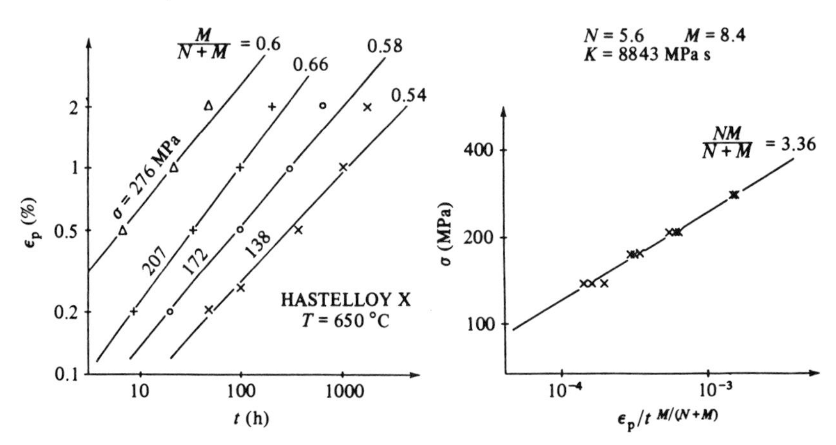

Cyclic loadings

The isotropic-hardening law may be used for cyclic loadings. In tension–compression the two laws described above can be written respectively as:

$$\dot{\varepsilon}_p = \left\langle \frac{|\sigma| - R(\tilde{\varepsilon}_p) - k}{K_a} \right\rangle^{N_a} \operatorname{Sgn}(\sigma)$$

$$\dot{\varepsilon}_p = \left(\frac{|\sigma|}{K \tilde{\varepsilon}_p^{1/M}} \right)^{N} \operatorname{Sgn}(\sigma)$$

where $\tilde{\varepsilon}_p$ is the accumulated plastic uniaxial strain defined by:

$$\tilde{\varepsilon}_p = \int_0^t |\dot{\varepsilon}_p(\tau)| \, d\tau.$$

After a first flow, the subsequent evolutions are identical in tension and compression (with the exception of sign). Fig. 6.31 illustrates this property schematically: at points B and B', we have the same value of $|\sigma|$ and the same value of accumulated plastic strain $\tilde{\varepsilon}_p$. Unfortunately, this property has not been well verified experimentally because of the Bauschinger effect.

Fig. 6.31. Isotropic hardening and cyclic load; the first tension followed by the first compression.

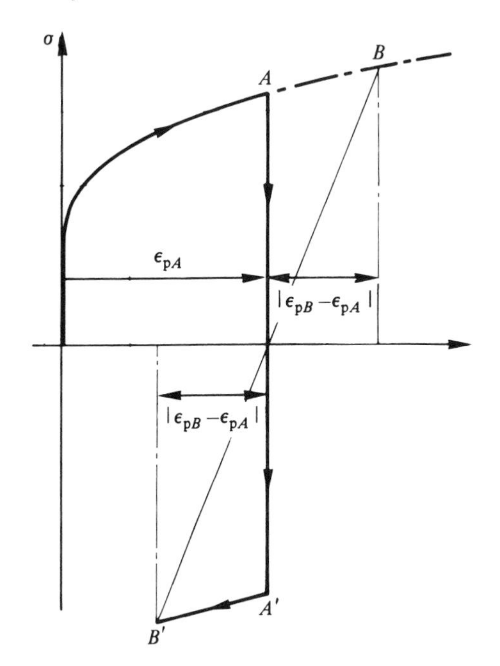

Under a cyclic load (with plastic deformation), the hardening variable increases indefinitely. This results either in a saturation of hardening according to the first law, if $Q_1 = 0$, or in a nonsaturation:

$$R(\tilde{\varepsilon}_p) = Q_1\tilde{\varepsilon}_p + Q_2[1 - \exp(-b\tilde{\varepsilon}_p)].$$

In the first case, this gives for the cyclic behaviour the properties of Norton's law (no more primary creep for example). In the second case, stabilization again occurs only when the behaviour is completely elastic. This shakedown, after a large number of cycles, is very seldom observed. The laws with isotropic hardening (as the laws without hardening) are therefore ill suited to the cyclic loading cases.

Method of extrapolation: 'relaxation–creep'

The problem of the extrapolation of short-duration test results to long periods is a fundamental problem in the analysis of structures whose lifetime is measured in years or decades. A reasonable duration for commonly performed tests is of the order of a day, a week, a month, even a year, but not longer.

If essentially time-dependent phenomena, like recovery and ageing occur, the problem of extrapolation is almost unsolvable. Since an increase in temperature, in general, accelerates these phenomena, we might use parameters with time–temperature equivalence, but these will have to be verified for each material for the maximum period of its use, so this procedure does not really help!

If, on the other hand, the material is stable and satisfies a viscoplastic law with isotropic hardening of the type:

$$\sigma = f(\varepsilon_p, \dot{\varepsilon}_p)$$

Fig. 6.32. Multiple relaxation procedure.

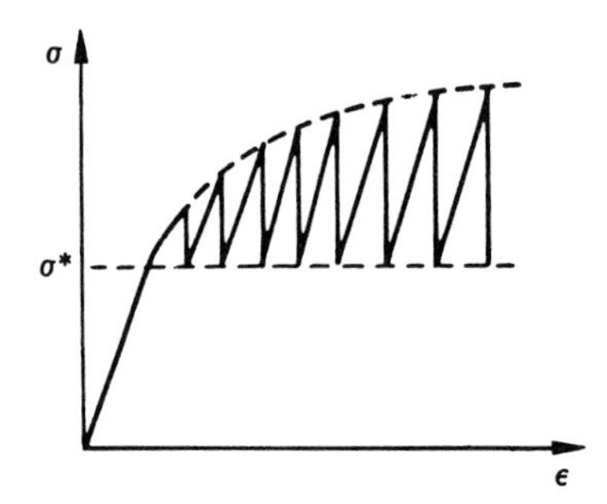

in simple tension, we may devise an experimental method which can be used to obtain, by relaxation, in considerably shorter times, the strain rates which correspond to those in creep tests. This makes it possible to construct long-term creep curves from short-term relaxation tests.

In principle, it is very simple: let $\varepsilon(t)$ be the creep curve to be determined

Fig. 6.33. Schematic illustration of the relaxation–creep method: (*a*) relaxation tests: (*b*) creep trajectory; (*c*) creep curve.

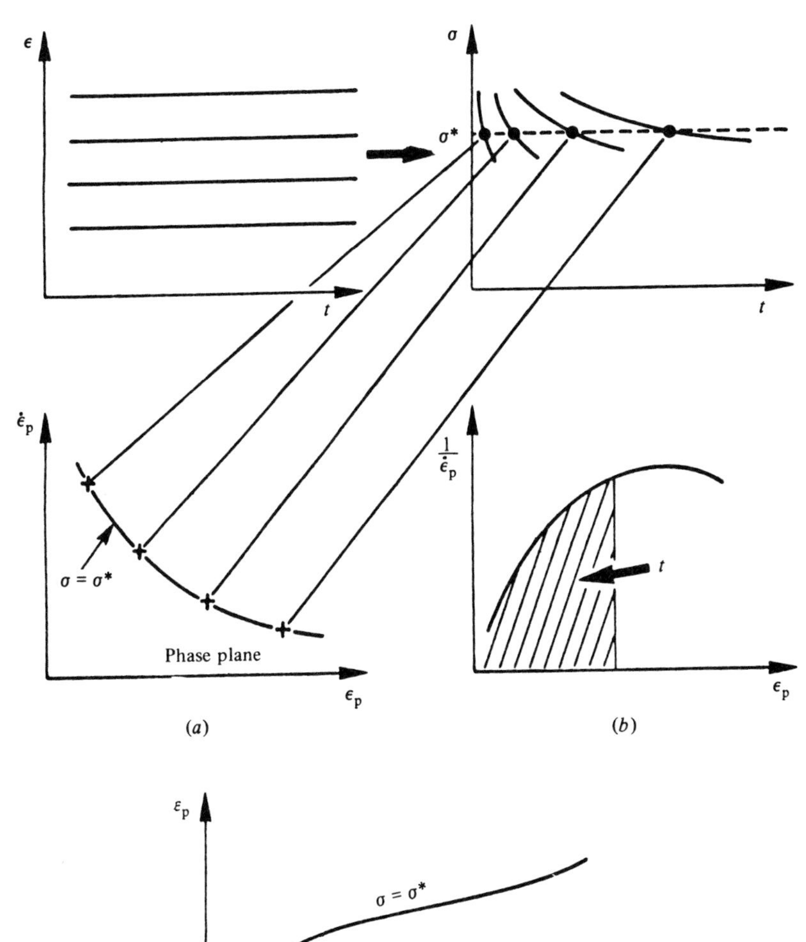

corresponding to a constant stress σ^* at temperature T^*. Let a series of relaxation tests, or better a multiple relaxation test (see Section 3.2.1) be performed with each relaxation reaching the stress σ^* (Fig. 6.32).

At every point of the relaxation curves $\sigma(t)$ where $\sigma = \sigma^*$, the material state is characterized by:

$$\sigma = \sigma^*$$
$$\varepsilon_p = \varepsilon - \sigma^*/E$$
$$\dot{\varepsilon}_p = -\dot{\sigma}(\sigma^*)/E.$$

The constitutive equation links σ, ε_p, $\dot{\varepsilon}_p$ for any loading history and so the above values are also representative of a creep test at $\sigma = \sigma^* =$ constant. We may therefore trace the trajectory of a creep test by plotting a graph $(\varepsilon_p, \dot{\varepsilon}_p)$ and by integrating the function whose curve is that which passes through these points. Denoting this function by

$$\dot{\varepsilon}_p = g(\varepsilon_p)$$

we obtain the creep curve $\varepsilon_p(t)$ where

$$t = \int_{\varepsilon_{p0}}^{\varepsilon_p} \frac{1}{g(\varepsilon_p)} \, d\varepsilon_p$$

with the given initial condition $\varepsilon_p(t=0) = \varepsilon_{p0}$. Fig. 6.33 schematically summarizes these different operations.

This calculation can be repeated for any stress value reached during the relaxation. Thus, a set of relaxation tests or a multiple relaxation test can be used to construct a set of creep curves at different stresses.

The most interesting feature of this method lies in the gain in time which results from the large nonlinearity of the behaviour of common metallic materials. The time necessary for attaining a state characterized by particular values of σ, ε_p, and $\dot{\varepsilon}_p$ is of the order of 5–50 times shorter for relaxation than for creep. This gain in time may be calculated for example, with the help of the viscosity-hardening law:

$$\sigma = E\varepsilon_e$$
$$\sigma = f(\varepsilon_p, \dot{\varepsilon}_p) = K\varepsilon_p^{1/M}\dot{\varepsilon}_p^{1/N}.$$

The time t_f^* necessary to reach the state σ^*, ε_p^*, $\dot{\varepsilon}_p^*$ in a creep test is obtained by integrating the law for $\sigma = \sigma^*$, with the initial condition $\varepsilon_p(t=0) = 0$:

$$t_f^* = \frac{M}{M+N}\left(\frac{K}{\sigma^*}\right)^N \varepsilon_p^{*(M+N)/M}.$$

The time t_r^* necessary for reaching the same state in a relaxation test is obtained by integrating the equation for $\varepsilon^* = \varepsilon_p^* + \varepsilon_e^* = \varepsilon_p^* + \sigma^*/E = \text{const}$-ant, with the same initial condition as above:

$$t_r^* = \left(\frac{K}{E}\right)^N \int_0^{\varepsilon_p^*} \frac{x^{N/M}}{(\varepsilon^* - x)^N}\,dx.$$

This integral can only be evaluated numerically.

The gain in time is expressed by

$$\frac{t_f^*}{t_r^*} = \frac{M}{M+N}\frac{\varepsilon_p^{*(M+N)/M}}{\varepsilon_e^{*N}}\left(\int_0^{\varepsilon_p^*} \frac{x^{N/M}}{(\varepsilon^* - x)^N}dx\right)^{-1}.$$

Setting

$$\Gamma = \frac{M}{M+N}\frac{\varepsilon_p^{*N/M}}{\varepsilon_e^{*N-1}}\left(\int_0^{\varepsilon_p^*} \frac{x^{N/M}}{(\varepsilon^* - x)^N}\right)^{-1}$$

it becomes apparent that for usual values of ε_e^* and ε_p^*, Γ varies only slightly with these variables, so that we may calculate Γ as a function of the parameters M and N only. The graph of Fig. 6.34 gives a parametric

Fig. 6.34. Numerical values of the function Γ.

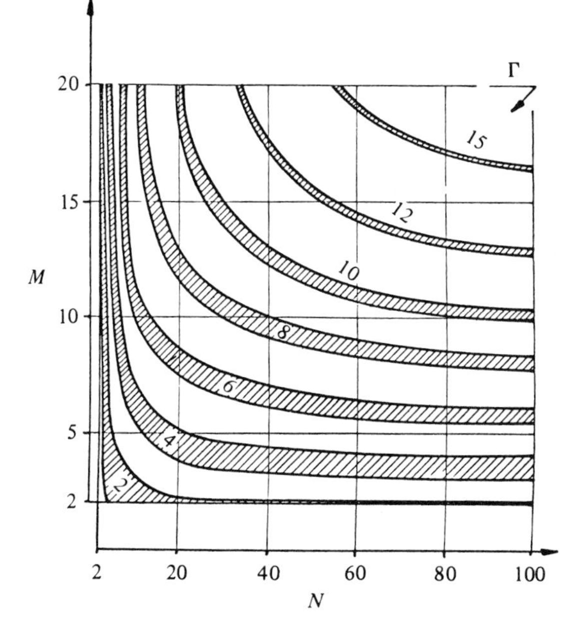

representation of Γ for

$$2 \leqslant M \leqslant 20$$

$$2 \leqslant N \leqslant 100$$

$$0.2 \times 10^{-2} \leqslant \varepsilon_e^* \leqslant 0.4 \times 10^{-2}$$

$$1 \leqslant \varepsilon_p^*/\varepsilon_e^* \leqslant 5$$

and the gain in time is equal to:

- $$t_f^*/t_r^* = \Gamma \varepsilon_p^*/\varepsilon_e^*.$$

Fig. 6.35 shows an example of the application of this method to creep tests of a little more than 1000 hours on a light alloy, AU2GN, at 130 °C, retraced by relaxation tests of no longer than 100 hours, with a gain in time by a factor of little more than 10 ($M = 13$, $N = 71$, $\varepsilon_p^*/\varepsilon_e^* \approx 1$).

Domain of validity

Table 6.2 gives a set of characteristic coefficients N, M, and K, of the multiplicative viscosity-hardening law for some materials at varying temperatures. These coefficients were determined by one of the methods described above.

For small variations in temperature, M and N can be considered as constants and we can limit ourselves to the introduction of a coefficient of

Fig. 6.35. Application of the relaxation–creep method: AU2GN alloy, $T = 130$ °C, $\sigma = 324$ MPa.

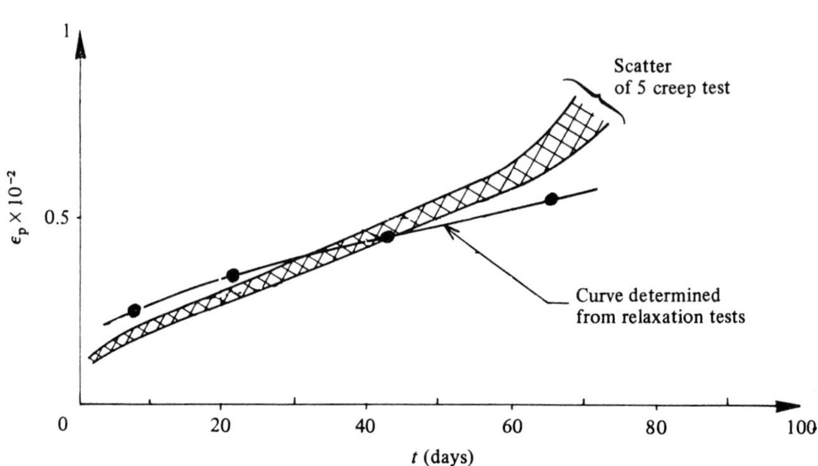

the activation energy such as the one described in Section 6.2.5:

$$\sigma = K'\varepsilon_p^{1/M}[\dot{\varepsilon}_p \exp(\Delta H/kT)]^{1/N}.$$

Such a term accounts for the decrease in the coefficient of resistance when the temperature increases. For significant variations in temperature (more

Fig. 6.36. Modelling of the monotonic behaviour of the wrought UDIMET 700 alloy by the viscoplasticity law with multiplicative hardening: (*a*) relaxation; (*b*) hardening.

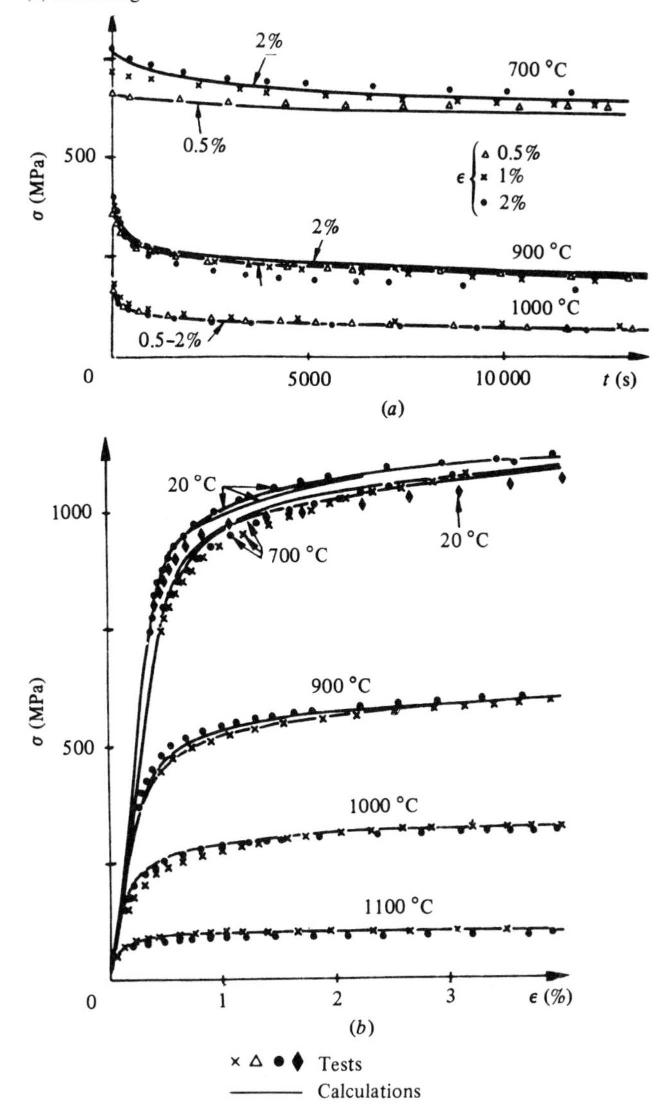

Fig. 6.37. Comparison of the two forms of laws with primary creep for IN 100 alloy (coated). These laws have been identified by hardening and relaxation; (*a*) additive form. (*b*) multiplicative form.

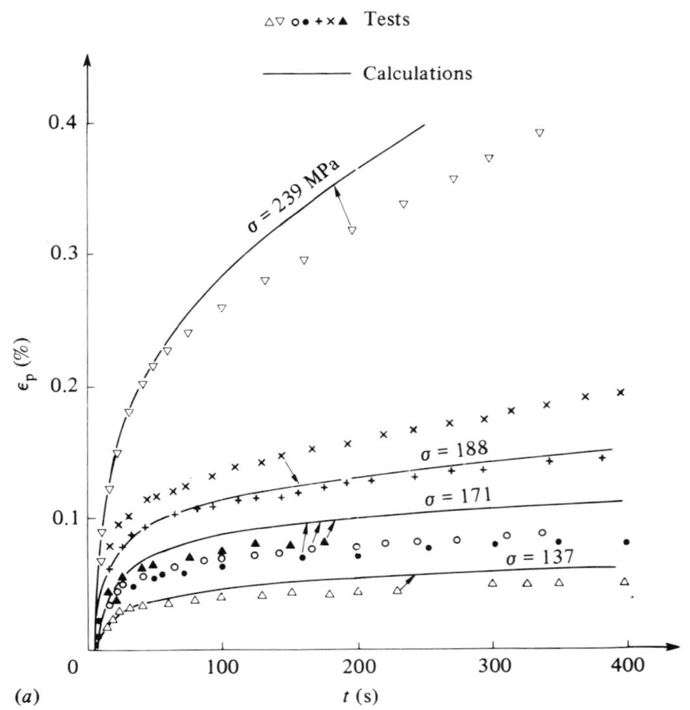

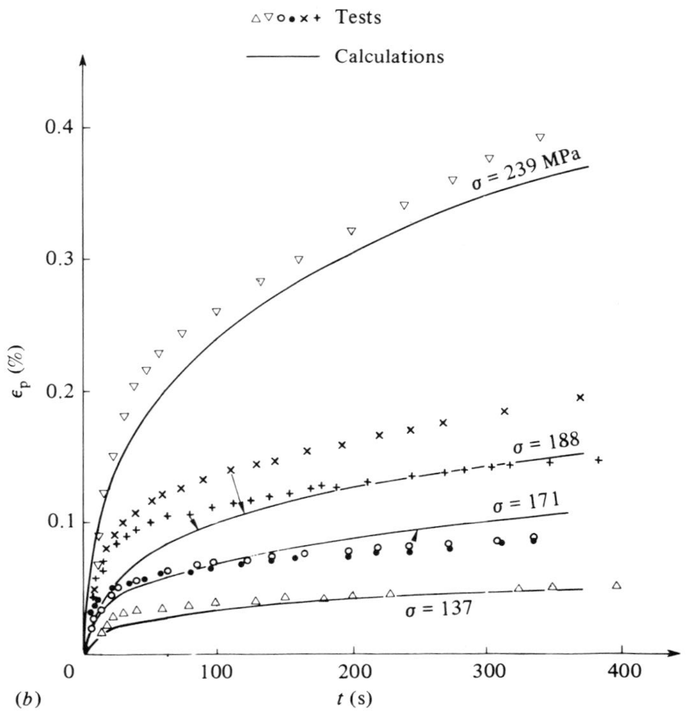

than 50 °C for example), we must treat the coefficients K, M and N as variables dependent on the temperature. Fig. 6.13 shows an example of such an evolution, interpolated piecewise by parabolas with continuity of tangents (see Chapter 3).

The two multiaxial, isotropic-hardening, viscoplasticity equations developed at the beginning of Section 6.4.2 (multiplicative and additive forms), satisfactorily account for the results from hardening, relaxation, and creep tests. Fig. 6.36 shows some examples of the comparison of the product law with the three types of tests, in addition to those shown in the preceding section. The two forms of the equations, multiplicative and additive, may be compared on Figs. 6.37 and 6.38 for the IN 100 alloy at 1000 °C.

The domain of validity of these two laws is approximately the same. The limitations are essentially the following:

> the range of the strain rate is approximately from $10^{-8}\,s^{-1}$ to $10^{-3}\,s^{-1}$. For low rates ($< 10^{-8}\,s^{-1}$) the exponent N sometimes evolves with the stress: we may then use an additional exponential term as in Norton's law (cf. Section 6.4.1) or introduce recovery

Fig. 6.38. Comparison of the additive and multiplicative equations with relaxation test on the IN 100 alloy at 1000 °C (coated).

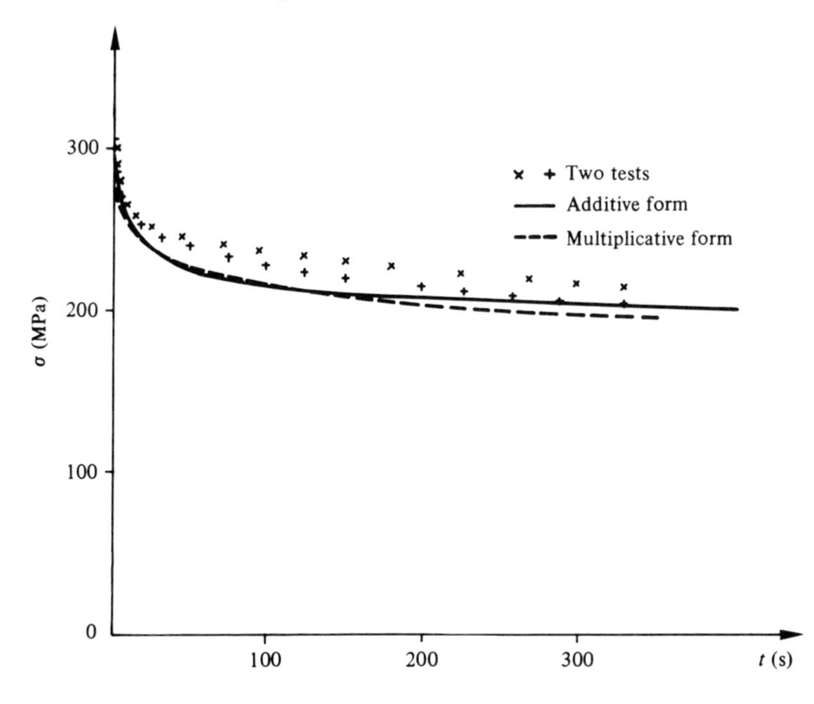

effects as shown in Section 6.4.4 (the additive model is more suitable for introducing such effects);

the range of strain depends on the tests used for identification. In a more general way, the range of validity can be between 10^{-4} and $(2\text{–}4) \times 10^{-2}$. It can reach more significant values with the additive model because of the linear hardening term $(Q_1 \varepsilon_p)$;

the three types of tests (hardening–relaxation–creep) can be described correctly by the two models provided the effects of time (recovery) are not very important;

the additive model represents secondary creep if the term $Q_1 \varepsilon_p$ is negligible (or when recovery is introduced). Secondary creep begins when the hardening process, represented by $R(\varepsilon_p)$, is saturated. From the additive equation we then find:

$$\dot{\varepsilon}_p = \left\langle \frac{\sigma - Q_2 - k}{k_a} \right\rangle^{N_a}.$$

If $\sigma < Q_2 + k$, then the creep has a logarithmic character (not strictly a secondary creep). With the multiplicative equation, we cannot get the secondary creep directly (see Section 6.2.4).

6.4.3 *Viscoplasticity law with kinematic hardening*

This equation is the result of several investigations of plasticity with reference to nonlinear kinematic hardening. Its aim is essentially to represent cyclic loading behaviour correctly by the incorporation of viscoplastic effects. It allows separation of different effects, and is directly related to the general thermodynamic concepts (cf. Section 6.3). Moreover, it should be noted that the kinematic character of hardening has been demonstrated quite well by experiments, as we shall see later in this section.

On the other hand, other methods are available for the representation of the nonlinearity of kinematic hardening: generalization of multilayer models mentioned in Chapter 5, superposition of linear kinematic hardening and recovery effects (see Section 6.4.4).

General formulation

To simplify the presentation, we first of all neglect the recovery effects; the way they can be incorporated will be studied in Section 6.4.4. Therefore, for the moment, the internal variable r is taken to be indistinguishable from the accumulated plastic strain p.

Pure kinematic hardening

Kinematic hardening corresponds to a translation of the elastic domain or of equipotential surfaces (in the stress space) during viscoplastic flow. Therefore, the tensorial internal variable $\boldsymbol{\alpha}$ and the associated thermodynamic force

$$\mathbf{X} = \rho(\partial\Psi/\partial\boldsymbol{\alpha})$$

are used in formulating the theory. The dissipation potential is expressed as a function of the associated thermodynamic variables and may depend upon the state variables, which are themselves treated as parameters:

$$\varphi^* = \Omega(\boldsymbol{\sigma}, \mathbf{X}; T, \boldsymbol{\alpha}).$$

In what follows the material is assumed to be initially isotropic. Kinematic hardening is the only anisotropy considered and the viscoplastic flow is assumed to be volume preserving. $\mathbf{X}$ is considered as the current centre of the elastic domain. The dissipation potential is therefore expressed as a function of the elementary invariants of $\boldsymbol{\sigma} - \mathbf{X}$, $\mathbf{X}$ and $\boldsymbol{\sigma}$.

In practice, the influence of the third invariant is often neglected. To simplify the development of the theory, the dissipation potential will therefore be taken in the form:

$$\Omega = \Omega(J_2(\boldsymbol{\sigma} - \mathbf{X}), J_2(\mathbf{X}); T, J_2(\boldsymbol{\alpha}))$$

with

$$J_2(\boldsymbol{\sigma} - \mathbf{X}) = [\tfrac{3}{2}(\boldsymbol{\sigma}' - \mathbf{X}'):(\boldsymbol{\sigma}' - \mathbf{X}')]^{1/2}.$$

The nonlinear kinematic hardening, initially introduced by Armstrong and Frederick for time-independent plasticity, and applied to viscoplasticity by Malinin and Khadjinsky, is based on an evolution equation for the variable $\mathbf{X}$ containing two terms: the (linear) kinematic hardening term and a recall term which provides a fading memory effect of the deformation path:

$$\dot{\mathbf{X}} = \tfrac{2}{3}C\dot{\boldsymbol{\varepsilon}}^{\mathrm{p}} - \gamma\mathbf{X}\dot{p}.$$

Note that the recall term involves the magnitude $\dot{p}$ of the plastic strain rate. It is homogeneous with the first term and therefore the effect of recall is time and rate independent. The model thus represents a nonlinear strain-hardening, without the effect of recovery with time.

In order to incorporate such a modelling in the general thermodynamic framework, it is necessary to keep $\boldsymbol{\alpha}$ in the expression for the potential. We now write:

$$\Omega = \Omega\left(J_2(\boldsymbol{\sigma} - \mathbf{X}) - k + \frac{\gamma}{2C} J_2^2(\mathbf{X}) - \frac{2\gamma C}{9} J_2^2(\boldsymbol{\alpha}); T \right)$$

in which the coefficients k, γ, C may also depend on the temperature.

The viscoplastic strain rate and the rate of change of the kinematic variable are obtained from the generalized normality law:

$$\dot{\boldsymbol{\varepsilon}}^{\mathrm{p}} = \frac{\partial \Omega}{\partial \boldsymbol{\sigma}} = \frac{3}{2} \frac{\partial \Omega}{\partial J_2} \frac{\boldsymbol{\sigma}' - \mathbf{X}'}{J_2(\boldsymbol{\sigma} - \mathbf{X})} = \frac{3}{2} \dot{p} \frac{\boldsymbol{\sigma}' - \mathbf{X}'}{J_2(\boldsymbol{\sigma} - \mathbf{X})}$$

$$\dot{\boldsymbol{\alpha}} = -\frac{\partial \Omega}{\partial \mathbf{X}} = \dot{\boldsymbol{\varepsilon}}^{\mathrm{p}} - \frac{3}{2} \frac{\gamma}{C} \mathbf{X} \dot{p}$$

where $\dot{p}$ is as usual the magnitude of the plastic strain rate $\dot{p} = (\frac{2}{3} \dot{\boldsymbol{\varepsilon}}^{\mathrm{p}} : \dot{\boldsymbol{\varepsilon}}^{\mathrm{p}})^{1/2} = \frac{2}{3} J_2(\dot{\boldsymbol{\varepsilon}}^{\mathrm{p}})$. Moreover, the free energy is assumed to include a term quadratic in $\boldsymbol{\alpha}$:

$$\rho \Psi = \rho \Psi_{\mathrm{e}} + \tfrac{2}{9} C J_2^2(\boldsymbol{\alpha}) = \rho \Psi_{\mathrm{e}} + \tfrac{1}{3} C \boldsymbol{\alpha} : \boldsymbol{\alpha}$$

so that

$$\mathbf{X} = \tfrac{2}{3} C \boldsymbol{\alpha}.$$

Using this relation we obtain:

$$\Omega = \Omega(J_2(\boldsymbol{\sigma} - \mathbf{X}) - k; T)$$

$$\dot{\boldsymbol{\varepsilon}}^{\mathrm{p}} = \frac{3}{2} \frac{\partial \Omega}{\partial J_2} \frac{\boldsymbol{\sigma}' - \mathbf{X}'}{J_2(\boldsymbol{\sigma} - \mathbf{X})} = \frac{3}{2} \dot{p} \frac{\boldsymbol{\sigma}' - \mathbf{X}'}{J_2(\boldsymbol{\sigma} - \mathbf{X})}$$

$$\dot{\boldsymbol{\alpha}} = \dot{\boldsymbol{\varepsilon}}^{\mathrm{p}} - \gamma \boldsymbol{\alpha} \dot{p}$$

$$\mathbf{X} = \tfrac{2}{3} C \boldsymbol{\alpha}$$

which completely define the constitutive equations. By combining the last two relations we easily recover the equation governing the variable $\mathbf{X}$:

$$\dot{\mathbf{X}} = \tfrac{2}{3} C \dot{\boldsymbol{\varepsilon}}^{\mathrm{p}} - \gamma \mathbf{X} \dot{p}.$$

Note the role played by the two additional terms in the initial potential: the first, dependent on $\mathbf{X}$, clearly introduces the recall term in the equation governing the kinematic variable; the second, which depends on the variable $\boldsymbol{\alpha}$ itself, acts by cancelling the effect of the first term on the elastic limit k with the help of the relation between $\mathbf{X}$ and $\boldsymbol{\alpha}$. This rather artificial procedure is, however, compatible with the general framework that we have set up.

In the above expressions the factor $\frac{2}{3}$ is used so that the corresponding relations in tension–compression can be written in a simple way. Moreover, it can be verified easily that if $\mathbf{X}(0)$ is deviatoric then the law of evolution implies that $\mathbf{X}$ remains deviatoric. Note that this is not an indispensable property as any theory of incompressible viscoplasticity is invariant in a translation of equipotential surfaces along the incompressibility axis (which is inclined equally to the three principal stress axes).

For a large range of variation of variables, the dissipation potential can be chosen to be a power function:

$$\Omega = \frac{K}{n+1}\left\langle \frac{J_2(\boldsymbol{\sigma} - \mathbf{X}) - k}{K} \right\rangle^{n+1}$$

where the coefficients k, K, n as well as C and γ depend on the temperature. The flow and hardening equations are then:

- $$\dot{\boldsymbol{\varepsilon}}^{\mathrm{p}} = \frac{3}{2}\dot{p}\frac{\boldsymbol{\sigma}' - \mathbf{X}'}{J_2(\boldsymbol{\sigma} - \mathbf{X})}$$
- $$\dot{p} = \left\langle \frac{J_2(\boldsymbol{\sigma} - \mathbf{X}) - k}{K} \right\rangle^{n}$$
- $$\mathbf{X} = \tfrac{2}{3}C\boldsymbol{\alpha}$$
- $$\dot{\boldsymbol{\alpha}} = \dot{\boldsymbol{\varepsilon}}^{\mathrm{p}} - \gamma\boldsymbol{\alpha}\dot{p}.$$

It is in this form that the law of viscoplasticity with nonlinear kinematic hardening is used. Note that this form is compatible with loading at variable temperature. For a rapid change in temperature, in which no flow occurs, the variable $\boldsymbol{\alpha}$ remains unchanged. On the other hand, the internal stress $\mathbf{X}$ is immediately generated since the coefficient C generally depends on the temperature.

Superposition of the isotropic hardening

It is possible to superpose isotropic hardening on nonlinear kinematic hardening, either to improve the modelling (in tension, for example), or because an expansion of the equipotential surfaces around their centres is observed, or to describe cyclic hardening or softening effects.

The isotropic hardening is introduced in the free energy with the help of the independent variable p and the associated thermodynamic force:

$$R = \rho(\partial\Psi/\partial p)$$

which appears in the expression for dissipation potential. This may also depend parametrically on the variable p:

$$\varphi^* = \Omega(\boldsymbol{\sigma}, \mathbf{X}, R; T, \boldsymbol{\alpha}, p).$$

Using the same assumptions as in the preceding section, we write the two potentials in the following forms:

$$\rho\Psi = \rho\Psi_e + \tfrac{1}{3}C\boldsymbol{\alpha}:\boldsymbol{\alpha} + h(p)$$

$$\Omega = \Omega\left(J_2(\boldsymbol{\sigma} - \mathbf{X}) - R - k + \frac{1}{2}\frac{\gamma(p)}{C} J_2^2(\mathbf{X}) - \frac{2}{9} C\gamma(p) J_2^2(\boldsymbol{\alpha}); T, p \right).$$

The associated thermodynamic variables can be deduced as:

$$\mathbf{X} = \tfrac{2}{3}C\boldsymbol{\alpha} \quad \text{and} \quad R = h'(p)$$

and the evolution laws as:

$$\dot{\boldsymbol{\varepsilon}}^{\mathrm{p}} = \frac{\partial\Omega}{\partial\boldsymbol{\sigma}} = \frac{3}{2}\frac{\partial\Omega}{\partial J_2}\frac{\boldsymbol{\sigma}' - \mathbf{X}'}{J_2(\boldsymbol{\sigma} - \mathbf{X})}$$

$$\dot{p} = -\frac{\partial\Omega}{\partial R} = \frac{\partial\Omega}{\partial J_2} = (\tfrac{2}{3}\dot{\boldsymbol{\varepsilon}}^{\mathrm{p}}:\dot{\boldsymbol{\varepsilon}}^{\mathrm{p}})^{1/2}$$

$$\dot{\boldsymbol{\alpha}} = -\frac{\partial\Omega}{\partial\mathbf{X}} = \dot{\boldsymbol{\varepsilon}}^{\mathrm{p}} - \frac{3}{2}\frac{\gamma(p)}{C}\mathbf{X}\dot{p}.$$

In place of the relation involving $\dot{\boldsymbol{\alpha}}$ we may use that with $\dot{\mathbf{X}}$:

$$\dot{\mathbf{X}} = \tfrac{2}{3}C\dot{\boldsymbol{\varepsilon}}^{\mathrm{p}} - \gamma(p)\mathbf{X}\dot{p}.$$

The role of each of the two additional terms in the potential is the same as described in the previous section. The relation between $\mathbf{X}$ and $\boldsymbol{\alpha}$ can be further used to reduce the potential to:

$$\Omega = \Omega(J_2(\boldsymbol{\sigma} - \mathbf{X}) - R - k; T, p).$$

The potentials can be specialized still further: the specific free energy by an exponential term in the expression for the function $h(p)$, as in Section 6.4.2, and a power function for the dissipation potential:

$$h(p) = Qp - \frac{Q}{b}[1 - \exp(-bp)]$$

$$\Omega = \frac{K(p)}{n+1}\left\langle \frac{J_2(\boldsymbol{\sigma} - \mathbf{X}) - R - k}{K(p)} \right\rangle^{n+1}$$

where the coefficients n, k, b, R, Q and the function $K(p)$ depend on the temperature.

The set of cyclic viscoplasticity equations with nonlinear kinematic hardening and isotropic hardening under isothermal conditions thus reduces to:

- $$\dot{\boldsymbol{\varepsilon}}^{\mathrm{p}} = \frac{3}{2}\dot{p}\frac{\boldsymbol{\sigma}' - \mathbf{X}'}{J_2(\boldsymbol{\sigma} - \mathbf{X})},$$
- $$\dot{p} = \left\langle \frac{J_2(\boldsymbol{\sigma} - \mathbf{X}) - R - k}{K(p)} \right\rangle^n,$$
- $$\dot{\mathbf{X}} = \tfrac{2}{3}C\dot{\boldsymbol{\varepsilon}}^{\mathrm{p}} - \gamma(p)\mathbf{X}\dot{p},$$
- $$\dot{R} = b(Q - R)\dot{p} \quad \text{with} \quad \mathbf{X}(0) = R(0) = 0.$$

The last equation may be replaced by its integrated form:

$$R = Q[1 - \exp(-bp)].$$

The linear term of the isotropic hardening relation of Section 6.4.2 has been omitted so as to allow cyclic stabilization: with an increasing number of plasticity cycles, p increases indefinitely but R becomes stabilized. It should be noted that there are three ways at our disposal for including the isotropic hardening effects in the constitutive law, by adjusting:

the size of the elastic domain and the equipotential surfaces by the function $R(p)$;
the viscous part of the stress (size of the equipotential surfaces) by the function $K(p)$;
the movement of the centre $\mathbf{X}$ of the elastic domain by the function $\gamma(p)$.

The choice between these different procedures or their combinations, as well as the determination of coefficients n, k, C constitutes one of the difficulties of the problem of identification (see the section dealing with tension–compression).

Study of dissipation
The mechanical power dissipated (or intrinsic dissipation) is given by:

$$\Phi_1 = \boldsymbol{\sigma}:\dot{\boldsymbol{\varepsilon}}^{\mathrm{p}} - \mathbf{X}:\dot{\boldsymbol{\alpha}} - R\dot{p}$$

where $\boldsymbol{\alpha}$ and p, which were used in the previous sections, are the only two independent internal variables.

In the framework adopted, invoking the generalized normality rule, the second principle of thermodynamics is easily verified when the dissipation potential Ω is a convex function of its variables and contains the origin, i.e., when it becomes zero at the origin. For a potential expressed in the form of a power function, for example proportional to:

$$\left\langle J_2(\boldsymbol{\sigma}-\mathbf{X}) - R - k + \frac{1}{2}\frac{\gamma(p)}{C} J_2^2(\mathbf{X}) - \frac{2}{9} C\gamma(p) J_2^2(\boldsymbol{\alpha}) \right\rangle^{n+1}$$

these two conditions follow immediately from the positive character of the coefficients k and C and the function $\gamma(p)$ (for $\boldsymbol{\sigma} = \mathbf{X} = 0$, $R = 0$ the quantity within the angle brackets is negative which then means that Ω is zero).

We may effectively introduce the constitutive equation in mechanical dissipation. We find:

$$\Phi_1 = (\boldsymbol{\sigma}-\mathbf{X}):\frac{3}{2}\dot{p}\frac{\boldsymbol{\sigma}'-\mathbf{X}'}{J_2(\boldsymbol{\sigma}-\mathbf{X})} - R\dot{p} + \frac{\gamma(p)}{C} J_2^2(\mathbf{X})\dot{p}$$

or, in still another form:

$$\Phi_1 = \left[J_2(\boldsymbol{\sigma}-\mathbf{X}) - R - k + \frac{1}{2}\frac{\gamma(p)}{C} J_2^2(\mathbf{X}) - \frac{2}{9} C\gamma(p) J_2^2(\boldsymbol{\alpha}) \right]\dot{p}$$

$$+ \left[k + \frac{1}{2}\frac{\gamma(p)}{C} J_2^2(\mathbf{X}) + \frac{2}{9} C\gamma(p) J_2^2(\boldsymbol{\alpha}) \right]\dot{p}.$$

In view of the relation between $\mathbf{X}$ and $\boldsymbol{\alpha}$, each of the two square bracketed terms can be simplified and each can be shown to be positive separately:

$$\Phi_1 = [J_2(\boldsymbol{\sigma}-\mathbf{X}) - R - k]\dot{p} + \left[k + \frac{\gamma(p)}{C} J_2^2(\mathbf{X}) \right]\dot{p}.$$

The first term is positive or zero since $\dot{p}$ is zero when the term inside the first square bracket is negative, and the second is positive since k, γ and C are positive.

Identification from tension–compression tests

Properties of the nonlinear kinematic hardening rule in tension–compression

The general equations are easily specialized to the tension–compression for which the deviatoric stress tensor and the tensor of 'internal kinematic stress' are expressed by:

$$\sigma' = \begin{bmatrix} \frac{2}{3}\sigma & 0 & 0 \\ 0 & -\frac{1}{3}\sigma & 0 \\ 0 & 0 & -\frac{1}{3}\sigma \end{bmatrix}, \quad X = X' = \begin{bmatrix} \frac{2}{3}X & 0 & 0 \\ 0 & -\frac{1}{3}X & 0 \\ 0 & 0 & -\frac{1}{3}X \end{bmatrix}$$

so that we have : $J_2(\boldsymbol{\sigma}-\mathbf{X})=|\sigma-X|$ and the flow law becomes:

$$\dot{\varepsilon}_p = \left\langle \frac{|\sigma - X| - R - k}{K(\tilde{\varepsilon}_p)} \right\rangle^n \operatorname{Sgn}(\sigma - X)$$

$$\dot{X} = C\dot{\varepsilon}_p - \gamma(\tilde{\varepsilon}_p)X|\dot{\varepsilon}_p|$$

$$\dot{R} = b(Q - R)|\dot{\varepsilon}_p|$$

where $\tilde{\varepsilon}_p$ is the accumulated plastic strain defined by $\dot{\tilde{\varepsilon}}_p = |\dot{\varepsilon}_p|$. The different components of the strain rate tensor are defined by:

$$\dot{\varepsilon}_{p11} = \dot{\varepsilon}_p \qquad \dot{\varepsilon}_{p22} = \dot{\varepsilon}_{p33} = -\tfrac{1}{2}\dot{\varepsilon}_p$$
$$\dot{X}_{11} = \tfrac{2}{3}\dot{X} \qquad \dot{X}_{22} = \dot{X}_{33} = -\tfrac{1}{3}\dot{X}.$$

The two hardening variables for a given load can be made explicit as follows: if $\varepsilon_p = \varepsilon_{p0}$ and $X = X_0$ initially, and if γ is constant:

$$X = \nu\frac{C}{\gamma} + \left(X_0 - \nu\frac{C}{\gamma}\right)\exp(-\nu\gamma(\varepsilon_p - \varepsilon_{p0}))$$

where $\nu = \operatorname{Sgn}(\sigma - X) = \pm 1$ indicates the sense of viscoplastic flow. This shows that the relation between X and ε_p is not one-to-one, the kinematic variable itself describes hysteresis loops (Fig. 6.39(*a*)). In contrast for the isotropic variable, it is possible to integrate from the initial instant and obtain:

$$R = Q[1 - \exp(-b\tilde{\varepsilon}_p)].$$

The flow law may be inverted to yield:

$$\sigma = X(\varepsilon_p, X_0, \varepsilon_{p0}) + \nu R(\tilde{\varepsilon}_p) + \nu k + \nu K(\tilde{\varepsilon}_p)|\dot{\varepsilon}_p|^{1/n}$$

where the functions X and R are given by the above relations. This form makes apparent the additive decomposition between hardening and viscous stress (which is eventually influenced by the hardening):

$$\sigma_v(\tilde{\varepsilon}_p, \dot{\varepsilon}_p) = K(\tilde{\varepsilon}_p)|\dot{\varepsilon}_p|^{1/n}.$$

Among the important properties pertaining to cyclic loading, already referred to in the chapter on plasticity (Section 5.4.4), we point out the following which are valid when the cyclic hardening or softening is stabilized, i.e., when R, γ, and K are constants:

under controlled stress loading, response stabilization is obtained in less than two cycles;

Fig. 6.39. Decomposition of stress into internal and viscous parts: (*a*) cyclic loading; (b) relaxation.

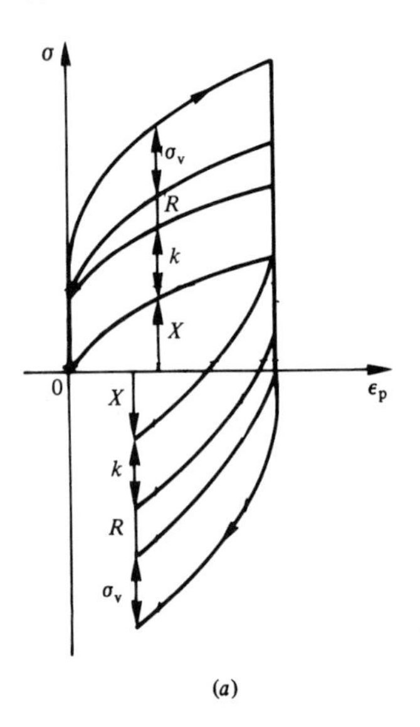

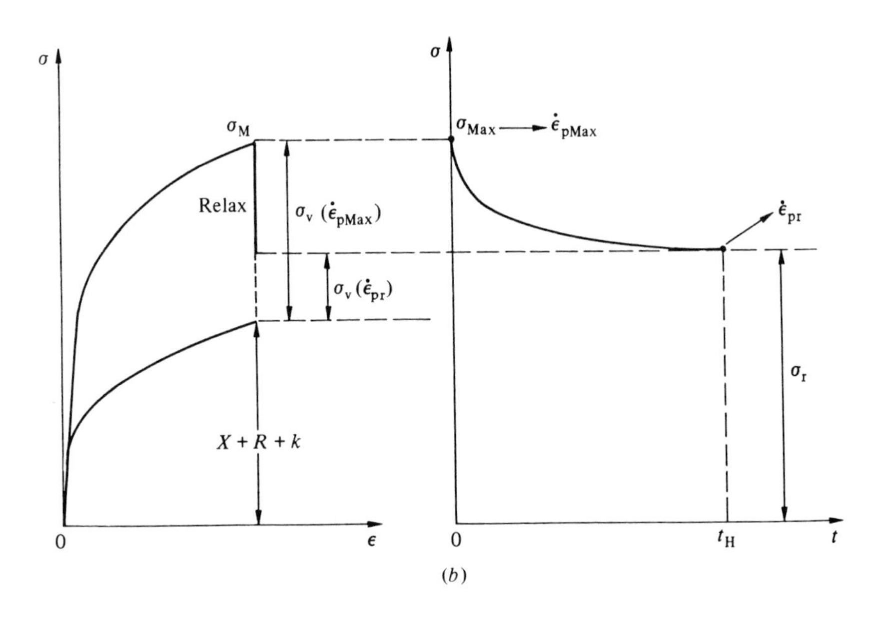

under controlled strain, stabilization can take a few cycles depending on the initial value of X and the values of C and γ; if the strain is controlled with the same rate of loading and unloading, the model gives the relaxation effect of the mean stress; stabilization (with flow) can occur when $X_{\min} = -X_{\mathrm{Max}}$. The

Fig. 6.40. Balance of dissipation; (*a*) in tension. (*b*) in tension–compression, when $X = 0$.

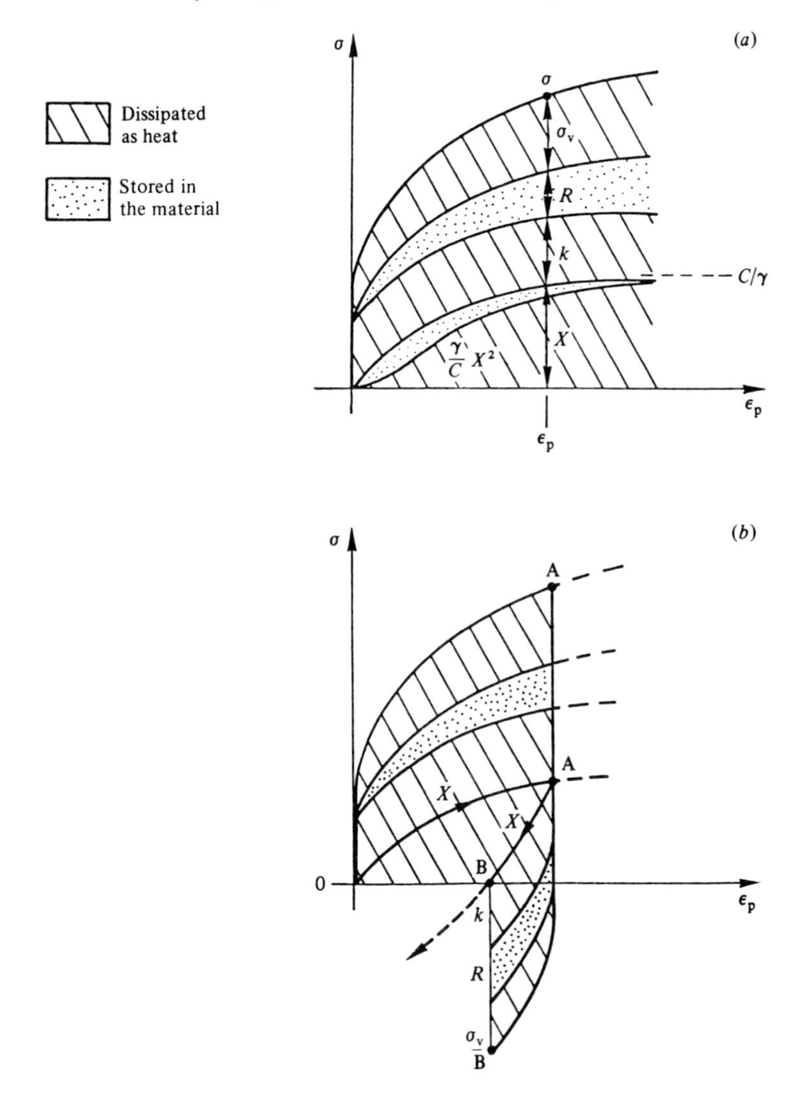

relation for X gives the maximum value:

$$X_{\text{Max}} = \frac{C}{\gamma} \tanh\left(\gamma \frac{\Delta\varepsilon_{\text{p}}}{2}\right)$$

where $\Delta\varepsilon_{\text{p}}$ is the strain range during the cycle;
in cyclic relaxation, i.e., with a strain, control including a holding time in tension, for example, the stress relaxed at each cycle

$$\delta\sigma = \sigma_{\text{Max}} - \sigma_{\text{r}}$$

evolves (or does not evolve) at the same time as the function $K(\tilde{\varepsilon}_{\text{p}})$. The following relation, wherein the index s stands for the stabilized cycle, can be shown to be approximately satisfied:

$$\frac{\delta\sigma_{\text{s}}}{\delta\sigma_1} = \frac{\sigma_{\text{Max s}} - \sigma_{\text{rs}}}{\sigma_{\text{Max}_1} - \sigma_{\text{r}_1}} \approx \frac{K_{\text{s}}\, \dot{\varepsilon}_{\text{p Max s}}^{1/n} - \dot{\varepsilon}_{\text{prs}}^{1/n}}{K_1\, \dot{\varepsilon}_{\text{p Max}_1}^{1/n} - \dot{\varepsilon}_{\text{pr}_1}^{1/n}} \approx \frac{K_{\text{s}}}{K_1}$$

using the notation defined in Fig. 6.39(*b*). The example of the 316 L steel, in Section 6.2.6 is a good example of the evolution of this relaxed stress.

Balance of dissipated power
The intrinsic dissipation is now expressed by:

$$\begin{aligned}\Phi_1 &= \sigma\dot{\varepsilon}_{\text{p}} - X\dot{\alpha} - R|\dot{\varepsilon}_{\text{p}}| \\ &= (|\sigma - X| - R - k)|\dot{\varepsilon}_{\text{p}}| + k|\dot{\varepsilon}_{\text{p}}| + [\gamma(\tilde{\varepsilon}_{\text{p}})/C]X^2|\dot{\varepsilon}_{\text{p}}|.\end{aligned}$$

The first term corresponds to viscous dissipation, the second to dissipation caused by the initial internal stresses, and the third to kinematic dissipation. Fig. 6.40 illustrates this decomposition of dissipation during the first tensile loading. We may write:

$$\sigma\dot{\varepsilon}_{\text{p}} = [\sigma - X - R + (\gamma/C)X^2]\dot{\varepsilon}_{\text{p}} + R\dot{\varepsilon}_{\text{p}} + [X - (\gamma/C)X^2]\dot{\varepsilon}_{\text{p}}.$$

Starting from the initial state, the dissipated plastic work:

$$\int_0^A \sigma\, \mathrm{d}\varepsilon_{\text{p}}$$

may be separated into the parts

Φ_1, released by the system (dissipated as heat for example),
$\int_0^A R\, \mathrm{d}\varepsilon_{\text{p}}$ which is blocked within the system,
$\int_0^A [X - (\gamma/C)X^2]\, \mathrm{d}\varepsilon_{\text{p}}$ 'blocked' temporarily, available for allowing fading of the kinematic stress X during the subsequent compression loading. It will be dissipated as heat at that time.

It should be kept in mind that the energy blocked in the system is nothing other than that part of the energy which corresponds to the internal variables:

$$\rho\Psi_p = \frac{1}{2}C\boldsymbol{\alpha}:\boldsymbol{\alpha} + Q\left[p + \frac{1}{b}\exp(-bp) - \frac{1}{b}\right]$$

and that we therefore have:

$$\int_0^A R\,\mathrm{d}\varepsilon_p = Q\left[\varepsilon_{p_A} + \frac{1}{b}\exp(-b\varepsilon_{p_A}) - \frac{1}{b}\right]$$

$$\int_0^A \left(X - \frac{\gamma}{C}X^2\right)\mathrm{d}\varepsilon_p = \frac{1}{2}C\alpha_A^2 = \frac{1}{2}\frac{X_A^2}{C}.$$

At the point B, when X is again equal to zero, we get back to the initial condition and the only energy stored is that due to the isotropic variable. It is then clear how calorimetric measurement recording in the order of 90% of the dissipated energy as heat, may be taken into consideration.

Identification of coefficients

The modelling of monotonic and cyclic behaviours requires the introduction of a large number of coefficients. Their identification for a given material (at a given temperature) is a difficult problem. The logical method has three steps:

choice of the characteristic tests and interpretation of the mechanical quantities ($\sigma(t)$, $\varepsilon_p(t)$). This choice depends on the envisaged field of application (initial loading, stabilized cycle, transient cyclic effects);

choice of the modelling; in particular the choice of the form of the isotropic hardening which can incorporate the effects of cyclic accommodation. The comparison of an initial tension curve with the stabilized hysteresis loops (for the same load) is extremely helpful for detecting changes in the size of the apparent elastic domain (variation of R) or changes in the apparent hardening modulus for the same strain (variation of γ). Fig. 5.43 illustrates the corresponding variations;

identification of the coefficients themselves: this is the most difficult step, but it is the one for which computer-aided identification may be used.

In most cases we limit ourselves first to the determination of the following coefficients:

n, K, k, C, γ for the initial behaviour,
n, K, $k+Q$, C, γ for behaviour at the stabilized cycle, and
b to eventually represent transient cyclic behaviour.

These seven coefficients are identified independently from tests performed in the corresponding domains: the response to the stablized cycle, which is usually independent of the previous loading, can be obtained directly by using the equation in the stabilized regime ($b=0$, $R=Q$). The calculation of 2–3 cycles is then sufficient, as shown in the previous section. Note also that the constant b has no influence on stabilized behaviour and little on the response during the first cycle. We thus have to determine successively five coefficients for the initial behaviour, five coefficients for the stabilized regime, and one coefficient for the transitory regime. Finally, the corresponding values of n, K, C, γ are averaged, unless it is considered necessary to also vary K or γ (as mentioned earlier). Note that the separation of coefficients k and $k+Q$ is not possible except by measuring dissipation (by calorimetric measurements). However, this distinction is unimportant for the mechanical response of the model.

Despite the above separations and the availability of numerical methods, the number of coefficients to be identified can prove too high, either because the available experiments are not sufficient to account correctly for all the coefficients, or because of poor convergence of the minimization algorithms. These two reasons are often found together. We may therefore resort to a preliminary treatment which consists in partially identifying the viscosity effects, i.e., the exponent n. Three complementary methods may be used.

(1) Start by determining the exponent N^* of Norton's law for the material if secondary creep tests at the temperature under consideration are available. Experiment shows that, in general, n is bounded by:

$$0.6N^* < n < N^*.$$

The value of n can then be set more or less arbitrarily with the help of similar identifications performed at other temperatures: the relation $n(T)$ may generally be assumed to be monotonic (decreasing) and homothetic to the relation $N^*(T)$. This first process is used mostly when sufficient accurate experimental information is lacking in the expected range of strain rates (not long enough relaxation, no tests at sufficiently different rates or frequencies).

(2) Use one (or several) of the relaxation tests showing a range of at least three orders of magnitude for the rates to establish an experimentally based

relationship between the exponent and the internal stress σ_i of the relation

$$\dot{\varepsilon}_p = \left(\frac{\sigma - \sigma_i}{K_\infty}\right)^n$$

for different (arbitrary) values of the parameter σ_i which globally represents:

$$\sigma_i = X(\varepsilon_p) + R(\tilde{\varepsilon}_p) + k.$$

This procedure is applicable only if the hardening $X + R$ does not vary too much during the relaxation, that is, if the relaxation is performed with an already high initial plastic strain (0.2% for example). Fig. 6.24 illustrates this.

(3) Define an average experimental point $(\sigma^*, \varepsilon_p^*, \dot{\varepsilon}_p^*)$ of an average test through which the corresponding calculated curves must be made to pass.

The numerical methods available for facilitating the identification are of two kinds (see Chapter 3):

algorithms for calculating (step-by-step in time) the response to an imposed load when the solution is not explicit,
methods of error minimization (between the model and the experiment) which are either entirely automatic or based on an interactive graphics system. These methods can be used only when the number of unknowns is not too large.

The operation of identification itself then consists in using these numerical methods after going through the preliminary treatment mentioned above ((1)–(3)). The number of independent coefficients can be reduced by 2 by using the relation obtained in (2) and (3); thus there remains a maximum of three coefficients for each of the steps mentioned above.

Material properties – domain of validity

A few characteristic coefficients are given in Table 6.3, where one part describes stabilized cyclic behaviour (the first five coefficients) and the other describes the first cycle and transient evolutions (the last four coefficients). As for the isotropic-hardening law, these coefficients are highly dependent on the temperature. Fig. 6.41 for example, enables us to compare the evolution with temperature of three viscosity coefficients for the IN 100 alloy: N^* corresponding to Norton's law (without hardening), N to the multiplicative isotropic-hardening law of Section 6.4.2., and n to the nonlinear kinematic(additive)-hardening law.

Table 6.3. *Coefficients of the nonlinear kinematic-hardening and isotropic-hardening laws for cyclic evolution. The function $\gamma(p)$ is of the form:* $\gamma(p) = \gamma_s + (\gamma_0 - \gamma_s)\exp(-\beta p)$.

Material	T(K)	n	K	k	C	γ_0	$k+Q$	γ_s	b	β
316 strainless steel	20	24	151	82	162 400	2 800	142	2 800	8	0
316 stainless steel	600	12	150	6	24 800	300	86	300	10	0
TA6V alloy	350	40	190	200	91 000	350	139	350	1.6	0
INCO 718 alloy	550	40	140	500	210 000	420	320	420		0
IN 100 alloy	700	28	580	110	678 000	1 750	90	2 060	400	1 500
IN 100 alloy	800	14	630	80	615 000	1 530	50	1 870	400	1 200
IN 100 alloy	900	9	490	60	362 500	1 200	25	1 540	200	1 000
IN 100 alloy	1 000	7.5	450	30	139 500	940	15	1 200	100	400
IN 100 alloy	1 100	6.5	420	17	41 000	700	7	830	60	300

Fig. 6.41. Evolution of the three types of viscosity exponents as functions of temperature for the IN 100 alloy.

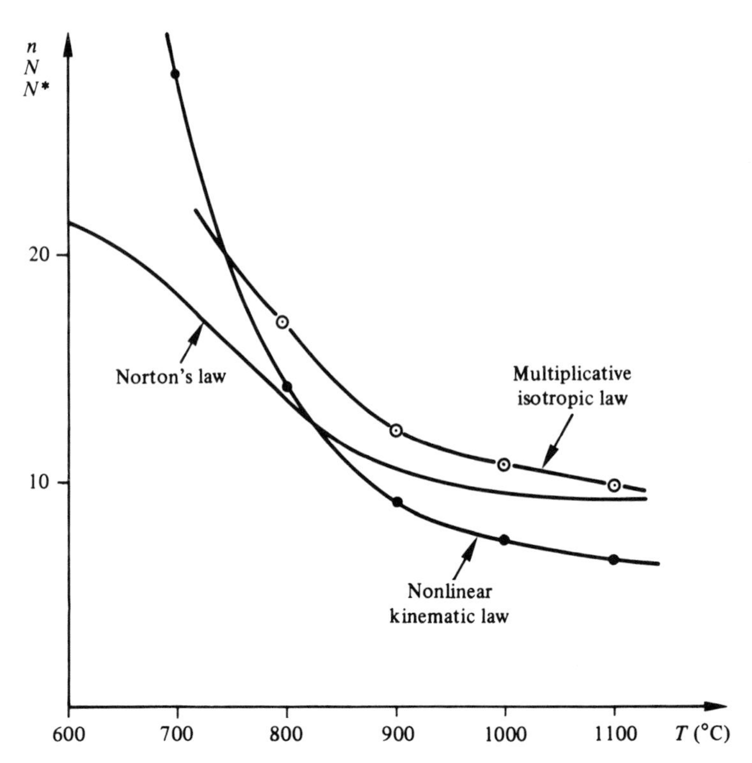

Fig. 6.42. Description of the first cycles on the IN 100 alloy (coated); (*a*), (*c*) for controlled force; (*b*), (*d*) for controlled displacement. Temperature 900 °C for (*a*), (*b*), and 1000 °C for (*c*), (*d*).

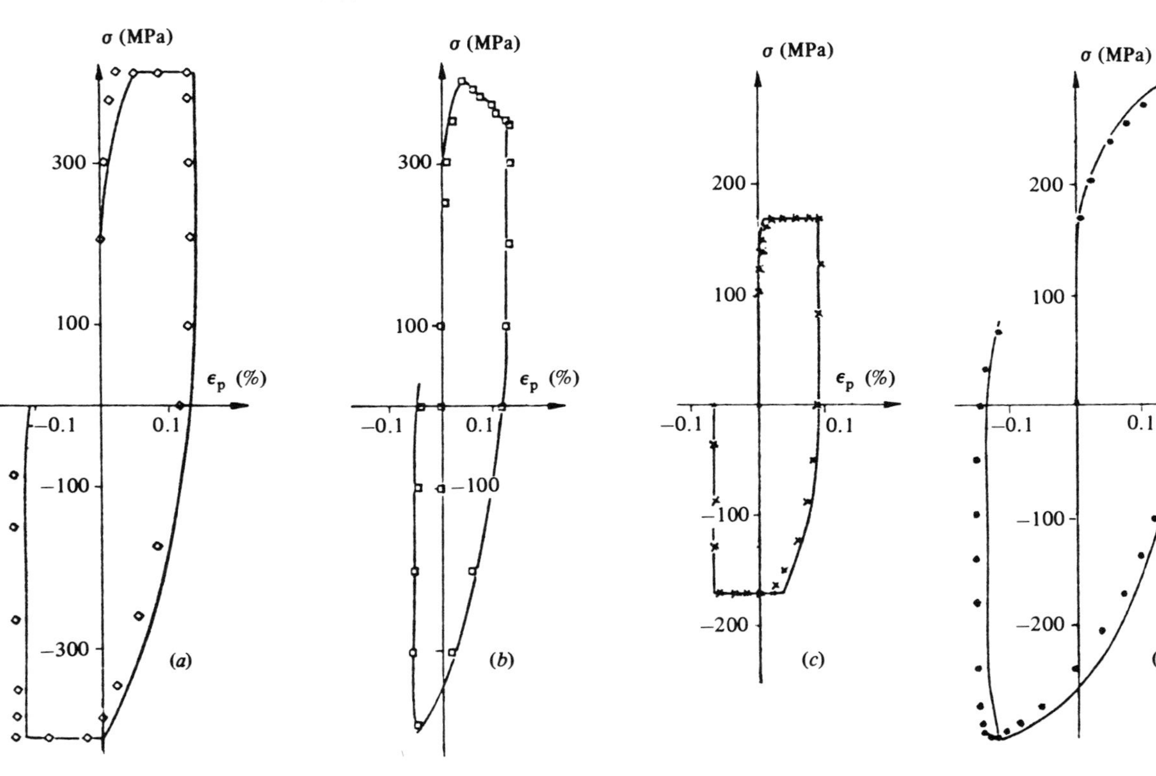

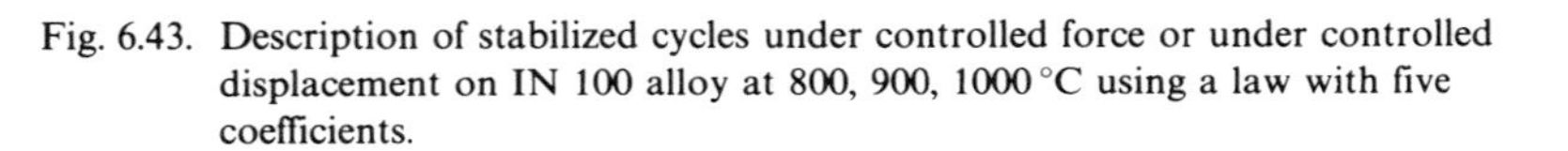

Fig. 6.43. Description of stabilized cycles under controlled force or under controlled displacement on IN 100 alloy at 800, 900, 1000 °C using a law with five coefficients.

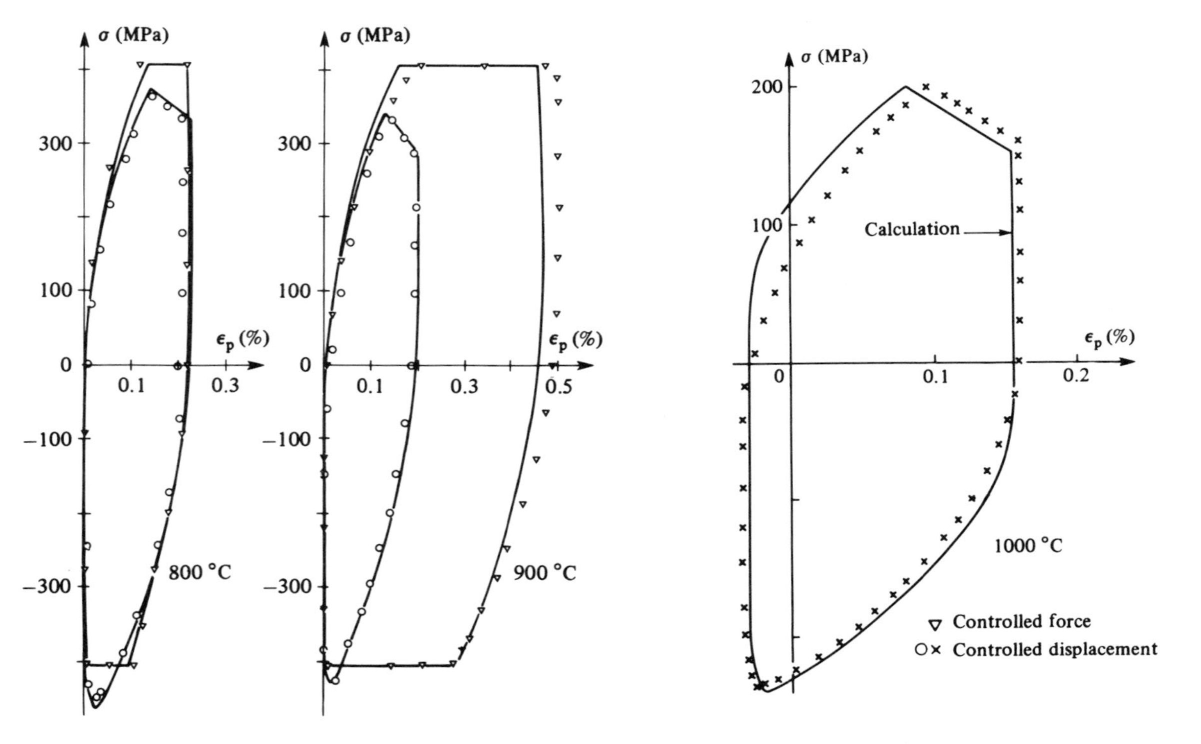

Figs. 6.42–6.44 show examples of comparison between experimental results and those obtained by modelling the IN 100 alloy. In particular, we note that the model accounts for:

the Bauschinger effect (reduction of the elastic limit in compression after the first tension);
primary creep and relaxation during the first tension;
the first cycle (Fig. 6.42) and the stabilized cycle in a strain or stress-controlled test (Fig. 6.43), with or without holding time. Holding at a constant force produces creep, i.e., an increase (or decrease) of strain at constant stress. Holding at constant displacement produces stress relaxation: the relation between $\delta\sigma$ and $\delta\varepsilon_p$ is then linear, it corresponds to a decrease in elastic elongation.
different cyclic curves as functions of the holding time or the frequency (Fig. 6.44). Note that the frequency intervals going from 0.006 to 5 Hz are sufficiently well represented.

The validity of the nonlinear kinematic model is, of course, as in all other cases, limited by certain considerations:

Fig. 6.44. Description of cyclic curves on the IN 100 alloy at 1000 °C (coated). Influence of holding time or frequency.

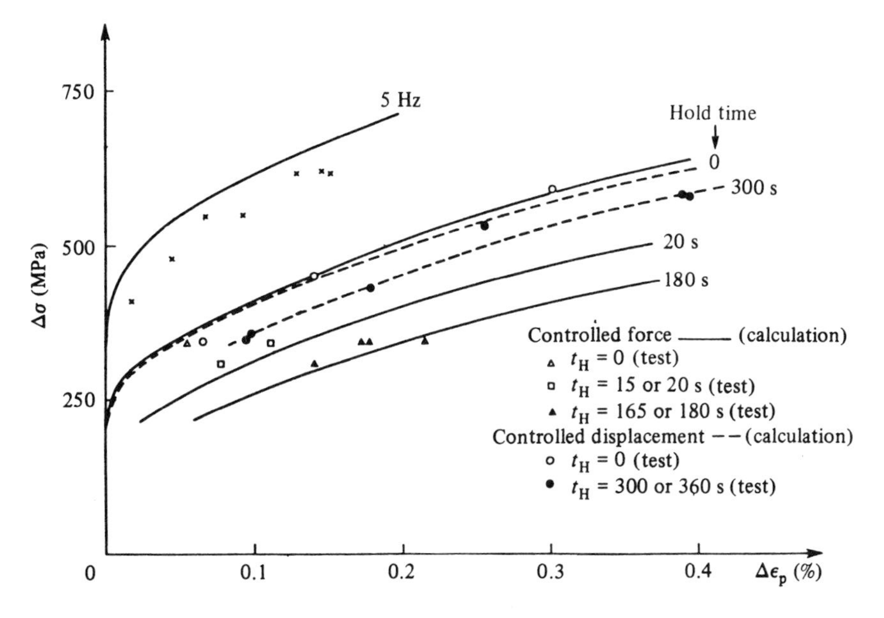

the range of temperature must correspond to a sufficiently high viscosity of the material,

the range of strain rate must be between 10^{-8} and $10^{-3}\,s^{-1}$: for the lowest strain rates, mainly in creep, it is necessary to account for recovery effects (see Section 6.4.4),

the model exhibits secondary creep at high stresses, the secondary creep rate is given by:

$$\dot{\varepsilon}_p^* = \left\langle \frac{\sigma - C/\gamma - Q - k}{K} \right\rangle^n$$

where Q is the value of R at saturation. To represent secondary creep at low stresses, it is necessary to introduce recovery effects (we shall see in Chapter 7 that creep damage is also liable to affect secondary creep).

The principal limitation is linked to the strain. The model predicts a saturation of kinematic hardening under increasing strain, the speed of the saturation being fixed by the coefficient γ. If it is chosen so as to represent small strains ($<0.5\%$ for example) correctly, the hardening effects are not well represented above 1% (at least in the stabilized regime). Comparison between IN 100 and INCO 718 illustrates this difficulty well: for the former the strains are limited to 0.5% and the modelling is therefore excellent; for the latter where the range goes up to 2%, the agreement is not so good.

As for plasticity (cf. Section 5.4.4), the range of validity in strain may be enlarged by introducing a second kinematic variable which represents hardening in the range where the first variable is saturated. The centre of the elastic domain can then be written as

$$\mathbf{X} = \sum_i \mathbf{X}_i$$

where each of the kinematic variables is governed by a similar equation (in a decoupled way):

$$\dot{\mathbf{X}}_i = \tfrac{2}{3} C_i \dot{\boldsymbol{\varepsilon}}^p - \gamma_i \mathbf{X}_i \dot{p}.$$

The introduction of the additional coefficients C_i, γ_i allows the refinement of the modelling as shown in Fig. 6.45 for the INCO 718 alloy. With a single kinematic variable, the calculations were totally wrong for the test with the higher range ($\Delta\varepsilon_p = 2.5\%$). The $\sigma - \varepsilon_p$ cycles in the figure have indefinite origins; they are made to coincide at the point $\sigma = 0$ in passing from compression to tension.

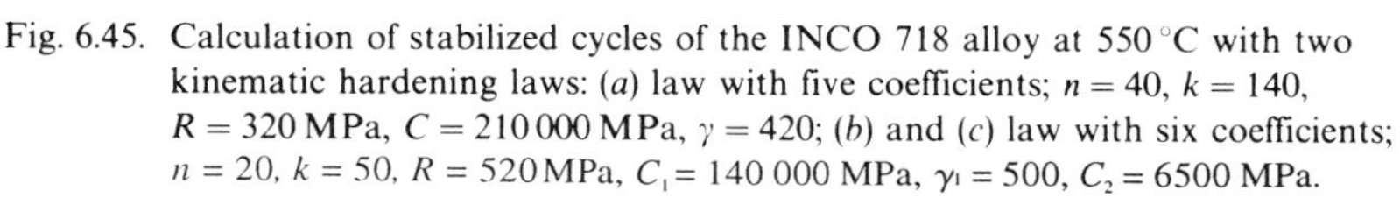

Fig. 6.45. Calculation of stabilized cycles of the INCO 718 alloy at 550 °C with two kinematic hardening laws: (*a*) law with five coefficients; $n = 40$, $k = 140$, $R = 320$ MPa, $C = 210\,000$ MPa, $\gamma = 420$; (*b*) and (*c*) law with six coefficients; $n = 20$, $k = 50$, $R = 520$ MPa, $C_1 = 140\,000$ MPa, $\gamma_1 = 500$, $C_2 = 6500$ MPa.

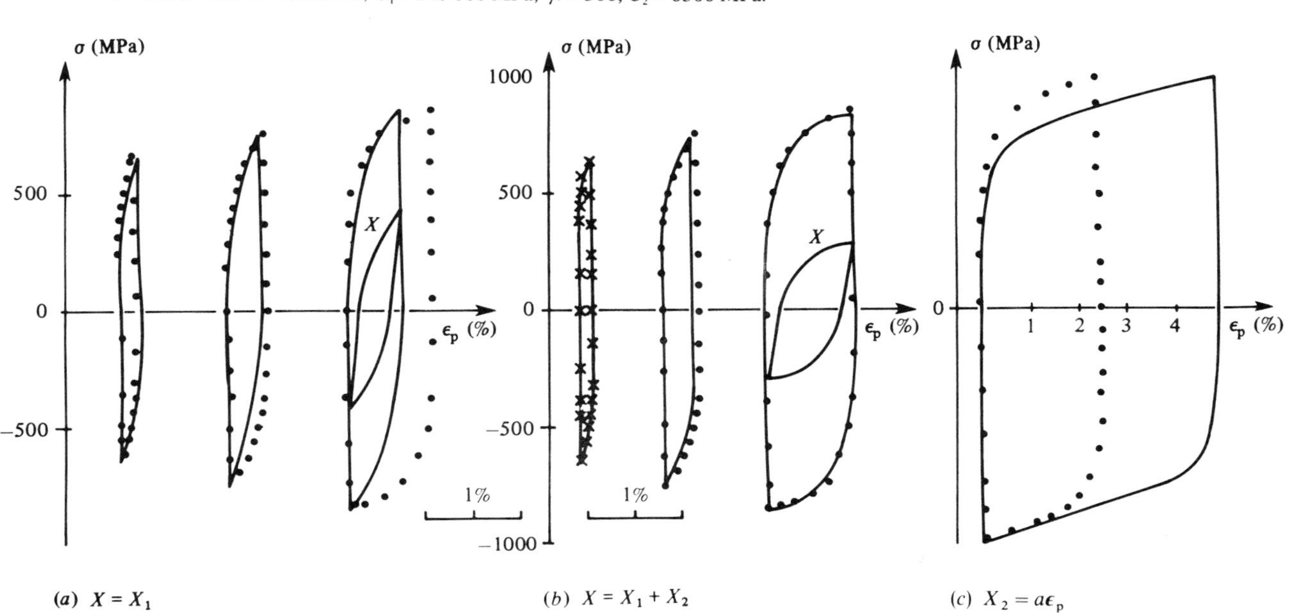

6.4.4 *Modelling of particular effects*

In the high-temperature range, we observe a large variety of behaviours: different phenomena which often correspond to an instability of the physical structure of the material can interact with viscoplasticity and produce complex effects such as effects of temperature history, recovery effects, strain ageing effects, and combinations of plasticity and viscoplasticity.

Effects of temperature history

The viscoplastic laws studied in the preceding sections are applicable to varying temperatures for stable materials. The expressions determined from isothermal tests involve coefficients which depend on the temperature in a unique way. This hypothesis has generally been well verified as long as the temperature varies at a rate lower than approximately 30 °C per second.

However, in some materials (and in the high-temperature range), the

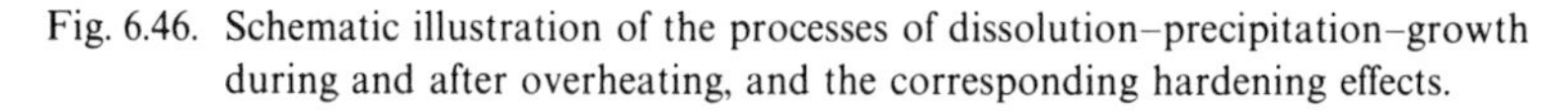

Fig. 6.46. Schematic illustration of the processes of dissolution–precipitation–growth during and after overheating, and the corresponding hardening effects.

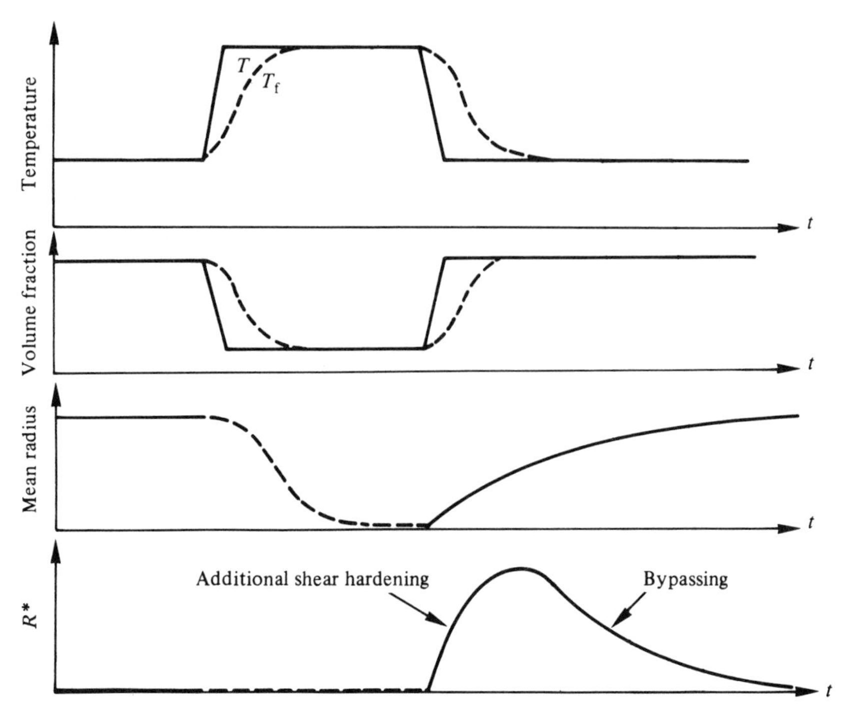

microstructure can evolve in an unstable way with the variation in temperature. In a biphase alloy such as the IN 100 (phase γ and precipitates of phase γ'), a rise in temperature from T_0 to T_1 produces a partial dissolution of average precipitates (initially stable), which corresponds to a decrease in the volume fraction of precipitates (Fig. 6.46); when the temperature returns to T_0, there is again a precipitation of a much more refined phase γ' which induces a higher resistance to deformation than exists under normal conditions. Fig. 6.47 illustrates this process. Later, at the temperature T_0, there is a coalescence of the precipitates which return to their normal size.

Such instability effects which are induced by the evolution of the temperature can be described by adding two internal variables T_f and r^* which correspond globally to the volume fraction and to the mean radius of the hardened precipitates. The associated thermodynamic variables are S_f and R^*. A few remarks will enable us to choose satisfactory evolution equations.

When the temperature varies, the fraction volume does not immediately reach its equilibrium state; there is a retardation effect which may be described globally by:

$$\dot{T}_f = (T - T_f)/\alpha.$$

We see that T_f acts as a retarded fictitious temperature.

A simple modelling consists in decomposing the dissipation potential into the sum of a viscoplastic potential Ω_p (of the same form as that for the stable case) and an instability potential Ω_i dependent upon the variables associated with the instability effect:

$$\Omega_i = \Omega_i(S_f, R^*).$$

The size of precipitates influences the viscoplastic flow. Only the difference between the actual state (in the course of instability) and the normal state (the metastable state) is taken into account and the variable R^* corresponds to an additional hardening. We replace R by $R + R^*$ in the viscoplastic potential since it is reasonable to consider the effect as an isotropic one:

$$\Omega_p = \Omega_p(J_2(\boldsymbol{\sigma} - \mathbf{X}) - R - R^* - k)$$

and R^* is assumed to be always positive. The volume fraction does not introduce any hardening effect; therefore Ω_p does not depend on S_f.

Fig. 6.47 Microstructure of the IN 100 (× 20 000); (*a*) at room temperature after pretreatment; (*b*) after 4 h at 900 °C; (*c*) after 4 h at 900 °C and overheating 10 min at 1000 °C; (*d*) after 4 h at 900 °C 10 min at 1000 °C and 10 min at 900 °C.

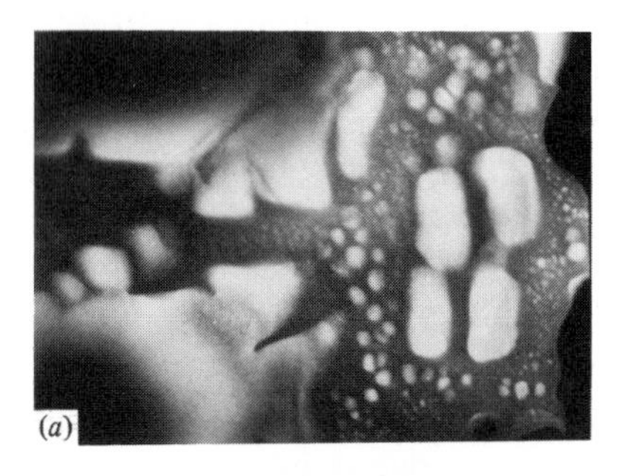

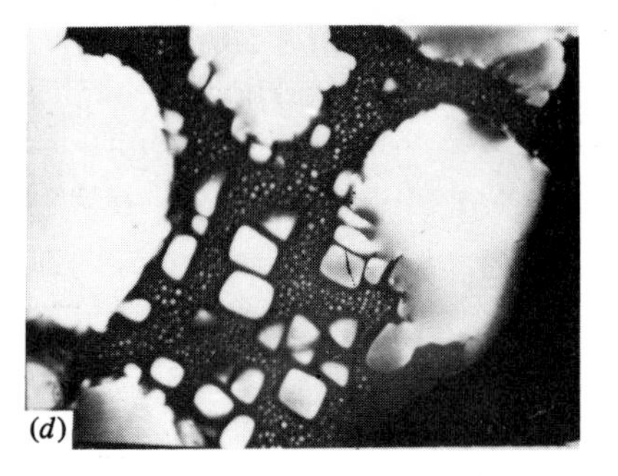

The evolution of R^* during and after a decrease in temperature produces two phases during which there is first an increase and then a decrease in hardening and a return to the normal state (Fig. 6.46). Observations as well as physical models indicate that these phases correspond to different mechanisms of dislocations crossing over the precipitates (shearing and then bypassing).

A qualitatively correct model, which gives satisfactory results, can be established within the framework of general thermodynamics. A simplified version of this model can be written as:

$$\dot{T}_\mathrm{f} = (T - T_\mathrm{f})/\alpha$$
$$\dot{R}^* = a(T_\mathrm{f} - T) - R^*/\beta, \quad R^* \geqslant 0.$$

The coefficients a, α, β are strongly temperature dependent like the phenomena of dissolution, precipitation, and precipitate coalescence. Fig. 6.48 can be used to ascertain the validity of this model for the IN 100 alloy.

Fig. 6.48. Hardening and evanescence periods of the IN 100 alloy. The loading and temperature are as indicated, $\Delta\varepsilon_{\mathrm{p\,sta}}$ is the plastic strain range before overheating. The overheating temperature is indicated on each curve (after Cailletaud).

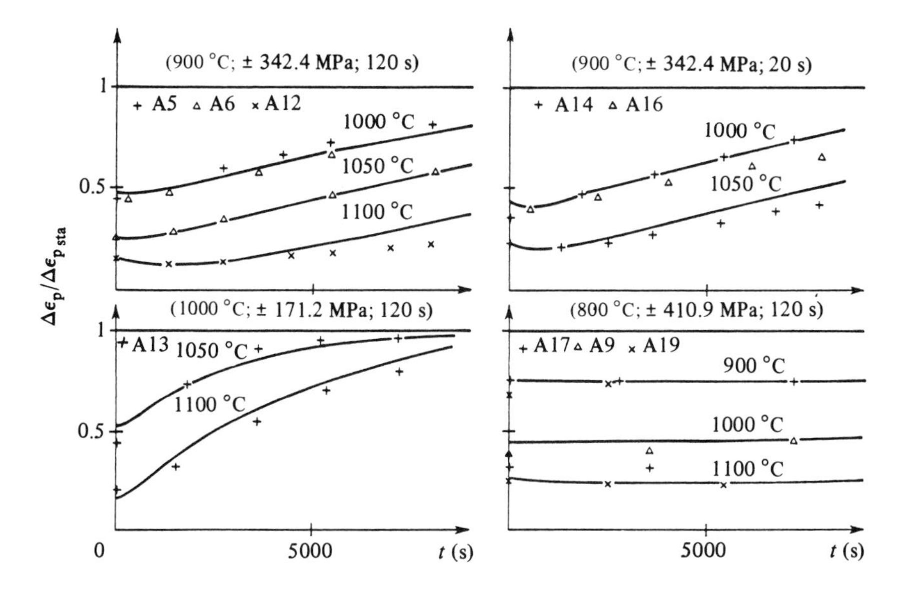

Restoration and recovery effects

The partial restoration of plastic strain can be demonstrated at high temperature for a highly viscous material: after deformation, in creep for example, and then elastic unloading, there is a progressive decrease in the residual strain at zero force while the temperature is held constant. This effect is taken into account by a nonlinear kinematic-hardening model as shown in Fig. 6.49: if the elastic domain $(R + k)$ is small, there is negative flow at zero stress. The recoverable strain (after an infinite time) is given by the position of point C such that $\sigma = X - R - k = 0$.

On the other hand, the recovery effect, which also occurs at high temperature, implies a partial loss of the hardening effect: under the action of thermal agitation, there is a slow recovery of the crystalline structure of the metal by annihilation of the dislocations and redistribution (or relaxation) of internal stresses. This recovery therefore produces a decrease in the mechanical resistance.

The equations which describe time recovery are, at the level of internal stresses, similar to those which describe relaxation of macroscopic stress in

Fig. 6.49. Description of recovery by kinematic hardening.

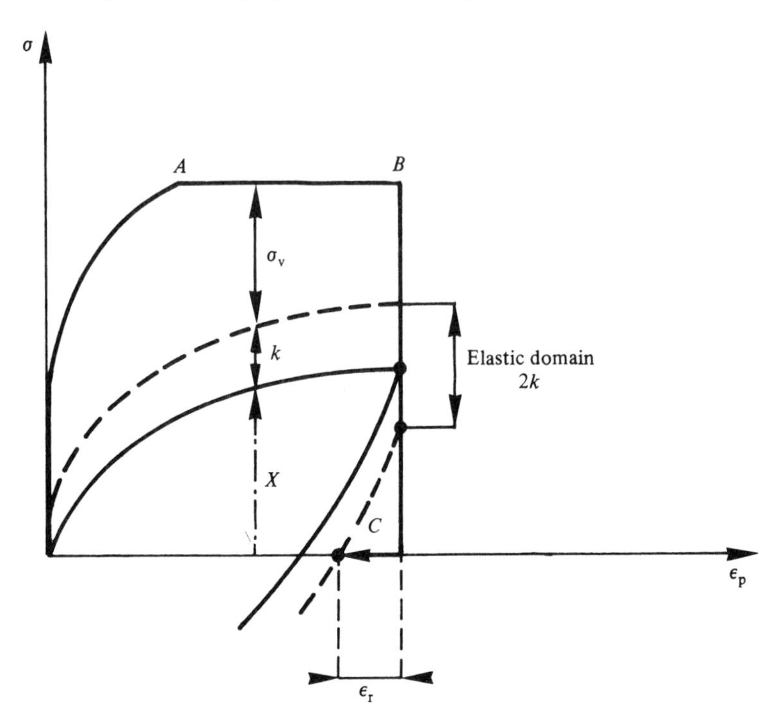

the relaxation test (described here by Norton's law):

$$\dot{\sigma} = E(\dot{\varepsilon} - \dot{\varepsilon}_p) = E[\dot{\varepsilon} - (\sigma/\lambda^*)^{N^*}] = -E(\sigma/\lambda^*)^{N^*}.$$

For internal stresses, we write an equation of the same form:

$$\dot{\sigma}_i = E_p[\dot{\varepsilon}_p - (\sigma_i/\lambda_i)^{N_i}]$$

where the first term $E_p\dot{\varepsilon}_p$ gives the hardening effect due to deformation. There is a decrease in σ_i when $\dot{\varepsilon}_p$ is zero or small:

$$\dot{\sigma}_i = -E_p(\sigma_i/\lambda_i)^{N_i}.$$

These recovery equations may be written for kinematic as well as isotropic variables. By using the dissipation potential in the form:

$$\varphi^* = \Omega_p(\boldsymbol{\sigma}, \mathbf{X}, R; T, \boldsymbol{\alpha}, r) + \Omega_r(\mathbf{X}, R)$$

where Ω_p is the flow potential explicitly mentioned in Section 6.4.3 and Ω_r is a recovery potential which can be written, for example, as:

$$\Omega_r = \frac{A}{a+1}\left[\frac{J_2(\mathbf{X})}{A}\right]^{a+1} + \frac{B}{b+1}\left(\frac{R}{B}\right)^{b+1}.$$

Application of the general equations then gives:

- $$\dot{\boldsymbol{\alpha}} = \dot{\boldsymbol{\varepsilon}}_p - \frac{3}{2}\frac{\gamma}{C}\mathbf{X}\dot{p} - \left[\frac{J_2(\mathbf{X})}{A}\right]^a \frac{3}{2}\frac{\mathbf{X}}{J_2(\mathbf{X})},$$
- $$\dot{r} = \dot{p} - (R/B)^b.$$

The free energy may be chosen in the same manner as in the previous section:

$$\rho\Psi = \rho\Psi_e + \tfrac{1}{3}C\boldsymbol{\alpha}:\boldsymbol{\alpha} + h(r)$$

so that we then have:

$$\mathbf{X} = \tfrac{2}{3}C\boldsymbol{\alpha} \quad \text{and} \quad R = h'(r).$$

By letting $g(R) = h''(r) = h''[h'^{-1}(R)]$, the above equations can be expressed in forms which involve only the variables $\mathbf{X}$ and R:

- $$\dot{\mathbf{X}} = \tfrac{2}{3}C\dot{\boldsymbol{\varepsilon}}^p - \gamma\mathbf{X}\dot{p} - C\left[\frac{J_2(X)}{A}\right]^a \frac{\mathbf{X}}{J_2(\mathbf{X})},$$

  $$\dot{R} = g(R)\dot{p} - g(R)(R/B)^b.$$

These equations describe not only the time recovery effects of hardening, but also some of the effects of the load history, such as those observed in the creep tests at two levels of stress (i.e., creep acceleration and delay). Fig. 6.50 illustrates in a schematic way tests with increasing stress levels: for the same plastic strain in the basic tests under σ_1 and σ_2, the state of internal stress is not the same due to the fact that the ratio between the hardening term in $\dot{\varepsilon}^p$ or $\dot{p}$ and the recovery term is not the same. The second stress level

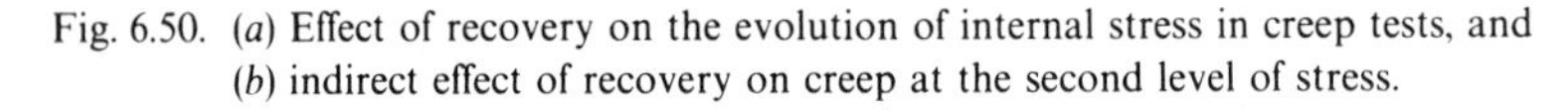

Fig. 6.50. (*a*) Effect of recovery on the evolution of internal stress in creep tests, and (*b*) indirect effect of recovery on creep at the second level of stress.

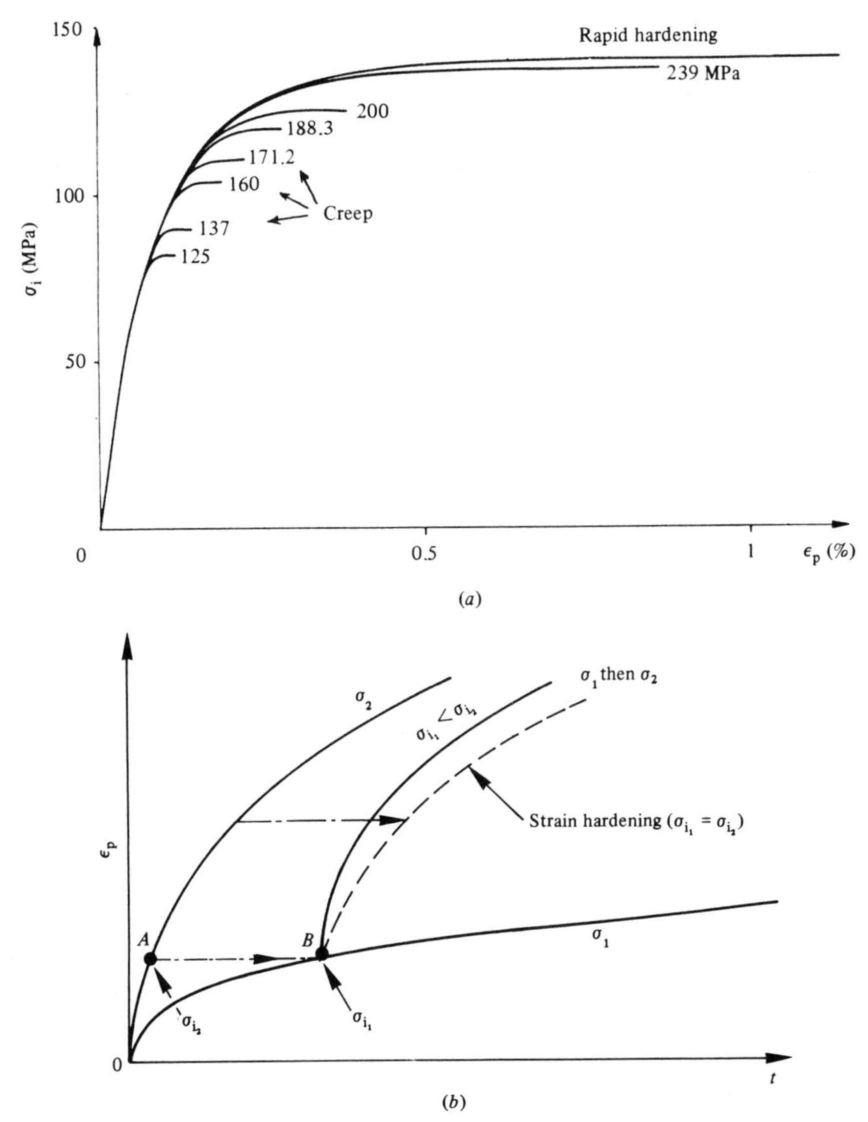

therefore starts with an internal stress σ_{i_1}, lower than the one normally present, which results in the acceleration of the effect as compared to strain-hardening (without recovery).

Creep delay for $\sigma_2 < \sigma_1$ can be described as being due to the same reasons: at the start, since $\sigma_{i_1} > \sigma_{i_2}$, a blockage of flow occurs which progressively disappears due to the recovery effect (σ_i decreases progressively from σ_{i_1} to σ_{i_2}). It is to be noted that the recovery effect also produces a decrease of the viscosity exponent for low rates.

Recovery effects are also observed in cyclic tests with holding periods (see Section 6.2.6). For a long holding time t_h, the stress amplitude of the stabilized cycle decreases when the strain amplitude and the strain rate are kept constant during the transitions (Fig. 6.16(*b*)). Only recovery effects on the isotropic variable can explain such observations. By using, for example, a simplified equation of the form:

$$\dot{R} = b(Q - R)\dot{p} - \beta(R - Q_r)$$

we find R for the stabilized cycle:

$$R_s = \frac{2bQ\Delta\varepsilon_p + \beta Q_r t_h}{2b\Delta\varepsilon_p + \beta t_h}$$

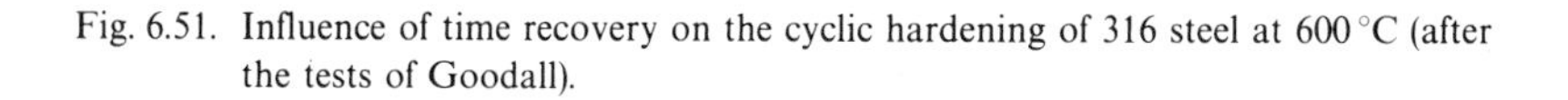

Fig. 6.51. Influence of time recovery on the cyclic hardening of 316 steel at 600 °C (after the tests of Goodall).

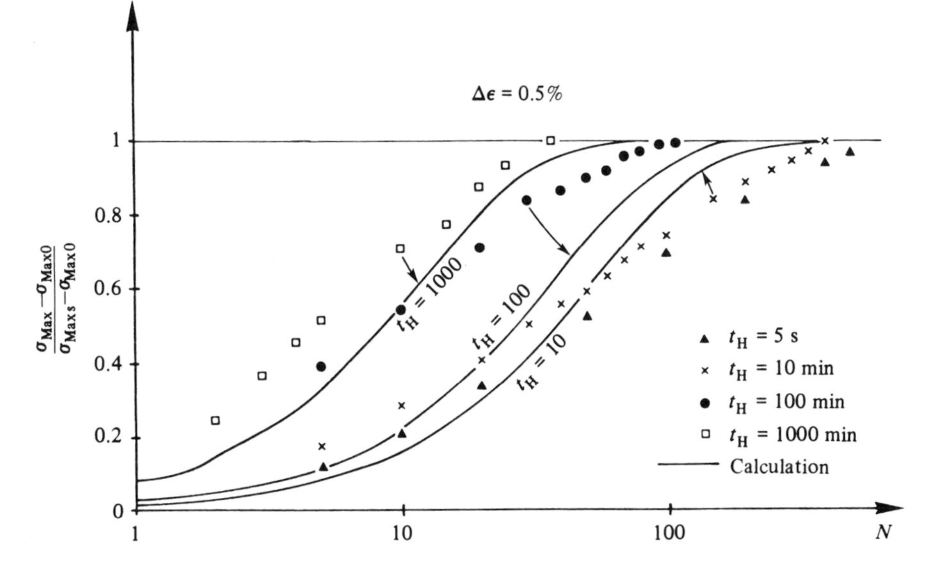

and the evolution of the maximum stress:

$$\frac{\sigma_{\text{Max}} - \sigma_{\text{Max}\,0}}{\sigma_{\text{Max}\,s} - \sigma_{\text{Max}\,0}} \approx \frac{R}{R_s} = 1 - \exp(-(2b\Delta\varepsilon_p + \beta t_h)N).$$

With this simple form, we immediately obtain evolutions consistent with the experimental results as shown in Fig. 6.51 (without the recovery effect, the evolution curve of the maximum stress would be unique).

Effects of ageing

We distinguish these effects from those of recovery for three main reasons:

- they can occur without previous deformation of the material,
- they really correspond to a microstructural instability, the structure of the material evolving as a function of time, and
- quite often this evolution produces a hardening of the alloy.

Ageing phenomena are thermally activated and occur more rapidly at high temperatures. Their influence on the mechanical properties (varying with time) is, however, significant only in the intermediate-temperature range in which evolution of the microstructure occurs in a period of time which is of the same order as the duration of the mechanical loads on the material or the structure.

The physical causes of these instabilities can be numerous: phase changes, precipitations, precipitation at the grain boundaries, growth and coalescence of precipitates, recrystallization. They can occur coupled with the mechanisms of deformation or uncoupled. The case in which the hardening effects are highly dependent on strain or on strain rate, is referred to as dynamic ageing.

An example of the initial ageing effect is shown in Fig. 6.52 for the 304 stainless steel: the reduction of creep strain as a function of the preheating time is a clear manifestation of induced hardening.

Very few studies of the mechanical modelling of a material undergoing ageing are available. We restrict ourselves here to the following model dealing with variable isotropic hardening:

$$\dot{R} = b(Q_1 - R)\dot{p} + \beta(Q_2 - R)$$

with superposed strain hardening and time hardening. If the strain rate is significant, the first term is dominant and only strain-hardening occurs; if it is low or zero, only time-hardening occurs.

Depending on the relative values of Q_1, Q_2 and R_0 (the initial value of R), we obtain either recovery or ageing effects (time-hardening). For example, for $R_0 < Q_2 < Q_1$, Fig. 6.53 schematically illustrates the case of a rapid tension test (strain-hardening evolution of $0A$) followed by a creep test (combination of the two effects is followed by the predominance of recovery as a function of time since $R_A > Q_2$ (curve AA')). For a previous ageing at the same temperature (curve $0B$), followed by a rapid tension test, the initial hardening is given by the curve BB' (in contrast to the curve $0A$).

Coupling of plasticity and viscoplasticity

To represent deformations which occur rapidly during the application of the load, together with those which occur slowly due to creep, we may

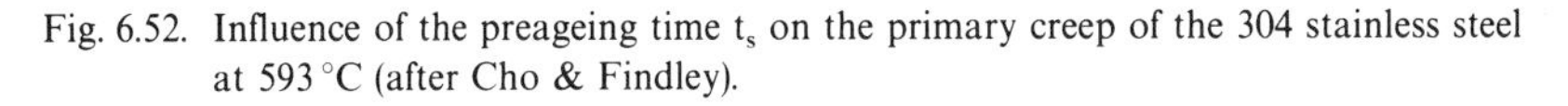

Fig. 6.52. Influence of the preageing time t_s on the primary creep of the 304 stainless steel at 593 °C (after Cho & Findley).

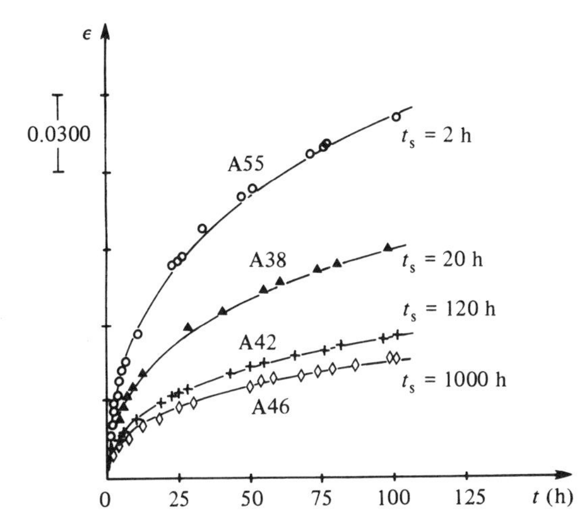

Fig. 6.53. Evolution of the hardening variable in case of a rapid tension test followed by the creep test, (*a*) as a function of strain, (*b*) as a function of time.

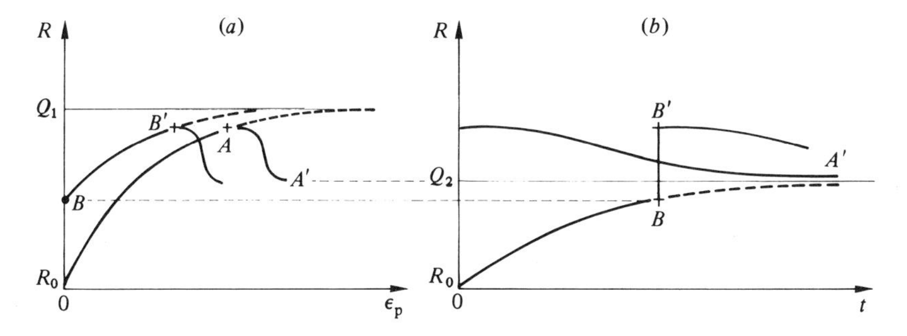

introduce a decomposition of the strain such as:

$$\boldsymbol{\varepsilon} = \boldsymbol{\varepsilon}^{e} + \boldsymbol{\varepsilon}^{in} = \boldsymbol{\varepsilon}^{e} + \boldsymbol{\varepsilon}^{p} + \boldsymbol{\varepsilon}^{vp}$$

where $\boldsymbol{\varepsilon}^{e}$ is the elastic strain, $\boldsymbol{\varepsilon}^{p}$ the instantaneous plastic strain, and $\boldsymbol{\varepsilon}^{vp}$ the viscoplastic (or creep) strain. Fig. 6.54 shows an example which explains why such a decomposition is useful. After a significant strain during the application of the load, even a creep test of very long duration produces only very small additional strains. There seem to be different time scales corresponding to different processes. On the other hand, a number of experiments show that the two deformation processes are coupled.

Plastic strain is governed by constitutive equations such as those presented in Chapter 5, while viscoplastic strain is governed by those described in the present chapter. Without going into details, and noting that V_k are the hardening variables and A_k the associated thermodynamic variables, the general equations of coupled plasticity and viscoplasticity can be expressed in the form:

$$\dot{\boldsymbol{\varepsilon}}^{in} = \lambda(\partial f/\partial \boldsymbol{\sigma}) + \partial \Omega/\partial \boldsymbol{\sigma}$$
$$\dot{V}_k = -\lambda(\partial f/\partial A_k) - \partial \Omega/\partial A_k$$

where the potentials f and Ω depend on $\boldsymbol{\sigma}$ and A_k. The hardening induced

Fig. 6.54. Rapid tension and long-duration creep test on 316 SPH stainless steel at 550 °C.

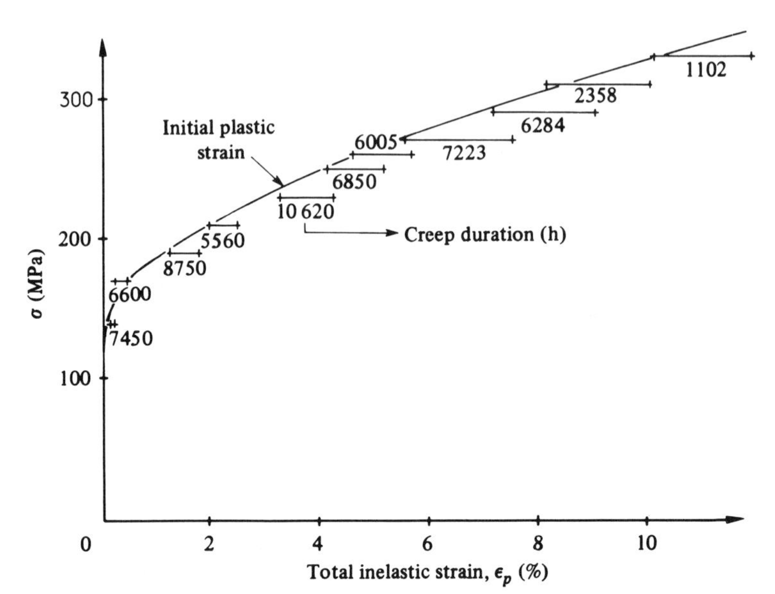

by plastic strain influences the viscoplastic strain and vice versa. The plastic multiplier λ is determined from the consistency condition of instantaneous plasticity: $f = \dot{f} = 0$; it depends on the viscoplastic strain rate by virtue of the term $\partial\Omega/\partial A_k$:

$$\lambda = \frac{\left\langle \dfrac{\partial f}{\partial \boldsymbol{\sigma}} : \dot{\boldsymbol{\sigma}} - \dfrac{\partial f}{\partial A_i} \dfrac{\partial A_i}{\partial V_k} \dfrac{\partial \Omega}{\partial A_k} \right\rangle}{\dfrac{\partial f}{\partial A_i} \dfrac{\partial A_i}{\partial V_k} \dfrac{\partial f}{\partial A_k}}.$$

The effect of such a coupling can be illustrated in tension (Fig. 6.55). The uppermost curve corresponds to instantaneous plastic deformations: it is the tension curve at infinite strain rate. Thus, there is a double saturation effect occurring as a function of the rate of deformation: for rates closer to zero, the limiting curve is that of the internal stress (see Section 6.4.2), while for those closer to infinity, it is along the instantaneous plasticity curve.

There is another way of combining the two time scales mentioned above. Instantaneous plastic deformation is not considered, but the two scales are introduced in the hardening law by a combination of strain-hardening and time-hardening:

$$\dot{\boldsymbol{\varepsilon}}^{\text{in}} = \dot{\boldsymbol{\varepsilon}}^{\text{vp}} = \partial\Omega_{\text{vp}}/\partial\boldsymbol{\sigma}$$
$$\dot{V}_k = -\partial\Omega/\partial A_k = -\partial\Omega_{\text{vp}}/\partial A_k - \partial\Omega_t/\partial A_k.$$

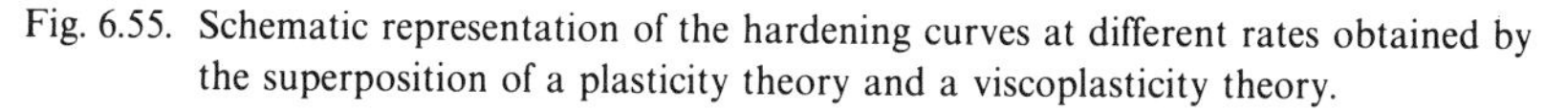

Fig. 6.55. Schematic representation of the hardening curves at different rates obtained by the superposition of a plasticity theory and a viscoplasticity theory.

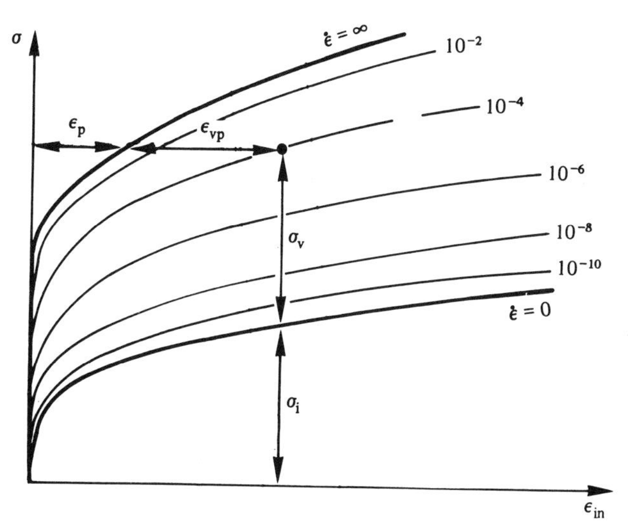

Thus, we are led to models such as those referred to in Sections 6.4.2 and 6.4.3. The ageing effects are opposed to the viscosity effects, and so in some domains, the influence of the rate is cancelled. Fig. 6.56(*a*) illustrates this situation for a tension test performed at a constant rate (upto a particular value of strain): the internal stress, which increases with time, is greater since the strain rate is low. The total stress is the sum of the internal stress and the viscous stress (a power fouction of $\dot{\varepsilon}$ for example). The settling of the tension curves for an increasing rate may only be apparent. Even negative strain rate sensitivities (negative apparent viscosity exponent) may be obtained, as observed for some materials (Fig. 6.56(*b*)).

6.5 Elements of the methods of viscoplastic structural analysis

These methods of analysis are based on those used in linear elasticity, such as the finite element method (see Section 4.1.5), modified to include the differential equations of viscoplasticity, and permitting solution despite the nonlinearity and time dependence.

We may recall that the use of the finite element method for elastic problems with initial stress amounts to solving a linear system of equations

$$Kq = Q + Q_0$$

where q is the column of nodal unknowns, in general, the nodal displacements of the structure discretized in finite elements.† K is the stiffness matrix of the structure: it depends on the geometry of the structure and the elastic law of the material. Q is the column of external forces, and Q_0 represents the nodal forces equivalent to the initial stresses.

After solving the linear system we can determine the displacements, strains and stresses at any point by using the matrix relations:

$$u = Nq \quad \varepsilon = Bq \quad \sigma = a\varepsilon + \sigma_0$$

where N and B are matrices which are dependent on the geometry and on the discretization into finite elements, and a is the elasticity matrix of the material.

6.5.1 *General scheme of viscoplastic analysis*

A viscoplasticity problem is solved by linearization starting from an initial solution assumed to be known at the instant t. The elasticity law becomes:

$$\sigma = a\varepsilon^{e} = a\varepsilon - a\varepsilon^{p} - a\varepsilon^{d} = a\varepsilon + \sigma_0$$

† The brackets of the matrix symbol have been omitted in this section.

Fig. 6.56. (*a*) Combination of viscosity and ageing effects may lead to an apparently rate-independent behaviour (schematic illustration), (*b*) inverse strain rate effect observed for an 18–10 azotized stainless steel (from Rabbe).

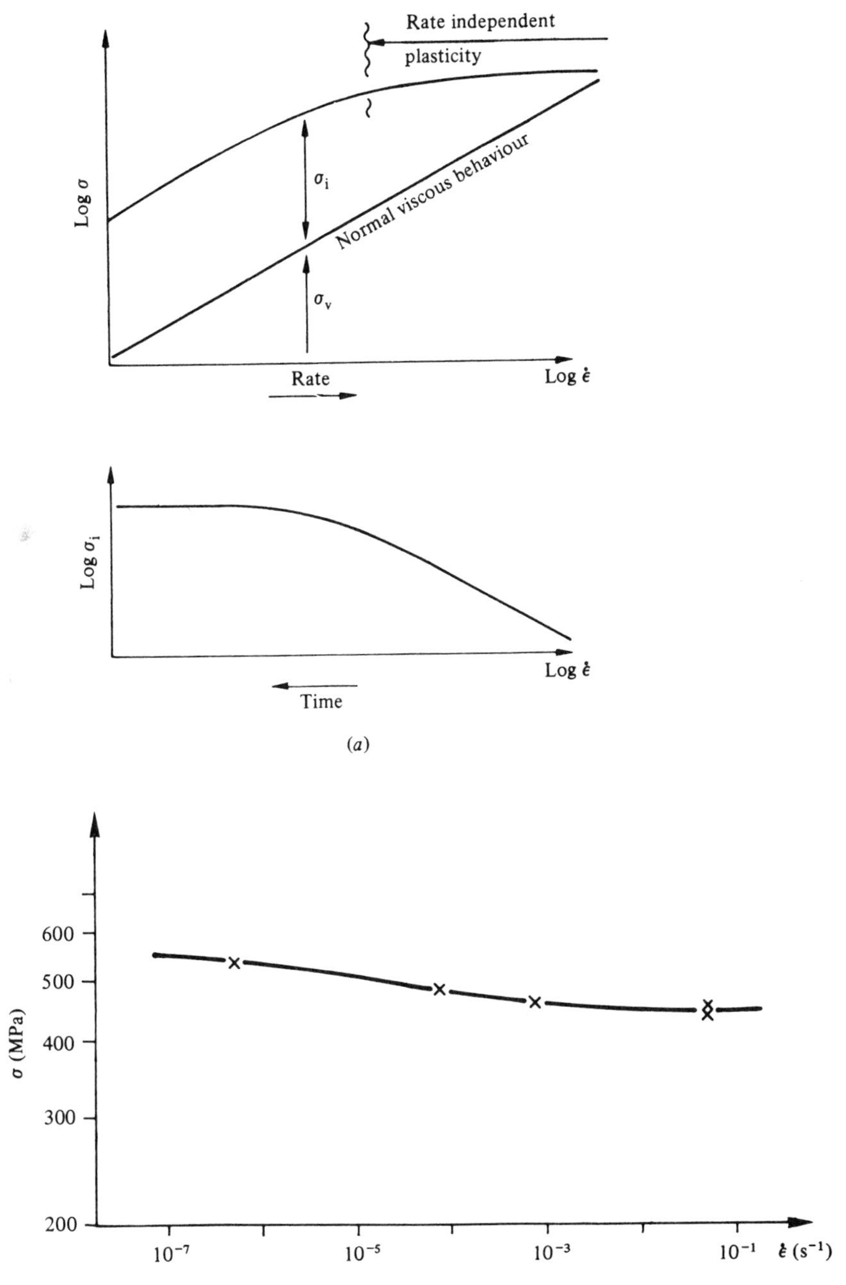

where a depends on the temperature and ε^{d} denotes the thermal dilatation

The viscoplastic flow law may be written in the form

$$\dot{\varepsilon}^{p} = f(\sigma, V_k, T) \qquad \dot{V}_k = g(\sigma, V_k, T)$$

where V_k denotes the hardening variables (p or $\boldsymbol{\alpha}$ for example). The analysis is then performed according to the following scheme (see Fig. 6.57):

Fig. 6.57. Schematic illustration of the algorithm for step-by-step calculation (first and second order methods).

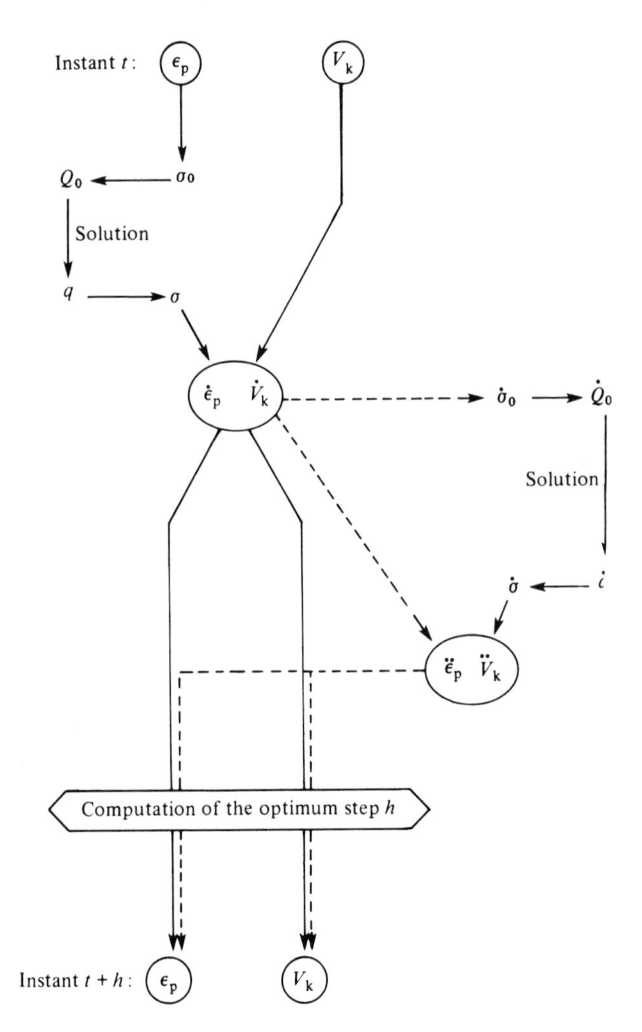

the values of ε_p and V_k are assumed to be known at the instant t, while the temperature and therefore the dilatation are given values for the problem,

σ_0 and then Q_0 can be deduced, and thus we have to solve an elasticity problem with initial stresses,

the solution of this problem gives the nodal unknowns and the stresses, which can be used to find the plastic strain rate and the hardening rate (at the same instant t).

So, starting from ε^p and V_k at the instant t, we can calculate $\dot{\varepsilon}_p$ and $\dot{V}_k$. The sequence of operations reduces the viscoplastic problem to the solution of a system of differential equations with nonseparable variables symbolically expressed as

$$\dot{y} = Y(y, t)$$

where y denotes all the principal unknowns (here ε^p and V_k). The linearization procedure which defines the values of these quantities at the instant $t + h$, knowing the value y_0 at the instant t, is still to be defined.

6.5.2 *Methods of step by step linearization*

These are classified as explicit methods (Euler, Euler–Cauchy, second order) and implicit methods which require an iteration.

Euler method:

$$\dot{y}_0 = Y(y_0, t) \qquad y = y_0 + \dot{y}_0 h.$$

Euler–Cauchy method:

$$\dot{y}_0 = Y(y_0, t) \qquad y_1 = y_0 + \dot{y}_0 h$$
$$\dot{y}_1 = Y(y_1, t + h) \qquad y = y_0 + (\dot{y}_0 + \dot{y}_1)h/2.$$

This latter method is numerically quite stable, and performs better than the Euler method (i.e., it provides a better cost–accuracy compromise). However, it is an algorithm of order 2 (two solutions per step) that does not take full advantage of the explicit nature of the viscoplasticity laws. For viscoplasticity, it is preferable to use the second order method:

Second order method:

$$y = y_0 + \dot{y}_0 h + \ddot{y}_0 h^2/2$$

where the second derivative $\ddot{y}_0$ is evaluated at the instant t after calculating $\dot{y}_0$ by explicit differentiation of the constitutive equation, which leads to solving a tangent problem:

$$K\dot{q} = \dot{Q} + \dot{Q}_0.$$

The nodal forces $\dot{Q}_0$ correspond to the 'initial stresses' of the tangent problem:

$$\dot{\sigma} = a\dot{\varepsilon} - a\dot{\varepsilon}^{\mathrm{p}} - a\dot{\varepsilon}^{\mathrm{d}} + a'a^{-1}\sigma\dot{T}$$

where a' denotes the derivative of a with respect to temperature.

The advantage of this second method consists in the evaluation of the third order terms in $(\varepsilon^{\mathrm{p}}, V_{\mathrm{k}})$ which may be used to determine an optimal value of the time step h. This is defined by assuming that the third order term in the Taylor expansion must be negligible in comparison to the sum of the first and second order terms. We are thus led to solve an inequality of the second degree:

$$|\dddot{y}_0|h^2/6 \leqslant \eta|\dot{y}_0 + \ddot{y}_0 h/2|$$

where η is the precision factor which is selected in advance, for example:

$$\eta = 2\% \quad \text{or} \quad 5\%.$$

This (heuristic) method of determining the optimal step size can also be used in conjunction with Euler's method (where we neglect the second order term in comparison to the first). In that case we find that:

$$h \leqslant 2\eta|\dot{y}_0/\ddot{y}_0|.$$

Implicit method:

$$\left.\begin{aligned} \dot{y}_0 &= Y(y_0, t_0) \\ \dot{y}_j &= Y(y_j, t+h) \\ y_{j+1} &= y_0 + [\alpha\dot{y}_0 + (1-\alpha)\dot{y}_j]h \end{aligned}\right\} \qquad \text{for } j \geqslant 1.$$

The system must be solved, i.e., the structure must be analysed, at each iteration at the instant $t + h$. It is necessary to have a test for stopping the iterations, which is applied at each point of the structure, for example of the kind:

$$\| y_{j+1} - y_j \| \leqslant \mu \| y_j \|.$$

The coefficient of precision μ and the coefficient α are selected in advance. A value for α between 0 and $\frac{1}{2}$, in general, assures a good convergence. For $\alpha = 0$ we have a purely implicit scheme. For $\alpha = \frac{1}{2}$ and by limiting j to 1 this reduces to the Euler–Cauchy method. A value for μ of approximately 1% provides a good compromise between cost and accuracy.

This method is more stable than the preceding ones. However, it requires an increase in the step size and a limit on the number of iterations in order not to use too much computer time. It is often preferred on account of its stability and generally good convergence (4–5 iterations are enough). But, even with correct convergence, the solution obtained is close to the real solution only if the step size h is sufficiently small. It is therefore preferable to supplement the procedure by calculation of the optimal step size (with the precision factor η perhaps larger than in other methods). Thus, one of the difficulties of this method lies in controlling the step size and the number of iterations simultaneously.

Bibliography

Odqvist F. K. G. *Mathematical theory of creep and creep rupture.* The Clarendon Press, Oxford (1974).

Odqvist F. K. G. & Hult J. *Kriechfestigkeit metallischer Werkstoffe.* Springer-Verlag, Berlin (1962).

Rabotnov Y. N. *Creep problems in structural members.* North-Holland Publishing Company, Amsterdam (1969).

Lemaitre J. *Sur la détermination des lois de comportement des matériaux élasto-viscoplastiques.* Thèse d'Etat, Université Paris XI (Orsay) (1971).

Radenkovic D. & Salençon J. *Plasticité et viscoplasticité.* Ediscience – McGraw Hill, New York (1974).

Padmanabhan K. A. & Davies G. J. *Superplasticity.* Springer-Verlag, Berlin (1980).

7

DAMAGE MECHANICS

L'endommagement, comme le diable, invisible mais redoutable.

The phenomenon of damage, described from a physical point of view in Chapter 1, and then presented schematically in Chapter 3, represents surface discontinuities in the form of microcracks, or volume discontinuities in the form of cavities. It, therefore, involves a rheological process quite different from deformation, although the initial causes of the two phenomena are identical: movement and accumulation of dislocations in metals, modification of intermolecular bonds in organic materials, microdecohesion in minerals. Damage is marked by pronounced irreversibility; the traditional thermomechanical treatments can only partially remove the defects caused by it. Macroscopic fracture has been studied for a long time. Around 1500, Leonardo da Vinci was already preoccupied with the characterization of fracture by means of mechanical variables. A number of failure criteria, i.e., functions of components of stress or strain, characterizing the fracture of the volume element have been proposed (e.g., by Coulomb, Rankine, Tresca, von Mises, Mohr, Caquot). However, it is only quite recently that concern has been directed towards modelling the progressive deterioration of matter preceding the macroscopic fracture. The development of damage mechanics began in 1958. In that year, Kachanov published the first paper devoted to a continuous damage variable, conceived within the framework, limited indeed, of creep failure of metals under uniaxial loads. This concept was taken up again in the seventies, mainly in France (Lemaitre & Chaboche), Sweden (Hult), England (Leckie), Japan (Murakami) and extended to ductile fracture and fatigue fracture. It has been generalized to the multiaxial isotropic case within the framework of the thermodynamics of irreversible processes described in Chapter 2; anisotropic damage remains to be 'finished' in the eighties!

7.1 Domain of validity and use

First of all, it should be stated what is meant by the beginning and the end of the process of damage; in other words what is the scale of the phenomena under consideration? A material is said to be free of any damage if it is devoid of cracks and cavities at the microscopic scale, or from a more pragmatic point of view, if its deformation behaviour is that of the material formed under the best conditions. Let us recall what was said in Chapter 2: the initial state of a material cannot be objectively defined. Usually, it is the state starting from which the history of the loads is known. The final damage state is that of the fracture of the volume element, i.e., the existence of a macroscopic crack of the size of the representative volume element as estimated in Chapter 3: 0.1–1 mm for metals or polymers, of the order of 1 cm for wood, and 10 cm for concrete. Beyond this, is the domain of crack mechanics (Chapter 8).

The theory of damage therefore, describes the evolution of the phenomena between the virgin state and macroscopic crack initiation. This evolution, which is not always easily distinguishable from the deformation phenomena which usually accompany it, is due to several mechanisms which have been classified in Chapter 3:

ductile plastic damage accompanying large plastic deformations of metals at ambient as well as average temperatures;
brittle viscoplastic (or creep) damage, a function of time which, for metals at average and high temperatures, corresponds to intergranular decohesions accompanying viscoplastic strains;
fatigue (or microplastic) damage, caused by stress repetitions and identified as a function of the number of cycles;
macrobrittle damage produced by monotonic loads without appreciable irreversible deformations, as in the case of concrete for example.

Other phenomena can be considered as damage: the oxidation process, corrosion, irradiation; but there thermomechanical modelling is yet to be done!

It is therefore clear that the theory of damage is concerned with all materials at low as well as high temperatures under any kind of load. It is from the nature of the evolution models that we can represent these different phenomena that can accumulate or interact with each other. Knowing the stress and strain history for a given volume element of a structure, the damage laws provide, by integration with respect to time, the damage evolution in the element up to the point of macroscopic crack initiation. Thus, the theory furnishes the time or the number of cycles corresponding to the initiation of such a crack at the most stressed point of the structure. This is the modern principle of analysing the resistance of structures used at the design level and in the verification and control of the service conditions of structures. This also provides a better means of optimizing the process of metal forming by plastic deformations, to avoid or reduce manufacturing defects. And, in any case, it is the means of knowing *a priori* the modifications of the product's mechanical properties which result from the forming operations.

7.2 Phenomenological aspects

The definition of a mechanical damage variable itself presents a difficult problem. There is nothing (almost nothing!) that macroscopically distinguishes a highly damaged volume element and a virgin one. It therefore becomes necessary to imagine internal variables which are representative of the deteriorated state of the matter. There are several possibilities for this choice depending on the school of thought and the types of damage measurement envisaged:

measurements at the scale of microstructure (density of microcracks or cavities) lead to microscopic models that can be 'integrated' over the macroscopic volume element with the help of mathematical homogenization techniques. We may thus obtain the properties of the damaged volume element; but from these it is difficult to define a macroscopic damage variable and a law of its evolution which is easy to use in continuum mechanics analysis;

global physical measurements (density, resistivity etc.) require the definition of a global model to convert them into properties which characterize mechanical resistance;

another type of damage evaluation is linked to the remaining lifetime, but this concept does not directly lead to a damage constitutive law;

global mechanical measurements (of modification of elastic, plastic, or viscoplastic properties) are easier to interpret in terms of damage variables using the concept of effective stress introduced by Rabotnov. This is the approach we will be following.

7.2.1 *Damage variable*

Definition

Consider a damaged solid in which an element of finite volume has been isolated, of a sufficiently large size with respect to the inhomogeneities of the medium, and imagine that this element has been grossly enlarged (Fig. 7.1).

Let S be the area of a section of the volume element identified by its normal $\vec{n}$. On this section, cracks and cavities which constitute the damage leave traces of different forms. Let $\tilde{S}$ be the effective area of resistance ($\tilde{S} < S$) taking account of the area of these traces, stress concentrations in the neighbourhood of geometric discontinuities, and the interactions between the neighbouring defects. Let S_D be the difference:

$$S_D = S - \tilde{S}.$$

S_D is the total area of the defect traces corrected for stress concentrations and interactions. We will see in Section 7.2.2 that the concept of effective stress associated with the hypothesis of strain – equivalence enables us to avoid the calculation of S_D (or $\tilde{S}$) which would be extremely difficult to do because of the lack of knowledge of the precise geometry of the defects and because of the doubts one might have regarding the applicability of continuum mechanics on this scale.

By definition:

- $$S_D/S = D_n$$

is the mechanical measure of local damage relative to the direction $\vec{n}$. From a physical point of view, the damage variable is therefore the relative (or corrected) area of cracks and cavities cut by the plane normal to the direction $\vec{n}$. From a mathematical point of view, as S tends to 0, the variable

Fig. 7.1. Damaged element.

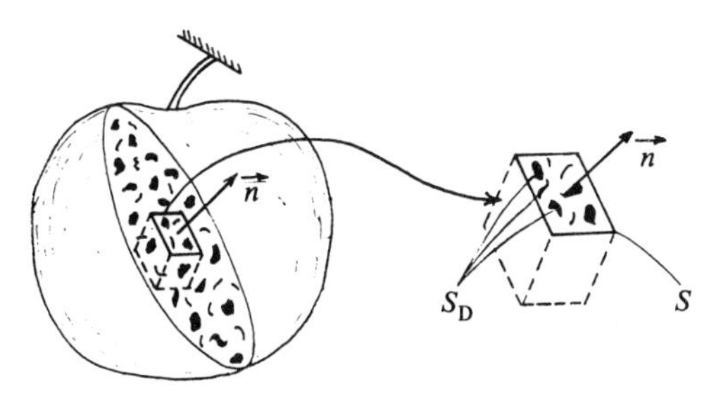

D_n is the (corrected) surface density of the discontinuities of the matter in the plane normal to $\vec{n}$.

$D_n = 0$ corresponds to the nondamaged or virgin state,
$D_n = 1$ corresponds to the breaking of the volume element into two parts along a plane normal to $\vec{n}$,
$0 \leqslant D_n < 1$ characterizes the damaged state.

In the general case of anisotropic damage consisting of cracks and cavities with preferred orientations, the value of the scalar variable D_n depends on the orientation of the normal. It will be seen in Section 7.3.1 that the corresponding intrinsic variable can be represented by a second or a fourth order tensor.

Hypothesis of isotropy

Isotropic damage consists of cracks and cavities with an orientation distributed uniformly in all directions. In this case, the variable does not depend on the orientation $\vec{n}$ and the damaged state is completely characterized by the scalar D. In this section we will limit ourselves to the case of isotropic damage:

$$D_n = D \qquad \forall \vec{n}.$$

7.2.2 Effective stress

Definition

The introduction of a damage variable which represents a surface density of discontinuities in the material leads directly to the concept of effective stress, i.e., to the stress calculated over the section which effectively resists the forces.

In the uniaxial case, if F is the applied force on a section of the representative volume element, $\sigma = F/S$ is the usual stress satisfying the equilibrium equation. In the presence of isotropic damage, D, the effective area of resistance is:

$$\tilde{S} = S - S_{\mathrm{D}} = S(1 - D)$$

and by definition the effective stress $\tilde{\sigma}$ is taken to be:

$$\tilde{\sigma} = \sigma S/\tilde{S} \quad \text{or} \quad \tilde{\sigma} = \sigma/(1 - D).$$

Evidently $\tilde{\sigma} \geqslant \sigma$

$\tilde{\sigma} = \sigma$ for a virgin material,

$\tilde{\sigma} \to \infty$ at the moment of fracture.

In the case of multiaxial isotropic damage, the ratio $S/\tilde{S}$ does not depend on the orientation of the normal and the operator $(1 - D)$ can be applied to all the components. We will therefore write for the effective stress $\tilde{\boldsymbol{\sigma}}$:

- $\tilde{\boldsymbol{\sigma}} = \boldsymbol{\sigma}/(1 - D).$

Principle of strain-equivalence

We assume that the deformation behaviour of the material is only affected by damage in the form of effective stress:

> any deformation behaviour, whether uniaxial or multiaxial, of a damaged material is represented by the constitutive laws of the virgin material in which the usual stress is replaced by the effective stress (Fig. 7.2).

For example, the uniaxial linear elastic law of a damaged material is written as:

$$\varepsilon_e = \frac{\tilde{\sigma}}{E} = \frac{\sigma}{(1 - D)E}$$

where E is the Young's modulus. This constitutes a nonrigorous hypothesis

Fig. 7.2. Effective stress and equivalence in strain.

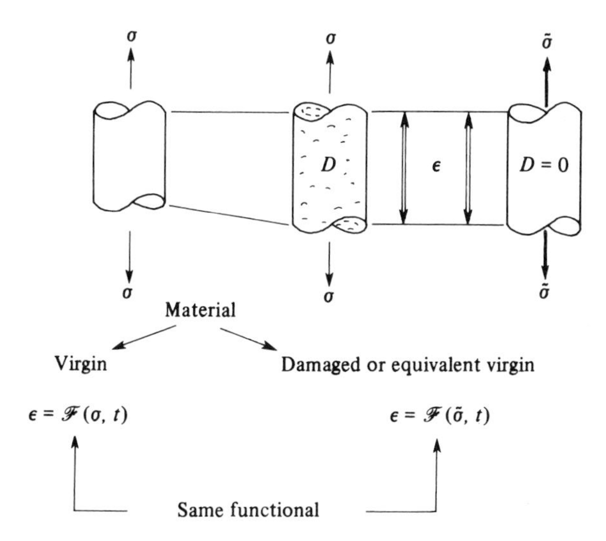

which assumes that all the different behaviours (elasticity, plasticity, viscoplasticity) are affected in the same way by the surface density of the damage defects. However, its simplicity allows the establishment of a coherent and efficient formalism.

Critical damage at fracture

By applying the concept of effective stress at the instant of fracture by interatomic decohesion, we define the critical value of damage D_c, as that corresponding to the occurrence of this phenomenon.

If $\tilde{\sigma}_u$ is the uniaxial stress at fracture by decohesion and σ_u is the usual ultimate fracture stress, we have:

$$\tilde{\sigma}_u = \sigma_u/(1 - D_c)$$

or

$$D_c = 1 - (\sigma_u/\tilde{\sigma}_u).$$

The physics of solids shows that $\tilde{\sigma}_u$ is of the order of $E/50$–$E/20$; for common materials $\tilde{\sigma}_u$ is of the order of $E/100 - E/250$, and D_c is therefore of the order of 0.5–0.9. This allows us to neglect $(1 - D_c)^x$ (with $x \gg 1$), a term which often appears in calculations, in comparison to 1. Multiaxial criteria are studied in Section 7.2.5.

7.2.3 ***Measurement of damage***

Damage is not directly accessible to measurement. Its quantitative evaluation, as for any physical value, is linked to the definition of the variable chosen to represent the phenomenon. Having chosen a definition based on the concept of effective stress, associated to the hypothesis of strain-equivalence, the measurements which result naturally from it, are essentially linked to the coupling between deformation and damage, i.e., to the modification of the mechanical properties caused by damage. These measurements are those of the properties of elasticity, plasticity, viscoplasticity, to which we also add the measurement of variation in resistivity.

Variation of the modulus of elasticity

Static method

Recall the uniaxial elastic damage law, which was mentioned in the previous Section, and which will be justified on thermodynamical grounds in Section 7.3.2:

$$\tilde{\sigma} = \sigma/(1 - D) = E\varepsilon_e$$

or

$$\sigma = E(1 - D)\varepsilon_e$$

where E is Young's modulus, i.e., the elasticity modulus of the material free from any damage; $E(1 - D) = \tilde{E}$ can then be interpreted as the elastic modulus of the damaged material. If Young's modulus E is known, any measurement of elastic stiffness can be used to determine the damage by $D = 1 - \sigma/(E\varepsilon_e)$ and with $\sigma = \tilde{E}\varepsilon_e$, we find that:

- $D = 1 - \tilde{E}/E.$

Although very simple in principle, this measurement is rather tricky to perform for the following reasons:

any measurement of the modulus of elasticity requires precise measurement of very small strains, as we have seen in Chapter 4; damage is usually very localized, which requires that the measurements be made on a very small base of the order of 0.5–5 mm; at the required level of the precision, the best straight line in the (σ, ε) graph representing an elastic loading or unloading is difficult to define.

In view of the above reasons, the following procedure is recommended:

(i) use of specimens with a weakened central section so as to localize the damage there; an example is given in Fig. 7.3;

Fig. 7.3. Method of measuring damage.

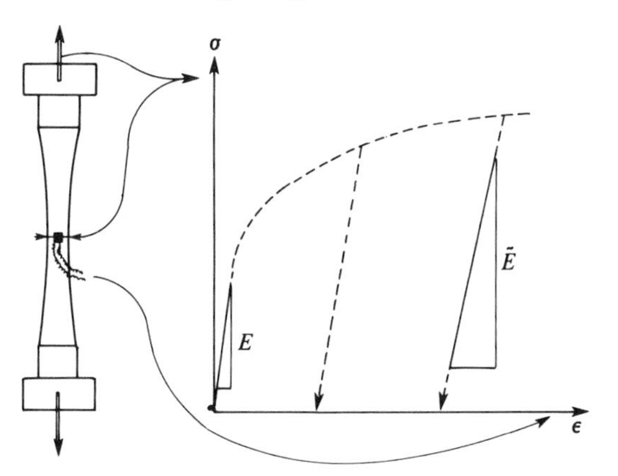

Fig. 7.4. Examples of evolution of damage: (*a*) ductile plastic damage: copper Cu/Al $T = 20\,^{\circ}\mathrm{C}$, $\dot{\varepsilon} \approx 10^{-4}\,\mathrm{s}^{-1}$; (*b*) fatigue damage: 316L stainless steel $T = 20\,^{\circ}\mathrm{C}$

$$\varepsilon = \begin{cases} +0.7 \times 10^{-2} \\ 0 \end{cases} \quad N_F(\text{fracture}) = 70\,450 \text{ cycles};$$

(*c*) cyclic creep damage: 316L stainless steel, $T = 550\,^{\circ}\mathrm{C}$

$$\varepsilon = \begin{cases} +10^{-2} \\ -10^{-2} \end{cases} \quad N_c(\text{fracture}) = 218 \text{ cycles}.$$

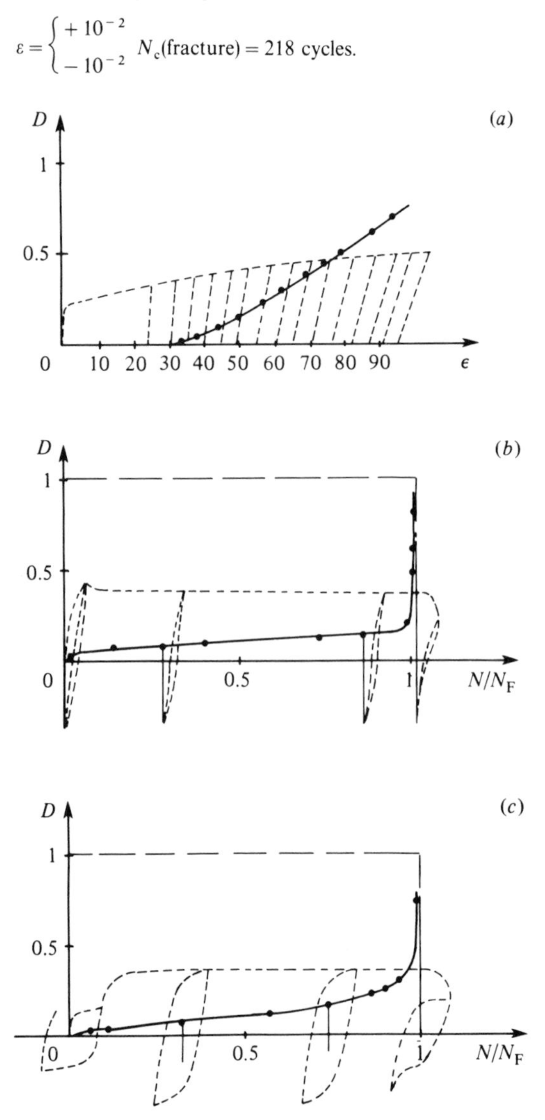

(ii) measurement of strain with small gauges: 0.5 mm × 0.5 mm when the temperature allows it, or by displacement transducers attached to as small a measurement base as possible, when the temperature is above 200 °C;

(iii) evaluation of the elasticity modulus during elastic unloading by avoiding zones of stronger nonlinearities: this is achieved by performing the evaluation in a range of stress defined as follows (see Fig. 7.3):

$$\sigma_{\text{Max}} < \sigma < \sigma_{\text{min}}$$
$$\sigma_{\text{Max}} = 0.85\sigma(\varepsilon)$$
$$\sigma_{\text{min}} = 0.15\sigma(\varepsilon).$$

If these precautions are taken, damage can be evaluated with a relative precision of the order of $\pm$ 5%, and the method can easily be applied for any kind of damage. Fig. 7.4 shows, as examples, evolution of damage in ductile plasticity, cyclic creep, and fatigue obtained by this technique.

Ultrasonic dynamic method

The same principle may be used in dynamic situations, particularly with regard to propagation of ultrasonic waves. A measurement of the speed or propagation time of plane waves in a specimen or a cylinder of damaged material leads to the determination of the elasticity modulus $\tilde{E}$ of the damaged material and of the damage by $D = 1 - \tilde{E}/E$. Expressed in terms of the velocity of longitudinal waves $\tilde{v}_{\text{L}}$ and transverse waves $\tilde{v}_{\text{T}}$ (see Section 4.1.3):

$$\tilde{E} = \rho\tilde{v}_{\text{T}}^2\frac{3\tilde{v}_{\text{L}}^2 - 4\tilde{v}_{\text{T}}^2}{\tilde{v}_{\text{L}}^2 - \tilde{v}_{\text{T}}^2}.$$

In the framework of the isotropic damage hypothesis (constant Poisson's coefficient) and neglecting the variation in ρ (which does not result in a relative error in D of more than 5%), we have:

$$v_{\text{L}}^2 = \frac{E}{\rho}\frac{1-\nu}{(1+\nu)(1-2\nu)} \quad \text{and} \quad \tilde{v}_{\text{L}}^2 = \frac{\tilde{E}}{\rho}\frac{1-\nu}{(1+\nu)(1-2\nu)}$$

and we obtain:

- $D = 1 - \tilde{v}_{\text{L}}^2/v_{\text{L}}^2.$

An example of damage measurement using the ultrasonic technique, applied to a concrete specimen loaded in compression, is given in Fig. 7.5.

Variation of plasticity characteristics

Monotonic hardening characteristics

This method is of interest, especially for the characterization of ductile plastic damage as a complement of the method of measurement by means of the variation of the elasticity modulus. Ductile plastic damage accompanies large plastic strains in metals and in polymers and only becomes important as the necking condition is approached. In hardening tests, it is apparent through a drop in the stress commencing from the point of instability defined by $d\sigma/d\varepsilon = 0$. This drop in stress is therefore due to a combination of the geometric effect (necking) and the damage effect (reduction of the effective area $\tilde{S} = S - S_D$).

The hardening law of a nondamaged material is expressed by the

Fig. 7.5. Measurement of damage in concrete by an ultrasonic technique (after S. Benounich): (*a*) evolution of the traverse time of the wave $\Delta t/\Delta t_0 = v_L/\tilde{v}_L$ as a function of the stress expressed as a fraction of the fracture stress; (*b*) evolution of the damage.

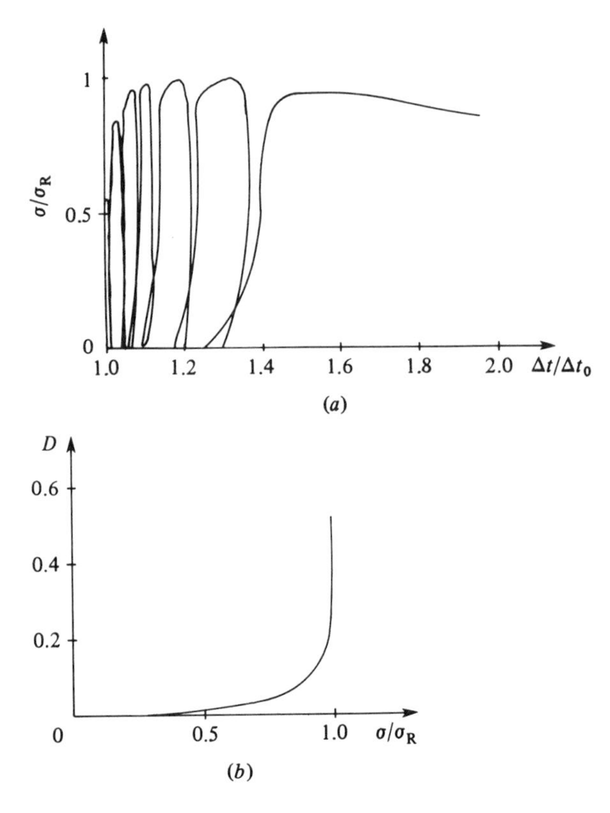

equation introduced in Section 2.1 and written here for large strains ($\varepsilon_{pv} = \ln(1 + \varepsilon_p)$):

$$\varepsilon_{pv} = \left\langle \frac{\sigma_v - \sigma_Y}{K_Y} \right\rangle^{M_Y}.$$

As soon as the damage becomes perceptible ($\varepsilon_v > \varepsilon_v^*$), the hypothesis of strain equivalence associated with the concept of effective stress can be invoked to write:

$$\varepsilon_{pv} = \frac{1}{K_Y^{M_Y}} \left\langle \frac{\sigma_v}{1 - D} - \sigma_Y \right\rangle^{M_Y}.$$

This expression can be used to measure D indirectly from the $(\sigma_v, \varepsilon_v)$ graph as soon as K_Y and M_Y are known (these latter quantities can be determined from the same graph for $\varepsilon_{pv} < \varepsilon_{pv}^*$):

$$D = 1 - \frac{\sigma_v}{K_Y \varepsilon_{pv}^{1/M_Y} + \sigma_Y}.$$

Cyclic hardening characteristics

The method discussed here is particularly suited to the measurement of low-cycle fatigue damage. The basic cyclic plasticity law, described in Chapter 5, is used in the form of a relation between the maximum plastic strain at the stabilized cycle (when it exists) and the maximum stress:

$$\varepsilon_{p\,Max}^* = \left[\frac{\sigma_{Max}^*}{K_c} \right]^{M_c}.$$

We generally assume that the fatigue damage is negligible in the stabilized cycle. The further evolution of the maximum plastic strain in a stress controlled test ($\sigma_{Max} = \sigma_{Max}^*$) is then attributed to damage. The hypothesis of strain-equivalence associated with the concept of effective stress leads to writing for each cycle:

$$\varepsilon_{pMax} = \left[\frac{\sigma_{Max}^*}{K_c(1 - D)} \right]^{M_c}$$

and the damage is obtained by combining the two relations:

$$D = 1 - \left(\frac{\varepsilon_{p\,Max}^*}{\varepsilon_{p\,Max}} \right)^{1/M_c}.$$

The relative precision of this method is of the order of 10–20%, the major difficulty being in the evaluation of the stabilized state. Fig. 7.6 shows the case of two alloys subject to fatigue at 5 Hz. The continuous curves correspond to the model of Section 7.4.3.

In contrast, under controlled plastic strain ($\varepsilon_{pMax} = \varepsilon^*_{pMax}$) loading, the

Fig. 7.6. Damage evolution in pure fatigue: (*a*) IN 100 at 1000 °C; (*b*) 316L at 20 °C.

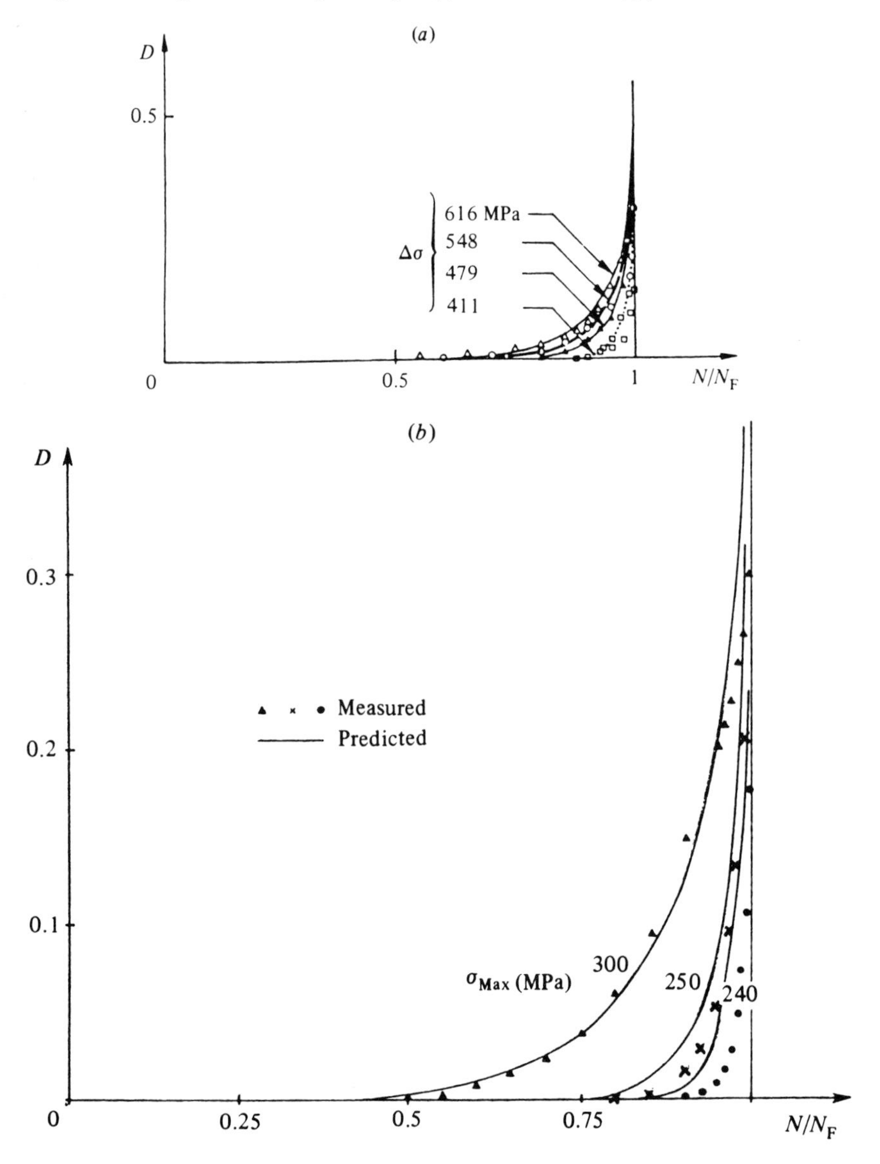

same method leads to:

$$D = 1 - \sigma_{\text{Max}}/\sigma^*_{\text{Max}}.$$

For a controlled total strain, the interpretation is more complex since in such a case both the elasticity and the cyclic laws are affected by the damage.

For a fatigue test under asymmetric loading (repeated tension, for example), we could use the same method but include the progressive ratchetting effect with a constitutive equation of the form (N being the number of cycles):

$$\frac{\delta\varepsilon_{\text{pMax}}}{\delta N} = \left[\frac{\sigma_{\text{Max}}}{(1-D)K_{\text{r}}}\right]^{M_{\text{r}}}.$$

Finally, note that this method is applicable only in the regime of low-cycle fatigue (measurable plastic strains) for materials not subjected to cyclic hardening (or softening) for most of their lives. Moreover, the fatigue damage that can thus be measured is the one that occurs in the last stage of the life of the volume element when microcracks have already been initiated (stage I) and are spreading on a microscopic scale (stage II).

Variation of viscoplastic characteristics

This method demonstrates creep damage, which in metal corresponds to the nucleation and growth processes of mainly intercrystalline microcracks. This damage increases as a function of time under constant or slowly increasing loads. This phenomenon becomes more pronounced as the temperature rises.

In a uniaxial creep test at constant stress, it manifests itself, especially during tertiary creep, by an increase in the creep velocity which becomes very high as rupture is approached. In fact in creep tests at constant load, tertiary creep also results from a reduction in the cross-sectional area of the speciman. Fig. 6.3 shows an example of this quite well known phenomenon.

As for ductile plastic damage, we may use the coupling between damage and strain behaviour (here viscoplastic) to identify the damage. By assuming the damage to be zero or negligible in primary creep, we may account for secondary creep by Norton's law (see Chapter 6):

$$\dot{\varepsilon}^*_{\text{p}} = (\sigma_{\text{v}}/\lambda^*)^{N^*}$$

and for tertiary creep by the concept of effective stress:

$$\dot{\varepsilon}_{\text{p}} = [\sigma_{\text{v}}/(1-D)\lambda^*]^{N^*}.$$

Knowing the viscosity exponent N^*, determined when $D = 0$, the damage can be expressed as a function of the secondary creep rate $\dot{\varepsilon}_p^*$ and the tertiary creep rate $\dot{\varepsilon}_p$ measured as functions of time on the creep curve. From the two relations above, it follows that when σ_v is constant:

$$D = 1 - (\dot{\varepsilon}_p^*/\dot{\varepsilon}_p)^{1/N^*}.$$

This method of evaluating damage can be used to plot the graph of $D(t/t_c)$ in tertiary creep as shown by Fig. 7.7 (where t_c is the rupture time). In practice, the test is performed at constant load and it is necessary to take into account the variation of σ_v. Noting that ε^* is the strain at the beginning of the secondary creep, we have:

$$D = 1 - \frac{1+\varepsilon}{1+\varepsilon^*}\left(\frac{\dot{\varepsilon}_p^*}{\dot{\varepsilon}_p}\right)^{1/N^*}.$$

Fig. 7.7. Creep damage evolution for the alloys IN 100 and AU2GN. The continuous curves correspond to the model of Section 7.4.2.

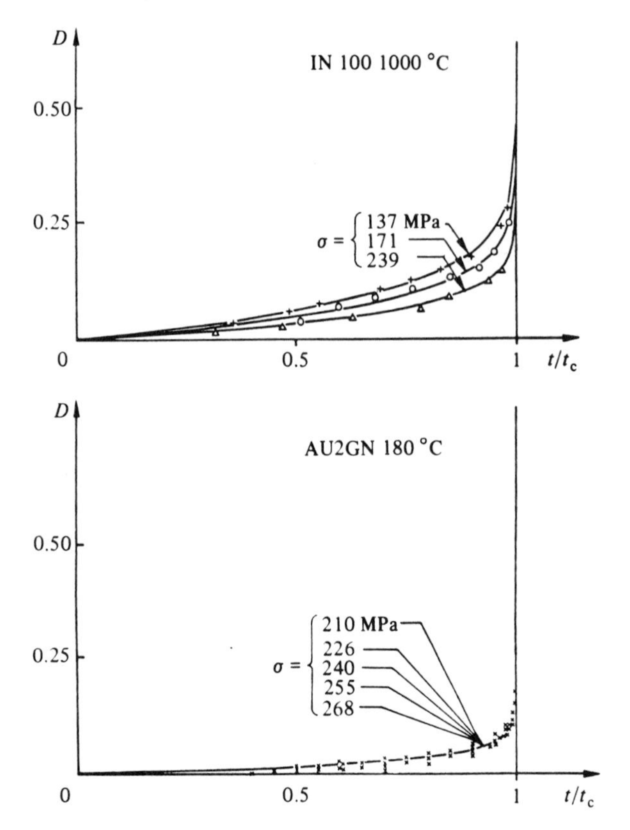

Variation of electrical resistance

The concept of effective stress is replaced here by the concept of effective current density:

$$\tilde{i} = i/(1 - D).$$

Consider an elementary tube of current (Fig. 7.8) and assume that this volume element is subjected to a simple tension in the direction of the current. The influence of the damage D is assumed to be distributed uniformly in the electrical resistance. The problem is complicated by the influence of deformations on the resistance. For an apparently constant current, the potential differences for the virgin material and the damaged one are expressed (with self-evident notation) by:

$$\mathrm{d}v = \rho \frac{\mathrm{d}l}{\mathrm{d}S} \mathrm{d}i, \qquad \mathrm{d}v' = \rho' \frac{\mathrm{d}l'}{\mathrm{d}S'} \mathrm{d}\tilde{i}.$$

We assume that the influence of damage is entirely represented by the density of the effective current $\mathrm{d}\tilde{i}$ and that the modifications of resistivity ρ, $\mathrm{d}l$ and $\mathrm{d}S$ are dependent only on the geometric changes resulting from elastic and plastic strains. In simple tension, the axial and transverse strains are:

$$\varepsilon_\mathrm{e} + \varepsilon_\mathrm{p} \quad \text{and} \quad -\nu\varepsilon_\mathrm{e} - \tfrac{1}{2}\varepsilon_\mathrm{p}$$

which can be used to write:

$$\mathrm{d}l' = \mathrm{d}l(1 + \varepsilon_\mathrm{e} + \varepsilon_\mathrm{p}), \qquad \mathrm{d}S' = \mathrm{d}S(1 - 2\nu\varepsilon_\mathrm{e} - \varepsilon_\mathrm{p}).$$

Traditionally the strain is considered to influence the resistivity only through the geometric effect of the variation in volume (Bridgman's law).

Fig. 7.8. Elementary volume.

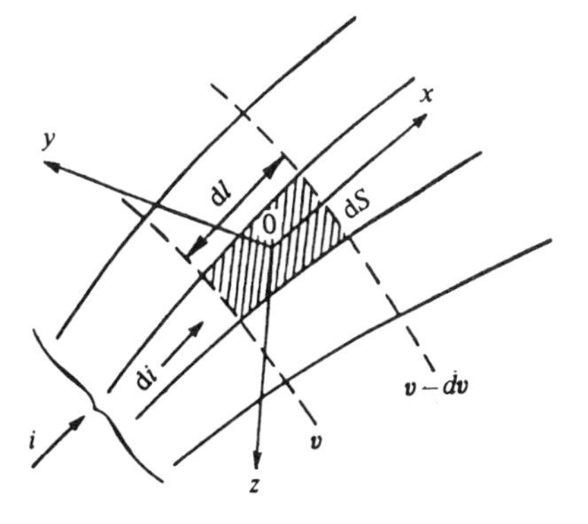

We therefore write:

$$\rho' = \rho(1 + C_1\varepsilon_e - 2\nu C_1\varepsilon_e).$$

In the framework of small strains, these assumptions lead to the relative variation of potential:

$$\frac{\Delta dv}{dv} = \frac{dv' - dv}{dv} = \frac{K_e\varepsilon_e + K_p\varepsilon_p + D}{1 - D}$$

where

$$K_e = 1 + 2\nu + C_1(1 - 2\nu), \quad K_p = 2.$$

The damage is therefore obtained from measurements of the relative variation in electrical potential, taking into account the presence of strains:

$$D = \frac{\Delta dv/dv - K_e\varepsilon_e - K_p\varepsilon_p}{1 + \Delta dv/dv}.$$

Fig. 7.9. Comparison between measurements from electrical and mechanical methods on the 316 steel at 20 °C ((*a*), (*c*)), and the IN 100 alloy at 1000 °C ((*b*), (*d*)). Mechanical measurements: drop in modulus of elasticity ((*a*), (*c*)), increase in creep rate ((*b*)), increase in maximum plastic strain ((*d*)).

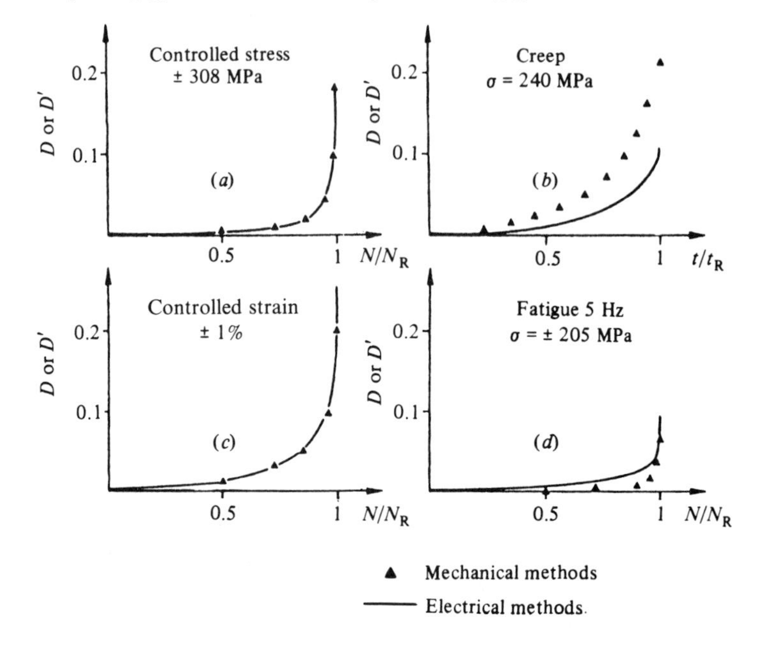

Tests performed on tension–compression specimens under very different conditions (stainless steel and aluminium alloy at 20 °C, refractory alloy at high temperature) show that the damage values D obtained by the electrical method are close to those from mechanical measurements. Fig. 7.9 shows some results obtained in fatigue and in creep. It is seen that in pure creep, where the damage manifests itself by the appearance of cavities, electrical measurements give lower values than mechanical ones. The comparison between the two types of measurement is more favourable in the case of pure fatigue, where the physical process consists in the initiation of microcracks, and for this case the electrical method is much more sensitive. However, on the whole, the agreement between the methods is good in view of the large uncertainty of this type of measurement which relies on the difference between two neighbouring mechanical or electrical states.

7.2.4 *Elementary damage laws*

We describe here some uniaxial phenomenological equations relative to three types of damage (ductile, creep and fatigue). It will be interesting to note the quality/complexity ratio.

Ductile plastic damage equation, linear in terms of strain

From a microscopic point of view this consists in the nucleation, growth and coalescence of cavities induced by large plastic strains. We have seen in the Section 7.2.3 how the variable D is measured from variations in the elasticity modulus or the monotonic hardening characteristics.

For many metallic materials subjected to a uniaxial monotonically increasing load, the damage D varies linearly with the strain (as in Fig. 7.4(a)). A simple law, well in accord with the experimental results, is therefore:

- $$D = D_c \left\langle \frac{\varepsilon_v - \varepsilon_{vD}}{\varepsilon_{vR} - \varepsilon_{vD}} \right\rangle,$$

where ε_{vD} is the true strain at the damage threshold, before which the damage is zero or negligible, and ε_{vR} is the true strain at fracture at which the damage is equal to its critical value D_c. This empirical equation contains three coefficients which have been identified for some materials, their values are listed in Table 7.1. The equation will be justified by the formalism of thermodynamics in Section 7.4.1.

Table 7.1. *Values of the coefficients of ductile plastic damage*

Material	T(°C)	ε_{vD}	ε_{vR}	D_c
99.9% copper	20	0.35	1.04	0.85
2024 alloy	20	0.03	0.25	0.23
E 24 steel	20	0.50	0.88	0.17
XC 38 steel	20	0	0.56	0.22
30 CD 4 steel	20	0.02	0.37	0.24
INCO 718 alloy	20	0.02	0.29	0.24

Table 7.2. *Values of the creep damage coefficients*

Material	T(°C)	A_0(h MPa)	r	k	A
AU2GN alloy	180	317	28	43	322
Magnesium alloy	260	162	7.5	5.4	156
304 steel	593	520	7	15	575
304 steel	650	555	5.6	15	650
316 L steel	600	1160	6.5	5	1120
L605 alloy	980	102	7.5		
UDIMET 560 alloy	980	190	5		
UDIMET 700 alloy	900	662	6.3		
UDIMET 700 alloy	1000	520	4.8		
IN 100 alloy	800	1010	10	15	1050
IN 100 alloy	900	768	6.3	15	870
IN 100 alloy	1000	433	5.2	15	520
IN 100 alloy	1100	244	3.6	15	345
0.4 Mn–0.12 C steel	450	1660	3.5		
321 Laminated steel	500	2915	3.9		
310 Laminated steel	550	1766	4		
	700	675	3.5		

Kachanov's creep damage law

This is concerned with brittle viscoplastic damage which was mentioned in Section 7.2.3 and an example of which is shown in Fig. 7.7.

In 1958, Kachanov proposed the modelling of this type of evolution by:

- $$\dot{D} = \left[\frac{\sigma_v}{A_0(1-D)}\right]^r,$$

where A_0 and r are the two characteristic creep damage coefficients for the material. This expression yields not zero but very low values for primary and secondary creep. A number of identifications have since shown the correctness of this modelling, at least for cases of sufficiently simple load histories. Table 7.2 gives the value of A_0 and r for some materials.

The time to rupture t_c, in a creep test under constant true stress, is obtained from the solution of the differential equation of the model for a damage value equal to the critical value, $D = D_c$. With the initial condition $D = 0$ at $t = 0$:

$$t_c = \frac{1 - (1 - D_c)^{r+1}}{r + 1} \left(\frac{\sigma_v}{A_0} \right)^{-r}.$$

For usual values of r we may neglect $(1 - D_c)^{r+1}$ in comparison to 1, and express the damage evolution in a particularly simple form by integrating the differential equation between 0 and D:

$$D = 1 - \left(1 - \frac{t}{t_c} \right)^{1/(r+1)} \quad \text{with} \quad t_c = \frac{1}{r+1} \left(\frac{\sigma_v}{A_0} \right)^{-r}.$$

Fig. 7.10 shows that the power relation between the stress and the time to rupture is verified satisfactorily by experiment and that the coefficients A_0 and r are strongly temperature dependent.

Fatigue damage equations

Fatigue damage in metals corresponds to nucleation and growth of microcracks, generally intracrystalline, under the action of cyclic loading,

Fig. 7.10. Time to reach creep rupture as a function of stress and temperature: IN 100 alloy (uncoated).

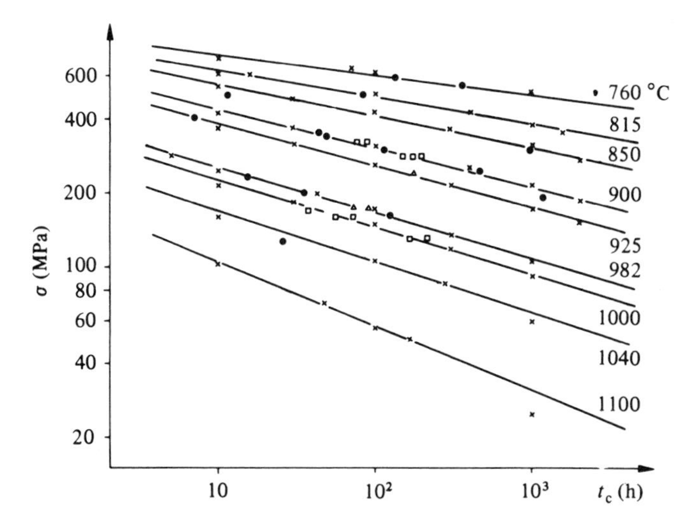

until the initiation of a macroscopic crack. This phenomenon can occur at low stress levels, lower than the conventional elastic limit. We thus distinguish:

> fatigue with a large number of cycles which almost certainly involves elastic deformations and which corresponds to a number of cycles to failure higher than approximately 50 000; beyond 10^6–10^7 or even 10^9 cycles, depending on the intended field of application, the life-time is generally considered to be unlimited, i.e., when the loads are lower than a conventional fatigue limit;
> fatigue with a low number of cycles (or low-cycle fatigue) which corresponds to less than 50 000 cycles in which plastic deformations occur; at high temperatures the fatigue phenomenon is usually coupled with creep damage and can be isolated only by performing tests at a sufficiently high frequency ($f > 5$–10 Hz).

The classical laws are expressed as functions of quantities defined with respect to a cycle, see Fig. 7.11:

> the stress amplitude $\Delta\sigma/2$ or strain amplitude $\Delta\varepsilon/2$,
> the mean value (over a cycle) of stress $\bar{\sigma}$ or strain $\bar{\varepsilon}$,
> the ratio of the minimum stress (or strain) to the maximum stress (or strain) $R_\sigma = \sigma_m/\sigma_{Max}$ (or $R_\varepsilon = \varepsilon_m/\varepsilon_{Max}$).

Linear and nonlinear accumulation

Cumulative effects, whether linear or nonlinear, are of great importance in fatigue. We have seen in Chapter 3 (Section 3.2.3) that the rule of linear

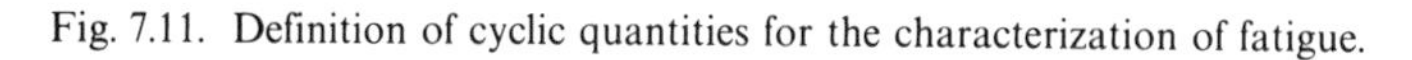

Fig. 7.11. Definition of cyclic quantities for the characterization of fatigue.

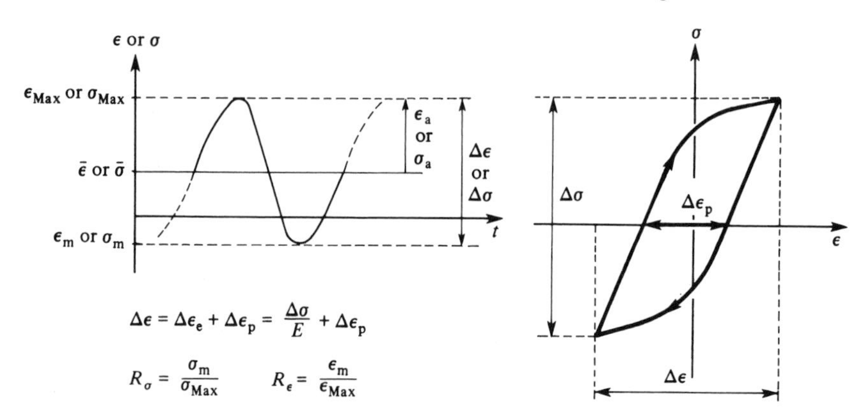

accumulation is in fact a property of any linear or nonlinear differential equation with separable variables.

The Palmgreen – Miner linear rule is based on the assumption that damage is accumulated additively when it is defined by the associated life ratio N_i/N_{F_i} where N_i is the number of cycles applied under a given load for which the number of cycles to fracture (under periodic conditions) would be N_{F_i}. The fracture criterion is:

$$\sum_i (N_i/N_{F_i}) = 1.$$

Therefore, in periodic tests, damage evolution is considered to be linear in that:

$$D = N/N_F.$$

In fact, the linear accumulation rule can be applied even to a damage which evolves nonlinearly. For this it is sufficient that a one-to-one relationship between D and N/N_F exists, or even that the damage evolution curve be a unique function (independent of the applied cycle) of the life ratio N/N_F. For a test at two stress levels, the evolution is as shown schematically in Fig. 7.12(*a*).

There are, therefore, two ways of defining a damage incremental law incorporating the linear accumulation rule. Following well-established rules this increment is introduced per cycle by:

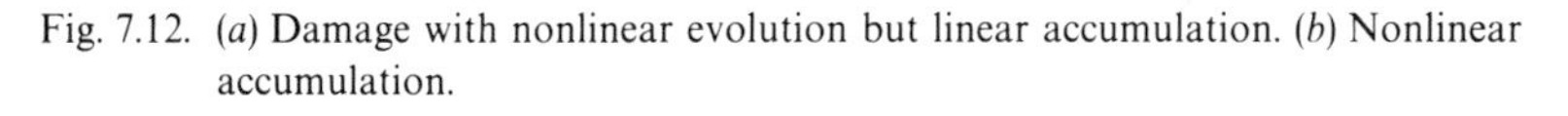

Fig. 7.12. (*a*) Damage with nonlinear evolution but linear accumulation. (*b*) Nonlinear accumulation.

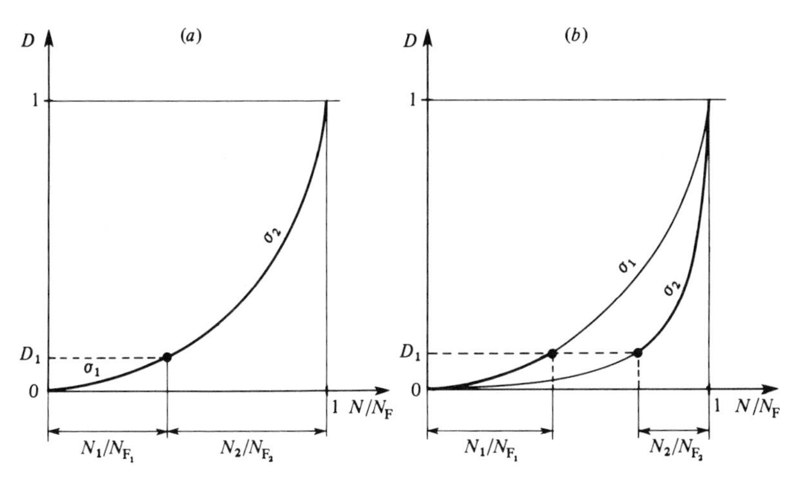

using simply an equation of the form:

$$\delta D = \delta N / N_\text{F}(\ldots)$$

where N_F is the number of cycles to failure defined by the chosen parametric relation;

employing a nonlinear evolution of damage which, however, is a unique function of N/N_F, for example:

$$\delta D = \frac{(1 - D)^{-k}}{k + 1} \frac{\delta N}{N_\text{F}(\ldots)}.$$

We recall (cf. Chapter 3) that such an equation, in which the loading and the damage D are separable variables, always leads to linear accumulation like that shown in Fig. 7.12(*a*).

In contrast, if the damage evolution curve, as a function of the life ratio N/N_F, depends on the applied loading we have the effect of nonlinear accumulation as shown in Fig. 7.12(*b*). D_1 represents the state of internal damage at the end of the first level σ_1. Evolution at the second level σ_2 continues from the same state, and it is clear that the sum of the life ratios is less than 1. From the point of view of the damage law, this nonlinearity

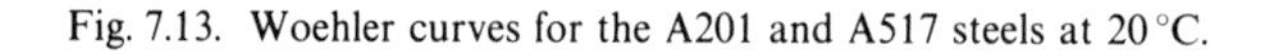

Fig. 7.13. Woehler curves for the A201 and A517 steels at 20 °C.

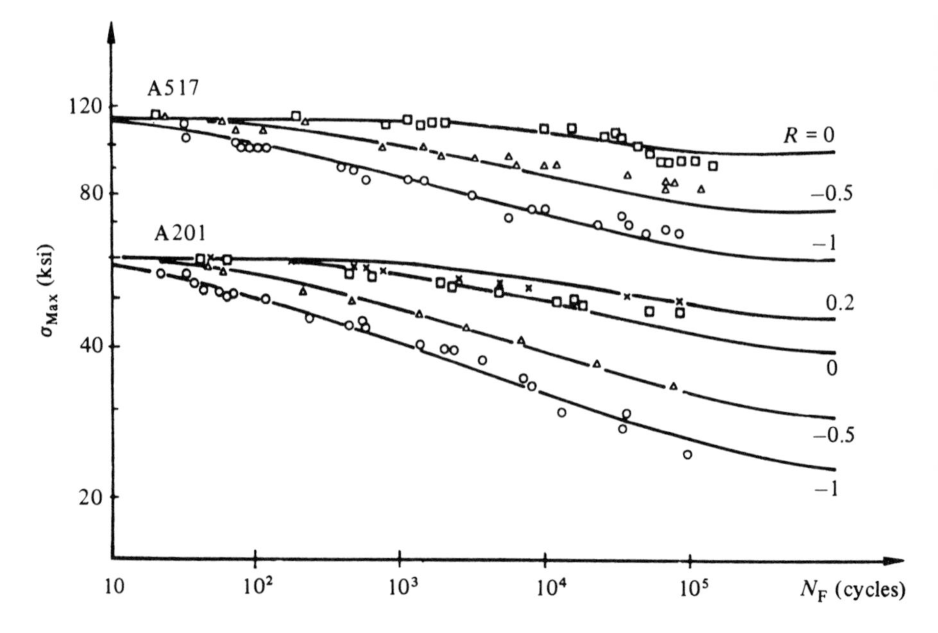

always corresponds to the case where the variables which represent the load and the damage variable D are not separable.

The Palmgreen–Miner linear accumulation law gives good results only for loads for which there is little variation in the amplitude and mean value of the stress. More sophisticated models for loads with complex histories will be described in Section 7.4.3.

Woehler–Miner law

This rule essentially applies to the fatigue of metals under periodic loading. Woehler curves express the experimental results of uniaxial fatigue tests. They represent the relations between the number of cycles to failure, the value of the maximum stress σ_{Max} and the mean stress $\bar{\sigma}$ (or the ratio $R_\sigma = \sigma_{\mathrm{m}}/\sigma_{\mathrm{Max}}$). A typical example is given in Fig. 7.13.

In general, each curve exhibits two limits: one, when σ_{Max} tends towards the static fracture stress σ_{u} (fracture in a quarter of the cycle), and the other, when σ_{Max} tends towards the fatigue limit σ_{l}. The influence of the mean stress on the fatigue limit is satisfactorily represented by Goodman's linear relation:

$$\sigma_{\mathrm{l}}(\bar{\sigma}) = \bar{\sigma} + \sigma_{\mathrm{l}_0}(1 - \bar{\sigma}/\sigma_{\mathrm{u}})$$

where σ_{l_0} is the fatigue limit at $\bar{\sigma} = 0$, or in a modified form by:

$$\sigma_{\mathrm{l}}(\bar{\sigma}) = \bar{\sigma} + \sigma_{\mathrm{l}_0}(1 - b\bar{\sigma})$$

in which b is a characteristic coefficient of the material and σ_{l} is the fatigue limit expressed in terms of maximum stress. Fig. 7.14 shows that this relation is applicable for a large range.

One way of writing a damage law which expresses these experimental results is to assume that the damage per cycle is a function of the maximum and the mean values of the stress:

$$\delta D/\delta N = f(\sigma_{\mathrm{Max}}, \bar{\sigma}).$$

In order to recover, after integration, one of the many forms proposed to represent the Woehler curves, we let:

$$\bullet \qquad \frac{\delta D}{\delta N} = \frac{\sigma_{\mathrm{Max}} - \sigma_{\mathrm{l}}(\bar{\sigma})}{\sigma_{\mathrm{u}} - \sigma_{\mathrm{Max}}}\left(\frac{\sigma_{\mathrm{Max}} - \bar{\sigma}}{B(\bar{\sigma})}\right)^{\beta}$$

with:

$$\sigma_{\mathrm{l}}(\bar{\sigma}) = \bar{\sigma} + \sigma_{\mathrm{l}_0}(1 - b\bar{\sigma})$$
$$B(\bar{\sigma}) = B_0(1 - b\bar{\sigma}).$$

The number of cycles to failure is obtained by an obvious integration, with the conditions:

$$N = 0 \rightarrow D = 0$$
$$N = N_F \rightarrow D = 1$$

so that:

$$\bullet \quad N_F = \frac{\sigma_u - \sigma_{Max}}{\sigma_{Max} - \sigma_l(\bar{\sigma})} \left(\frac{\sigma_{Max} - \bar{\sigma}}{B(\bar{\sigma})} \right)^{-\beta}.$$

By identifying this expression with Woehler's curves, one can determine (possibly as functions of temperature) the coefficients σ_u, σ_{l_0}, b, B_0 and β for each material. Some examples of the values of these coefficients are given in Table 7.3. Fig. 7.13 illustrates the application of this expression to two steels (the coefficients a and M_0 correspond to a generalization to be studied in Section 7.4.3).

For periodic loading (σ_{Max} = constant, σ_m = constant), damage evolves as a linear function of the number of cycles:

$$D = \frac{\sigma_{Max} - \sigma_l(\bar{\sigma})}{\sigma_u - \sigma_{Max}} \left(\frac{\sigma_{Max} - \bar{\sigma}}{B(\bar{\sigma})} \right)^{\beta} N = \frac{N}{N_F}.$$

This property is, however, very poorly verified by experiment, as we shall see in Section 7.4.3.

Coffin–Manson law

In low-cycle fatigue, plastic strains become dominant. To include certain time or frequency related effects in such a case, we may assume that the

Fig. 7.14. Effect of the mean stress on fatigue limit in tension–compression.

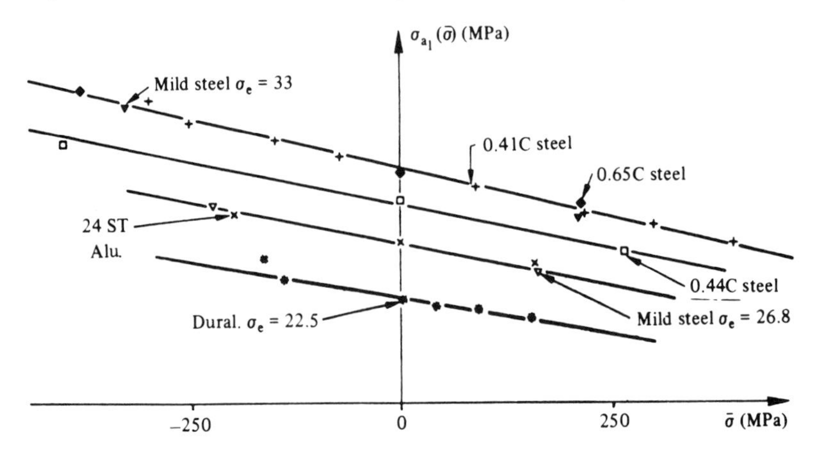

Table 7.3. *Coefficients of the fatigue damage law (in MPa units)*

Material	$T(^\circ\mathrm{C})$	σ_{l_0}	σ_u	β	B_0	b	a	M_0
IN 100	700	300	950	2.6	6560		0.120	4750
IN 100	900	240	715	3	4430		0.072	2925
IN 100	950	180	600	4.8	1637	0.0013	0.108	1485
IN 100	1000	140	470	6	880		0.138	875
IN 100	1110	70	250	4.7	645		0.162	635
INCO 718	550	600	1150	10.3	1586		0.015	1335
A 201	20	150	412	6	932	0.0070		
A 517	20	400	790	7.4	1528	0.0038	0.710	1945
C 35	20	255	458	4	5434			
316 L	20	222	760	5	1213		0.900	1700
Maraging	20	310	2028	4	5820			
AISI 4130 S	20	240	890	5.5	1590			
AISI 4130 H	20	292	1420	5	2460			
AISI 304	20	240	950	5.8	2000			
52100 steel	20	360	2005	3.3	6320			
Cu 5456 H311	20	76	400	4.7	832			
AISI 304	593	150	700	3	4207	0.005	0.100	3100
AISI 304	650	140	580	3.7	1728		0.200	1700
316 L (ICL 167)	600	200	650	5.5	1144		0.200	1200

damage per cycle is a power function of the plastic strain range:

$$\delta D/\delta N = f(\Delta\varepsilon_p) = (\Delta\varepsilon_p/C_1)^{\gamma_1}.$$

The integration of this relation for periodic loading where the cycle is assumed to be stabilized, from the first to the last cycle, gives the Coffin–Manson relation:

$$N_F = (\Delta\varepsilon_p/C_1)^{-\gamma_1}$$

which corresponds to the conditions:

$$N = 0 \rightarrow D = 0$$

$$N = N_F \rightarrow D = 1.$$

In the low stress range where $\Delta\varepsilon_p$ is small, we may complement this relation by a damage law expressed as a function of the stress amplitude $\Delta\sigma$:

$$\delta D/\delta N = (\Delta\sigma/C_2)^{\gamma_2}$$

or, by integrating with the same initial and final conditions as before:

$$N_F = (\Delta\sigma/C_2)^{-\gamma_2}.$$

By using the linear elasticity law $\Delta\sigma = E\Delta\varepsilon_e$, solving for $\Delta\varepsilon_e$, and then

Table 7.4. *Coefficients of the Coffin–Manson relations*

Material	T(°C)	γ_1	C_1	γ_2	C_2
TA6V	350	1.818	0.38	20	1450
INCO 718	550	1.615	0.407	18.9	2580
IN 100	700	2	0.244	6.5	3168
IN 100	900	2	0.037	7.1	2459
IN 100	1000	2	0.031	8.75	1268
IN 100	1110	2	0.02	7.8	750
AISI 304	20			10.5	2120
AISI 304	430	2.2	0.189		
Maraging, hardened	20	1.22	1.634	11.1	2940
35 CD 4 Air					
Cooled steel	20	1.54	0.662	8.3	3565
X 20 GM$_0$ V 12 1	350	1.83	0.356	13.5	1660
316 ICL 167 CN	20	2.07	0.34	5.7	3280
IMI 550	20	1.3	2.63	13.3	2940
COTAC 744	1000	1.32	0.408	11.1	1560
AISI 4130 soft	20			9.6	1925
AISI 4130 hard	20			9.2	2805
AISI 52100	20			6.5	5100
ALU 5456 H 311	20			7.2	1140
UDIMET 700	20	1.67	0.407	8.3	4830

adding $\Delta\varepsilon_p$ as determined from the Coffin–Manson relation, we obtain:

$$\Delta\varepsilon_e + \Delta\varepsilon_p = (C_2/E)N_F^{-1/\gamma_2} + C_1 N_F^{-1/\gamma_1}.$$

The coefficients C_1, γ_1, C_2, γ_2 depend on the material and the temperature. Table 7.4 and Fig. 7.15 give a few examples. This relation is interesting because after a large number of tests on many materials it has been possible to write it in the following form called 'universal slopes':

- $$\Delta\varepsilon = \Delta\varepsilon_e + \Delta\varepsilon_p = 3.5\frac{\sigma_u}{E}N_F^{-0.12} + D_u^{0.6}N_F^{-0.6}$$

where σ_u is the ultimate static fracture stress, E the Young's modulus, and D_u the ductility which is expressed as a function of the reduction of the area RA at necking: $D_u = -\ln(1 - RA)$. Fig. 7.15(*b*) illustrates the case of the UDIMET 700 alloy at room temperature.

The Coffin–Manson relation itself also contains the rule of linear accumulation and can therefore be used for cases of varying amplitude. Nevertheless, its range of application is limited to low variations in strain range and to the quite low temperatures for which the creep damage is small.

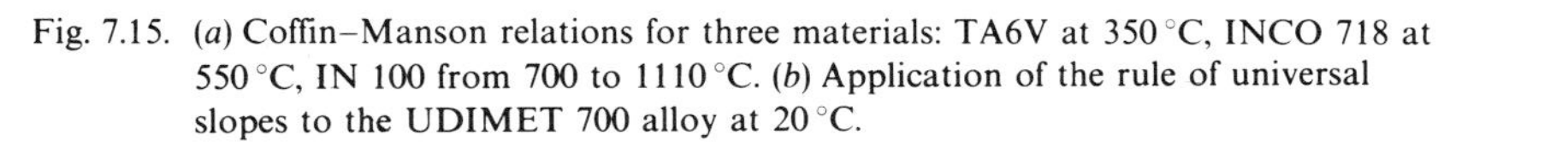

Fig. 7.15. (*a*) Coffin–Manson relations for three materials: TA6V at 350 °C, INCO 718 at 550 °C, IN 100 from 700 to 1110 °C. (*b*) Application of the rule of universal slopes to the UDIMET 700 alloy at 20 °C.

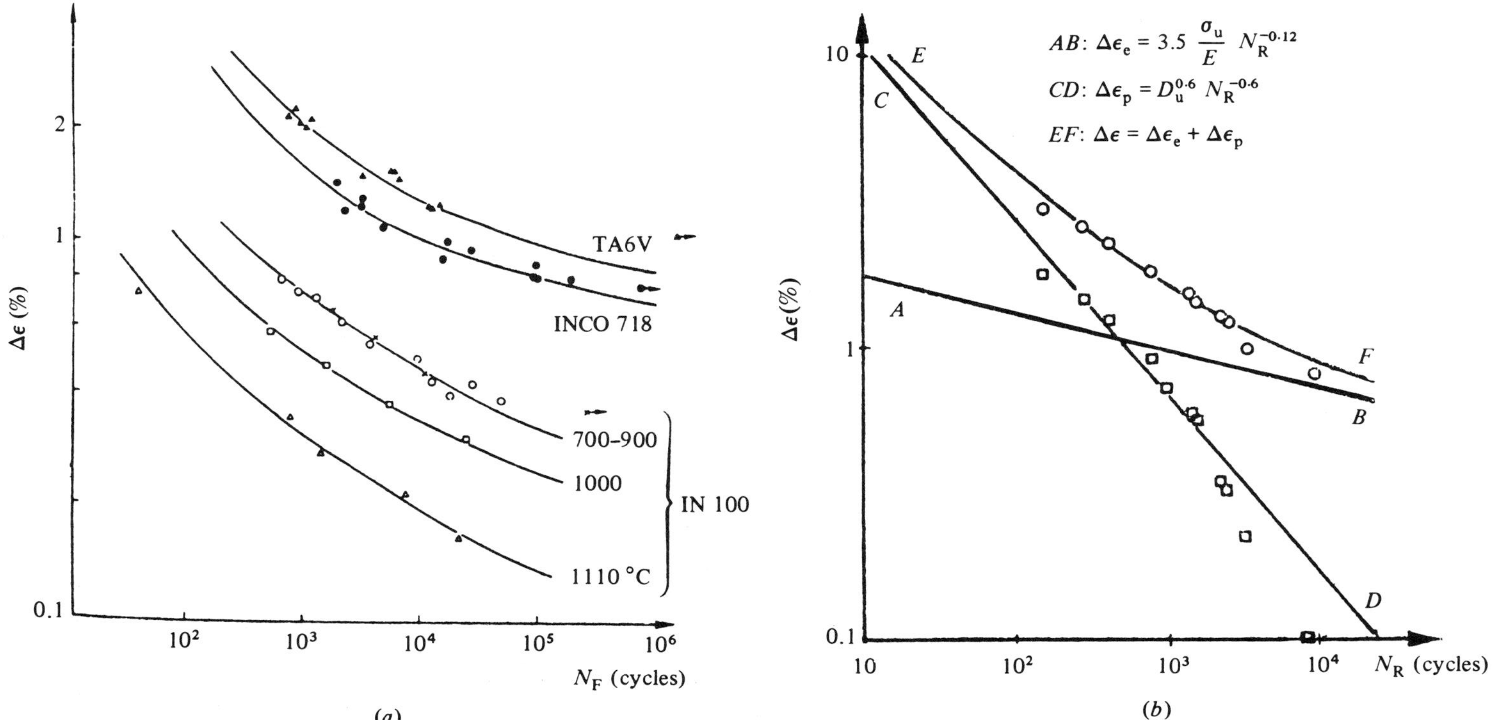

Energy criterion

An interesting parameter is Δw the energy dissipated per cycle, i.e., the area of the stress–strain hysteresis loop. This can be determined approximately from knowledge of the cyclic hardening coefficient by:

$$\Delta w = \frac{M_c - 1}{M_c + 1}\Delta\sigma\Delta\varepsilon_p.$$

The total dissipation energy up to failure is, unfortunately, not a constant intrinsic to the material; it has been observed to depend on the number of cycles to failure, generally as a power function (see, for example, Fig. 7.16):

$$w_F = N_F \Delta w = \alpha N_F^{-\delta}.$$

The laws of plasticity and their formulation within the thermodynamic framework can be used to replace Δw by the stored energy of the volume element (the rest being dissipated as heat). In the case of time-independent plasticity with isotropic hardening and superposed kinematic hardening with a yield function

$$f = J(\boldsymbol{\sigma} - \mathbf{X}) - R - k$$

where J is the second invariant of the deviator of $\boldsymbol{\sigma} - \mathbf{X}$, the power dissipated as heat can be written as (see Chapter 5):

$$\Phi = \boldsymbol{\sigma}:\dot{\boldsymbol{\varepsilon}}_p - \mathbf{X}:\dot{\boldsymbol{\alpha}} - R\dot{p} = \dot{p}[J(\boldsymbol{\sigma} - \mathbf{X}) - R] = \dot{p}(f + k).$$

When there is no flow we have $\dot{p} = 0$. On the other hand, during flow we

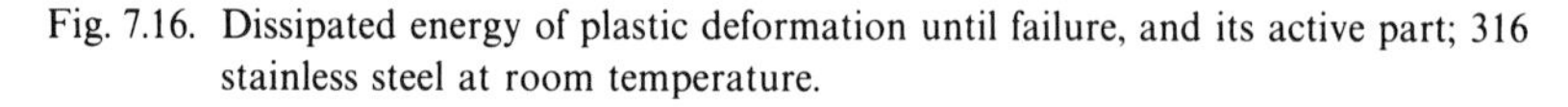

Fig. 7.16. Dissipated energy of plastic deformation until failure, and its active part; 316 stainless steel at room temperature.

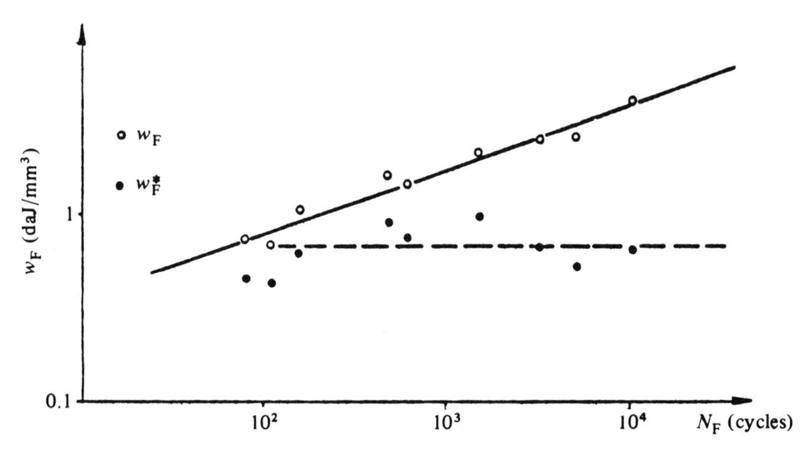

have $f = 0$, and the dissipated power is:

$$\Phi = k\dot{p} = k(\tfrac{2}{3}\dot{\varepsilon}_p : \dot{\varepsilon}_p)^{1/2}.$$

During a tension–compression cycle, the active energy stored in the volume element is therefore:

$$\Delta w^* = \Delta w - \Delta\Phi = \Delta w - 2k\Delta\varepsilon_p = \left(\frac{M_c - 1}{M_c + 1}\Delta\sigma - 2k\right)\Delta\varepsilon_p$$

where k represents the initial yield stress when

$$\mathbf{X}(0) = \varepsilon_p(0) = R(0) = 0.$$

The interesting feature of this parameter Δw^* is illustrated in Fig. 7.16; the energy to failure given by:

$$w_F^* = w_F - 2kN_F\Delta\varepsilon_p$$

is approximately constant and may be considered intrinsic to the material (the value $k = 220$ MPa used for this calculation corresponds to the limit of elasticity at 0.02% plastic strain for the 316 steel).

Strain range partitioning method

This method was developed to describe high temperature fatigue taking into account the effects of time. The basic assumption is that a complex cycle can always be decomposed into four parts corresponding to the four damage relations which can be determined independently.

Two types of damage can occur depending on whether the load generates plastic or creep deformation. A tensile load produces damage of a different type than a compressive load, and these two can be further distinguished to bring the total to four types, corresponding to the four basic cycles of Fig. 7.17.

In alternate tension–compression fatigue cycles, there is always:

plastic damage in both tension and compression:

D_{pp}, $\Delta\varepsilon_{pp}$;

plastic damage in tension, creep damage in compression:

D_{pc}, $\Delta\varepsilon_{pc}$;

creep damage in tension, plastic damage in compression:

D_{cp}, $\Delta\varepsilon_{cp}$;

Fig. 7.17. The four basic cycles of strain range partitioning.

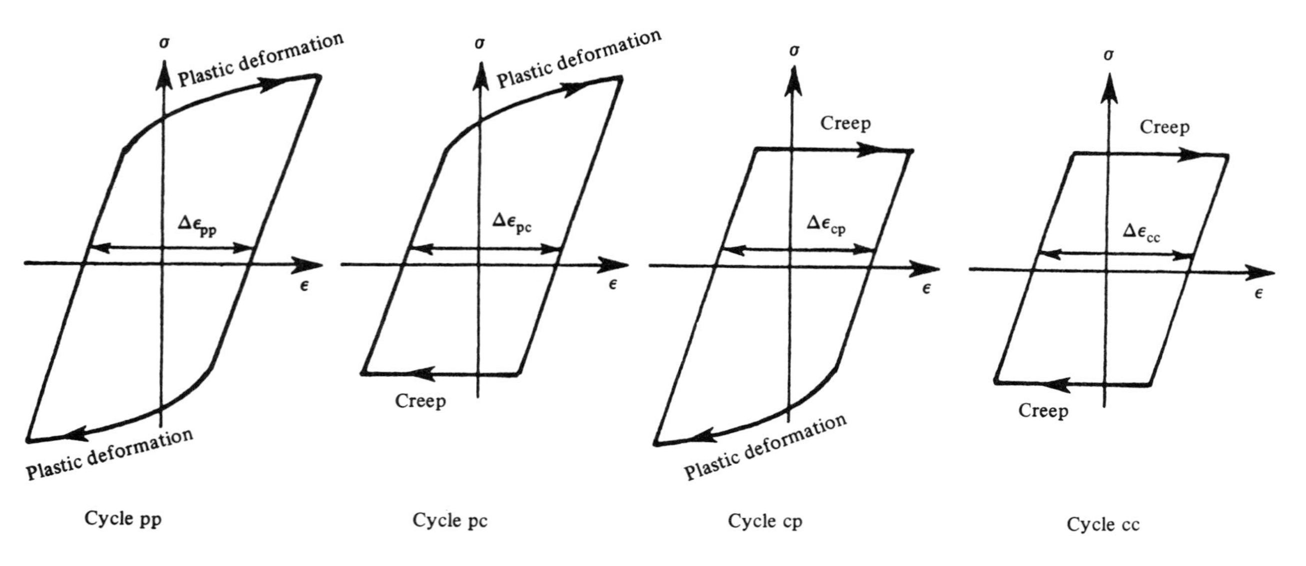

creep damage in both tension and compression:

D_{cc}, $\Delta\varepsilon_{cc}$.

Each damage is modelled by a differential relation with respect to the number of cycles. The form generally used reduces to assuming for each kind of damage:

$$\frac{\delta D_{ij}}{\delta N} = \left(\frac{\Delta\varepsilon_{in}}{C_{ij}}\right)^{\gamma_{ij}} \frac{\Delta\varepsilon_{ij}}{\Delta\varepsilon_{in}}.$$

Assuming that these damages are additive:

$$\frac{\delta D}{\delta N} = \left(\frac{\Delta\varepsilon_{in}}{C_{pp}}\right)^{\gamma_{pp}} \frac{\Delta\varepsilon_{pp}}{\Delta\varepsilon_{in}} + \left(\frac{\Delta\varepsilon_{in}}{C_{pc}}\right)^{\gamma_{pc}} \frac{\Delta\varepsilon_{pc}}{\Delta\varepsilon_{in}} + \left(\frac{\Delta\varepsilon_{in}}{C_{cp}}\right)^{\gamma_{cp}} \frac{\Delta\varepsilon_{cp}}{\Delta\varepsilon_{in}} + \left(\frac{\Delta\varepsilon_{in}}{C_{cc}}\right)^{\gamma_{cc}} \frac{\Delta\varepsilon_{cc}}{\Delta\varepsilon_{in}}.$$

In each of the four basic cases, the total inelastic strain $\Delta\varepsilon_{in}$ is equal to the strain $\Delta\varepsilon_{ij}$:

$$\Delta\varepsilon_{ij}/\Delta\varepsilon_{in} = 1.$$

The damage law is integrated assuming that in each case the loading is periodic with a constant strain range: $\Delta\varepsilon_{ij} = \Delta\varepsilon_{in} = \text{constant}$. Let N_{ij} be the solution of $\delta D_{ij}/\delta N = (\Delta\varepsilon_{in}/C_{ij})^{\gamma_{ij}}$ subject to the conditions

$$N = 0 \rightarrow D_{ij} = 0$$
$$N = N_{ij} \rightarrow D_{ij} = 1:$$
$$N_{pp} = (\Delta\varepsilon_{pp}/C_{pp})^{-\gamma_{pp}}$$
$$N_{pc} = (\Delta\varepsilon_{pc}/C_{pc})^{-\gamma_{pc}}$$
$$N_{cp} = (\Delta\varepsilon_{cp}/C_{cp})^{-\gamma_{cp}}$$
$$N_{cc} = (\Delta\varepsilon_{cc}/C_{cc})^{-\gamma_{cc}}.$$

These relations can be used to express, in a particularly simple manner, the equation of total damage under periodic loads:

$$\frac{\delta D}{\delta N} = \frac{1}{N_{pp}} \frac{\Delta\varepsilon_{pp}}{\Delta\varepsilon_{in}} + \frac{1}{N_{pc}} \frac{\Delta\varepsilon_{pc}}{\Delta\varepsilon_{in}} + \frac{1}{N_{cp}} \frac{\Delta\varepsilon_{cp}}{\Delta\varepsilon_{in}} + \frac{1}{N_{cc}} \frac{\Delta\varepsilon_{cc}}{\Delta\varepsilon_{in}}$$

and for

$$D = 1 \rightarrow N = N_R$$

● $$\frac{1}{N_R} = \frac{1}{N_{pp(\Delta\varepsilon_{in})}} \frac{\Delta\varepsilon_{pp}}{\Delta\varepsilon_{in}} + \frac{1}{N_{pc(\Delta\varepsilon_{in})}} \frac{\Delta\varepsilon_{pc}}{\Delta\varepsilon_{in}} + \frac{1}{N_{cp(\Delta\varepsilon_{in})}} \frac{\Delta\varepsilon_{cp}}{\Delta\varepsilon_{in}} + \frac{1}{N_{cc(\Delta\varepsilon_{in})}} \frac{\Delta\varepsilon_{cc}}{\Delta\varepsilon_{in}}.$$

This is the relation of the strain range partitioning method. Fig. 7.18 illustrates the shape of the four curves $N_{ij}(\Delta\varepsilon_{ij})$ for the particular case of 316 stainless steel.

Use of the partitioning method requires knowledge of the four basic curves and of the strain components $\Delta\varepsilon_{pp}$, $\Delta\varepsilon_{pc}$, $\Delta\varepsilon_{cp}$, $\Delta\varepsilon_{cc}$; $\Delta\varepsilon_{in}$ is the total inelastic strain range. The numbers N_{pp}, N_{pc}, N_{cp}, N_{cc} are the abscissae of the points of intersection of a straight line of ordinate $\Delta\varepsilon_{in}$ with the four basic curves (Fig. 7.19(*a*)).

The decomposition of a real cycle into 'pp, pc, cp, cc' basic cycles is very difficult because in some cases it assumes a subjective character. Consider the real cycle *ABCDEF* of Fig. 7.19(*b*):

> *AB* is a 'plastic' loading in tension, assumed to be rapid;
> *BC* is a 'creep' part;

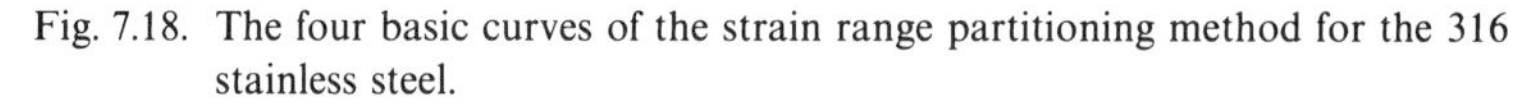

Fig. 7.18. The four basic curves of the strain range partitioning method for the 316 stainless steel.

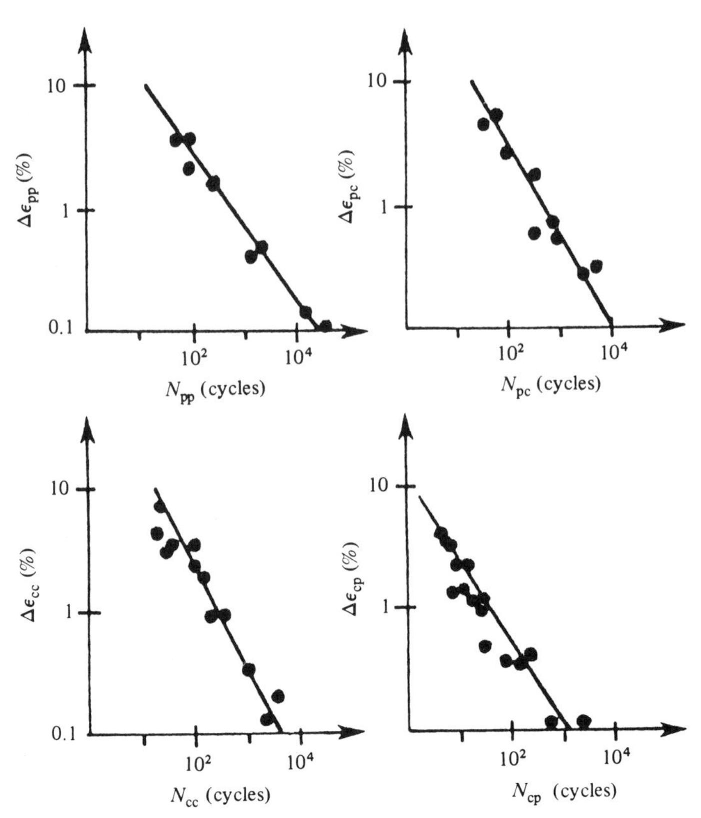

Fig. 7.19. (*a*) Obtaining the basic number of cycles to failure. (*b*) Partition of the actual cycle.

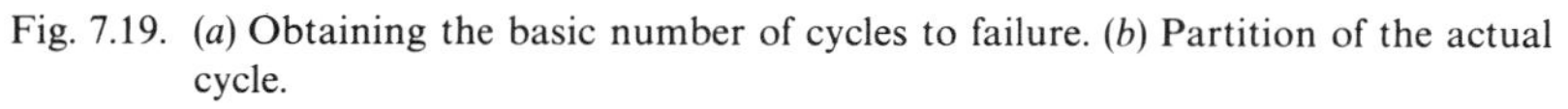

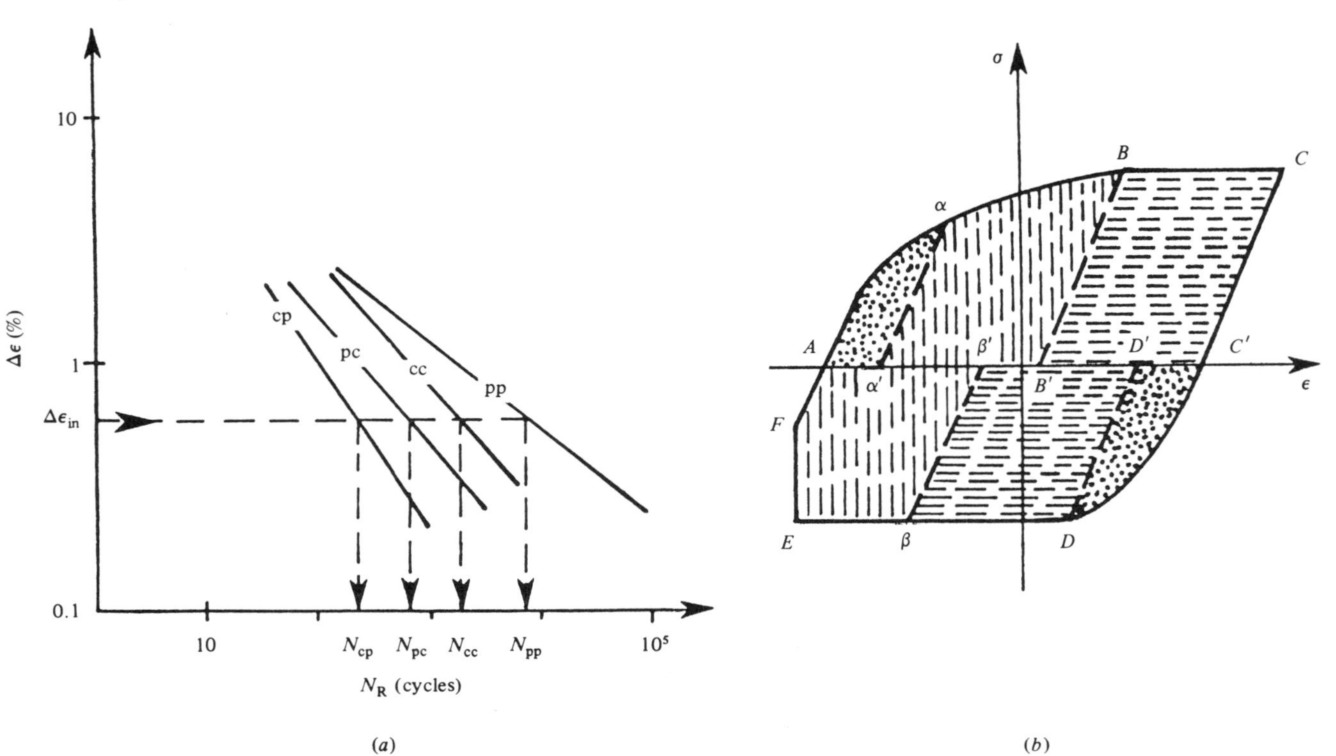

CD is an unloading and a plastic loading in compression, assumed to be rapid;
DE is a creep under compression;
EF is a relaxation in compression;
FA is an elastic loading.

(1) We search for the smallest plastic part, say *CD*. We associate to it a part which corresponds to a plastic strain of equal magnitude but of opposite sign, say $A\alpha$; we thus obtain the pp cycle with:

$$\Delta\varepsilon_{pp} = A\alpha' = D'C'.$$

(2) We search the smallest creep part, say *BC*. We associate to it a part which corresponds to a creep strain of equal magnitude but of opposite sign, say $D\beta$; we thus obtain the cc cycle with:

$$\Delta\varepsilon_{cc} = B'C' = \beta' D'.$$

(3) There remains the 'plastic' portion αB, the creep portion βE, and the relaxation portion *EF*. Relaxation and creep are governed by the same (viscoplastic) damage law.

The portion αB corresponds to a strain $\Delta\varepsilon_{p} = \alpha' B'$.
The portion βEF corresponds to a strain $\Delta\varepsilon_{c} = A\beta'$.

The partition is done in a way which equalizes these two strains. We may then associate these two portions to a pc cycle

$$\Delta\varepsilon_{pc} = \alpha' B' = A\beta'.$$

(4) In this particular case, the cp cycle does not exist: $\Delta\varepsilon_{cp} = 0$. We, of course, verify the fact that:

$$\sum_{ij} \Delta\varepsilon_{ij} = \Delta\varepsilon_{in} = AC'.$$

This method gives a unique decomposition for the cycle of Fig. 7.18(*b*), but it is conceivable that it could lead to different interpretations for cycles in which the transient loadings are so slow that there is creep (under variable load). In this case, which is quite common in practice, $\Delta\varepsilon_{pp}$ must be determined by a tension–compression cyclic test performed at a sufficiently high frequency (5–10 Hz) and with the same stress range. Another case where the partition is difficult is when the temperature changes during the cycle. The strain range partitioning method is a simple one which can be used for predictive purposes provided it is applied to cases which represent interpolations between the four basic curves and not extrapolations. Cases of varying temperatures, very small plastic strains, ratchetting effects, generalizations to multiaxial loading pose difficult, even unsolvable problems.

7.2.5 *Multiaxial damage criteria*

In Section 7.2.4 we introduced damage thresholds such as the stress or strain below which the damage is zero or negligible. This concept can be generalized to three dimensions, in the same manner as the plasticity criteria. Similarly, for time-dependent damage, we can use surfaces similar to the equipotentials introduced for viscoplasticity in Chapter 6.

In one dimension, the damage threshold (in terms of stress) defines the range of resistance of the material:

$$-\sigma_D < \sigma < \sigma_D \rightarrow \dot{D} = 0.$$

When $|\sigma|$ is higher than the threshold σ_D, there is damage.

In three dimensions, this concept is generalized by a damage threshold (yield) surface:

$$f_D(\boldsymbol{\sigma}, D) = 0.$$

When $f_D < 0$, no damage results but when $f_D \geqslant 0$, damage occurs. The following sub sections give some examples of such surfaces described in terms of the stress tensor invariants. This damage is assumed here to be isotropic. In some cases we will also employ a law in terms of strain and the limiting surface will then be expressed by:

$$f_D(\boldsymbol{\varepsilon}, D) = 0.$$

The elastic energy density release rate criterion

The von Mises plasticity criterion is concerned with the shear energy, since slip is the main mechanism of plastic deformation. Damage is equally sensitive to shear energy but also to the volumetric deformation energy since the growth of cavities and cracks is very sensitive to hydrostatic stress. To formulate an isotropic criterion, we may therefore postulate that the mechanism of damage is governed by the total elastic strain energy: distortion energy + volumetric energy:

$$w^e = w^e_d + w^e_H$$

$$w_e = \int_0^{\varepsilon^e} \boldsymbol{\sigma} : d\boldsymbol{\varepsilon}^e = \int_0^{\varepsilon^{e'}} \boldsymbol{\sigma}' : d\boldsymbol{\varepsilon}^{e'} + 3 \int_0^{\varepsilon^e_H} \sigma_H \, d\varepsilon_H \, .$$

$\boldsymbol{\sigma}'$ and $\boldsymbol{\varepsilon}^{e'}$ are, as before, the deviators of $\boldsymbol{\sigma}$ and $\boldsymbol{\varepsilon}^e$; σ_H and ε_H are the hydrostatic stress and strain respectively.

By introducing the linear elasticity law coupled with damage through the concept of effective stress:

$$\boldsymbol{\varepsilon}^{e'} = \frac{1+\nu}{E}\frac{\boldsymbol{\sigma}'}{1-D} \qquad \varepsilon_H = \frac{1-2\nu}{E}\frac{\sigma_H}{1-D}$$

and by assuming that the damage does not vary within the elastic domain, we obtain:

$$w_e = \frac{1}{2}\left(\frac{1+\nu}{E}\frac{\boldsymbol{\sigma}':\boldsymbol{\sigma}'}{1-D} + 3\frac{1-2\nu}{E}\frac{\sigma_H^2}{1-D}\right)$$

or, with the equivalent von Mises stress $\sigma_{eq} = (\frac{3}{2}\boldsymbol{\sigma}':\boldsymbol{\sigma}')^{1/2}$:

$$w_e = \frac{1}{2E(1-D)}[\tfrac{2}{3}(1+\nu)\sigma_{eq}^2 + 3(1-2\nu)\sigma_H^2].$$

Similar to the equivalent stress in plasticity, we define the equivalent damage stress σ^* by stating that this energy in a multiaxial state is equal to that in an equivalent uniaxial state defined by σ^*, for which ($\sigma_{eq} = \sigma^*$, $\sigma_H = \frac{1}{3}\sigma^*$):

$$\frac{\sigma^{*2}}{2E(1-D)} = \frac{1}{2E(1-D)}[\tfrac{2}{3}(1+\nu)\sigma_{eq}^2 + 3(1-2\nu)\sigma_H^2]$$

or

- $$\sigma^* = \sigma_{eq}[\tfrac{2}{3}(1+\nu) + 3(1-2\nu)(\sigma_H/\sigma_{eq})^2]^{1/2}.$$

The ratio σ_H/σ_{eq} in this relation expresses the triaxiality of the state of stress.

We shall see in Section 7.3.2 in the discussion on the thermodynamic formulation of damage that this equivalent stress σ^* is related to Y, the dual variable of D, in a simple way (such that $Y\dot{D}$ represents the dissipation); Y represents, if we do not consider the sign, the elastic energy release rate during damage growth at constant temperature and stress:

$$\sigma^* = (1-D)(-2EY)^{1/2} \quad \text{with} \quad Y = -\frac{1}{2}\frac{\mathrm{d}w_e}{\mathrm{d}D}\bigg|_{\sigma,T}$$

from which ensues the name the elastic energy density release rate criterion.

Fig. 7.20 gives the graphical representation of this criterion for (*a*) two-dimensional biaxial loading, and (*b*) uniaxial loading–torsion.

Fig. 7.20. Representation of the criterion of elastic energy density release rate: (*a*) two-dimensional biaxial loading and (*b*) uniaxial loading–torsion (after D. Baptiste).

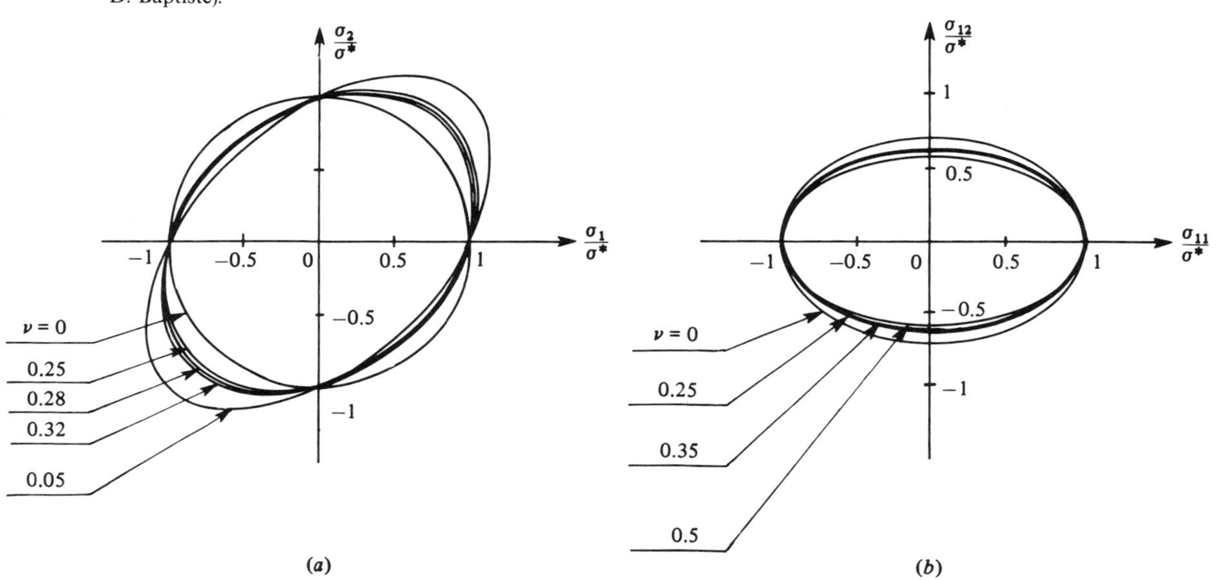

Three invariants criterion

The elastic energy density release rate criterion is very simple and is formally justified by thermodynamics; no material coefficient, other than Poisson's ratio, appears in this criterion. However, in certain cases, it can be at fault, particularly, for creep damage where it gives the same answer in tension and in compression.

Within the framework of isotropy, a more general, but more complex, form consists in expressing the damage criterion in terms of the three basic invariants of the stress tensor:

$$\sigma_H = \tfrac{1}{3}J_1(\boldsymbol{\sigma}) = \tfrac{1}{3}\mathrm{Tr}(\boldsymbol{\sigma})$$

$$\sigma_{eq} = J_2(\boldsymbol{\sigma}) = [\tfrac{3}{2}\mathrm{Tr}(\boldsymbol{\sigma}'^2)]^{1/2}$$

$$J_3(\boldsymbol{\sigma}) = [\tfrac{9}{2}\mathrm{Tr}(\boldsymbol{\sigma}'^3)]^{1/3}.$$

Instead of $J_3(\boldsymbol{\sigma})$ it is more practical to choose:

$$J_0(\boldsymbol{\sigma}) = \mathrm{Sup}(\sigma_i).$$

Fig. 7.21. Isochronous curves for creep rupture times (after Hayhurst).

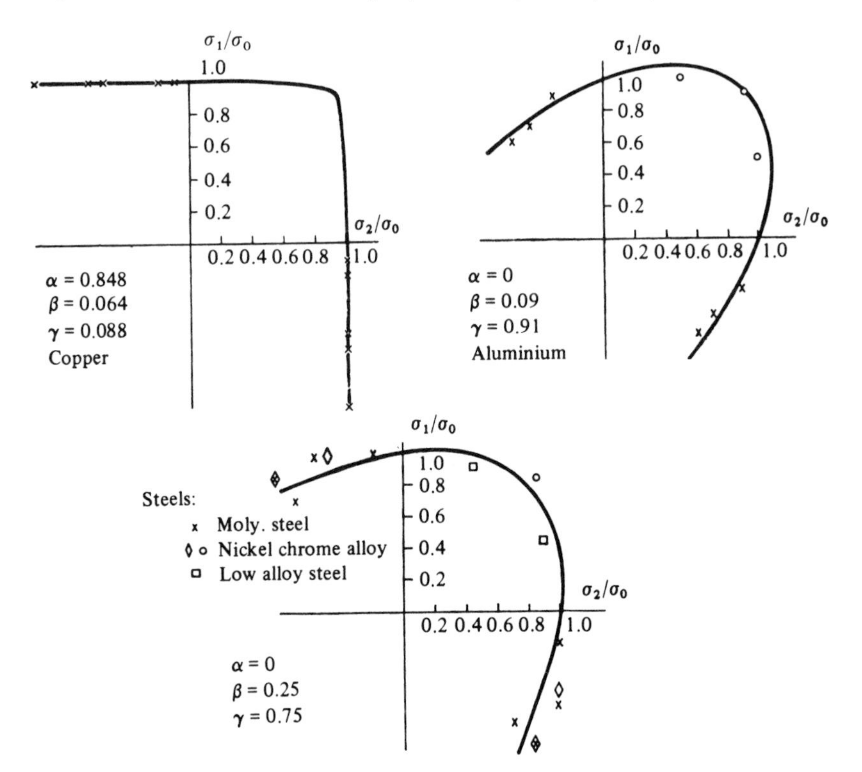

If a linear combination is chosen with α and β as phenomenological coefficients, then:

- $$\chi(\boldsymbol{\sigma}) = \alpha J_0(\boldsymbol{\sigma}) + \beta J_1(\boldsymbol{\sigma}) + (1 - \alpha - \beta) J_2(\boldsymbol{\sigma}).$$

For a uniaxial state we regain $\chi(\boldsymbol{\sigma}) = \sigma$.

Note that the three invariants correspond to the three basic mechanisms of nucleation and growth of defects, and that any isotropic surface in stress space can be described by this expression. $\chi(\boldsymbol{\sigma})$ may be negative for certain stress states (and certain values of α and β): these states then do not result in any damage growth.

Fig. 7.21 illustrates the possibilities of describing isochronic creep surfaces (loci of stress states that require the same time to failure in creep). Fig. 7.22 gives the shapes of these surfaces for different choices of the coefficients α and β.

A particularly interesting case is that where $\alpha = 0$:

$$\sigma_\beta^* = \beta J_1(\boldsymbol{\sigma}) + (1 - \beta) J_2(\boldsymbol{\sigma})$$

or

- $$\sigma_\beta^* = \sigma_{eq}[(1 - \beta) + 3\beta\sigma_H/\sigma_{eq}].$$

Here β acts as sensitivity coefficient of the material to the stress triaxiality. Fig. 7.22(*a*) shows graphs related to this equivalent stress.

Nonsymmetric criterion in terms of strain

This criterion has been developed for concrete, which has the property of being considerably more resistant to damage in compression than in tension. Microscopic observations reveal that microcracks always have a preferential orientation normal to the direction of the maximum principal extension. This suggests the idea of introducing a criterion dependent only upon the positive part of the principal strains.

Denoting the three principal strains by ε_1, ε_2 and ε_3, this criterion is expressed by:

- $$\varepsilon^* = (\langle \varepsilon_1 \rangle^2 + \langle \varepsilon_2 \rangle^2 + \langle \varepsilon_3 \rangle^2)^{1/2}$$

with

$$\langle \varepsilon_i \rangle = \varepsilon_i \quad \text{if} \quad \varepsilon_i > 0 \quad \text{for extension,}$$
$$\langle \varepsilon_i \rangle = 0 \quad \text{if} \quad \varepsilon_i \leqslant 0 \quad \text{for compression.}$$

Fig. 7.23 indicates the form of such a criterion compared to the

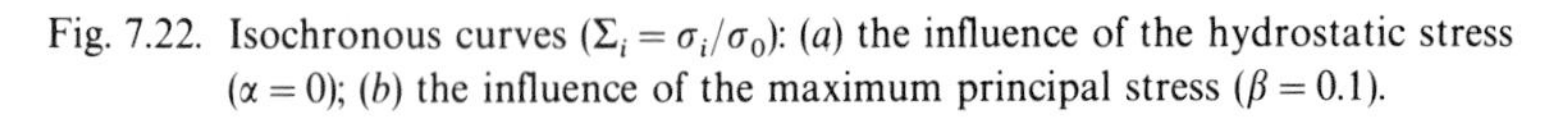
Fig. 7.22. Isochronous curves ($\Sigma_i = \sigma_i/\sigma_0$): (*a*) the influence of the hydrostatic stress ($\alpha = 0$); (*b*) the influence of the maximum principal stress ($\beta = 0.1$).

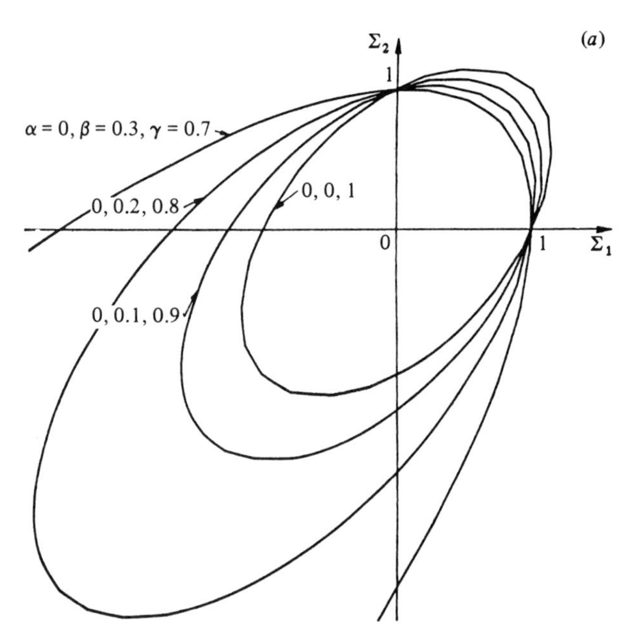

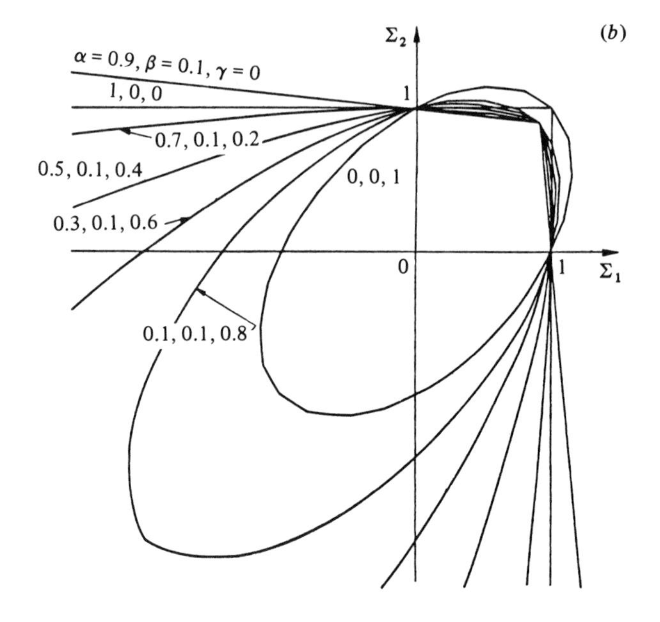

experimental results for biaxial tension, the stress is normalized to -1 for the failure stress in compression.

Fatigue limit criteria

The case of fatigue is more complex for two reasons:

> the loading is represented schematically by introducing the concept of a cycle: the parameters introduced are then defined globally for a cycle (amplitude, maximum value, mean value, etc.),
> the fatigue limit criteria expressed in terms of amplitude also depend on the average value during the cycle (see Section 7.2.4).

The concept of a uniaxial fatigue limit is therefore generalized by introducing at least two loading parameters. Numerous experiments have shown that the mean shear stress has no influence on the fatigue limit in either tension or torsion (Fig. 7.24).

On the other hand, the mean tensile stress influences the fatigue limit in tension (Fig. 7.14) and in torsion (Fig. 7.25) in a linear fashion. The two parameters which may be used to translate these effects are:

Fig. 7.23. Graphical representation of the strain criterion for concrete (after J. Mazars).

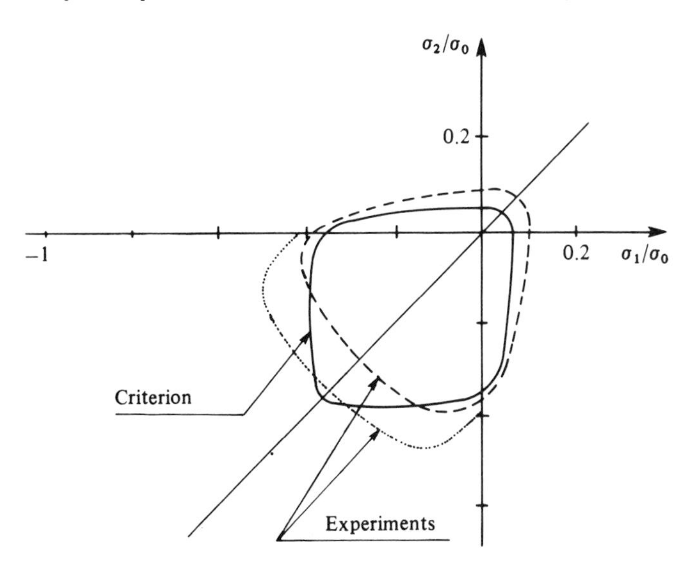

Fig. 7.24. Effect of mean torsion on the fatigue limit: (*a*) in torsion, and (*b*) in tension-compression. (τ_{a_1} = fatigue limit amplitude in torsion, $\bar{\tau}$ = mean stress in torsion, τ_{Max} = maximum stress in torsion; τ_e = elastic limit in torsion, σ_{a_1} = fatigue limit amplitude in bending.)

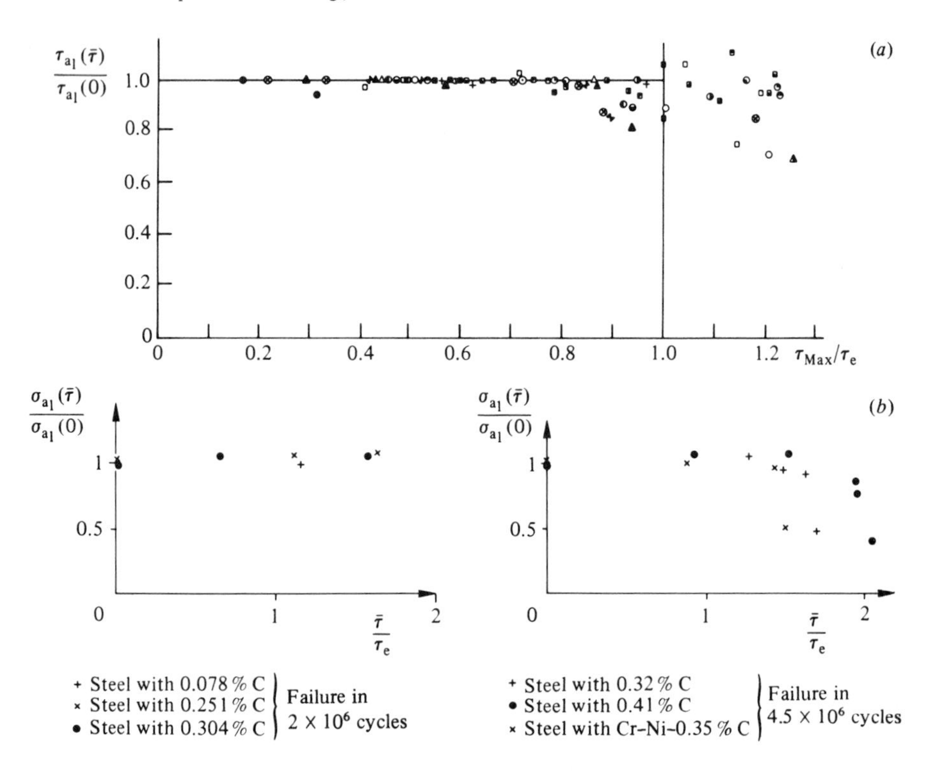

Fig. 7.25. Influence of mean tension on the fatigue limit in torsion.

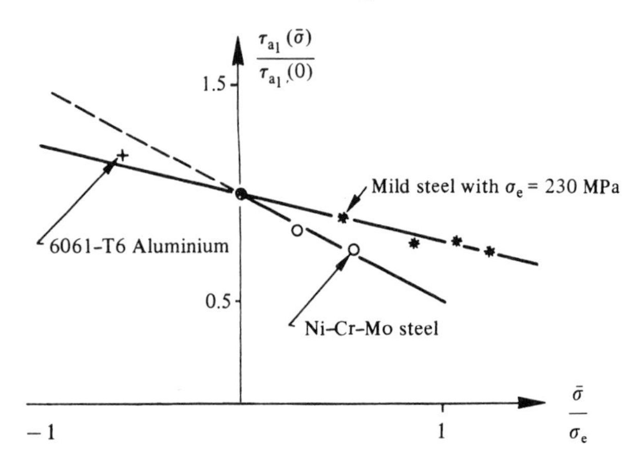

an octahedral shear stress amplitude, defined in proportional loading by:

$$A_{\mathrm{II}} = \tfrac{1}{2}\left[\tfrac{3}{2}(\sigma'_{ij\mathrm{Max}} - \sigma'_{ij\mathrm{m}})(\sigma'_{ij\mathrm{Max}} - \sigma'_{ij\mathrm{m}})\right]^{1/2}$$
$$= \left\{\tfrac{1}{2}\left[(a_1 - a_2)^2 + (a_2 - a_3)^2 + (a_3 - a_1)^2\right]\right\}^{1/2}$$

where σ'_{Max} and σ'_{m} are the maximum and minimum values of each component of the deviator (during the cycle) and a_1, a_2, a_3 represent the principal stress amplitudes $a_i = \Delta\sigma_i/2$.
the mean hydrostatic pressure during the cycle $\bar{\sigma}_{\mathrm{H}} = \tfrac{1}{3}$ mean $J_1(\boldsymbol{\sigma})$ in the Sines criterion, or the maximum hydrostatic pressure $\sigma_{\mathrm{H}_{\mathrm{Max}}} = \tfrac{1}{3}\mathrm{Max}\, J_1(\boldsymbol{\sigma})$ as in the Crossland criterion.

Fig. 7.26. Criteria of Sines, Crossland, Dang-Van in proportional biaxial loading, reversed or repeated.

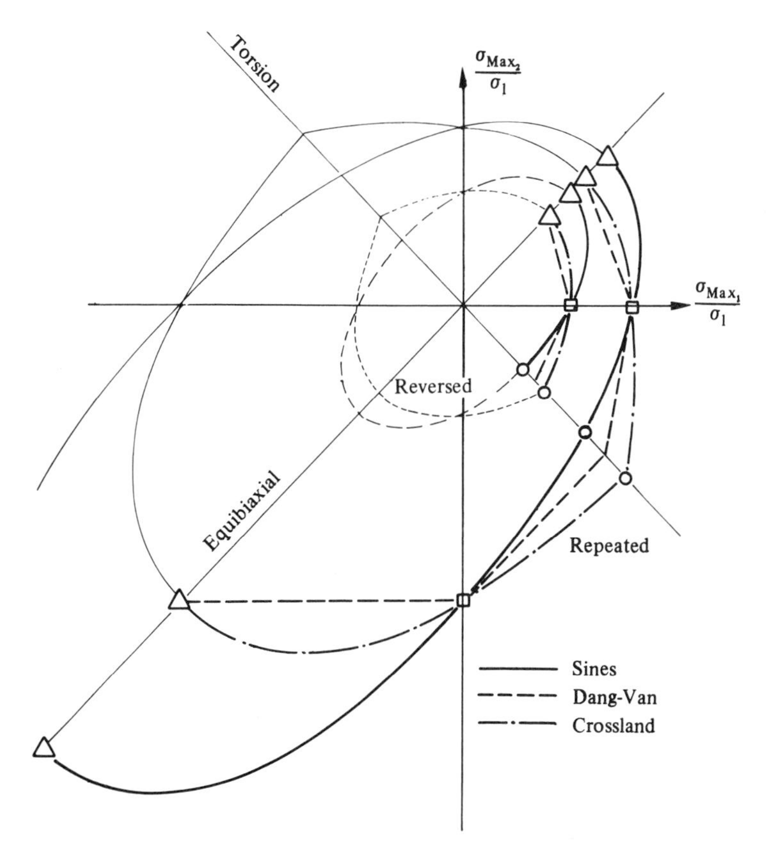

These two fatigue limit criteria can be written as:

Sines criterion:

- $$A_{\text{II}} = \sigma_{l_0}(1 - 3b\bar{\sigma}_{\text{H}});$$

Crossland criterion:

- $$A_{\text{II}} = \sigma_{l_0}\frac{1 - 3b\sigma_{\text{II}_{\text{Max}}}}{1 - b\sigma_{l_0}}.$$

In the first case, under alternate proportional loading, the fatigue limit surface in the plane $\sigma_3 = 0$ corresponds to the von Mises ellipse (Fig. 7.26): the Crossland criterion corresponds to a portion of an ellipse with a different eccentricity and centre.

Depending on the material, one or the other of the two criteria is the more

Fig. 7.27. Reversed tension–torsion: Sines and Crossland criteria for (*a*) 'silal' cast iron and (*b*) Cr–Va steel.

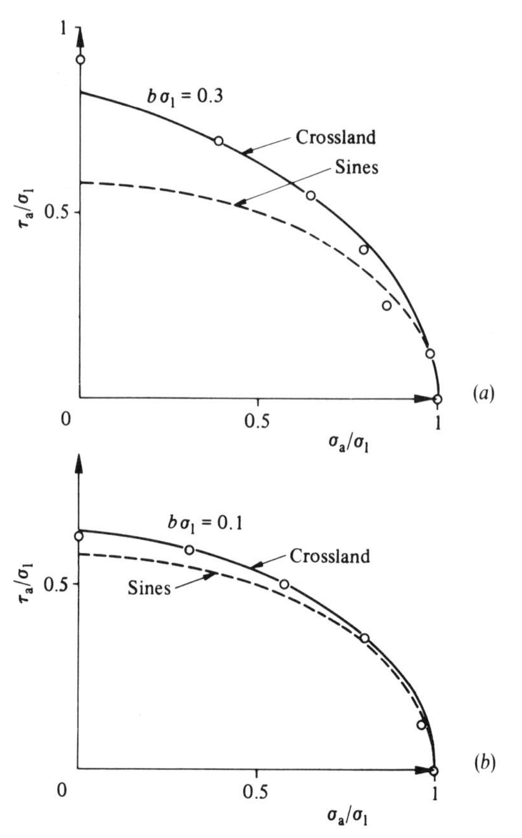

Table 7.5. *Fatigue limits in terms of the amplitude of* σ_{a_1} *for particular proportional loadings* (σ_l *means* $\sigma_{l_0} = \sigma_l(0)$)

Loading		Sines	Crossland
Reversed tension	$\sigma_2 = 0$	σ_l	
Repeated tension		$\sigma_l/(1 + bs_l)$	
Repeated compression		$\sigma_l/(1 - bs_l)$	
Reversed equibiaxial tension	$\sigma_2 = \sigma_1$	σ_l	$\sigma_l/(1 + b\sigma_l)$
Repeated equibiaxial tension		$\sigma_l/(1 + 2b\sigma_l)$	$\sigma_l/(1 + 3b\sigma_l)$
Repeated equibiaxial compression		$\sigma_l/(1 - 2b\sigma_l)$	$\sigma_l/(1 - b\sigma_l)$
Repeated and reversed torsion	$\sigma_2 = -\sigma_1$	$\sigma_l/\sqrt{3}$	$\sigma_l/\sqrt{3}(1 - b\sigma_l)$

Sines: $(\sigma_{a_1}^2 + \sigma_{a_2}^2 - \sigma_{a_1}\sigma_{a_2})^{1/2} = \sigma_l[1 - b(\bar{\sigma}_1 + \bar{\sigma}_2)]$

Crossland: $(\sigma_{a_1}^2 + \sigma_{a_2}^2 - \sigma_{a_1}\sigma_{a_2})^{1/2} = \dfrac{\sigma_l}{1 - b\sigma_l}[1 - b\,\mathrm{Max}(\bar{\sigma}_1 + \bar{\sigma}_2 + \sigma_{a_1} + \sigma_{a_2})]$

realistic (Fig. 7.27). Table 7.5 gives different expressions for the fatigue limit in terms of amplitude $\sigma_{a_1} = \Delta\sigma_1/2$ for different types of loading. The fatigue limit surface depends on the mean stress during the cycle. Fig. 7.28 shows, for example, the cases of reversed and repeated loadings.

Under nonproportional loadings, these criteria are always applicable but the definition of equivalent amplitude must be replaced by an intrinsic formula:

$$A_{II} = \tfrac{1}{2} \underset{t_0}{\mathrm{Max}} \underset{t}{\mathrm{Max}} J_2(\boldsymbol{\sigma}(t) - \boldsymbol{\sigma}(t_0))$$

where J_2 denotes the second invariant of the deviator. This generalization is illustrated in Fig. 7.29(*a*) in the deviatoric plane for any loading.

An intermediate criterion, with a stronger physical basis, has been introduced by Dang-Van. In order to define the amplitude parameter, he considers the plane which suffers the maximum shear amplitude during the cycle. On this plane, represented by its normal $\vec{n}$, we consider at each instant the difference $\vec{\tau}(t) - \vec{\bar{\tau}}$ where $\vec{\tau}(t)$ is the shear $\boldsymbol{\sigma}\cdot\vec{n} - (\boldsymbol{\sigma}\cdot\vec{n}\cdot\vec{n})\vec{n}$ and where $\vec{\bar{\tau}}$ represents the mean shear stress on this plane during the cycle. Usually, $\vec{\bar{\tau}}$ is defined by the centre of the smallest circle containing the path of $\vec{\tau}(t)$ in the plane. The radius of this circle is taken to be the maximum shear amplitude (Fig. 7.29(*b*)):

$$\Delta\tau/2 = \tfrac{1}{2} \underset{t}{\mathrm{Max}} \underset{t_0}{\mathrm{Max}} \| \vec{\tau}(t) - \vec{\tau}(t_0) \|$$

where $\|\ \ \|$ denotes the magnitude. The quantity $\vec{\tau}(t) - \vec{\bar{\tau}}$ is combined with the hydrostatic pressure $\sigma_H(t)$ to define the critical instant and the critical

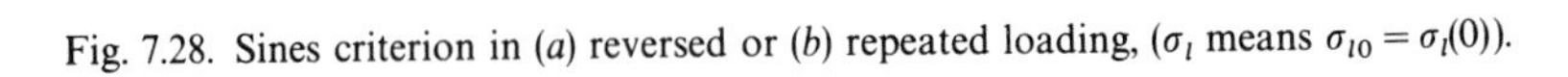

Fig. 7.28. Sines criterion in (*a*) reversed or (*b*) repeated loading, (σ_l means $\sigma_{l0} = \sigma_l(0)$).

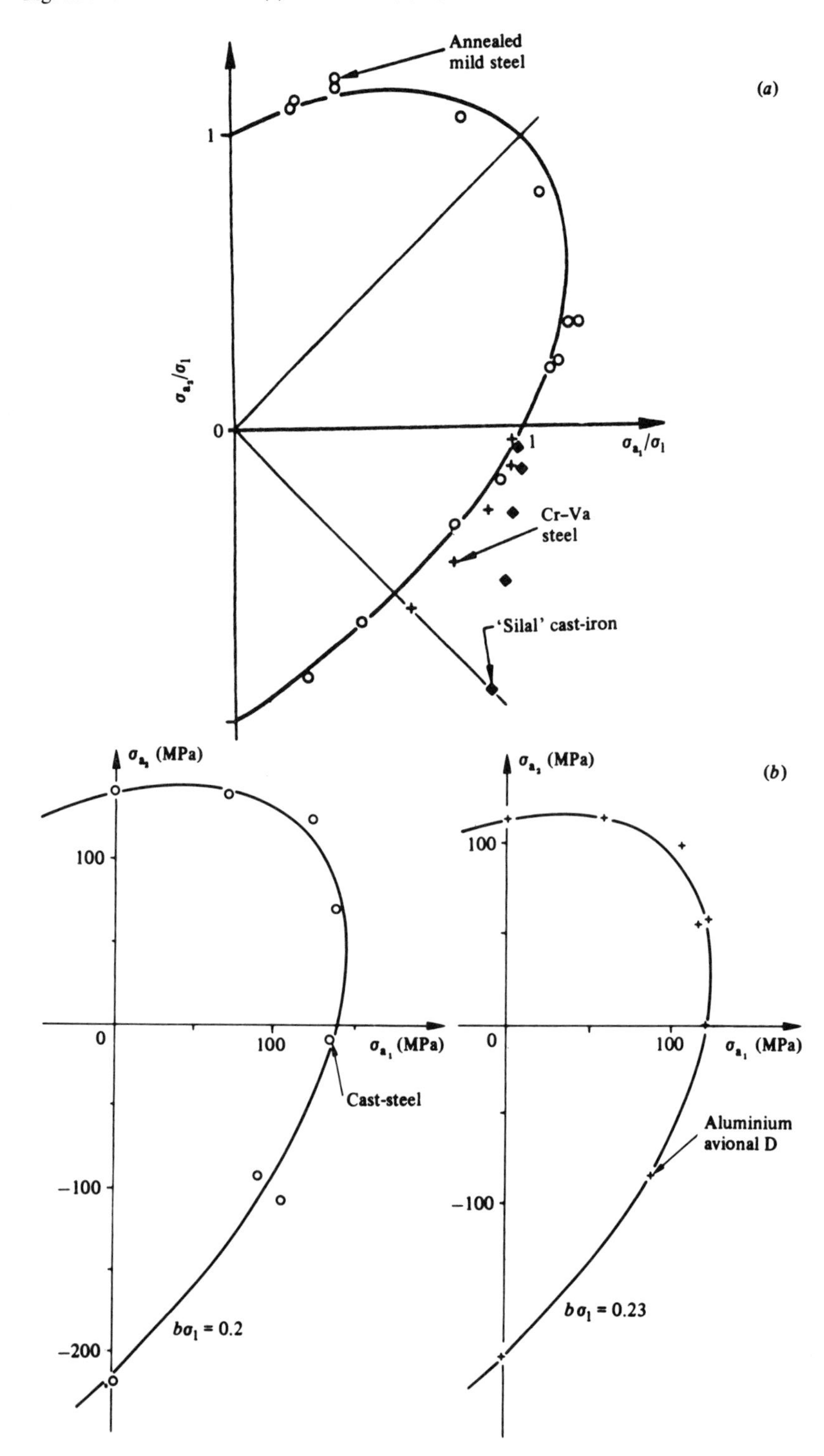

parameter. The criterion is then expressed as:

$$\underset{\vec{n}}{\text{Max}}\ \underset{t}{\text{Max}}\left[\ \|\vec{\tau}(t)-\vec{\bar{\tau}}\| + \frac{1}{2}\frac{3b\sigma_{l_0}}{1-b\sigma_{l_0}}\sigma_{\mathrm{H}}(t)\ \right] = \frac{\sigma_{l_0}}{2(1-b\sigma_{l_0})}.$$

Fig. 7.30 illustrates the recentring and the determination of the critical instant in the shear–pressure diagram (for the case where shear has a fixed

Fig. 7.29. Determination of the amplitude of the cycle in the general case: (*a*) Sines and Crossland criteria; (*b*) Dang-Van criterion.

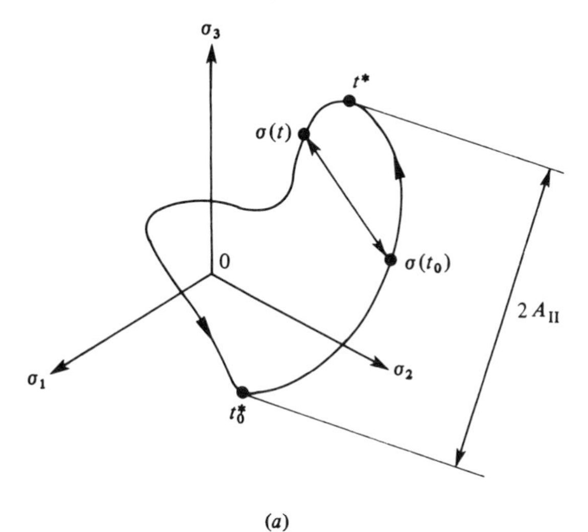

(*a*)

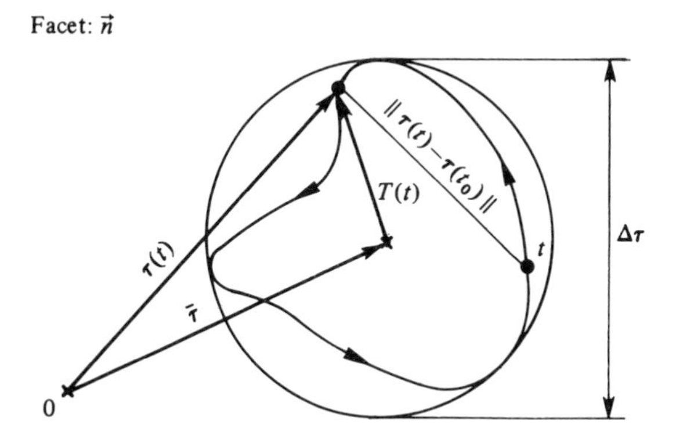

(*b*)

direction). The essential difference with the two previous criteria is that the quantities appearing in the criterion (shear and pressure) are defined at the same instant (point A). In contrast, in the Crossland criterion, these quantities are defined at different instants (A' and C). We therefore have a more accurate description of the real process. Fig. 7.31 shows different cases of proportional loading after recentring the cycle. In this case, the critical plane is really the plane that minimizes the shear amplitude and we can write the criterion in the simplified form:

$$\frac{\Delta\tau}{2} = \frac{\sigma_{10}}{2(1 - b\sigma_{10})}(1 - 3b\sigma_{\mathrm{HMax}}).$$

In torsional proportional loading, the Dang-Van criterion gives results which are between those given by the criterion of Sines and that of Crossland:

$$\tau_{\mathrm{a}} = \sigma_{\mathrm{a}} = \frac{\sigma_{1_0}}{2(1 - b\sigma_{1_0})}.$$

For other proportional loadings (tension–compression, reversed and repeated equibiaxial), it is identical to the Crossland criterion as can be seen from Fig. 7.26.

Fig. 7.30. Application of the Dang-Van criterion when the shear has a fixed direction during the cycle. A: critical instant. B and C: instants corresponding to the choice of the hydrostatic pressure in the criteria of Sines and Crossland (σ_1 means $\sigma_1(0)$).

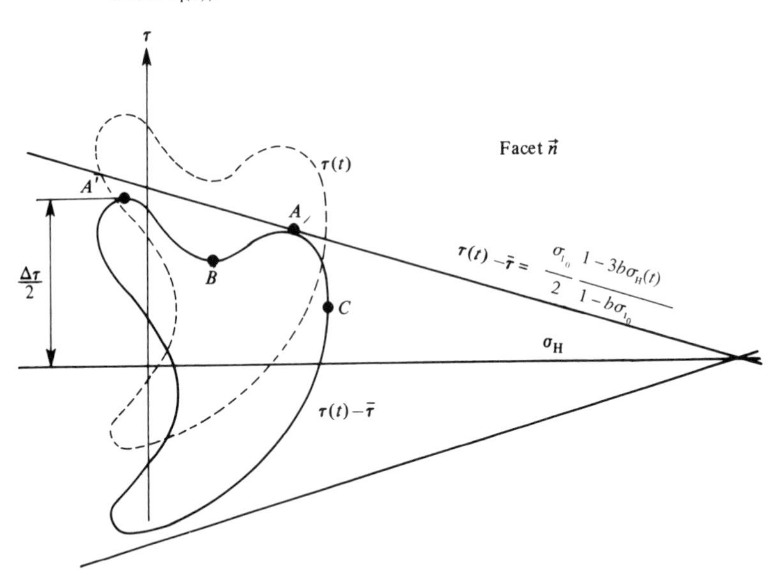

Note that this criterion implicitly gives the direction of defects (critical planes) but that, in the general case, it is much more complex to use because it involves three maximizations: twice as a function of t and t_0 at each plane, and then as a function of the direction of the normal $\vec{n}$.

Instead of using the shear amplitude $\Delta\tau/2$, we may also use the range of octahedral shear:

$$\Delta J_2 = \underset{t_1}{\text{Max}}\left[\underset{t_2}{\text{Max}}\, J_2(\boldsymbol{\sigma}(t_1) - \boldsymbol{\sigma}(t_2))\right].$$

This maximum being reached at a time t_i, the equivalent shear stress is defined by:

$$J_2^*(t) = J_2(\boldsymbol{\sigma}(t) - \boldsymbol{\sigma}(t_i)) - \Delta J_2/2.$$

It is then sufficient to replace $(\vec{\tau}(t), \sigma_H(t))$ by $(J_2^*(t), \sigma_H(t))$ in the criterion and we obtain an expression, equivalent to the Crossland criterion in proportional loading, which is simpler to use than the Dang-Van criterion.

Fig. 7.31. Application of the Dang-Van criterion for particular proportional loadings.

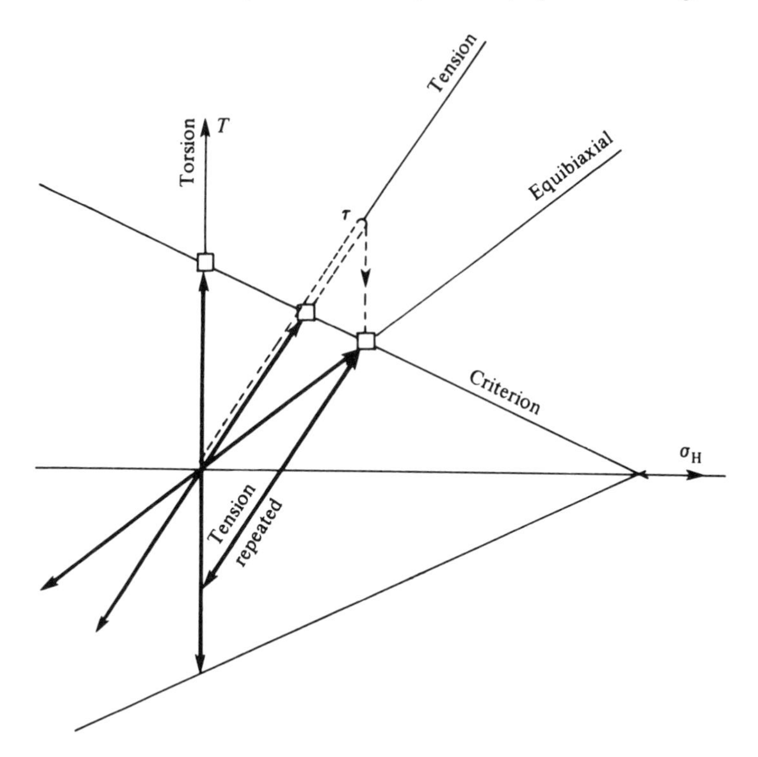

7.3 Thermodynamic formulation

7.3.1 *Multiaxial representation of damage*

Within the isotropic framework, the damage variable is a scalar D which, for the purpose of defining the effective stress $\tilde{\boldsymbol{\sigma}}$, affects all components of the stress tensor $\boldsymbol{\sigma}$ identically:

$$\tilde{\boldsymbol{\sigma}} = \boldsymbol{\sigma}/(1 - D).$$

In fact, the surface microcracks and the volume cavities that make up damage usually have preferred orientations and thus constitute anisotropic damage. For example, microcracks in metals have a tendency to propagate in a direction perpendicular to that of the maximum stress. The damage operator, which applied to the usual stress tensor gives the effective stress tensor, is then tensorial and of an order which differs depending on the adopted theoretical scheme: a second order tensor, if a stress vector is defined on a damaged oriented surface, but this concept does not necessarily satisfy the second principle of thermodynamics; a fourth order tensor if the damage operator is defined on the basis of the elastic potential as we show below; eighth order and higher tensors are also possible!

If only geometrical effects of cavities and cracks are taken into account, we introduce, for each element of area, represented by its normal $\vec{n}$, an area reduction $\Omega(\vec{n})$, and the state of damage is characterized by a second order symmetric tensor expressed in principal axes representation by:

$$\boldsymbol{\Omega} = \sum_{j=1}^{3} \Omega_j \cdot \vec{n}_j \otimes \vec{n}_j.$$

The net stress tensor (referred to the noncorrected resistant sections), symmetrized according to Murakami and Ohno can be deduced from it by:

$$\tilde{\boldsymbol{\sigma}} = \tfrac{1}{2}[\boldsymbol{\sigma} \cdot (\mathbf{1} - \boldsymbol{\Omega})^{-1} + (\mathbf{1} - \boldsymbol{\Omega})^{-1} \cdot \boldsymbol{\sigma}].$$

In fact this effective stress cannot be used directly in the constitutive equations because the behaviour of the damaged material depends not only upon the area reduction caused by cracks and cavities, but also upon their arrangement and the effects of their interactions. It is necessary to introduce another effective stress derived from the Cauchy stress by a fourth order tensor. This tensor also characterizes the damage, but it is preferable to introduce it directly from a theory based on the concept of strain equivalence.

Definition of effective stress by strain equivalence

This theory, introduced by Chaboche and Lemaitre, is applicable to materials whose initial behaviour is undefined (elastic or plastic, isotropic or anisotropic). In practice, it is sufficient to examine the case of isothermal elasticity to define the effective stress tensor. Let, therefore

$$\boldsymbol{\sigma} = \mathbf{a}:\boldsymbol{\varepsilon}^{\mathrm{e}}$$

be the constitutive equation of the virgin material, and

$$\tilde{\boldsymbol{\sigma}} = \mathbf{a}:\boldsymbol{\varepsilon}^{\mathrm{e}}$$

be the constitutive equation of the damaged material, $\mathbf{a}$ being the fourth order elasticity tensor. The strain energy of the damaged material can be defined directly by using an elasticity tensor $\tilde{\mathbf{a}}$ different from $\mathbf{a}$ but possessing the same thermodynamic symmetry properties:

$$\tilde{a}_{ijkl} = \tilde{a}_{ijlk} = \tilde{a}_{jikl} = \tilde{a}_{klij}$$
$$\psi_{\mathrm{e}} = \tfrac{1}{2}\tilde{\mathbf{a}}:\boldsymbol{\varepsilon}^{\mathrm{e}}:\boldsymbol{\varepsilon}^{\mathrm{e}}.$$

The constitutive equation of the damaged material is then deduced by:

$$\boldsymbol{\sigma} = \partial\psi_{\mathrm{e}}/\partial\boldsymbol{\varepsilon}^{\mathrm{e}} = \tilde{\mathbf{a}}:\boldsymbol{\varepsilon}^{\mathrm{e}}.$$

According to the strain equivalence principle (Section 7.2.2), $\tilde{\boldsymbol{\sigma}}$ is that stress tensor which, when applied to the virgin material, produces the same strain tensor as that obtained by applying the tensor $\boldsymbol{\sigma}$ to the damaged material. The combination of the two laws easily leads to:

● $$\tilde{\boldsymbol{\sigma}} = (\mathbf{a}:\tilde{\mathbf{a}}^{-1}):\boldsymbol{\sigma} = \boldsymbol{\Delta}:\boldsymbol{\sigma}$$

where $\boldsymbol{\sigma}$ and $\tilde{\boldsymbol{\sigma}}$ are second order tensors: it follows that the damage operator $\boldsymbol{\Delta} = \mathbf{a}:\tilde{\mathbf{a}}^{-1}$ is a fourth order tensor whose coefficients can be determined from the elasticity matrices of the virgin and damaged materials.

We can obtain the same result by using the mathematical technique of homogenization, at least, for the idealized cases of volume elements damaged by ellipsoidal cavities or periodically distributed parallel cracks. The elasticity matrix of the damaged material is expressible as:

$$\tilde{a}_{ijkl} = \left(\frac{\tilde{V}}{V}\delta_{ir}\delta_{js} - \frac{1}{V}\int_{\tilde{V}} b_{ijrs}\mathrm{d}V\right)a_{rskl}$$

where b_{ijrs} represents the matrix of the stress-concentration coefficients. V is the apparent volume of the element and $\tilde{V}$ is the solid volume (equal to the apparent volume minus the volume of voids). We then introduce $\mathbf{D}$, a 4th order operator with components:

$$D_{ijrs} = \left(1 - \frac{\tilde{V}}{V}\right)\delta_{ir}\delta_{js} + \frac{1}{V}\int_{\tilde{V}} b_{ijrs}\,\mathrm{d}V$$

such that:

$$(\mathbf{1} - \mathbf{D})^{-1} = \boldsymbol{\Delta}.$$

where $\mathbf{1}$ is the fourth order unity tensor.
This operator is not symmetric, for example $D_{1122} \neq D_{2211}$. We can write:

$$\tilde{\mathbf{a}} = (\mathbf{1} - \mathbf{D}):\mathbf{a}$$

$$\tilde{\boldsymbol{\sigma}} = (\mathbf{1} - \mathbf{D})^{-1}:\boldsymbol{\sigma}.$$

We will stop here in defining the three-dimensional damage variable $\mathbf{D}$ which can represent the most general states of anisotropic damage. It should, however, be noted that an energy equivalence, instead of the strain equivalence, allows us to introduce a second order damage tensor (still written as $\boldsymbol{\Omega}$) and an effective stress of the form:

$$(\mathbf{1} - \boldsymbol{\Omega})^{-1/2}\cdot\boldsymbol{\sigma}\cdot(\mathbf{1} - \boldsymbol{\Omega})^{-1/2}$$

but this theory suffers from the disadvantage that it cannot be identified with any general elastic anisotropy.

The two problems of damage multiaxiality

It is important to distinguish correctly between two aspects of damage which often cause confusion:

> the first is linked to the multiaxiality of the stress state in the form of the damage criterion mentioned in Section 7.2.5;
> the second is related to the isotropy or anisotropy of the current state of damage and its evolution.

To illustrate the first aspect, we may imagine different isotropic damages induced by states of stress with different directions. For example, in the same material, a tensile loading σ may produce an isotropic damage $D = 0.2$ (which is the same for all material directions), but a compressive loading $-\sigma$ may result in an isotropic damage $D = 0.02$. In other words, the directions of the stress state that results in damage should not be confused with those linked to the material. These latter ones define the directions of anisotropy of the current state of damage which will be explained below.

The second aspect may be illustrated by the case of a radial loading (where the principal directions of the stress tensor remain constant) which produces an anisotropic damage with isotropic evolution; for example the case of a tensile loading which produces:

$D = D(t)$ in the material direction parallel to that of the tension; $D = \alpha D(t)$ in other directions, for example with $\alpha = 0.1$ in a perpendicular direction.

The properties of isotropy and anisotropy can be applied independently to the current state and to the damage growth.

For radial (or proportional) loading, the principal directions of the damage tensor coincide with those of the stress tensor. A simplified form of the damage evolution equation is then:

$$\dot{\mathbf{D}} = \mathbf{Q}^* \dot{D}$$

where $\mathbf{Q}^*$ is a tensor dependent on the material, and possibly on the temperature. All nonlinearities of the evolution phenomenon are assumed to be incorporated in the scalar $\dot{D}$, the rate of damage, representative of the quantity of defects per unit volume, independent of their orientation.

In the general case, the evolution law depends on the principal directions of the effective stress tensor:

$$\dot{\mathbf{D}} = \mathbf{Q}(\tilde{\sigma})\dot{D}.$$

The relation between $\mathbf{Q}(\tilde{\sigma})$ and $\mathbf{Q}^*$ involves rotations to change from the principal system to the reference system.

In what follows, we will limit ourselves to the very simple case of isotropic damage, or to the more general case studied above in which the anisotropy of the damage evolution does not depend upon any particular instant during the life of the volume element.

7.3.2 *Theory of isotropic damage*

Thermodynamic potential

We will only consider damageable materials exhibiting elastic, elastoplastic or elastoviscoplastic behaviour (the viscoelastic case would require a slightly different formal treatment). The damage variable is the scalar D which will be considered as a state variable amenable to the thermodynamic representation described in Chapter 2. With the notations of the preceding section, this amounts to treating the damage as a spherical tensor $\mathbf{1}D$.

The free energy thermodynamic potential is therefore a convex function of all the state variables, and, in particular, of the damage

$$\Psi = \Psi(\varepsilon^{e}, T, D, V_{k})$$

where V_k denotes the internal variables such as the hardening variables.

As in elastoplasticity, we may adopt the hypothesis of decoupling between hardening and other effects, represented by the variable V_k, and the effects of elasticity associated with the damage:

$$\Psi = \Psi_{e}(\varepsilon^{e}, T, D) + \Psi_{p}(T, V_{k})$$

where Ψ_e is the thermoelastic potential of a damageable material. We have seen in Chapter 4 that the linear thermoelasticity can be derived from a quadratic potential in ε^e and T. Remaining within this framework, the fundamental hypothesis of the effective stress and the definition chosen for D (see Section 7.3.1) require that Ψ_e depends linearly on D:

$$\rho\Psi_{e} = \tfrac{1}{2}(1-D)\mathbf{a}:(\varepsilon^{e} - \mathbf{k}\Delta T):(\varepsilon^{e} - \mathbf{k}\Delta T) + C\Delta T^{2}$$

where $\mathbf{k}\Delta T$ represents the thermal dilatation.

The specification of this thermodynamic potential furnishes the thermoelastic law of the damaged material:

$$\boldsymbol{\sigma} = \rho\,\partial\Psi_{e}/\partial\varepsilon^{e} = (1-D)\mathbf{a}:(\varepsilon^{e} - \mathbf{k}\Delta T) = \tilde{\mathbf{a}}:(\varepsilon^{e} - \mathbf{k}\Delta T)$$

which is indeed of the form:

$$\tilde{\boldsymbol{\sigma}} = \frac{\boldsymbol{\sigma}}{1-D} = \mathbf{a}:(\varepsilon^{e} - \mathbf{k}\Delta T).$$

The variable associated with the damage variable is the scalar:

$$Y = \rho\,\partial\Psi_{e}/\partial D = -\tfrac{1}{2}\mathbf{a}:(\varepsilon^{e} - \mathbf{k}\Delta T):(\varepsilon^{e} - \mathbf{k}\Delta T).$$

It is interesting to note that this quantity can be identified, except for the sign, with one-half of the variation in elastic energy generated by a damage variation at constant stress and temperature, $\mathrm{d}w_e/\mathrm{d}D|_{\sigma,T}$. To show this, let us calculate: $\mathrm{d}w_e = \boldsymbol{\sigma}:\mathrm{d}\varepsilon^e$ by obtaining $\mathrm{d}\varepsilon^e$ from the expression $\mathrm{d}\boldsymbol{\sigma} = 0$ deduced from the thermoelasticity law with $\Delta T = 0$:

$$\mathrm{d}\sigma|_{T=\text{constant}} = (1-D)\mathbf{a}:\mathrm{d}\varepsilon^{e} - \mathbf{a}:\varepsilon^{e}\mathrm{d}D = 0$$

$$\mathrm{d}w_{e} = \boldsymbol{\sigma}:\mathrm{d}\varepsilon^{e} = \boldsymbol{\sigma}:\varepsilon^{e}\mathrm{d}D/(1-D) = \mathbf{a}:\varepsilon^{e}:\varepsilon^{e}\mathrm{d}D$$

$$\mathrm{d}w_{e}/\mathrm{d}D|_{\sigma,T} = \mathbf{a}:\varepsilon^{e}:\varepsilon^{e} = a_{ijkl}\varepsilon^{e}_{ij}\varepsilon^{e}_{kl}$$

and indeed, as expected, we find the stated result:

- $$-Y = \tfrac{1}{2}(\mathrm{d}w_{e}/\mathrm{d}D)|_{\sigma,T}.$$

$-Y$ is called the elastic energy release rate.

Thus, for isotropic damage and when the material is elastically isotropic, there exists a simple relation between the variable associated with the damage and the equivalent damage stress introduced in Section 7.2.5.

$$\bullet \qquad -Y = \frac{\sigma^{*2}}{2E(1-D)^2} = \frac{1}{2E}\tilde{\sigma}^{*2}.$$

This relation justifies the elastic energy density release rate criterion by giving a thermodynamic meaning to the damage equivalent stress: the variables $-Y$ and $\tilde{\sigma}^*$ are equivalent variables.

Dissipation potential

Under the hypothesis of decoupling between mechanical and thermal dissipations, the second principle of thermodynamics requires the mechanical dissipation to be positive:

$$\boldsymbol{\sigma}:\dot{\boldsymbol{\varepsilon}}^{\mathrm{p}} - Y\dot{D} - A_{\mathrm{k}}\dot{V}_{\mathrm{k}} \geqslant 0.$$

The phenomenon of plastic flow can occur without damage; similarly the damage phenomenon can occur without noticeable macroscopic flow; we must therefore have separately:

$$\boldsymbol{\sigma}:\dot{\boldsymbol{\varepsilon}}^{\mathrm{p}} - A_{\mathrm{k}}\dot{V}_{\mathrm{k}} \geqslant 0 \qquad -Y\dot{D} \geqslant 0$$

$-Y$ being a positive definite quadratic form, the second principle requires that $\dot{D} \geqslant 0$. The damage defined in this way can therefore only increase or remain constant.

In accordance with the phenomenological thermodynamic method presented in Chapter 2, the damage growth law can be derived from a dissipation potential whose existence is postulated:

$$\varphi(\dot{\boldsymbol{\varepsilon}}^{\mathrm{p}}, \dot{V}_{\mathrm{k}}, \dot{D}, \vec{q}; \boldsymbol{\varepsilon}^{\mathrm{e}}, T, V_{\mathrm{k}}, D).$$

This convex function with a scalar value is a function of all the flux variables, with observable and internal variables acting as parameters.

Using the Legendre–Fenchel transformation, we may construct another equivalent dual potential so as to be able to express $\dot{D}$ as a function of Y rather than the opposite

$$\varphi^*(\boldsymbol{\sigma}, A_{\mathrm{k}}, Y, \vec{g}; \boldsymbol{\varepsilon}^{\mathrm{e}}, T, V_{\mathrm{k}}, D).$$

The generalized normality law that results from this is expressed by:

$$\dot{D} = -\partial\varphi^*/\partial Y.$$

Using the hypothesis of thermomechanical decoupling φ^* can be expressed as a sum of two functions, one which depends only on $\vec{g} = \overrightarrow{\text{grad}}\ T$, and the other which depends on all the other variables, so that $\dot{D}$ does not depend on $\vec{g}$.

The balance of the dissipated power can be illustrated by a tensile stress cycle, shown schematically in Fig. 7.32, in which the increase in damage is assumed to occur at a constant maximum stress. $0A$ corresponds to the plastic flow during loading; AB represents the plastic flow accompanying the damage process; BC represents the increase in elastic strain produced by the increase in damage. The curve $0A'B'$ represents the evolution of the hardening variable A_k related to the plastic flow $\delta\varepsilon_p$. The total energy dissipated during the cycle is the area $0ABCD0$. The energy dissipated plastically corresponding to the integration of $\sigma d\varepsilon_p$ is the area $0ABD0$; from this we subtract (1), the energy blocked in dislocations, the integral of $A_k d V_k$, and obtain the energy dissipated as heat during the flow. The energy dissipated by the process of decohesion (eventually as heat) corresponds to the integral of $\Phi_D = -Y\dot{D}$. This is represented by (3).

In effect, we have:

$$-Y\,dD = \frac{1}{2}\frac{dw_e}{dD}\bigg|_\sigma dD = \frac{1}{2}\sigma\frac{d\varepsilon_e}{dD}\bigg|_\sigma dD = \frac{1}{2}\sigma\, d\varepsilon_e.$$

Fracture criterion for the volume element

From what we have just said, an energetic definition of the criterion of macrocrack initiation is justified. The variable $-Y$, the elastic energy

Fig. 7.32. Balance of dissipated energies during plastic flow and damage.

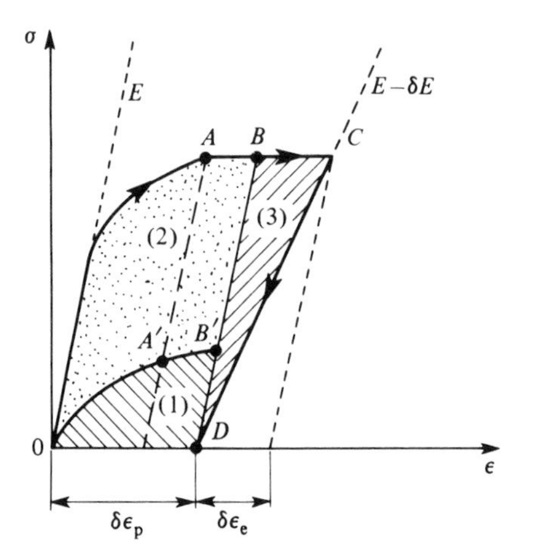

release rate, represents the elastic energy of the equivalent virgin material (with ΔT taken here as zero):

$$-Y = \tfrac{1}{2}\mathbf{a}:\boldsymbol{\varepsilon}^{\mathrm{e}}:\boldsymbol{\varepsilon}^{\mathrm{e}}.$$

The criterion consists in postulating that the initiation of a macroscopic crack corresponds to a critical value of this energy, intrinsic to every material:

$$-Y = |Y| = Y_{\mathrm{c}} \leftrightarrow \text{fracture}.$$

This critical energy is identified with the decohesion energy of the material which may be evaluated by performing a uniaxial tension test

$$-Y = \tfrac{1}{2}E\varepsilon_{\mathrm{e}}^{2}$$

and by replacing ε_{e} by its expression taken from the elasticity law of the damaged material:

$$\varepsilon_{\mathrm{e}} = \frac{\sigma}{(1-D)E}$$

$$|Y| = \frac{1}{2}\frac{\sigma^2}{(1-D)^2 E}$$

and, for the fracture conditions:

$$\left.\begin{aligned} |Y| &= Y_{\mathrm{c}} \\ \sigma &= \sigma_{\mathrm{u}} \\ D &= D_{\mathrm{c}} \end{aligned}\right\} \qquad Y_{\mathrm{c}} = \frac{1}{2}\frac{\sigma_{\mathrm{u}}^2}{E(1-D_{\mathrm{c}})^2}$$

or

- $$D_{\mathrm{c}} = 1 - \sigma_{\mathrm{u}}/(2EY_{\mathrm{c}})^{1/2},$$

a criterion which is identified with the critical damage at failure calculated in Section 7.2.2:

$$D_{\mathrm{c}} = 1 - \sigma_{\mathrm{u}}/\tilde{\sigma}_{\mathrm{u}}, \qquad \tilde{\sigma}_{\mathrm{u}} = (2EY_{\mathrm{c}})^{1/2}.$$

7.3.3 *A nonisotropic damage theory*

In order to simplify the notation, we limit ourselves here to the isothermal case, $\Delta T = 0$. The theory consists in using the damage tensor $\mathbf{D}$, introduced in Section 7.3.1, conjointly with a scalar measure D which will be taken as the trace of the tensor $\mathbf{D}$. These two variables furnish respectively the orientation of defects and their density. The anisotropy of the growth law is assumed to depend only on the material and on the principal stress directions.

Thermodynamic potential

The specific free energy depends on the tensor $\mathbf{D}$. As usual we assume decoupling between the hardening effects and the law of elasticity:

$$\Psi = \Psi_e(\boldsymbol{\varepsilon}^e, T, \mathbf{D}) + \Psi_p(T, V_k)$$

where Ψ_e depends linearly on the damage:

$$\rho\Psi_e = \tfrac{1}{2}(\mathbf{1} - \mathbf{D}):\mathbf{a}:\boldsymbol{\varepsilon}^e:\boldsymbol{\varepsilon}^e.$$

The elasticity law is:

$$\boldsymbol{\sigma} = \rho\, \partial\Psi/\partial\boldsymbol{\varepsilon}^e = (\mathbf{1} - \mathbf{D}):\mathbf{a}:\boldsymbol{\varepsilon}^e = \tilde{\mathbf{a}}:\boldsymbol{\varepsilon}^e$$

and the effective stress is:

$$\tilde{\boldsymbol{\sigma}} = (\mathbf{1} - \mathbf{D})^{-1}:\boldsymbol{\sigma}.$$

The variable associated with damage is expressed by:

$$\mathbf{Y} = \rho\, \partial\Psi_e/\partial\mathbf{D} = -\tfrac{1}{2}(\mathbf{a}:\boldsymbol{\varepsilon}^e)\boldsymbol{\varepsilon}^e \qquad \text{or} \qquad Y_{ijkl} = -\tfrac{1}{2}a_{ijrs}\varepsilon^e_{rs}\varepsilon^e_{kl}.$$

If we introduce the trace of $\mathbf{D}$ as a scalar measure of damage:

$$D = c\,\mathrm{Tr}\,(\mathbf{D}) = c\mathbf{D}::\mathbf{1}$$

where the symbol : : denotes the tensorial product contracted on four indices (the coefficient c will be specified later), we may consider that Ψ_e also depends on D and obtain:

$$Y = \mathrm{Tr}(\mathbf{Y}) = -\tfrac{1}{2}\mathbf{a}:\boldsymbol{\varepsilon}^e:\boldsymbol{\varepsilon}^e = \rho\frac{\partial\Psi_e}{\partial\mathbf{D}}::\mathbf{1} = \rho c\frac{\partial\Psi_e}{\partial D}.$$

We can still write these variables as functions of effective stress:

$$-\mathbf{Y} = \tfrac{1}{2}\tilde{\boldsymbol{\sigma}}\boldsymbol{\varepsilon}^e = \tfrac{1}{2}\tilde{\boldsymbol{\sigma}}\mathbf{a}^{-1}:\tilde{\boldsymbol{\sigma}}$$

$$-Y = \tfrac{1}{2}\tilde{\boldsymbol{\sigma}}:\boldsymbol{\varepsilon}^e = \tfrac{1}{2}\mathbf{a}^{-1}:\tilde{\boldsymbol{\sigma}}:\tilde{\boldsymbol{\sigma}}.$$

Dissipation potential

Using the same assumptions as before, we consider a dissipation potential such as:

$$\varphi^*(\boldsymbol{\sigma}, A_k, \mathbf{Y}; \boldsymbol{\varepsilon}^e, T, V_k, \mathbf{D}).$$

The generalized normality law resulting from this is expressed by:

$$\dot{\boldsymbol{\varepsilon}}^p = \partial\varphi^*/\partial\boldsymbol{\sigma}, \quad \dot{V}_k = -\partial\varphi^*/\partial A_k, \quad \dot{\mathbf{D}} = -\partial\varphi^*/\partial\mathbf{Y}.$$

For simplification, we assume that the dissipations due to the deformation process (and hardening) and the damage process are independent of each other. The potential φ^* can then be decomposed as:

$$\varphi^* = \varphi^*_{\mathrm{p}}(\tilde{\sigma}, \tilde{A}_{\mathrm{k}}; V_{\mathrm{k}}, T) + \varphi^*_{\mathrm{D}}(\mathbf{Y}; \overset{\mathrm{e}}{\varepsilon}, T, \mathbf{D}).$$

Again, for simplification, the damage potential is assumed to depend linearly on $\mathbf{Y}$:

$$\varphi^*_{\mathrm{D}} = -F(\boldsymbol{\varepsilon}^{\mathrm{e}}, T, D)\mathbf{Q}: :\mathbf{Y}$$

where $\mathbf{Q}$ is a fourth order tensor defining the anisotropy of the damage growth law. This anisotropy expressed with reference to the principal axes of the effective stress $\tilde{\boldsymbol{\sigma}}$, is thus taken as constant, the whole nonlinearity of the damage process being contained in the function F. The normality law gives:

$$\dot{\mathbf{D}} = \mathbf{Q}F(\boldsymbol{\varepsilon}^{\mathrm{e}}, T, \mathbf{D}).$$

The growth of the scalar D is obtained by assuming that $D = c\,\mathrm{Tr}(\mathbf{D})$ and $Y = \mathrm{Tr}(\mathbf{Y})$, with $c = 1/\mathrm{Tr}(\mathbf{Q})$. We find:

$$\dot{D} = c\,\mathrm{Tr}(\dot{\mathbf{D}}) = -c\,\mathrm{Tr}(\partial\varphi^*/\partial\mathbf{Y}) = -c(\partial\varphi^*/\partial Y) = F(\boldsymbol{\varepsilon}^{\mathrm{e}}, T, \mathbf{D}).$$

These equations can also be written as functions of the effective stress with the help of the elasticity law. We will later use a law of the form:

$$\dot{D} = G(\tilde{\boldsymbol{\sigma}}, T, D)$$
$$\dot{\mathbf{D}} = \mathbf{Q}\dot{D}.$$

Fig. 7.33. Examples of microcracks in two-dimensional media calculated by the homogenization technique: (*a*) rectangular arrangement; (*b*) staggered arrangement.

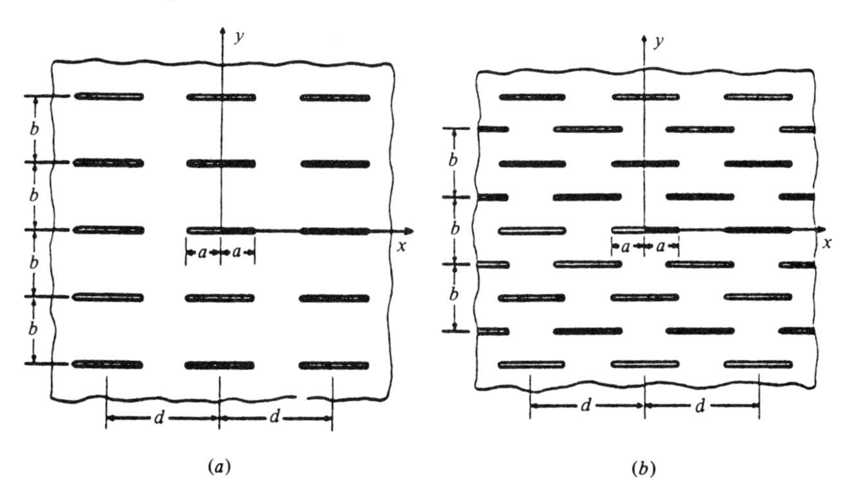

Definition of the nonisotropic relation

The inconvenience of a general theory is often linked to the difficulty of the identification of the material parameters. In order to limit the number of degrees of freedom as far as possible, we may, for the definition of the operator $\mathbf{Q}$, use elementary linear elasticity solutions obtained for particular arrangements of defects. Thus the case of all parallel cracks (Fig. 7.33) may be homogenized and interpreted by a damage tensor $\mathbf{D} = \mathbf{1} - \tilde{\mathbf{a}}:\mathbf{a}^{-1}$ with the matrix representation:

$$D = \begin{bmatrix} D_1 & 0 & 0 & & & \\ \frac{\nu}{1-\nu} D_1 & 0 & 0 & & & \\ \frac{\nu}{1-\nu} D_1 & 0 & 0 & & & \\ & & & 0 & & \\ & & & & D_5 & \\ & & & & & D_5 \end{bmatrix}$$

The symmetry of the stress tensor limits it to a 6×6 matrix. Direction 1 is perpendicular to the plane of the cracks. The effective stress tensor is obtained as in Section 7.3.1:

$$\tilde{\boldsymbol{\sigma}} = (\mathbf{1} - \mathbf{D})^{-1} : \boldsymbol{\sigma}$$

which written in a matrix form is:

$$\begin{Bmatrix} \tilde{\sigma}_{11} \\ \tilde{\sigma}_{22} \\ \tilde{\sigma}_{33} \\ \tilde{\sigma}_{23} \\ \tilde{\sigma}_{31} \\ \tilde{\sigma}_{12} \end{Bmatrix} = \begin{bmatrix} \frac{1}{1-D_1} & 0 & 0 & & & \\ \frac{\nu}{1-\nu}\frac{D_1}{1-D_1} & 1 & 0 & & & \\ \frac{\nu}{1-\nu}\frac{D_1}{1-D_1} & 0 & 1 & & & \\ & & & 1 & & \\ & & & & \frac{1}{1-D_5} & \\ & & & & & \frac{1}{1-D_5} \end{bmatrix} \begin{Bmatrix} \sigma_{11} \\ \sigma_{22} \\ \sigma_{33} \\ \sigma_{23} \\ \sigma_{31} \\ \sigma_{12} \end{Bmatrix}.$$

Study of this particular case shows that it can be described by two damage variables D_1 and D_5. If we admit a law of defect growth perpendicular to the maximum principal stress, it is sufficient to make the following choice: $\mathbf{Q} = \mathbf{\Gamma}$, with:

$$\mathbf{\Gamma} = \begin{bmatrix} 1 & 0 & 0 & & & \\ \dfrac{\nu}{1-\nu} & 0 & 0 & & & \\ \dfrac{\nu}{1-\nu} & 0 & 0 & & & \\ & & & 0 & & \\ & & & & \xi & \\ & & & & & \xi \end{bmatrix}.$$

In fact, each material is sensitive to a greater or lesser degree to this anisotropy effect. A linear combination with the isotropic case then offers the simplest way of describing possible cases with only two material-dependent coefficients (plus Poisson's ratio):

$$\mathbf{Q} = (1-\gamma)\mathbf{\Gamma} + \gamma\mathbf{1}.$$

The two characteristic coefficients in the above relation are ξ in the operator $\mathbf{\Gamma}$, and the coefficient γ which can be used to represent all the degrees of anisotropy: $\gamma = 1$ if the material is isotropic, and $\gamma = 0$ in the completely anisotropic case.

It is possible to verify that the results obtained for the isotropic damage law remain valid; the second principle requires that:

$$\dot{D} \geqslant 0.$$

The power dissipated during the process of damage is written as:

$$\Phi_D = -\mathbf{Y}::\dot{\mathbf{D}} = -\mathbf{Y}::\mathbf{Q}\dot{D} = -[(1-\gamma)\mathbf{Y}::\mathbf{\Gamma} + \gamma Y]\dot{D}.$$

The second term in the square brackets is always negative since $-Y$ is a positive definite form in thermoelastic strains. The quantity $\mathbf{Y}::\mathbf{\Gamma}$ is also negative as can be checked by replacing $\mathbf{Y}$ with its expression as a function of strains.

Anisotropy effectively develops only when the loading is nonproportional. In this case the anisotropy tensor depends on the principal direction of $\tilde{\boldsymbol{\sigma}}$. This is schematically illustrated in Fig. 7.34 for a simple example for which $\gamma = 0$. Under the action of a first tension load (I), the damage has grown and the defects are arranged in planes perpendicular to the maximum principal stress. We have:

$$\mathbf{D}_1 = \mathbf{Q}D_1.$$

If an equibiaxial tension is now applied ($\sigma_1 = \sigma_2$, $\sigma_3 = 0$), the maximum principal effective stress remains such that $\tilde{\sigma}_1 > \tilde{\sigma}_2$. The defects continue to grow in the same plane (II). If a tension is now applied in direction 2 ($\sigma_1 = \sigma_3 = 0$), $\tilde{\sigma}_2$ becomes the maximum principal effective stress and further growth of defects takes place in planes perpendicular to it. We will therefore write (III):

$$\mathbf{D}_2 = \mathbf{D}_1 + \mathbf{Q}_2\mathbf{D}_2 = \mathbf{Q}D_1 + \mathbf{V}{:}\mathbf{Q}{:}\mathbf{V}^{\mathrm{T}}D_2$$

with

$$\mathbf{V} = \mathbf{R} \otimes \mathbf{R}$$

where **R** denotes the rotation operator between axes 1 and 2. After a certain

Fig. 7.34. Schematic representation of anisotropic damage under complex loading: (*a*) applied loading; (*b*) corresponding effective stresses; (*c*) resulting damage from this loading.

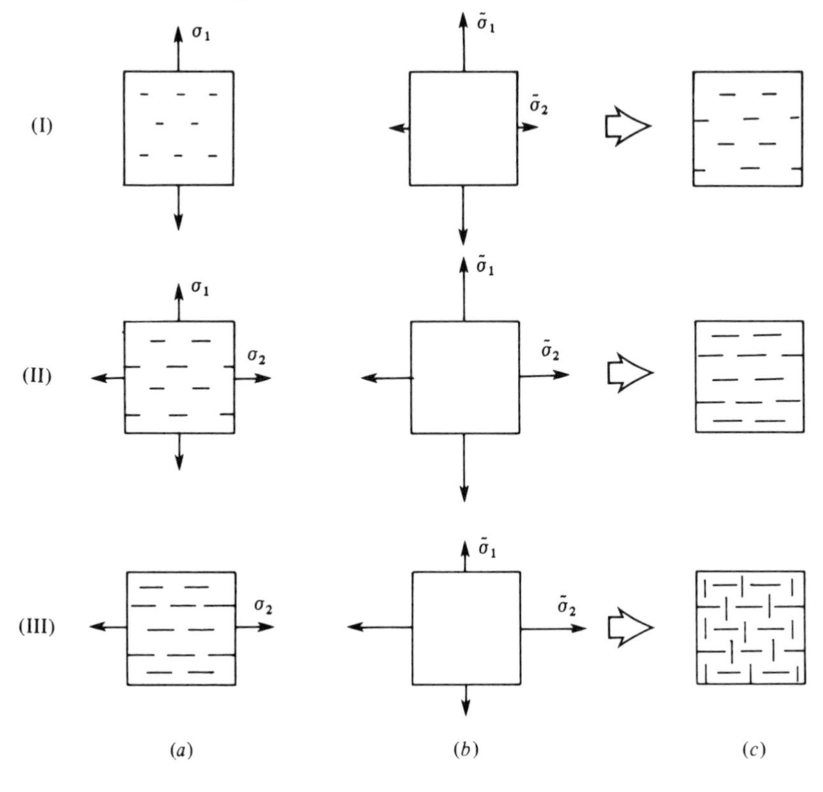

time there will be superposition of the two perpendicular systems of defects; when $D_1 = D_2$ the anisotropy will be much less, directions 1 and 2 exhibiting the same behaviour with only small variations for the intermediate directions. This illustrates quite well the point that anisotropy is a characteristic of the growth law and not of the current state of damage. Note that the choice of the expression $\dot{\mathbf{D}} = \mathbf{Q}\mathbf{D}$ implies that in simple tension (direction 1) the first component of $\mathbf{D}$ is always equal to D.

7.4 Particular models

The isotropic models described below assume that there is no coupling between damage and other dissipative phenomena. The evolution law of the variable D does not depend on other internal variables; this prevents taking into account some phenomena such as the increase in the fatigue lifetime produced by a prestrain. In order to account for this particular phenomenon, it is sufficient to introduce hardening variables in the damage law. This is possible with the formalism described in Section 7.3, but it is a complex process!

7.4.1 *Ductile plastic damage*

Ductile plastic damage accompanies large plastic deformations and, like plasticity, it is rate independent and does not involve time explicitly. If, on the other hand, we consider only isotropic damage and hardening, the only internal variable which appears, besides damage, is the accumulated plastic strain defined (see Chapter 5) by:

$$\dot{p} = (\tfrac{2}{3}\dot{\boldsymbol{\varepsilon}}^{\mathrm{p}} : \dot{\boldsymbol{\varepsilon}}^{\mathrm{p}})^{1/2}.$$

The two damage models described above are derived from a dissipation potential formulated with

$$f(\boldsymbol{\sigma}, R) \quad \text{and} \quad \varphi_{\mathrm{D}}^{*}(Y; T, D).$$

Coupling between plasticity and damage is studied in detail in Section 7.5.2. Here, let us use the loading function in the form:

$$f = f(\boldsymbol{\sigma}, R, D) = \sigma_{\mathrm{eq}}/(1 - D) - R - k \leqslant 0.$$

The damage potential is chosen as a power function of $-Y$:

$$\varphi_{\mathrm{D}}^{*} = \frac{S_0}{s_0 + 1} \frac{1}{1 - D} \left(-\frac{Y}{S_0} \right)^{s_0 + 1}$$

where s_0 and S_0 are characteristic coefficients for each material and depend on temperature. By applying the general method we get:

$$\dot{\boldsymbol{\varepsilon}}^{\mathrm{p}} = \dot{\lambda}\frac{\partial f}{\partial \boldsymbol{\sigma}} = \frac{3}{2}\frac{\dot{\lambda}}{1-D}\frac{\boldsymbol{\sigma}'}{\sigma_{\mathrm{eq}}}, \qquad \dot{p} = (\tfrac{2}{3}\dot{\boldsymbol{\varepsilon}}^{\mathrm{p}}:\dot{\boldsymbol{\varepsilon}}^{\mathrm{p}})^{1/2} = \frac{\dot{\lambda}}{1-D},$$

$$\dot{D} = -\dot{\lambda}\frac{\partial \varphi_{\mathrm{D}}^{*}}{\partial Y} = \frac{\dot{\lambda}}{1-D}\left(-\frac{Y}{S_0}\right)^{s_0},$$

$$\dot{D} = \left(-\frac{Y}{S_0}\right)^{s_0}\dot{p}.$$

Uniaxial stress model

The associated variable Y and the equivalent effective stress $\tilde{\sigma}^*$ are equivalent variables, as has been shown in Section 7.3.2:

$$-Y = (1/2E)\tilde{\boldsymbol{\sigma}}^{*2}.$$

Moreover, with a monotonically increasing uniaxial load, which is the assumption for the validity of this model, $\tilde{\sigma}^*$ is identified with the effective stress $\tilde{\sigma}$ and $\dot{p}$ with the magnitude of the plastic strain rate $|\dot{\varepsilon}_{\mathrm{p}}|$. We may therefore write:

$$\dot{D} = (\tilde{\sigma}^2/2ES_0)^{s_0}|\dot{\varepsilon}_{\mathrm{p}}|.$$

In order to obtain the model suggested by phenomenological considerations, it is necessary to replace $\dot{\varepsilon}_{\mathrm{p}}$ by its expression in the hardening law introduced in Section 5.4.2 and written here in terms of the effective stress and with $\sigma_{\mathrm{Y}} = 0$:

$$\varepsilon_{\mathrm{p}} = \left(\frac{\tilde{\sigma}}{K}\right)^{M} \quad \text{or} \quad \dot{\varepsilon}_{\mathrm{p}} = \frac{M}{K}\left(\frac{\tilde{\sigma}}{K}\right)^{M-1}\dot{\tilde{\sigma}}.$$

After an obvious change in the notation of the constants:

$$\dot{D} = \left(\frac{\tilde{\sigma}}{S_1}\right)^{s_1}\frac{\dot{\tilde{\sigma}}}{S_1} \quad \text{or} \quad \mathrm{d}D = \left(\frac{\tilde{\sigma}}{S_1}\right)^{s_1}\frac{\mathrm{d}\tilde{\sigma}}{S_1}.$$

A closely related form, closer to the experimental results because it introduces a damage threshold σ_{D}, is the following:

- $$\mathrm{d}D = \left\langle\frac{\sigma - \sigma_{\mathrm{D}}}{(1-D)S}\right\rangle^{s}\frac{\mathrm{d}\sigma}{S}.$$

The corresponding potential φ^* does not have an explicit form. The three characteristic coefficients of ductile plastic damage, which are possibly dependent on temperature, are determined through identification with the damage measurements conducted, for example, with the help of the method of elastic modulus variation described in Section 7.2.3.

Integration of this differential equation gives the value of the damage for any value of stress. With the initial condition $\sigma = \sigma_D \to D = 0$, we find:

$$D = 1 - \left(1 - \left\langle \frac{\sigma - \sigma_D}{S} \right\rangle^{s+1}\right)^{1/s+1}.$$

The final ductile fracture occurs when:

$$D = D_c \quad \text{or} \quad \sigma = \sigma_R.$$

Identification of the characteristic coefficients of the ductile damage can be performed by measuring variations of the elasticity modulus during the hardening tests. Two examples are:

Copper Cu/Al, $T = 20\,°\mathrm{C}$: $\sigma_D = 330$ MPa, $S = 445$ MPa, $s = 0.70$, $D_c = 0.85$;
AU4G – T_4 alloy, $T = 20\,°\mathrm{C}$: $\sigma_D = 400$ MPa, $S = 580$ MPa, $s = 0.58$, $D_c = 0.23$.

Multiaxial strain model

Starting from the same dissipation potential, we can develop a multiaxial model within the framework of the hypotheses of isotropic damage and isotropic hardening which is valid for any loading:

$$-\frac{\partial \varphi_D^*}{\partial Y} = \left(\frac{-Y}{S_0}\right)^{s_0} \dot{p}.$$

We may replace $-Y$ by its expression in terms of equivalent von Mises stress σ_{eq} and the hydrostatic stress σ_H (Sections 7.2.5 and 7.3.2):

$$-Y = \frac{\tilde{\sigma}^{*2}}{2E} = \frac{\sigma_{eq}^2}{2E(1-D)^2}\left[\frac{2}{3}(1+\nu) + 3(1-2\nu)\left(\frac{\sigma_H}{\sigma_{eq}}\right)^2\right].$$

The ductile damage occurs only when the strain-hardening is saturated ($R = R_\infty = \text{constant}$) and the material is then considered perfectly plastic. The expression of the plastic criterion $\sigma_{eq}/(1-D) - R - k = 0$ shows that

$$\frac{\sigma_{eq}}{1-D} = \tilde{\sigma}_{eq} = \text{constant} = K.$$

Then

- $\dot{D} = (K^2/2ES_0)^{s_0}[\frac{2}{3}(1+\nu) + 3(1-2\nu)(\sigma_H/\sigma_{eq})^2]^{s_0}\dot{p}.$

Evolution of the damage as a function of p is obtained by integrating this differential equation with the initial condition $p \leqslant p_D \rightarrow D = 0$, where p_D is the damage threshold in terms of strain.

If we restrict ourselves to radial loading for which the triaxiality ratio σ_H/σ_{eq} is constant, we obtain:

$$D = (K^2/2ES_0)^{s_0}[\tfrac{2}{3}(1+\nu) + 3(1-2\nu)(\sigma_H/\sigma_{eq})^2]^{s_0}\langle p - p_D\rangle.$$

We can simplify this expression by introducing the condition of fracture

$$p = p_R \rightarrow D = D_c$$

where p_R is the accumulated strain at fracture. Since D_c is a material constant, p_R depends on the triaxiality ratio σ_H/σ_{eq}

$$D_c = (K^2/2ES_0)^{s_0}[\tfrac{2}{3}(1+\nu) + 3(1-2\nu)(\sigma_H/\sigma_{eq})^2]^{s_0}\langle p_R - p_D\rangle.$$

The expression for D then may be written as:

$$D = D_c\left\langle \frac{p - p_0}{p_R - p_D}\right\rangle.$$

This expression generalizes to the multiaxial empirical model of Section 7.2.4 with the difference that p_R and p_D are functions of the triaxiality ratio. From the expression for D_c we get:

$$p_R - p_D = D_c\left(\frac{K^2}{2ES_0}\left[\frac{2}{3}(1+\nu) + 3(1-2\nu)\left(\frac{\sigma_H}{\sigma_{eq}}\right)^2\right]\right)^{-s_0}.$$

Assuming that the stress triaxiality affects the damage threshold p_D and the fracture strain p_R in the same manner, the ratio p_D/p_R is a constant for each material equal to its uniaxial value $\varepsilon_D/\varepsilon_R$ where no distinction is made between total strains and plastic strains. The model involves only two constants D_c and $\varepsilon_D/\varepsilon_R$, and a function $\dot{p}_R$ which can also be expressed as a function of the uniaxial fracture strain ε_R corresponding to a triaxiality ratio $\sigma_H/\sigma_{eq} = \frac{1}{3}$:

$$p_R\left(\frac{1}{3}\right) = \varepsilon_R = \left(\frac{2ES_0}{K^2}\right)^{s_0}\frac{D_c}{1 - \varepsilon_D/\varepsilon_R}$$

$$p_R\left(\frac{\sigma_H}{\sigma_{eq}}\right) = \varepsilon_R\left[\frac{2}{3}(1+\nu) + 3(1-2\nu)\left(\frac{\sigma_H}{\sigma_{eq}}\right)^2\right]^{-s_0}.$$

If the numerical values given by this expression are compared with those obtained by applying the cavity growth models of MacClintock, Rice and Tracey, we find that $s_0 = 1$ constitutes the best identification, also in agreement with the uniaxial stress model.

In summary, the multiaxial ductile plastic damage strain model is the following:

In differential form:

$$\dot{D} = \frac{D_c}{\varepsilon_R - \varepsilon_D}\left[\frac{2}{3}(1+\nu) + 3(1-2\nu)\left(\frac{\sigma_H}{\sigma_{eq}}\right)^2\right]\dot{p}\,.$$

In integrated form:

$$D \approx \frac{D_c}{\varepsilon_R - \varepsilon_D}\left\{p\left[\frac{2}{3}(1+\nu) + 3(1-2\nu)\left(\frac{\sigma_H}{\sigma_{eq}}\right)^2\right] - \varepsilon_D\right\}.$$

The three constants, intrinsic to each material, are determined by uniaxial tests and are those given in Table 7.1.

7.4.2 *Creep Damage*

The creep damage law commonly used is that described in Section 7.2.4. The relations developed below correspond to generalizations for complex loadings:

variable stress,
multiaxial proportional loading,
multiaxial nonproportional loading.

Nonlinear accumulation model

A form which is an improvement over Kachanov's law is obtained by introducing an extra coefficient k as suggested by Rabotnov:

$$\dot{D} = (\sigma/A)^r(1-D)^{-k}.$$

It always includes the rule of linear accumulation, and in general k is larger than r: the damage rate is influenced more strongly by the degree of damage than the global mechanical behaviour is. Integration of this law, for a constant stress, gives:

$$D = 1 - \left(1 - \frac{t}{t_c}\right)^{1/(k+1)} \quad \text{with} \quad t_c = \frac{1}{k+1}\left(\frac{\sigma}{A}\right)^{-r}.$$

Table 7.2 gives the values of coefficients r, k, $A = A_0[(k+1)/(r+1)]^{1/r}$. Fig. 7.7 shows the correlation with the damage measurements made on two materials. It also shows that the curves can depend on the stress level (the case of the IN 100 alloy) which is consistent with the existence of nonlinear accumulation effects and leads to the following generalization:

- $\dot{D} = (\sigma/A)^r(1-D)^{-k(\sigma)}$.

Since the exponent k depends on the stress level, the two variables σ and D are no longer separable. No theoretical justification exists for the introduction of the function $k(\sigma)$ except that it provides the simplest means of representing nonlinear accumulation.

Discrete values of $k(\sigma)$ are determined by identification with the damage growth measured during creep under constant stress. In fact, the evolution of D as a function of t/t_c, obtained by integration from 0 to D, depends on σ:

$$D = 1 - \left(1 - \frac{t}{t_c}\right)^{1/[k(\sigma)+1]}, \quad t_c = \frac{1}{k(\sigma)+1}\left(\frac{\sigma}{A}\right)^{-r}.$$

The function $k(\sigma)$ is an increasing one, because damage growth is faster for lower stress (Figs. 7.7 and 7.35). This is confirmed by the results of tests at

Fig. 7.35. Schematic diagram illustrating nonlinear accumulation in creep tests at two levels of stress.

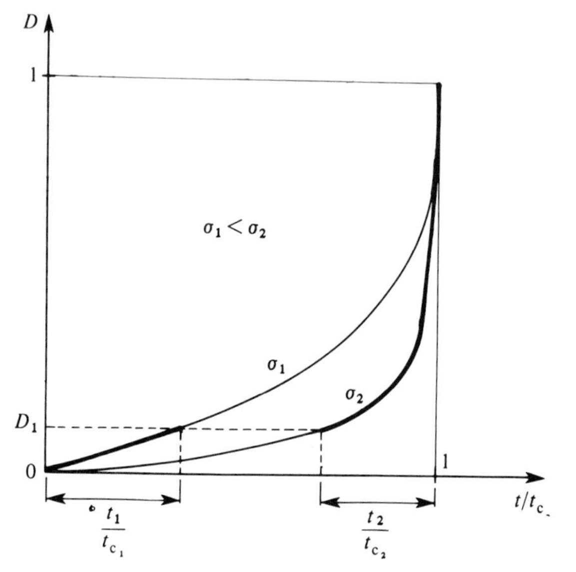

two levels of stress. The equation represents the effects of nonlinear accumulation quite well; by integrating from 0 to D_1, at the first level of stress, and then from D_1 to 1 at the second level, we find (Fig. 7.36) that:

$$\frac{t_2}{t_{c_2}} = \left(1 - \frac{t_1}{t_{c_1}}\right)^{[k(\sigma_2)+1]/[k(\sigma_1)+1]}.$$

The above discussion applies to creep under tension; few studies have been made regarding creep damage in compression: fracture is never observed as it is preceded by instability of geometric origin (buckling). However, there is nothing to prove that this damage does not exist or that compression does not influence damage in further tension.

Multiaxial isotropic model

The domain of validity of this model is restricted to loading which can be considered to be almost proportional. If such is not the case, it may become necessary to consider the effects of damage anisotropy such as those mentioned in the next subsection.

The writing of a multiaxial law therefore requires only the use of an equivalent stress in the sense of creep damage. Depending on the material it

Fig. 7.36. Example of prediction by the damage model for the IN 100 refractory alloy.

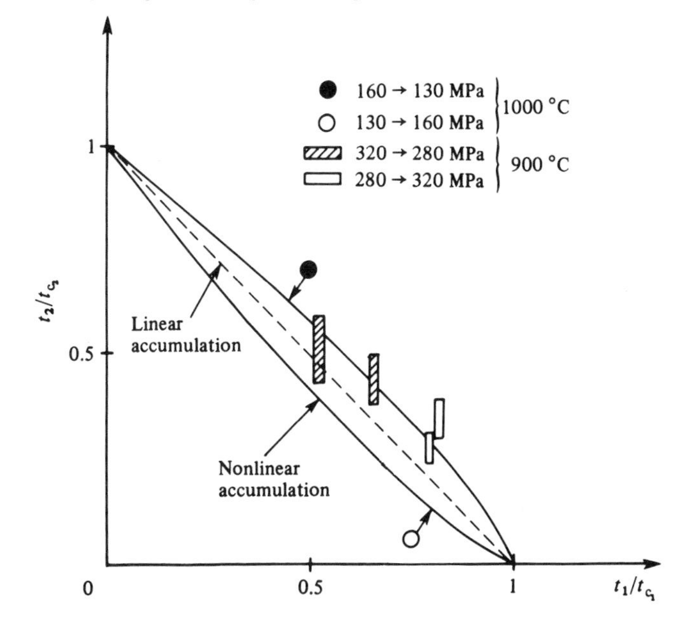

will be possible to use one of the criteria described in Section 7.2.5 and consequently to describe isochronous surfaces of failure in stress space (surfaces linking the states of stress which result in creep rupture in the same time):

the elastic energy density release rate criterion:

$$\sigma^* = \sigma_{eq}[\tfrac{2}{3}(1+\nu) + 3(1-2\nu)(\sigma_H/\sigma_{eq})^2]^{1/2},$$

the shear–dilatation sensitivity coefficient criterion:

$$\sigma_\beta^* = (1-\beta)\sigma_{eq} + 3\beta\sigma_H,$$

the more general criterion involving a combination of the three basic invariants:

$$\chi(\boldsymbol{\sigma}) = \alpha J_0(\boldsymbol{\sigma}) + \beta J_1(\boldsymbol{\sigma}) + (1-\alpha-\beta)J_2(\boldsymbol{\sigma})$$

where $J_1(\boldsymbol{\sigma}) = 3\sigma_H$, $J_2(\boldsymbol{\sigma}) = \sigma_{eq}$, and $J_0(\boldsymbol{\sigma})$ denotes the maximum principal stress.

In the isotropic case, the damage variable $\mathbf{D}$ is reduced to a scalar

$$\mathbf{D} = \mathbf{1}D$$

and the effective stress is expressed by:

$$\tilde{\boldsymbol{\sigma}} = \frac{\boldsymbol{\sigma}}{1-D}.$$

The multiaxial law of damage evolution is expressed as follows:

$$\dot{D} = \left\langle \frac{\chi(\tilde{\boldsymbol{\sigma}})}{A} \right\rangle^r (1-D)^{r-k}.$$

If $r = k$, we obtain Kachanov's law. For simple tension, we have again exactly the same expressions as those obtained in the previous section. If the exponent depends on the stress, the following expression can be chosen:

- $$\dot{D} = \left\langle \frac{\chi(\tilde{\boldsymbol{\sigma}})}{A} \right\rangle^r (1-D)^{r-k\langle\chi(\boldsymbol{\sigma})\rangle}.$$

Note that this can also be written as:

$$\dot{D} = \left\langle \frac{\chi(\boldsymbol{\sigma})}{A} \right\rangle^r (1-D)^{-k\langle\chi(\boldsymbol{\sigma})\rangle} \quad \text{or} \quad \dot{D} = \left\langle \frac{\chi(\tilde{\boldsymbol{\sigma}})}{A} \right\rangle^{k\langle\chi(\boldsymbol{\sigma})\rangle} \left[\frac{\chi(\boldsymbol{\sigma})}{A} \right]^{r-k\langle\chi(\boldsymbol{\sigma})\rangle}.$$

As usual the symbol $\langle\chi\rangle$ denotes the positive part of χ. In fact $\chi(\boldsymbol{\sigma})$ may be negative for certain stress states depending on the values of the coefficients α

and β. The damage rate is then zero, and there is no isochronous surface passing through these points.

Anisotropic damage model

In accordance with the theory introduced in Section 7.3.3, a simple way of incorporating damage anisotropy (already produced) in the damage law (in further evolution) consists in admitting a multiplicative decomposition of the damage rate tensor:

$$\dot{\mathbf{D}} = \mathbf{Q}\dot{D} = [(1-\gamma)\mathbf{\Gamma} + \gamma\mathbf{1}]\dot{D}.$$

The anisotropic part is contained in the tensor $\mathbf{Q}$. It plays a part, it should be recalled, only in cases of nonproportional loading. The effect of anisotropy is thus considered fixed in the current principal directions of the effective stress tensor $\tilde{\boldsymbol{\sigma}}$ (see Section 7.3.3). The nonlinearity of the damage evolution process is entirely contained in the equation governing D. We may choose an equation for this similar to that used in the isotropic case:

$$\dot{D} = \left\langle \frac{\chi^*(\tilde{\boldsymbol{\sigma}}, D)}{A} \right\rangle^{k\langle\chi(\boldsymbol{\sigma})\rangle} \left(\frac{\chi(\boldsymbol{\sigma})}{A}\right)^{r-k\langle\chi(\boldsymbol{\sigma})\rangle}$$

where the equivalent effective stress depends on D:

$$\chi^*(\tilde{\boldsymbol{\sigma}}, D) = \alpha J_0(\tilde{\boldsymbol{\sigma}}) + \frac{\beta}{1+2A} J_1(\tilde{\boldsymbol{\sigma}}) + \frac{1-\alpha-\beta}{1-A} J_2(\tilde{\boldsymbol{\sigma}})$$

with

$$A = \frac{\nu}{1-\nu}\frac{(1-\gamma)D}{1-\gamma D}.$$

This change in the expression of the equivalent effective stress is introduced in order to recover Kachanov's law in the simple tension case. Indeed, the first component of the tensor $\mathbf{D}$ is equal to D and we then have:

$$\tilde{\sigma}_1 = \frac{\sigma_1}{1-D},$$

$$\tilde{\sigma}_2 = \frac{\nu(1-\gamma)}{1-\nu}\frac{D}{1-\gamma D}\frac{\sigma_1}{1-D} = \tilde{\sigma}_3.$$

The multiaxial anisotropic model is therefore completely identified with the uniaxial model and the coefficients ξ and γ which fix the anisotropy (in addition to ν). In the isotropic case ($\gamma = 1$) we recover $\tilde{\sigma}_2 = \tilde{\sigma}_3 = 0$ and the expressions of the previous subsection.

This formulation therefore includes the two limiting cases (isotropic and purely anisotropic) and also the intermediate ones (by linear combination). It applies to any nonproportional loading and can be used to represent the effects of damage anisotropy on the damage law and on material behaviour (through the effective stress established in Section 7.3.1).

It has been applied to the case of creep under tension followed by creep under torsion. Fig. 7.37 shows a few examples of predictions of the total rupture time (in terms of time to failure under tension alone) as function of the time spent under tension. This is repeated for several values of the coefficients, and the results of Murakami's and Kachanov's theories are presented for comparison.

7.4.3 **Fatigue damage**

Definition of damage per cycle

The general theory of Section 7.3 uses time as a reference variable, and the increments of damages are written as functions of the increments of time, and of strain or stress. In the case of fatigue, we usually employ the concept of the loading cycle to evaluate the evolution of damage and to measure the fatigue life-time. The equations then depend on the load through globally

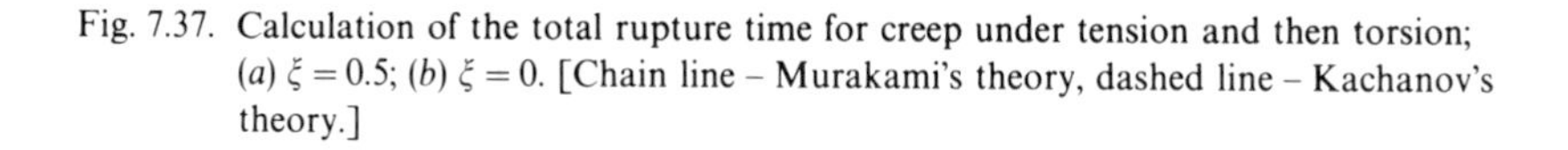

Fig. 7.37. Calculation of the total rupture time for creep under tension and then torsion; (*a*) $\xi = 0.5$; (*b*) $\xi = 0$. [Chain line – Murakami's theory, dashed line – Kachanov's theory.]

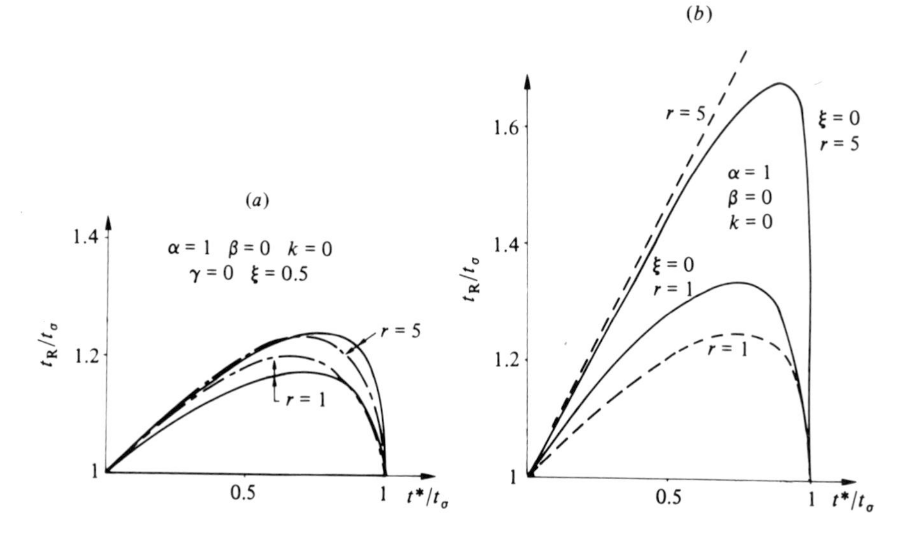

defined quantities over a cycle, such as amplitude, maximum value, mean value.

Therefore, we only have to imagine a formal integration, during each cycle, of an evolution law similar to that used for ductile plastic damage. The growth equation of fatigue damage is therefore taken in the form:

$$\delta D = f(\cdots)\delta N$$

in which the variables on which the function f depends are:

the state variables at the beginning of the cycle (temperature, damage D, and the hardening variables),
the maximum and the mean values over a cycle of the parameters which define the loading (stress, strain or plastic strain).

A particular problem occurs in representing the influence of the temperature when this changes during the cycle. This case will be studied later in this section.

Fatigue damage equations, which obey the rule of linear accumulation, have already been discussed as basic laws (Section 7.2.4). In contrast, the equation studied below allows us to describe the effects of nonlinear accumulation in the case of nonperiodic cyclic loads.

Uniaxial model

This model is derived from the macroscopic definition of damage with reference to two types of evaluation:

(1) damage evaluation in terms of the remaining life-time;
(2) damage evaluation by using the concept of effective stress.

We first examine the construction of a model based on the remaining life-time. Fatigue tests at two stress levels ($\Delta\sigma_1$ during N_1 cycles followed by $\Delta\sigma_2$ during N_2 cycles with $N_1 + N_2 = N_R$, the number of cycles to failure) demonstrate the nonlinearity of the damage accumulation and furnish an indirect measure of the current state of damage from knowledge of the remaining life–time N_2/N_{F_2}. These measurements are sufficient to prove that the damage evolution curves as functions of the life ratio N/N_F depend on the load level, as illustrated in Fig. 7.38(*a*): the remaining life at the second level N_2/N_F is then different from $1 - N_1/N_{F_1}$, the value given by the linear accumulation rule.

A simple way to introduce such effects in the damage growth equation consists in rendering the load and damage variables nonseparable. For example, we may take:

$$\delta D = D^{\alpha(\sigma_{\text{Max}},\bar{\sigma})}.\left(\frac{\sigma_{\text{Max}} - \bar{\sigma}}{C(\bar{\sigma})}\right)^{\beta} \delta N.$$

The exponent α depends on the loading $(\sigma_{\text{Max}}, \bar{\sigma})$, which results in nonseparability. Note that an equation similar to that used for creep damage is not suitable because of a reversal of the cumulative effect: in fatigue, the life-time at the second level, when the first load is higher, is shorter than predicted by the linear rule. Integrating from $D = 0$ to $D = 1$, the above equation gives the number of cycles to failure:

$$N_{\text{F}} = \frac{1}{1 - \alpha(\sigma_{\text{Max}}, \bar{\sigma})}\left(\frac{\sigma_{\text{Max}} - \bar{\sigma}}{C(\bar{\sigma})}\right)^{-\beta}.$$

Fig. 7.38. Nonlinear accumulation in fatigue and partial indetermination of the damage measured in terms of the remaining life-time.

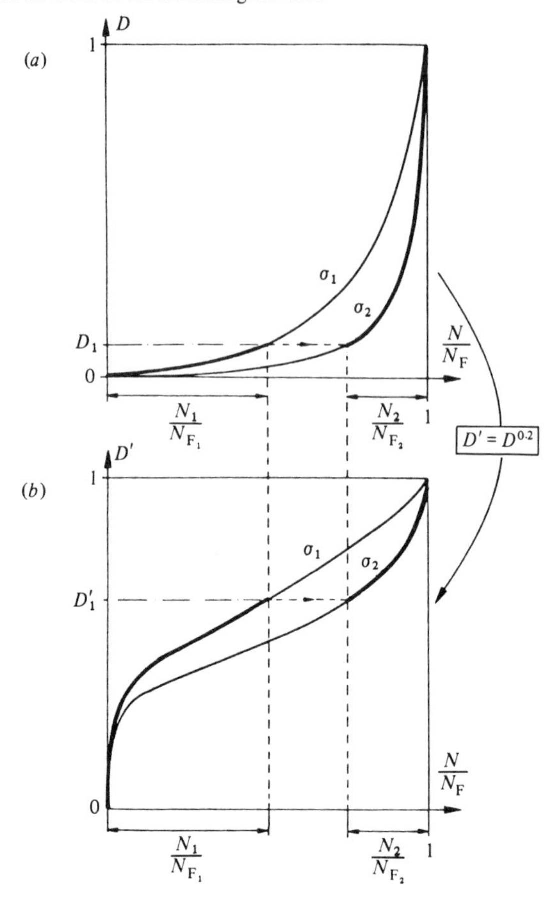

D evolves as a function of N/N_F and, in a two-level test, the life-time at the second level is given by:

$$\frac{N_2}{N_{F_2}} = 1 - \left(\frac{N_1}{N_{F_1}}\right)^{\eta} \quad \text{with} \quad \eta = \frac{1-\alpha_2}{1-\alpha_1} = \frac{1-\alpha(\sigma_{\text{Max}_2}, \bar{\sigma}_2)}{1-\alpha(\sigma_{\text{Max}_1}, \bar{\sigma}_1)}.$$

This formulation permits a qualitative description of most of the results of tests conducted at different levels on a number of materials. The evaluation of D in terms of the remaining life is, however, not sufficient to totally fix its value at each instant. A look at Fig. 7.38 is enough to be convinced of this; a simple variable mapping in fact shows that the life-time can be identical for any loading while the damage value is different. In other words the concept of remaining life-time provides only a relative evaluation of damage.

In order to avoid this indetermination, the second type of evaluation is used. The concept of effective stress applied to fatigue provides an indirect measure as we have seen in Section 7.2.3. The measured evolutions are extremely nonlinear as can be seen in Fig. 7.6. With this concept, damage can really be measured only in the last part of the life-time, when microscopic initiations have already occurred (this is the phase of micropropagation of defects).

In order to combine this evaluation with the one corresponding to the 'remaining life-time', it is sufficient to make a change of variable by replacing D in the previous equation by:

$$1-(1-D)^{\beta+1}.$$

The differential law can then be written as:

$$\bullet \qquad \delta D = [1-(1-D)^{\beta+1}]^{\alpha(\sigma_{\text{Max}},\bar{\sigma})}\left[\frac{\sigma_{\text{Max}}-\bar{\sigma}}{M(\bar{\sigma})(1-D)}\right]^{\beta}\delta N.$$

The form is more complex, but its properties are identical to the properties of the previous equation, except for the current value of damage. The number of cycles to failure, obtained by integrating the above is:

$$N_F = \frac{1}{(\beta+1)[1-\alpha(\sigma_{\text{Max}}, \bar{\sigma})]}\left(\frac{\sigma_{\text{Max}}-\bar{\sigma}}{M(\bar{\sigma})}\right)^{-\beta}$$

and we find that $M(\bar{\sigma}) = C(\bar{\sigma})(\beta+1)^{1/\beta}$. The damage, expressed as a function of N/N_F is:

$$D = 1 - \left[1-\left(\frac{N}{N_F}\right)^{1/(1-\alpha)}\right]^{1/(\beta+1)}.$$

Fig. 7.39. Predictions of two-level fatigue tests on (*a*) maraging steel at room temperature (amplitudes in ksi: treatment I) (*b*) AISI 4130 soft steel and maraging steel (Treatment I & II) at room temperature and (*c*) IN 100 alloy at high temperatures.

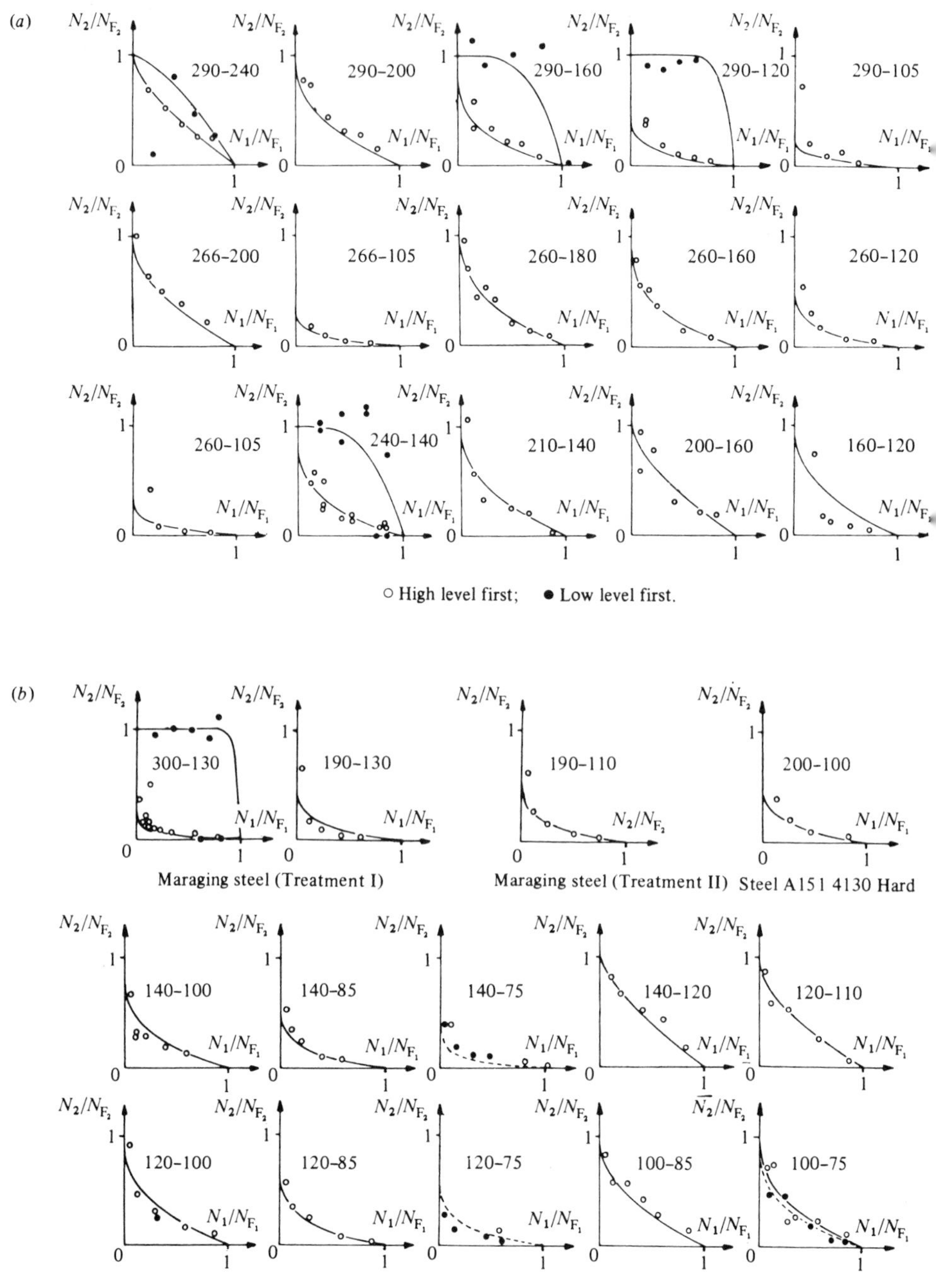

The calculated curves of Fig. 7.6 show that this expression is in good agreement with experimental results. The functions α and M still have to be chosen to represent at the same time the fatigue limit and the static fracture (in one cycle) and the effects of nonlinear accumulation:

- $$\alpha(\sigma_{\text{Max}}, \bar{\sigma}) = 1 - a\left\langle \frac{\sigma_{\text{Max}} - \sigma_l(\bar{\sigma})}{\sigma_u - \sigma_{\text{Max}}} \right\rangle,$$
- $$\sigma_l(\bar{\sigma}) = \sigma_{l_0} + (1 - b\sigma_{l_0})\bar{\sigma},$$
- $$M(\bar{\sigma}) = M_0(1 - b\bar{\sigma})$$

where σ_{l_0} is the fatigue limit at zero mean stress and σ_u is the ultimate tensile stress. The choice of a linear relation to express the influence of $\bar{\sigma}$ is justified by experimental observations on the fatigue limit. Examples of correlation with Woehler's curves have been shown in Fig. 7.13 (the coefficients β and M_0 are determined from these experimental curves).

The coefficient a can be identified only from the damage measurements mentioned above (Fig. 7.6), but the choice of the value of a is important only when fatigue damage is combined with another type of damage (creep for example). In pure fatigue, all loadings can be described by an arbitrary value of a. The relation between B_0 and M_0 of Table 7.3 is: $M_0 = B_0[a(\beta + 1)]^{1/\beta}$.

The tests at two stress levels are described by the same relations as before. The remaining life-time at the second level is:

$$\frac{N_2}{N_{F_2}} = 1 - \left(\frac{N_1}{N_{F_1}}\right)^\eta \quad \text{with} \quad \eta = \frac{1 - \alpha_2}{1 - \alpha_1} = \frac{\sigma_{\text{Max}_2} - \sigma_l(\bar{\sigma}_2)}{\sigma_{\text{Max}_1} - \sigma_l(\bar{\sigma}_1)} \frac{\sigma_u - \sigma_{\text{Max}_1}}{\sigma_u - \sigma_{\text{Max}_2}}.$$

Experimental results on a number of materials show correct correlations at room temperature and at high temperatures (Fig. 7.39). These nonlinear cumulative effects are not very different from those described with the equations used by Subramanyan and by Manson. The same relation

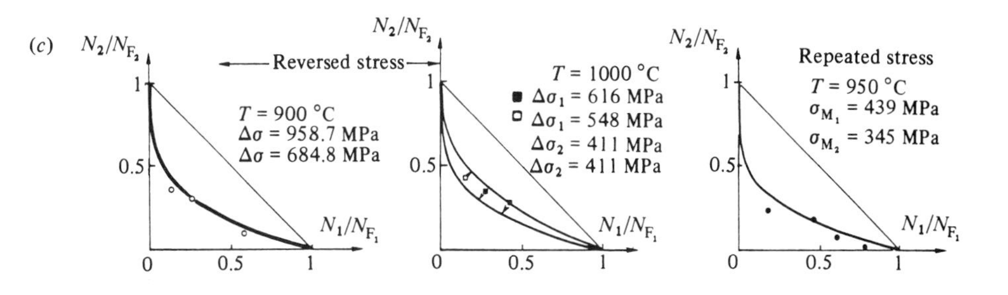

governs N_2/N_{F_2}, with

$$\eta = \frac{\sigma_{Max_2} - \sigma_l}{\sigma_{Max_1} - \sigma_l} \quad \text{and} \quad \eta = \left(\frac{N_{F_1}}{N_{F_2}}\right)^{0.45}.$$

respectively (Fig. 7.40, and Table 7.3).

Extension to the multiaxial case

The principle of generalization of the preceding model to the multiaxial case consists in the simultaneous use of:

the uniaxial law for correctly describing the damage evolution in tension and, later, the nonlinear accumulative effects,

Fig. 7.40. Comparison of three exponents for nonlinear accumulation of fatigue damage. Case of C 35 steel (stress ranges in M Pa).

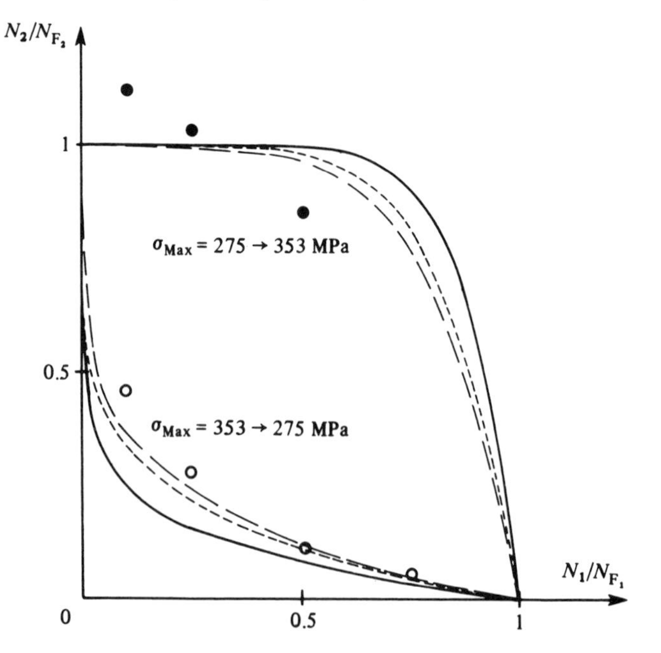

the multiaxial fatigue limit criteria as a guide in the choice of significant invariants for use in the law. Here, we choose the Sines criterion, but an analogous formulation is possible with the Crossland or the Dang-Van criterion (see Section 7.2.5),
the anisotropy of the damage law, with a method similar to that used in creep.

In the uniaxial case, it is the function α which is used to introduce the fatigue limit with the term $\sigma_{Max} - \sigma_l(\bar{\sigma})$ that can also be written in the form $\sigma_a - \sigma_{al}(\bar{\sigma})$. In the multiaxial case, a stress cycle corresponding to an unlimited life-time is such that:

$$A_{II} \leqslant A_{II}^*(\bar{\sigma}_H) = \sigma_{l_0}(1 - 3b\bar{\sigma}_H)$$

with

$$A_{II} = \tfrac{1}{2} \underset{t_0}{\text{Max}}\, \underset{t}{\text{Max}}\, J_2(\boldsymbol{\sigma}(t) - \boldsymbol{\sigma}(t_0)).$$

Similarly, static fracture is described by the von Mises criterion (for example) by replacing $\sigma_u - \sigma_{Max}$ in the function α by $\sigma_u - \sigma_{eqMax}$ where σ_{eqMax} is the maximum value of the second invariant of $\boldsymbol{\sigma}$ during the course of the cycle:

$$\sigma_{eq} = J_2(\boldsymbol{\sigma}) = \frac{1}{\sqrt{2}}[(\sigma_1 - \sigma_2)^2 + (\sigma_2 - \sigma_3)^2 + (\sigma_3 - \sigma_1)^2]^{1/2}.$$

The functions α and M are therefore expressed by:

$$\alpha(A_{II}, \bar{\sigma}_H, \sigma_{eqMax}) = 1 - a\left\langle \frac{A_{II} - A_{II}^*(\bar{\sigma}_H)}{\sigma_u - \sigma_{eqMax}} \right\rangle$$

$$M(\bar{\sigma}_H) = M_0(1 - 3b\bar{\sigma}_H).$$

In the uniaxial damage law, the amplitude of effective stress is $(\sigma_{Max} - \bar{\sigma})/(1 - D)$. In the multiaxial case we have instead:

$$\tilde{A}_{II} = \tfrac{1}{2} \underset{t_0}{\text{Max}}\, \underset{t}{\text{Max}}\, J_2(\tilde{\boldsymbol{\sigma}}(t) - \tilde{\boldsymbol{\sigma}}(t_0)).$$

The general isotropic damage law is then written as:

- $$\delta D = [1 - (1 - D)^{\beta + 1}]^{\alpha(A_{II}, \bar{\sigma}_H, \sigma_{eqMax})} \left[\frac{\tilde{A}_{II}}{M(\bar{\sigma}_H)}\right]^{\beta} \delta N.$$

It should be noted that for an isotropic damage theory:

$$\tilde{A}_{II} = A_{II}/(1 - D).$$

The anisotropy of the fatigue damage-growth law is introduced by expressing the rate of growth of the tensor $\mathbf{D}$ by:

$$\dot{\mathbf{D}} = \mathbf{Q}_\mathrm{F}\dot{D}$$

where $\mathbf{Q}_\mathrm{F}$ depends on the material and is defined with respect to the principal axes of the effective stress tensor $\tilde{\boldsymbol{\sigma}}$ at the instant considered most damaging in the cycle. There is not enough research or experimental results (in 1984 !) that could enable us to establish and validate completely such an anisotropic theory.

Case of variable temperature during the cycle

When the temperature changes slowly from one cycle to another, or from one group of cycles to another, it is sufficient to consider each cycle as isothermal, and take account of the temperature in the model coefficients by treating them as variables dependent on the number of cycles.

On the other hand, if the temperature varies greatly during each cycle, the problem is more tricky. It can be solved by introducing:

(i) A reduced stress: $S = \sigma/\sigma_\mathrm{u}(T)$ where $\sigma_\mathrm{u}(T)$ is the ultimate tensile stress

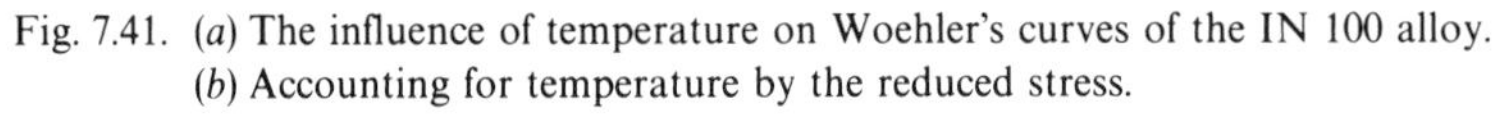

Fig. 7.41. (*a*) The influence of temperature on Woehler's curves of the IN 100 alloy. (*b*) Accounting for temperature by the reduced stress.

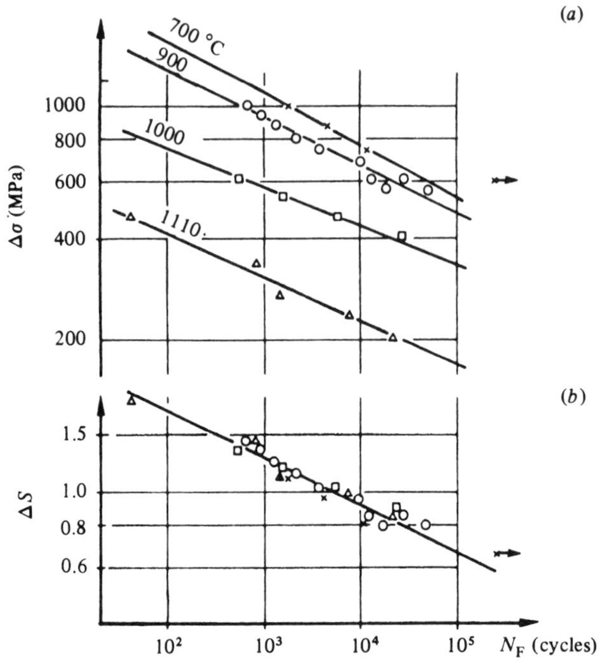

which depends on the temperature. Fig. 7.41 illustrates the case of the IN 100 alloy and shows quite well that a large part of the influence of the temperature is accounted for by this parameter. Note that a law expressed in terms of the plastic strain range (Coffin–Manson law described in Section 7.2.4) is also not much influenced by the temperature.

(ii) A mean equivalent temperature, as proposed by Taira, by linearly accumulating the temperature effects during a cycle. To define this equivalent temperature, we use a simplified (linearly accumulative) isothermally determined law such as:

$$\delta D = \frac{\delta N}{N_F(S_{\text{Max}}, \bar{S}, T)} = \frac{(S_{\text{Max}} - \bar{S})^{\beta(T)}}{A(\bar{S}, T)} \delta N$$

where S_{Max} and $\bar{S}$ are the maximum and mean values of S during the cycle. We then define the damage growth per cycle as:

$$\frac{\delta D}{\delta N} = \int_{\text{cycle}} \frac{\beta(T)}{2} \frac{\langle S - \bar{S} \rangle^{\beta(T)-1}}{A(\bar{S}, T)} \, \mathrm{d}S$$

where T depends on time, and therefore on S. For the isothermal case we recover:

$$\delta D / \delta N = 1 / N_F(S_{\text{Max}}, \bar{S}, T).$$

When the temperature varies, we obtain the equivalent temperature T^*, i.e., the temperature of an isothermal loading producing the same damage:

$$\frac{1}{N_F(S_{\text{Max}}, \bar{S}, T^*)} = \frac{1}{2} \int_{\bar{S}}^{S_{\text{Max}}} \frac{\beta(T(S)) \, \mathrm{d}S}{(S - \bar{S}) N_F(S, \bar{S}, T(S))}.$$

In practice, S_{Max} and $\bar{S}$ are known, as are the evolutions of S and T during the cycle. They can then be used to calculate the right hand side of the above equation. Assuming N_F to be a monotonic function of T, we can define the equivalent temperature T^*. It is this temperature which must be then used to define the coefficients of the damage law.

7.4.4 *Interaction effects of fatigue and creep damage*

At elevated temperatures, the processes of fatigue damage (due to cyclic loading) and creep damage (linked to the loading duration) can interact. The physical phenomena which occur at the microstructural scale are complex and the influence of the environment is not negligible. We limit ourselves here to a macroscopic description of these interaction effects through the damage growth models described in the preceding sections.

Accumulation of damages of different natures

The creep and fatigue damage models defined in Sections 7.4.2 and 7.4.3 consider separately two types of processes:

creep damage, D_c, characterized by intercrystalline defects, evolves as a function of time elapsed under stress; in the uniaxial case:

$$dD_c = f_c(\sigma, T, D_c, \ldots)\, dt,$$

fatigue damage D_F, for which the defects are generally initiated on the surface and progress across the crystals (transcrystalline); these evolve as functions of the loading cycles; in the uniaxial case:

$$\delta D_F = f_F(\sigma_{Max}, \bar{\sigma}, T, D_F, \ldots)\delta N.$$

When the two processes are present simultaneously, the interaction effects can be represented macroscopically by introducing the couplings:

$$dD_c = f_c(\sigma, T, D_c, D_F, \ldots)\, dt$$
$$\delta D_F = f_F(\sigma_{Max}, \bar{\sigma}, T, D_F, D_c, \ldots)\, \delta N.$$

How should D_F be introduced in the first equation and D_c in the second? In order to simplify and make possible a prediction from pure creep and pure fatigue, we are led to assume that the damage D_c and D_F have additive macroscopic effects:

$$dD_c = f_c(\sigma, T, D_c + D_F, \ldots)\, dt$$
$$\delta D_F = f_F(\sigma_{Max}, \bar{\sigma}, T, D_F + D_c, \ldots)\delta N$$

so that there is only one damage variable and its evolution can be described by:

$$dD = dD_c + dD_F = f_c(\sigma, T, D, \ldots)\, dt + f_F(\sigma_{Max}, \bar{\sigma}, T, D, \ldots)\, dN.$$

It is this form of the equation which will be used below.

Rule of linear accumulation and interaction

This rule was developed first by Robinson and then by Taira. It consists in adding together the creep and pure fatigue damages defined by the linear relation:

- $$dD = \frac{dt}{t_c(\sigma, T)} + \frac{dN}{N_F(\Delta\varepsilon, T)},$$

where $t_c(\sigma, T)$ is the time to rupture in creep under constant stress σ and temperature T and N_F is the number of cycles to failure in pure fatigue for a periodic loading of strain range $\Delta\varepsilon$ (in this method one uses $\Delta\varepsilon$ rather than the stress parameters).

The preceding relation when integrated over a cycle gives:

$$\delta D = \left[\int_0^{\Delta t} \frac{dt}{t_c(\sigma, T)} + \frac{1}{N_F(\Delta\varepsilon, T)}\right]\delta N$$

where Δt is the period of the cycle. When the load is periodic, the number of cycles to failure under combined fatigue–creep is found by integration to be:

$$\frac{1}{N_R} = \int_0^{\Delta t} \frac{dt}{t_c(\sigma, T)} + \frac{1}{N_F(\Delta\varepsilon, T)}.$$

If the temperature varies during the cycle, T in N_F is replaced by the equivalent temperature T^* defined in Section 7.4.3.

Although satisfactory in some cases, this method proves ineffective for many materials, in particular for the IN 100 alloy as shown in Fig. 7.42. This figure also shows the experimental values of the life ratios in fatigue

Fig. 7.42. Nonlinear interaction of fatigue and creep damages in the IN 100 alloy. Fatigue criterion expressed in terms of $\Delta\varepsilon$.

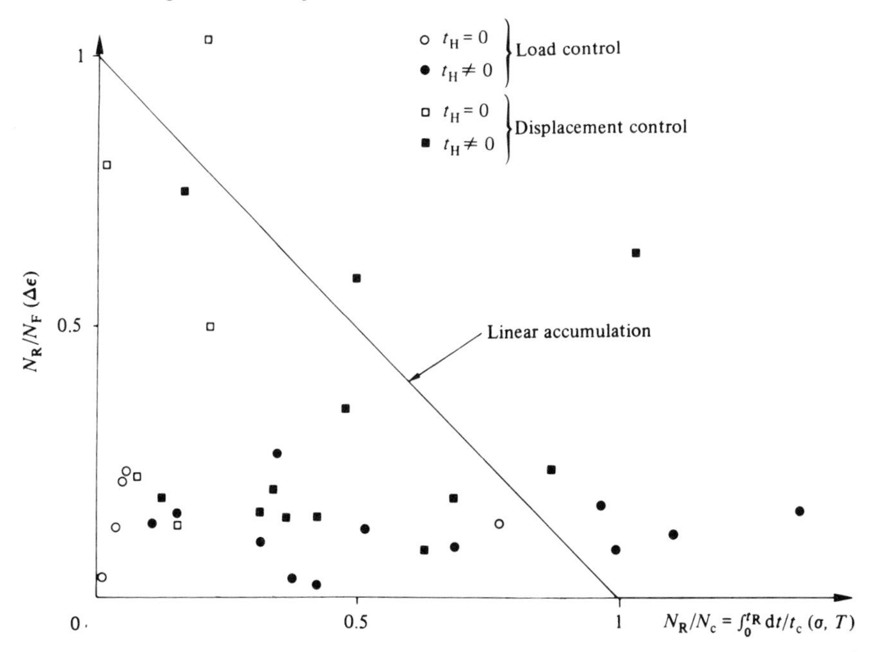

and creep. It should be noted that for materials which exhibit significant ductility in creep, it is necessary to perform geometric corrections; the creep tests must be interpreted in terms of the true stress as these data are then applied to cyclic tests in which the strains always remain small.

Model of nonlinear accumulation and interaction

The method presented here is derived directly from the damage equations with nonlinear accumulation:

for creep, Kachanov's law is used in which the characteristic coefficients are identified by pure creep tests;
for fatigue, the model presented in Section 7.4.3 is used in which damage evolution is highly nonlinear in accordance with the damage measurements.

These two models are coupled by the assumption of additivity of fatigue and creep damages. This enables us to take account of the following physical effects:

the presence of creep damage accelerates the growth of fatigue microcracks (the grain boundaries close to the tip of the crack are less resistant);
the presence of fatigue cracks increases the rate of growth of cavities by virtue of stress concentration effects.

In compression, the creep damage growth rate can be lower than in tension. This effect is incorporated by replacing σ by:

$$\chi(\sigma) = \langle c\sigma + (1-c)|\sigma| \rangle.$$

For $c = 0$, we have the symmetry property (compression identical to tension). In contrast, for $c = 1$ there is no creep damage in compression, since we then have $\chi = 0$. This is the application of the general criteria of Sections 7.2.5 and 7.4.2 to tension–compression.

Moreover, in the interest of simplicity, the exponent of Rabotnov's law is assumed to be constant. With these assumptions, the general evolution law under conditions of fatigue–creep interaction is written as:

$$\mathrm{d}D = \left[\frac{\chi(\sigma)}{A}\right]^r (1-D)^{-k}\,\mathrm{d}t + [1-(1-D)^{\beta+1}]^{\alpha(\sigma_{\mathrm{Max}},\bar{\sigma})}\left[\frac{\sigma_{\mathrm{Max}}-\bar{\sigma}}{M(\bar{\sigma})(1-D)}\right]^\beta \mathrm{d}N.$$

For life-times which are, at least, in the range of hundreds of cycles, as is the case during service, the damage variation during each isolated cycle may be neglected. We may thus simply integrate the effects of creep damage during a cycle and include them at the end of this cycle. Accordingly we obtain:

$$\delta D = \left\{ (1-D)^{-k} \int_0^{\Delta t} \left(\frac{\chi(\sigma)}{A} \right)^r \mathrm{d}t \right.$$
$$\left. + [1-(1-D)^{\beta+1}]^{\alpha(\sigma_{\mathrm{Max}},\bar{\sigma})} \left[\frac{\sigma_{\mathrm{Max}} - \bar{\sigma}}{M(\bar{\sigma})(1-D)} \right]^{\beta} \right\} \delta N$$

where Δt is the period of the cycle.

Denoting by N_{F} the number of cycles to failure for fatigue damage alone and by N_{c} the number for creep damage alone, we have:

$$\frac{1}{N_{\mathrm{c}}} = \int_0^{\Delta t} \frac{\mathrm{d}t}{t_{\mathrm{c}}(\chi(\sigma), T)} = (k+1) \int_0^{\Delta t} \left(\frac{\chi(\sigma)}{A} \right)^r \mathrm{d}t$$

$$\frac{1}{N_{\mathrm{F}}} = (\beta+1)[1-\alpha(\sigma_{\mathrm{Max}}, \bar{\sigma})] \left[\frac{\sigma_{\mathrm{Max}} - \bar{\sigma}}{M(\bar{\sigma})} \right]^{\beta}.$$

By substituting these expressions into the preceding equation, we are led to the following relation:

- $$\delta D = \left\{ \frac{(1-D)^{-k}}{(k+1)N_{\mathrm{c}}} + \frac{[1-(1-D)^{\beta+1}]^{\alpha(\sigma_{\mathrm{Max}},\bar{\sigma})}}{(\beta+1)N_{\mathrm{F}}[1-\alpha(\sigma_{\mathrm{Max}},\bar{\sigma})](1-D)^{\beta}} \right\} \delta N.$$

The number of cycles to failure N_{R} is found by integrating for D from 0 to 1, and is expressed by:

$$\frac{N_{\mathrm{R}}}{N_{\mathrm{F}}} = \int_0^1 \left(\frac{N_{\mathrm{F}}}{N_{\mathrm{c}}} \frac{(1-D)^{-k}}{k+1} + \frac{[1-(1-D)^{\beta+1}]^{\alpha(\sigma_{\mathrm{Max}},\bar{\sigma})}}{(\beta+1)[1-\alpha(\sigma_{\mathrm{Max}},\bar{\sigma})](1-D)^{\beta}} \right)^{-1} \mathrm{d}D.$$

The calculation of N_{R}, the number of cycles to theoretical fracture is performed in two steps:

calculation of N_{c} and N_{F}, the numbers of cycles to failure in creep alone and fatigue alone respectively. In tests under displacement control, evolution of the stress cycle is taken into account until stable conditions are reached;

knowing the maximum and the mean stress, $N_{\mathrm{F}}/N_{\mathrm{c}}$, and the coefficients r, k, β, we can easily calculate $N_{\mathrm{R}}/N_{\mathrm{F}}$ by performing the above integration which gives the value of N_{R}.

Comparison between predictions and experimental results is made on interaction diagrams; Fig. 7.43 gives an example of these for stress-

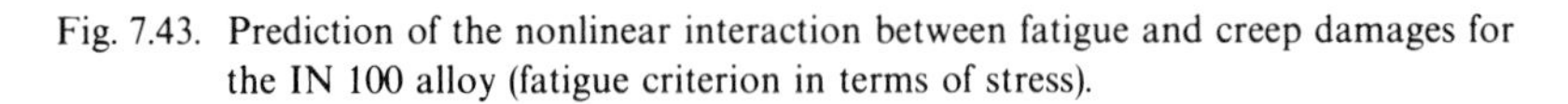

Fig. 7.43. Prediction of the nonlinear interaction between fatigue and creep damages for the IN 100 alloy (fatigue criterion in terms of stress).

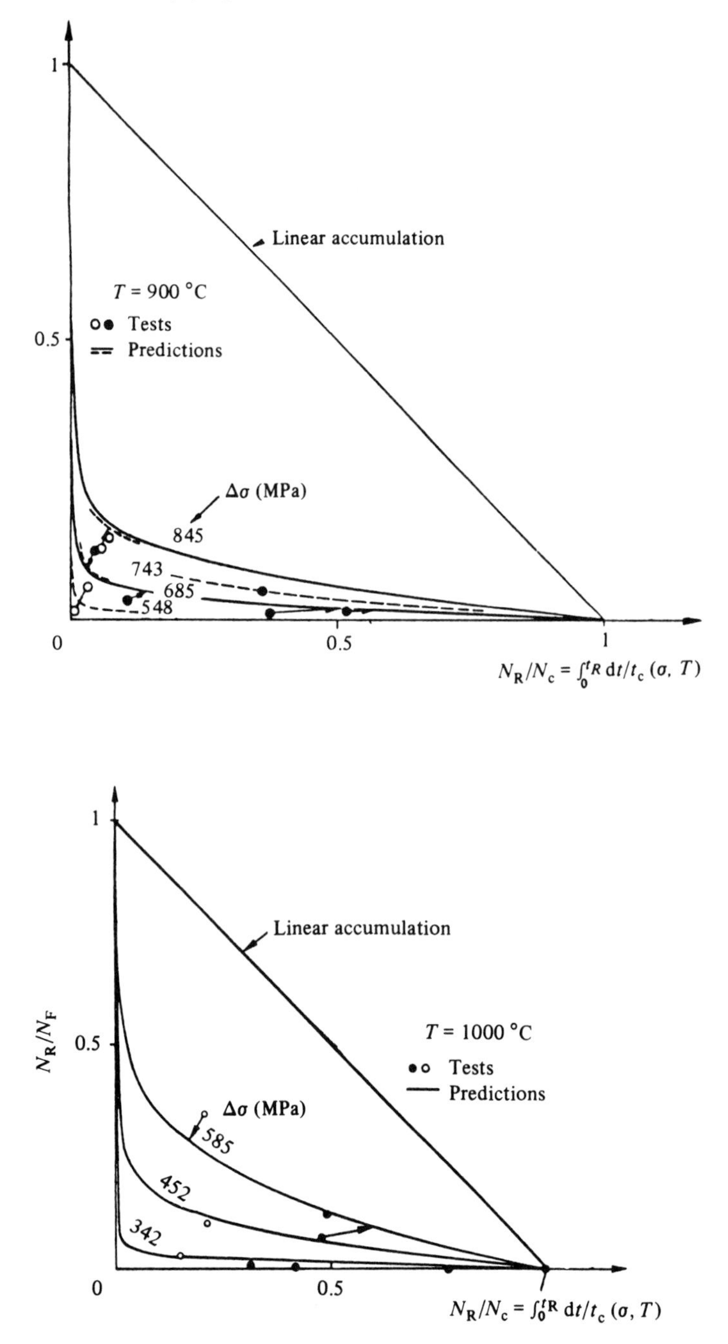

controlled tests. The experimental points correspond to the measured values of N_R divided by the calculated values of N_c and N_F.

For IN 100 and for many other materials, the nonlinearity of the interaction increases as the stress amplitude decreases, as predicted correctly by the theory, particularly when comparing to the prediction by the rule of linear accumulation.

Fig. 7.44 sums up the set of comparisons between predictions and the results of reference tests (pure fatigue and pure creep), stress and strain-controlled viscoplastic fatigue tests, viscoplastic tests at two temperature levels, pure fatigue tests at two stress levels, creep tests after fatigue (see

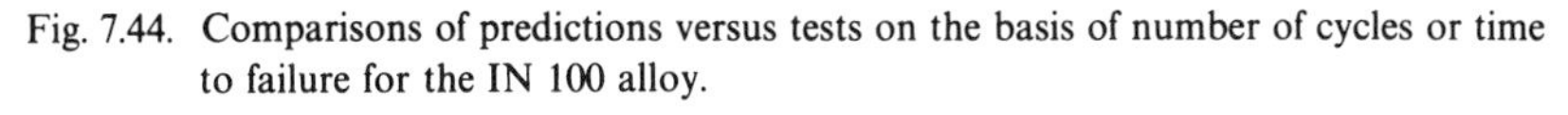
Fig. 7.44. Comparisons of predictions versus tests on the basis of number of cycles or time to failure for the IN 100 alloy.

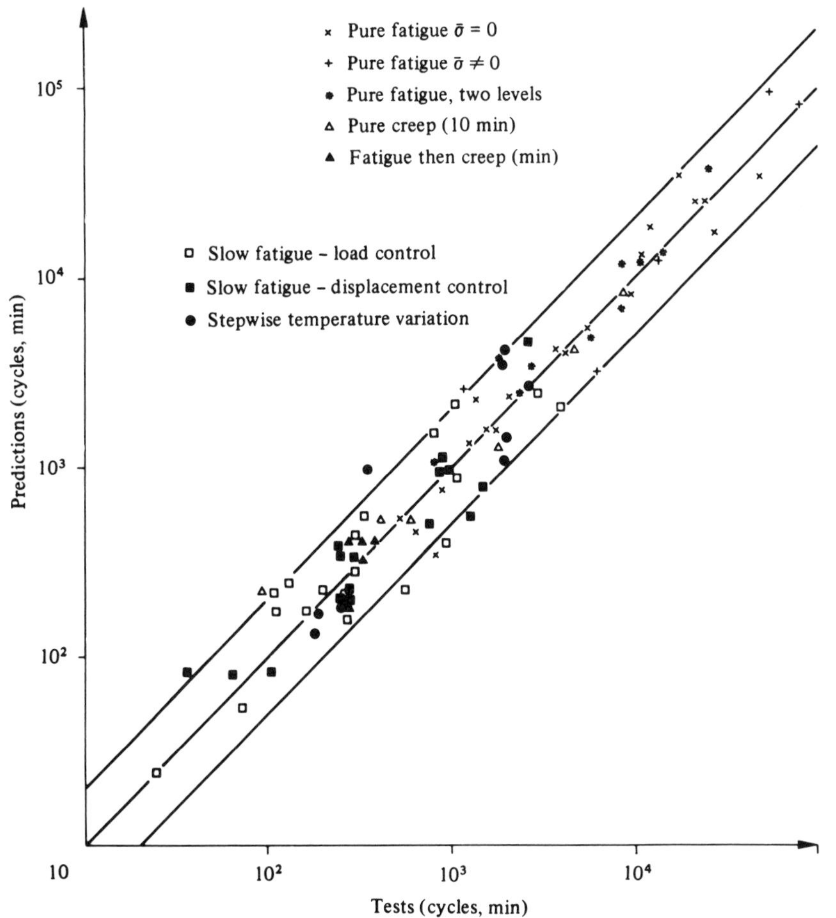

below), and thermomechanical fatigue tests. The differences vary within a factor of 2, which constitutes a good verification of the model as the rule of linear accumulation can lead to errors higher than a factor of 10.

The validity of this modelling derives from some well-known facts for this type of refractory alloy:

> creep damage evolves much faster than fatigue damage (relative to nominal life-time): this difference between the rates of evolution is correctly described by the chosen models;
> there is no cyclic consolidation effect: the effect of hardening on the damage evolution is negligible.

The first remark implies the existence of test conditions for which the effect of nonlinear accumulation is observed in the opposite sense, i.e., in a sense beneficial with respect to the rule of linear accumulation. This represents tests started in fatigue for N_1 cycles followed by creep for a duration t_c until rupture; we must then have:

$$\frac{N_1}{N_F} + \frac{t'_c}{t_c} > 1$$

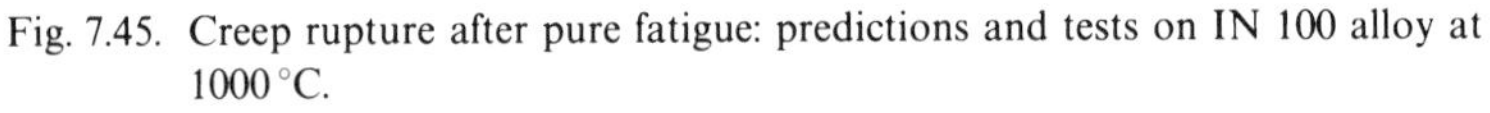

Fig. 7.45. Creep rupture after pure fatigue: predictions and tests on IN 100 alloy at 1000 °C.

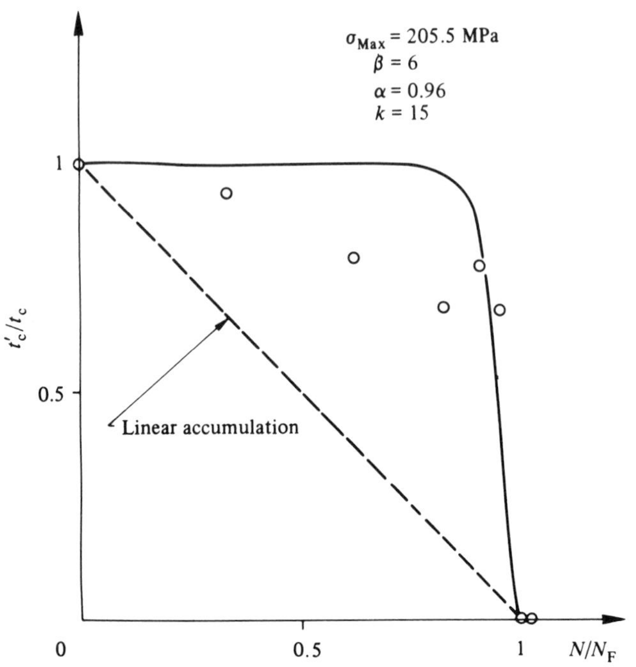

where N_F and t_c are respectively the number of cycles to failure in fatigue and the time to rupture in creep, corresponding to the applied stress.

Such tests have been conducted on the IN 100 alloy at 1000 °C and the predicted effect is indeed observed as can be seen from Fig. 7.45. The fatigue and creep damage models can be used for a more accurate prediction. In fatigue, after N_1 cycles, we have:

$$[1-(1-D_1)^{\beta+1}]^{1-\alpha}=N_1/N_F$$

where α denotes the value of the function $\alpha(\sigma_{Max},\bar{\sigma})$ corresponding to the maximum applied stress ($\bar{\sigma}=0$). The integration of the creep damage law between D_1 and 1 leads to:

$$(1-D_1)^{k+1}=t'_c/t_c$$

where k is the value of $k(\sigma)$. Hence we can deduce the relation predicting the time to rupture in creep as a function of the length of fatigue precycling:

$$\frac{t'_c}{t_c}=\left[1-\left(\frac{N_1}{N_F}\right)^{1/(1-\alpha)}\right]^{(k+1)/(\beta+1)}.$$

This relation, used with $\beta=6$, $\alpha=0.96$, $k=15$, which correspond to the loading under consideration, satisfactorily reproduces the experimental results considering the relatively significant dispersion among the creep rupture tests (Fig. 7.10).

7.5 Deformation and damage coupling

By deformation and damage coupling, we refer to the fact that the solution of a problem in terms of stresses and strains at the limit states depends on the state of damage of the structure under consideration. Generally, as we have seen, damage reduces the stiffness and strength of materials. For a given state of stresses, the larger are the strains the higher is the damage, and hence we recognize the importance of 'coupled' calculations for evolution problems in which stresses, strains and damage are calculated simultaneously.

The concept of effective stress, associated to the principle of strain equivalence, enables us to write the laws of coupled behaviour very simply since it is sufficient to replace the usual stress $\boldsymbol{\sigma}$ by the effective stress $\tilde{\boldsymbol{\sigma}}$.

We introduce here some basic elements of the general approach and apply them to a few examples of constitutive equations. For simplicity, we assume that the damage is isotropic.

7.5.1 *Elasticity coupled with damage*

This is important for materials for which plasticity effects are not very significant. A typical example is concrete in which nonlinear deformations, which are sometimes irreversible, are produced by the process of decohesion. This process may be analysed using a damage theory.

The theoretical study of this case issues directly from the choice of the effective stress. For isotropic elasticity and isotropic damage in the initial conditions, the elasticity law of the damaged material is written as:

$$\boldsymbol{\varepsilon} = \mathbf{A}:\tilde{\boldsymbol{\sigma}}$$

or

$$\bullet \qquad \boldsymbol{\varepsilon} = \frac{1+\nu}{E}\frac{\boldsymbol{\sigma}}{1-D} - \frac{\nu}{E}\frac{\mathrm{Tr}(\boldsymbol{\sigma})}{1-D}\mathbf{1}.$$

This relation must be coupled with the chosen damage law.

Let us take a very schematic example, specialized to uniaxial tension, by assuming that D evolves with (elastic) deformations and introducing a variable threshold ε_D:

$$\mathrm{d}D = \begin{cases} (\varepsilon/\varepsilon_0)^{s^*} & \text{when } \varepsilon = \varepsilon_D \quad \text{and} \quad \mathrm{d}\varepsilon = \mathrm{d}\varepsilon_D > 0 \\ 0 & \text{when } \varepsilon < \varepsilon_D \quad \text{or} \quad \mathrm{d}\varepsilon < 0. \end{cases}$$

By taking the initial conditions to be $D = \varepsilon_D = 0$, and integrating up to fracture, $D = 1$, we obtain:

$$D = (\varepsilon/\varepsilon_R)^{s^*+1} \qquad \sigma = E\varepsilon[1 - (\varepsilon/\varepsilon_R)^{s^*+1}]$$

where $\varepsilon_R = [(s^* + 1)\varepsilon_0^{s^*}]^{1/(s^*+1)}$ is the fracture strain. Fig. 7.46(*a*) illustrates the stress evolution obtained from such a theory ($s^* = 2$) of elasticity coupled with damage. During an eventual unloading–loading we again find linear elastic behaviour but with a lower elasticity modulus. The results obtained are similar to the experimental observations as shown in Fig. 7.46(*b*).

7.5.2 *Plasticity coupled with damage*

General formulation

In order to model the behaviour of a damaged and hardened plastic material, the strain equivalence principle is once more applied. The stress $\boldsymbol{\sigma}$ is replaced by the effective stress $\tilde{\boldsymbol{\sigma}}$ in the dissipation potential, all other variables remain unchanged (see Chapter 5). The associated variable of

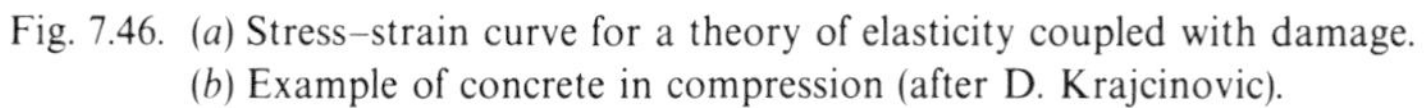
Fig. 7.46. (*a*) Stress–strain curve for a theory of elasticity coupled with damage. (*b*) Example of concrete in compression (after D. Krajcinovic).

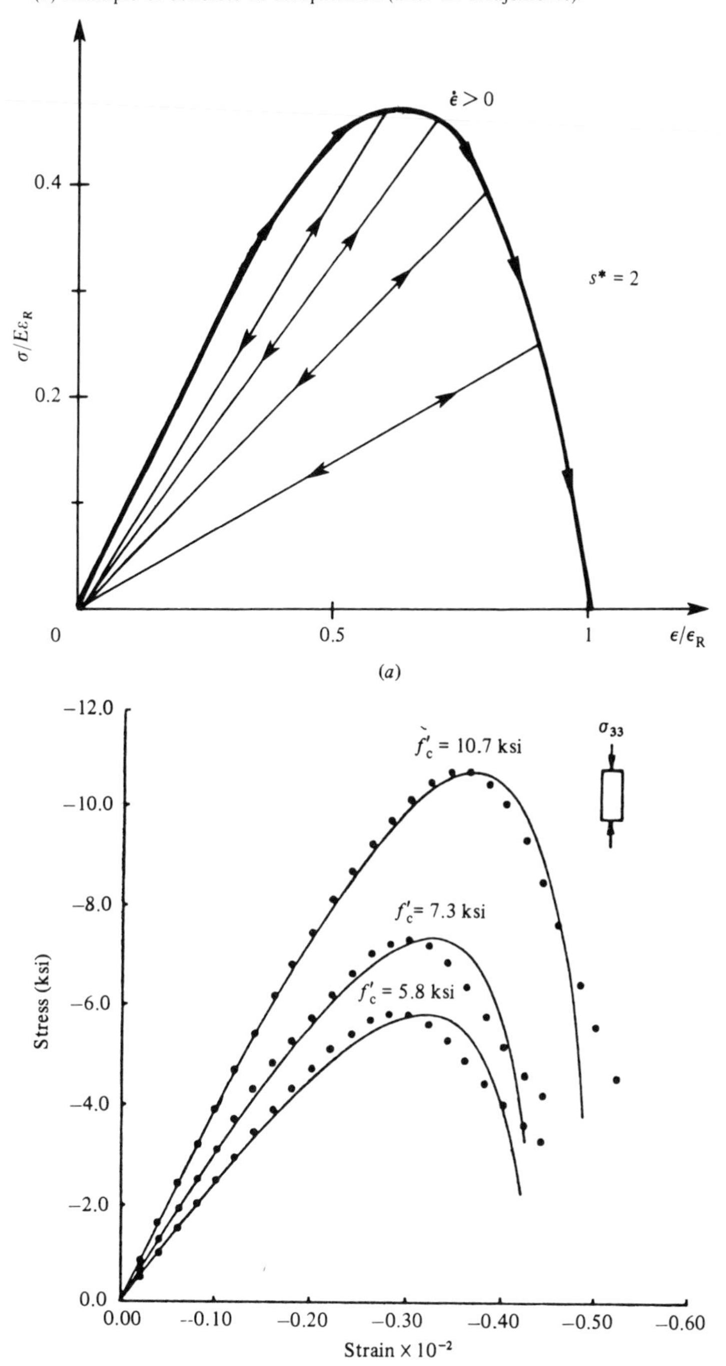

isotropic strain hardening remains R while the corresponding internal variable changes and is now denoted as r.

$$\boldsymbol{\sigma} \to \tilde{\boldsymbol{\sigma}} = \frac{\boldsymbol{\sigma}}{1-D}$$

$$\mathbf{X} \to \mathbf{X}$$

$$R \to R$$

Thus, the elastic domain is expressed by:

$$f = J_2(\tilde{\sigma} - \mathbf{X}) - R - k = J_2\left(\frac{\boldsymbol{\sigma}}{1-D} - \mathbf{X}\right) - R - k \leqslant 0.$$

The flow law and the evolutions of internal and damage variables are obtained by the law of normal dissipativity, as in the general case.

The dissipation potential φ^* is decomposed into:

$$f(\tilde{\boldsymbol{\sigma}}, \mathbf{X}, R) \qquad \text{and} \qquad \varphi_{\mathrm{D}}^*(Y; \boldsymbol{\varepsilon}^{\mathrm{e}}, \boldsymbol{\alpha}, r, D),$$

the first term corresponds to plastic dissipation, the second to dissipation by damage, assumed to be decoupled from the first.

The generalized normality rule provides the flow law and the evolutions of the internal variables:

$$\dot{\boldsymbol{\varepsilon}}^{\mathrm{p}} = \dot{\lambda}(\partial f / \partial \boldsymbol{\sigma})$$

$$\dot{\boldsymbol{\alpha}} = -\dot{\lambda}(\partial f / \partial \mathbf{X})$$

$$\dot{r} = -\dot{\lambda}(\partial f / \partial R).$$

$$\dot{D} = -\dot{\lambda}(\partial \varphi_{\mathrm{D}}^* / \partial Y).$$

The plastic multiplier $\dot{\lambda}$ is then determined by the plastic flow consistency condition: $\dot{f} = f = 0$ (see Chapter 5). Only the case of isotropic hardening will be studied below.

Coupling between isotropic ductile plastic damage and isotropic hardening

Assuming that the material obeys the von Mises criterion, the elastic domain is now expressed by $f = \sigma_{\mathrm{eq}}/(1-D) - R - k \leqslant 0$; φ_{D}^* depends only on Y and D. We, therefore, have:

$$\dot{\boldsymbol{\varepsilon}}^{\mathrm{p}} = \frac{3}{2}\frac{\dot{\lambda}}{1-D}\frac{\tilde{\boldsymbol{\sigma}}'}{\tilde{\sigma}_{\mathrm{eq}}} = \frac{3}{2}\frac{\dot{\lambda}}{1-D}\frac{\boldsymbol{\sigma}'}{\sigma_{\mathrm{eq}}}$$

$$\dot{r} = \dot{\lambda} = (\tfrac{2}{3}\dot{\boldsymbol{\varepsilon}}^{\mathrm{p}} : \dot{\boldsymbol{\varepsilon}}^{\mathrm{p}})^{1/2}(1-D) = \dot{p}(1-D)$$

which represents the corrected accumulated plastic strain rate by virtue of irreversible strain due to damage, assuming that the given hardening law of the material without damage is:

$$R = \rho \partial \Psi(r)/\partial r = R(r).$$

$\dot{\lambda}$ or $\dot{r}$ is expressed by the consistency condition:

$$\dot{f} = \frac{\partial f}{\partial \boldsymbol{\sigma}} : \dot{\boldsymbol{\sigma}} + \frac{\partial f}{\partial R}\dot{R} + \frac{\partial f}{\partial D}\dot{D} = 0$$

or

$$\frac{3}{2}\frac{\boldsymbol{\sigma}' : \dot{\boldsymbol{\sigma}}}{\sigma_{\mathrm{eq}}(1-D)} - R'(r)\dot{r} + \frac{\sigma_{\mathrm{eq}}}{(1-D)^2}\dot{D} = 0.$$

Then, with:

$$\dot{r} = \dot{\lambda}, \frac{\sigma_{\mathrm{eq}}}{1-D} - R = k \quad \text{and} \quad \dot{D} = -\frac{\partial \varphi_{\mathrm{D}}^*}{\partial Y}\dot{\lambda}$$

we get:

$$\dot{\lambda} = \frac{\dot{\sigma}_{\mathrm{eq}}}{(1-D)R'(r) + [k + R(r)]\partial \varphi_{\mathrm{D}}^*/\partial Y}.$$

The loading–unloading criterion is expressed by:

$$\dot{\lambda} = 0 \quad \text{if} \quad \dot{\sigma}_{\mathrm{eq}} \leqslant 0$$
$$\dot{\lambda} > 0 \quad \text{if} \quad \dot{\sigma}_{\mathrm{eq}} > 0.$$

Moreover, there is flow only if $f = 0$. Denoting the hardening modulus by $h(r, D, Y)$ we have:

$$\dot{p} = \frac{\dot{\lambda}}{1-D} = H(f)\frac{\langle \dot{\sigma}_{\mathrm{eq}} \rangle}{h(r, D, Y)}$$

with:

$$h(r, D, Y) = (1-D)^2 R'(r) + (1-D)[k + R(r)]\partial \varphi_{\mathrm{D}}^*/\partial Y.$$

Since $\partial \varphi_{\mathrm{D}}^*/\partial Y$ is negative, this hardening modulus may be negative. In fact it is positive when the strain is small and $R'(r)$ is large, and it becomes negative at the instability point:

$$h(r, D, Y) = 0.$$

In this case, the above description is no longer applicable and we must use the loading–unloading condition with $\dot{\sigma}_{\mathrm{eq}}$.

Thus, the Prandtl–Reuss equations in which the hardening law appears in the form of a power function

$$R(p) = K_Y p^{1/M_Y} = \sigma_{eq} - \sigma_Y$$

can now be written

- $$d\boldsymbol{\varepsilon} = d\boldsymbol{\varepsilon}^e + d\boldsymbol{\varepsilon}^p$$
- $$d\boldsymbol{\varepsilon}^e = \frac{1+\nu}{E}\frac{d\boldsymbol{\sigma}}{1-D} - \frac{\nu}{E}\frac{d(\mathrm{Tr}(\boldsymbol{\sigma}))}{1-D}\mathbf{1}$$
- $$d\boldsymbol{\varepsilon}^p = \tfrac{3}{2}H(f)\frac{\langle d\sigma_{eq}\rangle}{\dfrac{K_Y(1-D)^2}{M_Y}\left\langle \dfrac{\sigma_{eq}}{K_Y(1-D)} - \dfrac{\sigma_Y}{K_Y}\right\rangle^{1-M_Y} + \sigma_{eq}\dfrac{\partial \varphi_D^*}{\partial Y}}\frac{\boldsymbol{\sigma}'}{\sigma_{eq}}.$$

To these, the corresponding damage law should be adjoined (see Section 7.4.1):

$$\dot{D} = -\frac{\partial \varphi_D^*}{\partial Y}\dot{\lambda} = -\frac{\partial \varphi_D^*}{\partial Y}\dot{p}(1-D)$$

or

- $$dD = \frac{D_c}{\varepsilon_R - \varepsilon_D}\left[\tfrac{2}{3}(1+\nu) + 3(1-2\nu)\left(\frac{\sigma_H}{\sigma_{eq}}\right)^2\right]dp$$

and

$$\frac{\partial \varphi^*}{\partial Y} = -\frac{D_c}{(\varepsilon_R - \varepsilon_D)(1-D)}\left[\tfrac{2}{3}(1+\nu) + 3(1-2\nu)\left(\frac{\sigma_H}{\sigma_{eq}}\right)^2\right].$$

7.5.3 *Viscoplasticity coupled with damage*

Assuming that the attentive [*sic* !] reader has understood the method presented in Sections 7.5.1 and 7.5.2 well, we immediately write the coupled isotropic viscoplasticity law studied in Section 6.4.2:

- $$\dot{\boldsymbol{\varepsilon}} = \dot{\boldsymbol{\varepsilon}}^e + \dot{\boldsymbol{\varepsilon}}^p,$$
- $$\dot{\boldsymbol{\varepsilon}}^e = \frac{1+\nu}{E}\frac{\dot{\boldsymbol{\sigma}}}{1-D} - \frac{\nu}{E}\frac{\mathrm{Tr}(\dot{\boldsymbol{\sigma}})}{1-D}\mathbf{1},$$
- $$\dot{\boldsymbol{\varepsilon}}^p = \tfrac{3}{2}\dot{p}\frac{\boldsymbol{\sigma}'}{\sigma_{eq}}, \quad \dot{r} = \dot{p}(1-D),$$
- $$\dot{r} = \frac{1}{1-D}\left[\frac{\sigma_{eq}}{(1-D)Kr^{1/M}}\right]^N.$$

As with plasticity, the relation $\dot{r} = \dot{p}(1-D)$ comes from the choice of a viscoplastic potential of the form:

$$\varphi^* = \Omega = \frac{K}{N+1}\left\langle \frac{J_2(\tilde{\boldsymbol{\sigma}}) - R + h'(r)}{K} \right\rangle^{N+1} r^{-N/M}$$

and from the normality rule:

$$\dot{\boldsymbol{\varepsilon}}^{\mathrm{p}} = \partial\Omega/\partial\boldsymbol{\sigma}.$$

Only creep damage can occur simultaneously with viscoplasticity:

- $\dot{D} = \langle \chi(\boldsymbol{\sigma})/A \rangle^r (1-D)^{-k\langle \chi(\boldsymbol{\sigma})\rangle}.$

In applications, we may consider that the damage is neglected during primary creep and that strain-hardening is saturated during tertiary creep. For example, in tension:

- during primary creep:

$$\varepsilon_{\mathrm{p}} < \varepsilon_{\mathrm{p}}^*, \quad D = 0 \rightarrow \qquad \varepsilon_{\mathrm{p}} = \left[\frac{N+M}{M}\left(\frac{\sigma}{K}\right)^N t \right]^{M/(N+M)},$$

- during tertiary creep: $\varepsilon_{\mathrm{p}} \geqslant \varepsilon_{\mathrm{p}}^*$, $Kr^{1/M}$ is replaced by $K\varepsilon_{\mathrm{p}}^{*\,1/M}$ first, the damage equation then gives

$$D = 1 - \left(1 - \frac{t}{t_{\mathrm{c}}}\right)^{1/(k+1)} \qquad \text{with} \qquad t_{\mathrm{c}} = \frac{1}{k+1}\left(\frac{\sigma}{A}\right)^{-r}.$$

Then, the constitutive equation is written as:

$$\mathrm{d}\varepsilon_{\mathrm{p}} = \left(1 - \frac{t}{t_{\mathrm{c}}}\right)^{-(N+1)/(k+1)} \left(\frac{\sigma}{K}\right)^N \mathrm{d}t.$$

After integrating for $\sigma = $ constant, we find:

$$\varepsilon_{\mathrm{p}} = \varepsilon_{\mathrm{p}}^* + (\varepsilon_{\mathrm{pR}} - \varepsilon_{\mathrm{p}}^*)\left[1 - \left(\frac{1 - t/t_{\mathrm{c}}}{1 - t^*/t_{\mathrm{c}}}\right)^{(k-N)/(k+1)}\right]$$

$$\varepsilon_{\mathrm{pR}} = \varepsilon_{\mathrm{p}}^* + \frac{k-1}{k-N}\left(\frac{\sigma}{K\varepsilon_{\mathrm{p}}^{*\,1/M}}\right)^N t_{\mathrm{c}}\left(1 - \frac{t^*}{t_{\mathrm{c}}}\right)^{(k-N)/(k+1)}$$

with

$$t^* = \frac{M}{N+M}\left(\frac{\sigma}{K}\right)^{-N} \varepsilon_{\mathrm{p}}^{*(N+M)/M}$$

where the exponents are usually ordered as follows:

$$r \leqslant N \leqslant k.$$

Hence, it follows that the rupture strain is a decreasing function of the stress applied during creep. This is generally well confirmed by experiment. Fig. 7.47 shows good agreement for tertiary creep as well as for the rupture strain for IN 100 superalloy.

7.6 Prediction of crack initiation in structures

In this section, we present some basic elements on the use of damage models for predicting crack initiation. The use of these models to foresee the growth of cracks within the framework of the local approach will be discussed in Section 8.7.

7.6.1 *Initial damage*

The first problem that arises in the integration of differential damage models is that of the initial conditions. The processes of cold-working, forming, machining of mechanical parts, etc. can leave an initial damage, denoted by D_0, which must be taken into account in fracture prediction.

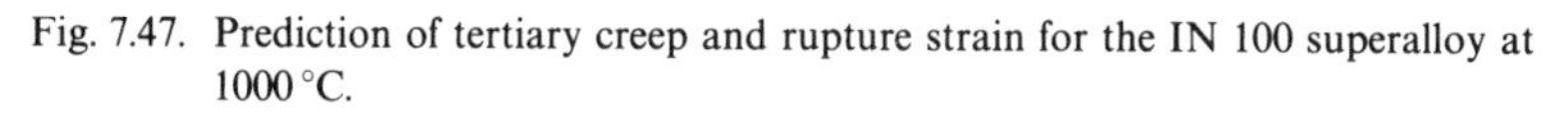

Fig. 7.47. Prediction of tertiary creep and rupture strain for the IN 100 superalloy at 1000 °C.

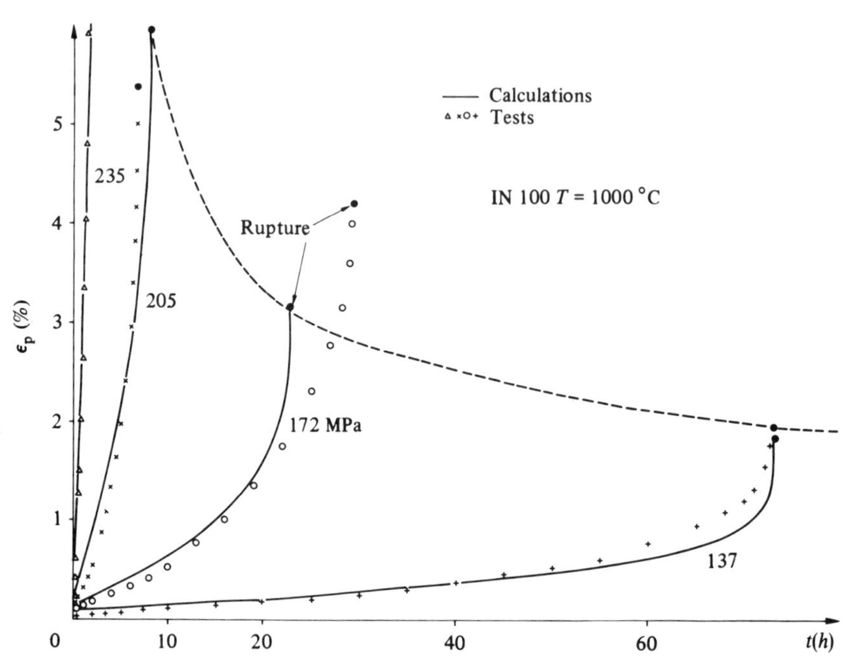

To write $D = D_0$ at $t = 0$ or $N = 0$ is easy enough! But to determine the value of the initial damage D_0 is not so. Ultrasonic methods can provide some indications, but, usually, it is necessary to resort to an assessment based on indirect information.

7.6.2 *Calculation of damage at critical points*

Classical method without coupling

In a number of applications, we may consider the influence of damage on the deformation behaviour to be weak and assume decoupling. In fatigue, for example, in the restricted deformation range, this is an acceptable assumption: the coupling effect does not come into play, as we have seen (in Section 7.2.3), until very late when one or several microcracks have already been initiated. In any case this is the simplest assumption, and is commonly made in most applications.

Calculation of fracture initiation is then performed in two steps:

(1) Calculation of the stress–strain response in the structure by the finite element method (see Chapters 4, 5 and 6 for a brief description of this method in elasticity, plasticity and viscoplasticity). Under enforced loading, assumed to be known, we have to solve a boundary value problem, which is made complex in many applications by the effects of plastic or viscoplastic flows. For cyclic loadings, we try to obtain the solution at the so-called stabilized cycle, that is to say for a constitutive equation of stable cyclic behaviour (the effects of deconsolidation or of cyclic hardening are not accounted for in this case), and for stabilized stress redistributions. To meet this last requirement, calculations have to be done successively for many cycles in order to take account of the stress redistribution in the structure.

(2) Knowing the stresses and strains at any point of the structure, we are able to determine the critical point or points (where there is a danger of a crack initiating) and a life-time of the structure, i.e., the time or the number of cycles corresponding to crack initiation. This step requires the use of an initiation criterion which may be derived from the damage evolution law. Generally it is sufficient to use a direct relation between the stresses and/or strains, calculated in the stabilized regime, and a time or a number of cycles to failure. In fatigue, for example, we will use:

$$N_R = N_F(\sigma_{Max}, \bar{\sigma}, T).$$

An additional problem arises when the loads are neither constant nor periodic. For certain applications, it is necessary to take into consideration the real service loads, which are usually idealized; but this involves consideration of the effects of accumulated damage. Staying within the context of a noncoupled theory, the structure is analysed independently for each loading, and the adopted rule is used to sum the damages which are obtained independently for each loading. For example, for the rule of linear accumulation in fatigue under a loading programme of repeated blocks (see Fig. 7.48), we have:

$$n_R \sum_{i=1}^{m} (N_i/N_{Fi}) = 1$$

where n_R is the number of sequences before initiation, N_i the number of cycles in each block and N_{F_i} is the number of cycles to failure corresponding to the periodic stresses and/or strains equal to those calculated for the ith loading condition.

Example of crack initiation calculation in a turbine blade

This involves the blades of a high pressure turbine of an aeroplane engine cooled by internal circulation of compressed air through ducts or cavities (Fig. 7.49). The material is the Nickel based IN 100 superalloy. It is protected against oxydization by an aluminizing process. The main loads are those from the start–stop cycles of the engine:

forces of centrifugal origin,

Fig. 7.48. Block loading programmes.

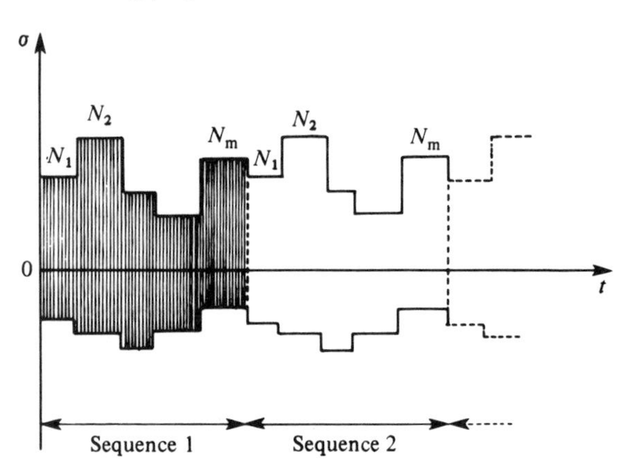

cyclic temperature changes producing thermal stresses responsible for 'thermal fatigue'.

The method described above for predicting crack initiation is applied as follows:

use of a cyclic viscoplasticity law with nonlinear kinematic hardening (Section 6.4.3) in the stabilized regime;

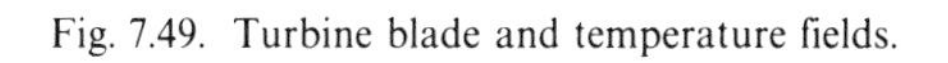
Fig. 7.49. Turbine blade and temperature fields.

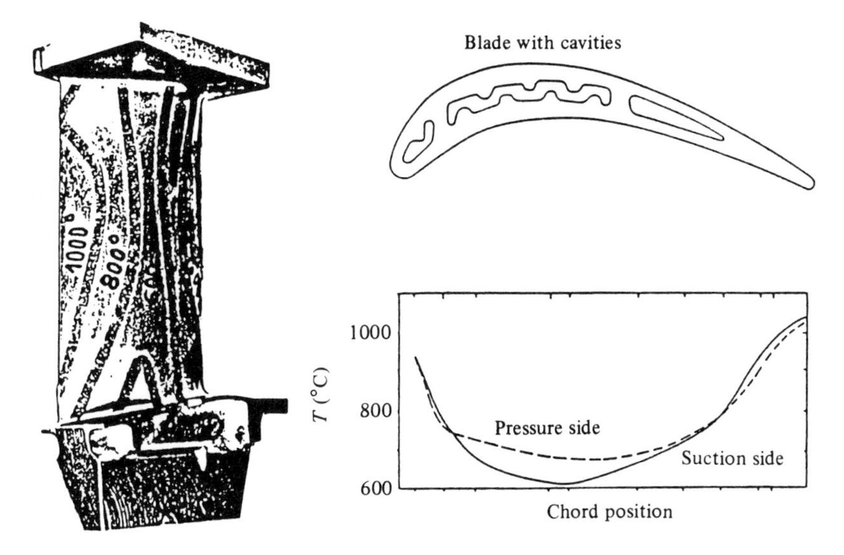

Fig. 7.50. Schematic illustration of the plane cross-section method.

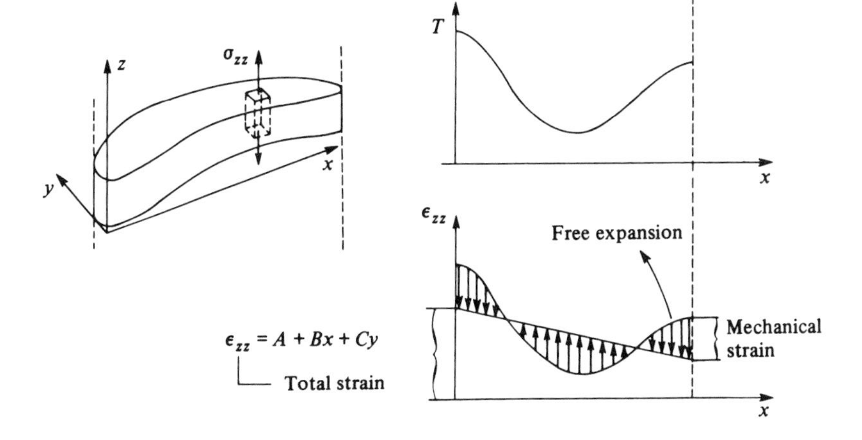

calculation of the stresses and strains by a method of plane cross-sections and uniaxial stresses, permitting three degrees of freedom (one of translation and two of bending). Fig. 7.50 shows schematically how the free expansion of the elements is inhibited, thus resulting in thermal stresses and elastoviscoplastic strains. The effects of stress redistribution due to viscoplasticity and successive cycles are taken into account by a step-by-step calculation as indicated in Section 6.5.2;

calculation of the number of cycles to crack initiation at each point, the critical initiation point obviously being that which corresponds to the minimum number of cycles to failure. These calculations are done with stresses and strains obtained at the first, second,..., sixth or seventh cycle in order to extrapolate the number of cycles at initiation, taking into account the stress redistribution at each cycle (Fig. 7.51). The nonlinear fatigue–creep interaction effects are incorporated by the models described in Section 7.4.4.

The predictions have been confirmed by experimental observations in practical laboratory tests on real blades under loadings close to those experienced by engines in service. Similar results have been obtained for plates under thermal gradients and for different types of anisothermal tension–compression tests, all with the same choice of coefficients (Fig. 7.51(*b*)).

This method is commonly used at the Société SNECMA in the final dimensioning phase of the turbine blades, with the help of the program

Fig. 7.51. (*a*) Prediction of number of cycles to crack initiation as function of the number of calculation cycles. (*b*) Comparison with tests.

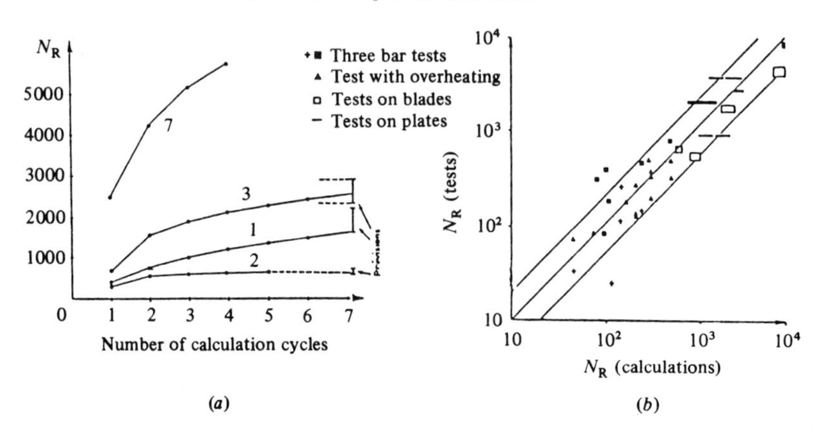

'CALIFAT' (CALcul d'Initiation des Fissures dans les Aubes de Turbine). The finite element method allows its generalization for use in other critical parts of the engines (discs, combustion chambers) and in other industrial sectors (nuclear industry, automobile industry, etc.).

Taking account of deformation–damage coupling

This effect can be important for certain materials or certain types of loading insofar as the calculated stresses and strains are modified by the introduction of such a coupling. Two examples are briefly described.

Creep damage

The influence of creep damage on viscoplastic behaviour is important. It becomes apparent, as we have seen, in tertiary creep. Even if a structure is loaded by specified forces, the influence of tertiary creep can result in an additional stress redistribution.

Such a calculation is not easy, even under a constant load. Fig. 7.52 shows the example of a notched plate with the calculation performed in plane stress by taking this coupling into consideration. The equations are those of perfect viscoplasticity (without hardening) and of isotropic damage specialized by:

$$\dot{\boldsymbol{\varepsilon}}^{\mathrm{p}} = \frac{3}{2}\left[\frac{\sigma_{\mathrm{eq}}}{\lambda^*(1-D)}\right]^{N^*}\frac{\boldsymbol{\sigma}'}{\sigma_{\mathrm{eq}}}, \quad \dot{D} = \left[\frac{\sigma_{\mathrm{eq}}}{A(1-D)}\right]^{r}.$$

Fig. 7.52. Damaged zone in a notched plate (after D. Hayhurst).

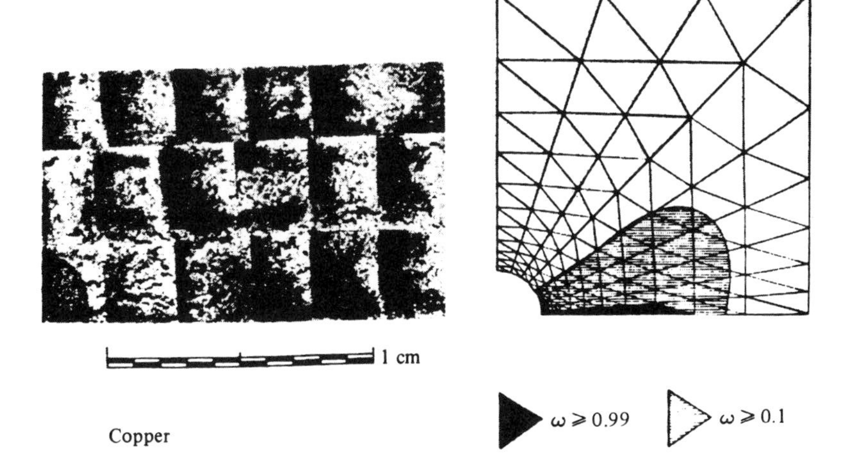

The calculation is done by the finite element method applied to viscoplasticity (see Chapter 6). The stiffness matrix of the most loaded elements is periodically updated to take into account the reduction of elastic moduli. The equations are integrated step-by-step in time.

Damage in concrete

Concrete is a material whose nonlinear behaviour in tension can be represented by coupling between the linear elastic law and a deformation

Fig. 7.53. Evolution of damage at the most critical point of a concrete beam loaded in bending (after J. Mazars).

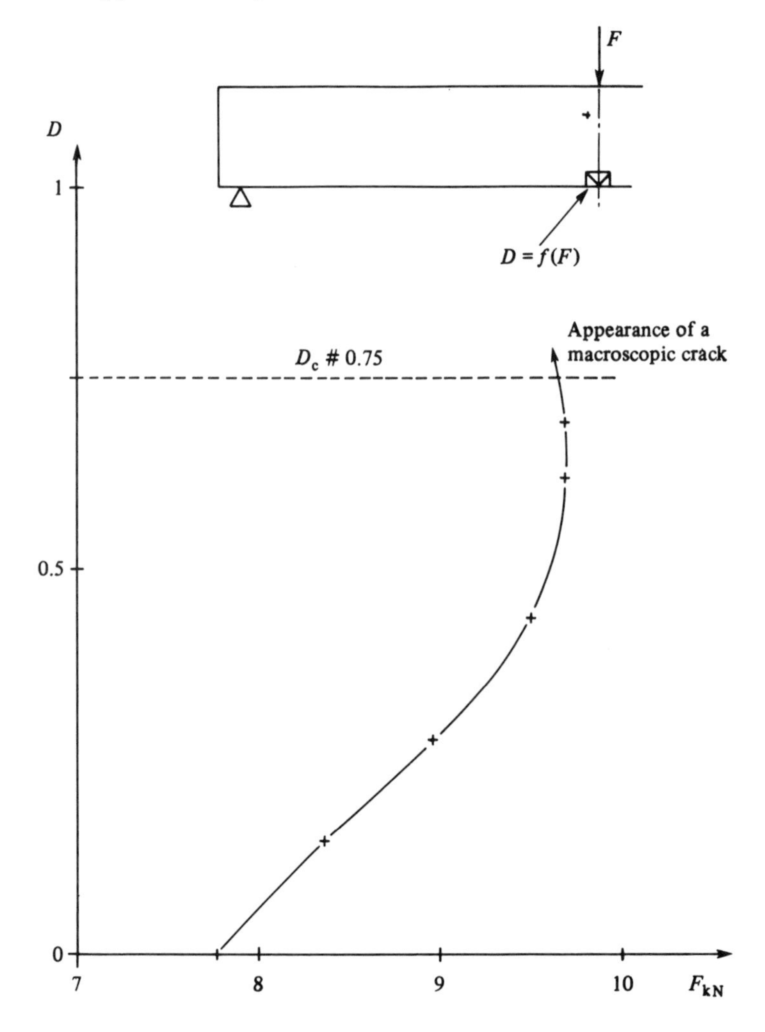

induced damage. In Section 7.5.1 we have seen a simple but realistic model for describing its uniaxial behaviour. A multiaxial criterion in terms of strain is added to differentiate tension from compression. As an example, we mention the criterion described in Section 7.2.5 which contains the invariant ε^* derived from the positive parts of the principal strains:

$$\varepsilon^* = (\langle \varepsilon_1 \rangle^2 + \langle \varepsilon_2^2 \rangle + \langle \varepsilon_3^2 \rangle)^{1/2}.$$

By assuming isotropic damage, we write:

$$\dot{D} = K H(f) \varepsilon^{*s^*} \langle \dot{\varepsilon}^* \rangle$$

where f is the damage criterion:

$$f = \varepsilon^* - \varepsilon_D \leqslant 0.$$

$H(f)$ is the Heaviside function as used in plasticity.

Structural analysis is performed by the finite element method. The damage equation is integrated step-by-step; the elastic stiffness of each finite element is modified at each step to take into account the effect of damage on the elasticity law:

$$\boldsymbol{\varepsilon} = \frac{1+\nu}{E} \frac{\boldsymbol{\sigma}}{1-D} - \frac{\nu}{E} \frac{\mathrm{Tr}(\boldsymbol{\sigma})}{1-D} \mathbf{1}.$$

We note that it is sufficient to introduce the elasticity modulus of the damaged material $\tilde{E} = E(1-D)$.

Fig. 7.53 shows an example of results obtained for a concrete beam subjected to bending. The damage law, integrated for monotonic loading, is here of the type:

$$D = 1 - \frac{\varepsilon_{D_0}(1-A)}{\varepsilon^*} - \frac{A}{\exp(B(\varepsilon^* - \varepsilon_{D_0}))}$$

where for the concrete under consideration, $E = 30\,000$ MPa $\nu = 0.2$; $\varepsilon_{D_0} = 10^{-4}$; $A = 0.8$; $B = 2 \times 10^4$.

The experiments performed in corresponding conditions have confirmed the location of the damage zones as well as the damage level for a given load (ultrasonic method of measurement). The prediction of crack propagation considered as the growth of a completely damaged zone is described in Section 8.7.

Bibliography

Kachanov L. M. *Introduction to continuum damage mechanics.* Martinus Nijhoff, Dordricht (1986).

Krajcinovic D. & Lemaitre J. *Continuum damage mechanics – theory and applications.* CISM course. Springer-Verlag, Berlin (1987).

Chaboche J. L. *Description thermodynamique et phénoménologique de la viscoplasticité cyclique avec endommagement.* Thèse d'Etat, Université P. et M. Curie (Paris VI) (1978).

Cazaud R., Janssen C., Pomey G. & Rabbe P. *La fatigue des métaux.* Dunod, Paris (1977).

Bathias C., Bailon J. P. *La fatigue des matériaux et des structures.* Maloine, Paris (1980).

Osgood C. C. *Fatigue design.* Pergamon Press, New York (1982).

Bazant Z. P. *Advanced topics in inelasticity and failure of concrete.* Cement Och. Betoninstituted, Stockholm (1979).

Wilshire B. & Owen D. R. J. *Engineering approaches of engineering design.* Pineridge Press, Swansea (1983).

8

CRACK MECHANICS

Une fissure instable est une crique qui craque!

All the previous chapters have treated the behaviour of a volume element by considering it as a continuous medium, represented by its state variables. The theory of cracking phenomena or fracture mechanics describes the behaviour of solids or structures with macroscopic geometric discontinuities at the scale of the structure. These discontinuities are line discontinuities in two-dimensional media (such as plates and shells) and surface discontinuities in three-dimensional media. The theory of damage can be used to predict the onset of a macroscopic crack. The theory of cracking phenomena, on the other hand, can be used to predict the evolution of the crack up to a complete failure of the structure.

In structural analysis these discontinuities must be taken into consideration as they modify the stress, strain and displacement fields on such a scale that the assumption of a homogeneous medium would no longer be meaningful. The theory of damage can also be used to study the evolution of cracks, as shown in Section 8.6.4, but more global methods allow consideration of these problems in a more synthetic and simplified way at least for quite simple structures and loads.

As early as 1920, Griffith showed that the failure of a brittle elastic medium could be characterized by a variable, later called the energy release rate, whose critical value, independent of the geometry of the structure, was a characteristic of the material. This approach, called the global approach, was not generalized and formally constructed starting from the thermodynamics of irreversible processes until the seventies (as evidenced by the works of Son, Lemaitre and Chaboche). It showed that in all cases the essential phenomena occur in the vicinity of the crack front and that it is possible to study the macroscopically cracked medium with the help of intrinsic variables. This is due to the high stress concentration present at the 'tip' of a crack which, for linear elasticity, can be represented by the singularity of the stress field. The study of these singularities led Irwin in 1956 to define stress intensity factors corresponding to the particular kinematics of the crack propagation. These semilocal parameters are used largely to study brittle fracture or fatigue fracture of two-dimensional media. Between these two approaches, the global and the semilocal, there exist the contour integrals of Rice (1968) and Bui (1973) which characterize the singularity from the energy point of view.

The years 1960–80 witnessed a great development in the mechanics of fracture by crack propagation. A large number of highly voluminous works have been published on this subject. The present chapter provides no more than a quick synopsis presented in the same spirit as the other chapters. The courageous reader will have to look into some 6000 pages representing the most significant works referred to in the bibliography of this chapter.

8.1 Domain of validity and use

The linear fracture mechanics is based on an elastic analysis of the stress field for small strains. It gives excellent results for brittle-elastic materials like high-strength steel, glass, and to a lesser extent, concrete and wood. The critical value of the energy release rate, the stress intensity factors, or the Rice integral represents a precise condition of fracture by instability of the cracked medium. Similarly the fatigue model of Paris and its derivatives, which express the crack growth rate per cycle as a function of the amplitude of one of the above three variables, can be used to predict correctly, that is within a factor of 2 of the rate, the propagation of cracks in two-dimensional structures subjected to periodic loading.

Three-dimensional problems for brittle-elastic materials can be dealt with by using the same concepts if all the external loads are proportional to only one parameter. This frequently occurs in practice. If, however, the load is not proportional, bifurcation criteria must be adjoined. The word criterion written in its plural form means that several theories with neighbouring domains of validity are possible.

With the occurrence of plasticity or viscoplasticity we enter into the field of nonlinear fracture mechanics. This is the case for ductile materials like low-carbon steel, stainless steel, certain aluminium alloys and polymers. The plasticity manifests itself in two ways: at the level of the plastic zone in front of the crack tip, being the source of a history effect by virtue of the development of residual stresses; and at the level of the mechanism of crack propagation by superposition of the mechanism of ductile fracture. As long as the load is low enough, whether monotonically increasing or periodic, these effects can be neglected; linear fracture mechanics continues to provide a good approximation to the physical reality. In contrast, for large and highly variable loads, the stable progression of ductile fracture cracks and the history effects due to overloads can be modelled only by taking plasticity into account in one way or another. Average and high temperatures induce, as in damage, creep–fatigue interaction effects.

8.2 Elements of analysis of cracked media

The analysis of stresses and strains in the vicinity of crack tips and fronts constitutes a necessary basis for studying crack behaviour. This behaviour is, in effect, governed by the transfer of strain energy into decohesion energy in these zones. Although a plastic or damaged zone is always present at the crack tip, it will be seen that linear elastic analysis provides a sufficiently accurate representation of the reality.

8.2.1 *Initial cracks*

Fracture mechanics assumes the existence of an initial crack in the structure under consideration, and a system of external loads. The initial crack is either a crack created by damage under the effect of these loads, or a defect created during the manufacture or forming of the material. In the first case, it will progress by continuation of the damage mechanism, without transition. In the second case, there is no relation between the crack orientation and the loading system. The crack's progression involves the initiation of a new direction which constitutes a transition phase. The crack is called an incompatible crack.

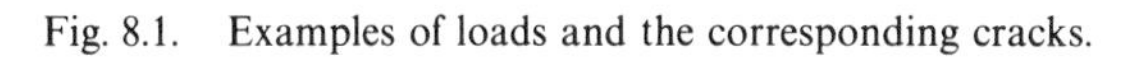
Fig. 8.1. Examples of loads and the corresponding cracks.

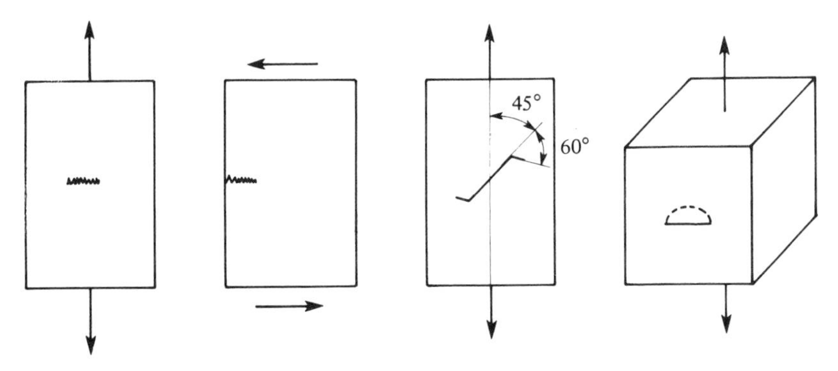

In order to know whether this phase should be studied or not in a given problem, the following definition will be used: a loading system is compatible with a crack front configuration $\Gamma(x_1, x_2, x_3)$ if the crack increment $\delta\Gamma$ which results from the load occurs without bifurcation in the sense of criteria stated in Section 8.4.5. $\Gamma(x_1, x_2, x_3)$ is the expression of the crack front line in a rectangular (x_1, x_2, x_3) frame of reference. Some examples are given in Fig. 8.1.

8.2.2 *Elastic analysis*

In fracture mechanics, the basic problem is the analysis of stress distribution in plane, linear elastic, cracked media. This is done for theoretical reasons (in plane elasticity there is the possibility of finding analytical solutions!) as well as for practical ones (many mechanical structures are built with metal sheets which are highly susceptible to crack phenomena: 'planes fly with many cracks, but we are not worried about them, thank you!').

Loading modes

Let us consider a crack in a plane medium. Depending on the direction of the load with respect to the direction of the crack, we distinguish three important mechanisms for the relative displacements of the crack lips; these are illustrated in Fig. 8.2 with reference to an orthogonal set of axes (x_1, x_2, x_3).

Mode I: or opening mode; the field of relative displacement of the crack lips is defined by the discontinuity:

$$[\![u_1]\!] = 0, \qquad [\![u_2(x_1)]\!] \neq 0, \qquad [\![u_3]\!] = 0.$$

Fig. 8.2. Loading modes of a crack (drawn by B. Bocquet).

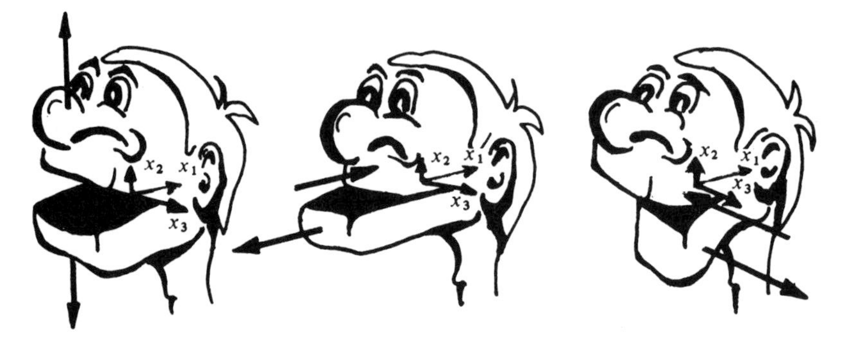

Mode II: or plane shear (sliding) mode:

$$[\![u_1(x_1)]\!] \neq 0, \qquad [\![u_2]\!] = 0, \qquad [\![u_3]\!] = 0.$$

Mode III: or antiplane shear (tearing) mode:

$$[\![u_1]\!] = 0, \qquad [\![u_2]\!] = 0, \qquad [\![u_3]\!] \neq 0$$

Westergaard's asymptotic solution

In accordance with the plane-stress assumption, we consider, for example, the problem of mode I deformation of a crack of length $2a$ in a thin, finite, plane medium subjected to a uniform stress field, $\sigma_{ij} = 0$ except $\sigma_{22} = \sigma_{22}^{\infty}$, sufficiently far from the crack.

The plane elasticity problem to be solved can be written as:

Stresses:

$$[\sigma] = \begin{bmatrix} \sigma_{11} & \sigma_{12} \\ \sigma_{12} & \sigma_{22} \end{bmatrix},$$

Strains:

$$[\varepsilon] = \begin{bmatrix} \dfrac{\partial u_1}{\partial x_1} & \dfrac{1}{2}\left(\dfrac{\partial u_1}{\partial x_2} + \dfrac{\partial u_2}{\partial x_1}\right) & \varepsilon_{13} \\ \dfrac{1}{2}\left(\dfrac{\partial u_1}{\partial x_2} + \dfrac{\partial u_2}{\partial x_1}\right) & \dfrac{\partial u_2}{\partial x_2} & \varepsilon_{23} \\ \varepsilon_{13} & \varepsilon_{23} & \varepsilon_{33} \end{bmatrix}.$$

Law of linear elasticity (Chapter 4):

$$[\varepsilon] = \frac{1}{E}\begin{bmatrix} \sigma_{11} - \nu\sigma_{22} & (1+\nu)\sigma_{12} & 0 \\ (1+\nu)\sigma_{12} & \sigma_{22} - \nu\sigma_{11} & 0 \\ 0 & 0 & -\nu(\sigma_{11} + \sigma_{22}) \end{bmatrix}.$$

The solution of the asymptotic problem in the vicinity of the crack tip is obtained by the following Airy stress function of the complex variable $z = x_1 + \mathrm{i}x_2$ (Re and Im denote the real and imaginary parts respectively):

$$\chi = \mathrm{Re}(\bar{\bar{Z}}) + x_2 \,\mathrm{Im}\,(\bar{Z})$$

with

$$\frac{\mathrm{d}\bar{\bar{Z}}}{\mathrm{d}z} = \bar{Z}, \quad \frac{\mathrm{d}\bar{Z}}{\mathrm{d}z} = Z, \quad \frac{\mathrm{d}Z}{\mathrm{d}z} = Z' \quad \text{and} \quad Z = \frac{K_{\mathrm{I}}(\sigma_{22}^{\infty}, a)}{[(\pi/a)(z^2 - a^2)]^{1/2}}$$

where K_{I} is a scalar parameter independent of the location and varies only with the load σ_{22}^{∞} and the crack half-length a.

The equilibrium equations are automatically satisfied and the Cauchy–Riemann conditions can be used to determine the stresses:

$$\sigma_{11} = \partial^2\chi/\partial x_2^2 = \operatorname{Re} Z - x_2 \operatorname{Im} Z'$$

$$\sigma_{22} = \partial^2\chi/\partial x_1^2 = \operatorname{Re} Z + x_2 \operatorname{Im} Z'$$

$$\sigma_{12} = -\partial^2\chi/\partial x_1 \partial x_2 = -x_2 \operatorname{Re} Z'.$$

With the change of variable $z - a = r\exp(\mathrm{i}\theta)$ we can represent the stresses in polar coordinates with the origin centred at the crack tip (Fig. 8.3). Then using the approximation $r/a \ll 1$ we obtain the following expression for the function Z, which is valid when r tends to zero (and for mode I)

$$Z = \frac{K_{\mathrm{I}}}{(2\pi r)^{1/2}}\left(\cos\frac{\theta}{2} - \mathrm{i}\sin\frac{\theta}{2}\right).$$

The stresses and displacements in the vicinity of the crack tip can be derived to be

• $$[\sigma] = \frac{K_{\mathrm{I}}(\sigma_{22}^{\infty}, a)}{(2\pi r)^{1/2}} \times \begin{bmatrix} \cos\frac{\theta}{2}\left(1 - \sin\frac{\theta}{2}\sin\frac{3\theta}{2}\right) & \sin\frac{\theta}{2}\cos\frac{\theta}{2}\cos\frac{3\theta}{2} \\ \sin\frac{\theta}{2}\cos\frac{\theta}{2}\cos\frac{3\theta}{2} & \cos\frac{\theta}{2}\left(1 + \sin\frac{\theta}{2}\sin\frac{3\theta}{2}\right) \end{bmatrix}$$

Fig. 8.3. Crack loaded in mode I.

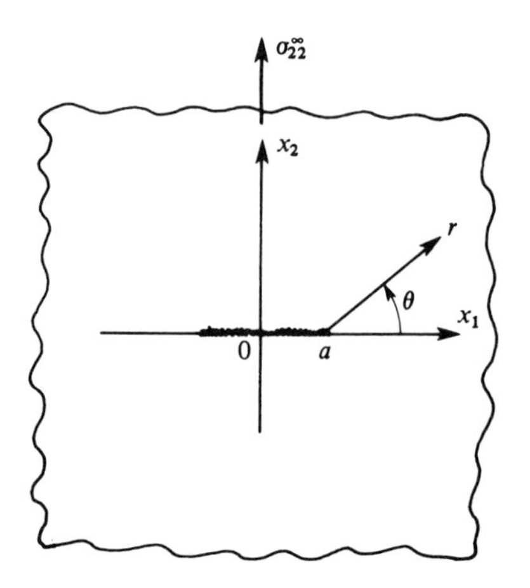

$$u_1 = \frac{K_I}{E}\left(\frac{r}{2\pi}\right)^{1/2} \cos\frac{\theta}{2}[3 - \nu - (1 + \nu)\cos\theta]$$

$$u_2 = \frac{K_I}{E}\left(\frac{r}{2\pi}\right)^{1/2} \sin\frac{\theta}{2}[3 - \nu - (1 + \nu)\cos\theta].$$

In plane-stress, the crack-opening displacement $[\![u_2]\!]$ is defined by

$$\bullet \qquad [\![u_2]\!] = u_2(r, \pi) - u_2(r, -\pi) = \frac{8K_I}{E}\left(\frac{r}{2\pi}\right)^{1/2}.$$

From the above it is seen that at the crack tip the stresses are singular in $r^{-1/2}$ and the crack-opening displacement tends to zero as $r^{1/2}$ tends to zero.

The same problem treated under the plane-strain assumption yields the same expressions for stresses with

$$\sigma_{33} = \nu(\sigma_{11} + \sigma_{22}) \neq 0$$

but a different displacement field:

$$u_1 = \frac{K_I}{E}\left(\frac{r}{2\pi}\right)^{1/2} (1 + \nu)\cos\frac{\theta}{2}(3 - 4\nu - \cos\theta)$$

$$u_2 = \frac{K_I}{E}\left(\frac{r}{2\pi}\right)^{1/2} (1 + \nu)\sin\frac{\theta}{2}(3 - 4\nu - \cos\theta)$$

and a crack opening expressed by:

$$\bullet \qquad [\![u_2]\!] = \frac{8K_I}{E}\left(\frac{r}{2\pi}\right)^{1/2} (1 - \nu^2).$$

Remark:

Note that the plane-stress solution contains an incompatibility with regard to the strain component ε_{33}. According to the plane-stress assumption $\varepsilon_{33} = -(\nu/E)(\sigma_{11} + \sigma_{22})$, which means that ε_{33} is singular in $r^{-1/2}$. On the other hand, the equations of compatibility $\varepsilon_{33,11} = \varepsilon_{33,22} = \varepsilon_{33,12} = 0$ require that ε_{33} be linear in x_1 and x_2, and hence regular! In plane-strain, this incompatibility does not exist.

Similar procedures can be used to study the behaviour under mode II loading (the plane problem) and mode III loading (the antiplane problem). The results are as follows.

Mode II

Stresses and displacements both for the plane-stress and plane-strain assumptions are given by:

- $$[\sigma] = \frac{K_{\rm II}(\sigma_{12}^{\infty}, a)}{(2\pi r)^{1/2}} \begin{bmatrix} -\sin\frac{\theta}{2}\left(2+\cos\frac{\theta}{2}\cos\frac{3\theta}{2}\right) & \cos\frac{\theta}{2}\left(1-\sin\frac{\theta}{2}\sin\frac{3\theta}{2}\right) \\ \cos\frac{\theta}{2}\left(1-\sin\frac{\theta}{2}\sin\frac{3\theta}{2}\right) & \sin\frac{\theta}{2}\cos\frac{\theta}{2}\cos\frac{3\theta}{2} \end{bmatrix}$$

$$u_1 = \frac{K_{\rm II}}{E}\left(\frac{r}{2\pi}\right)^{1/2}(1+v)\sin\frac{\theta}{2}(C_1 + 2 + \cos\theta)$$

$$u_2 = \frac{K_{\rm II}}{E}\left(\frac{r}{2\pi}\right)^{1/2}(1+v)\cos\frac{\theta}{2}(C_1 - 2 + \cos\theta)$$

where $C_1 = (3 - v)/(1 + v)$ for plane-stress, and $C_1 = 3 - 4v$ for plane-strain. The shear displacement of the crack lips is

$$[\![u_1]\!] = u_1(r, \pi) - u_1(r, -\pi)$$

which can be expressed in plane-stress as:

- $$[\![u_1]\!] = \frac{8K_{\rm II}}{E}\left(\frac{r}{2\pi}\right)^{1/2}$$

and in plane-strain as:

- $$[\![u_1]\!] = \frac{8(1-v^2)K_{\rm II}}{E}\left(\frac{r}{2\pi}\right)^{1/2}.$$

Mode III
Antiplane stress:

- $$[\sigma] = \frac{K_{\rm III}(\sigma_{13}^{\infty}, a)}{(2\pi r)^{1/2}} \begin{bmatrix} 0 & 0 & -\sin\frac{\theta}{2} \\ 0 & 0 & \cos\frac{\theta}{2} \\ -\sin\frac{\theta}{2} & \cos\frac{\theta}{2} & 0 \end{bmatrix}.$$

Antiplane displacements:

$$u_1 = u_2 = 0$$

$$u_3 = \frac{4(1+v)}{E}K_{\rm III}\left(\frac{r}{2\pi}\right)^{1/2}\sin\frac{\theta}{2}.$$

Antiplane shear displacement of crack lips:

$$[\![u_3]\!] = u_3(r, \pi) - u_3(r, -\pi)$$

- $$[\![u_3]\!] = \frac{8(1+\nu)K_{\text{III}}}{E}\left(\frac{r}{2\pi}\right)^{1/2}.$$

Muskhelishvili's exact solution

If the medium studied in the above is infinite in both directions, then an exact solution is available for mode I loading. This is obtained by superposing the trivial solution for the medium without the crack and the solution of a problem with a stress, zero at infinity and $-\sigma_{22}^{\infty}$ at the crack line $-a < x_1 < a$. This solution is analytic for $x_2 = 0$. For plane-stress, it is the following:

for $x_1 \geqslant a$:

$$\sigma_{22}(x_2 = 0) = \frac{\sigma_{22}^{\infty}}{[1 - (a/x_1)^2]^{1/2}}$$

$$\sigma_{11}(x_2 = 0) = \sigma_{22} - \sigma_{22}^{\infty}$$

$$\varepsilon_{22}(x_2 = 0) = \frac{\sigma_{22}^{\infty}}{E}\left\{\frac{1-\nu}{[1-(a/x_1)^2]^{1/2}} + \nu\right\}$$

for $0 \leqslant x_1 \leqslant a$:

$$[\![u_2]\!] = 2u_2 \frac{4\sigma_{22}^{\infty} a}{E}\left[1 - \left(\frac{x_1}{a}\right)^2\right]^{1/2}.$$

if in the expression for σ_{22} we make a change of variable $x_1 = a + r$ (Fig. 8.4) we find that

$$\sigma_{22} = \frac{\sigma_{22}^{\infty}}{[(r^2 + 2ar)/(a+r)^2]^{1/2}} = \sigma_{22}^{\infty}\frac{a + r}{[r(2a + r)]^{1/2}}$$

where, for r small in comparison to a:

$$\sigma_{22} \to \sigma_{\infty}(a/2r)^{1/2}.$$

This shows that the stress component σ_{22} $(x = 0)$ is indeed singular in $r^{-1/2}$ at the crack tip $r = 0$ (or $x_1 = a$). We recover Westergaard's approximate solution by putting

$$K_{\text{I}} = \sigma_{22}^{\infty}(\pi a)^{1/2}.$$

This expression for the factor K_I corresponds to a large plate containing a (small) central crack of length $2a$ in the direction perpendicular to the tension. In contrast to the approximate solution, the above solution leads to a stress field which at points remote from the crack is equal to the applied one. Moreover, the expression for crack-opening displacement shows that the loaded crack has the form of an ellipse; it is represented in Fig. 8.4 together with the graph of σ_{22} $(x_2 = 0)$.

Stress intensity factors

The functions $K_I(\sigma_{22}^{\infty}, a)$, $K_{II}(\sigma_{12}^{\infty}, a)$, $K_{III}(\sigma_{13}^{\infty}, a)$ introduced in Westergaard's stress function characterize in essence the singularity of the stress field at the crack tip and are proportional to the discontinuity in the displacements of the crack lips. We call them stress intensity factors, factors of mode I, mode II, and mode III respectively.

From a knowledge of these values we can completely determine the stress or the displacement field in a cracked structure which is assumed to behave elastically. Conversely, if we know the expressions for nonzero stress components and the displacement components, we can determine the stress intensity factors from the following expressions which, in fact, define these factors for plane, cracked media.

Fig. 8.4. Crack in an infinite plane medium loaded in mode I.

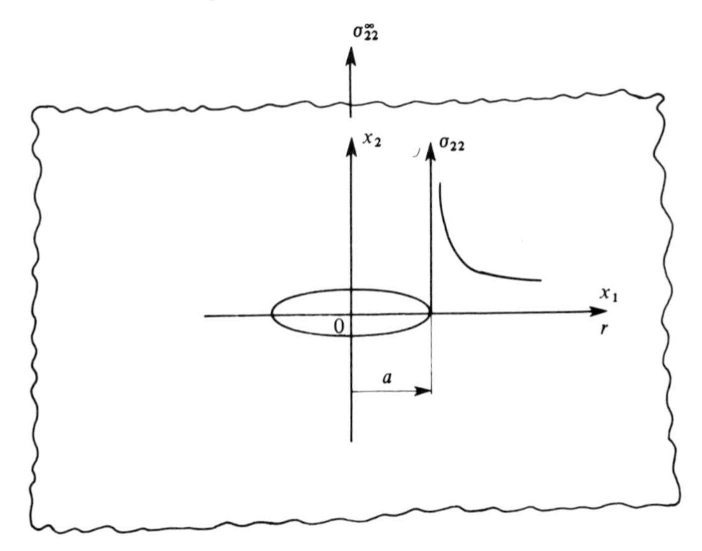

- $$K_{\text{I}} = \lim_{r\to 0} [\sigma_{ij}(2\pi r)^{1/2}] = \lim_{r\to 0}\left[\frac{E}{8C_2}\left(\frac{2\pi}{r}\right)^{1/2} [\![u_2]\!]\right],$$

- $$K_{\text{II}} = \lim_{r\to 0} [\sigma_{ij}(2\pi r)^{1/2}] = \lim_{r\to 0}\left[\frac{E}{8C_2}\left(\frac{2\pi}{r}\right)^{1/2} [\![u_1]\!]\right],$$

with

- $C_2 = 1$ in plane-stress,

- $C_2 = 1 - \nu^2$ in plane-strain,

- $$K_{\text{III}} = \lim_{r\to 0} [\sigma_{ij}(2\pi r)^{1/2}] = \lim_{r\to 0}\left[\frac{E}{8(1+\nu)}\left(\frac{2\pi}{r}\right)^{1/2} [\![u_3]\!]\right].$$

If, for example, we apply the definition of K_{I} to Muskhelishvili's solution for a plane, infinite medium loaded in mode I we recover

- $$K_{\text{I}} = \sigma_{22}^{\infty}(\pi a)^{1/2}.$$

In the same fashion, for a plane, infinite medium subjected to mode II loading, we find

- $$K_{\text{II}} = \sigma_{12}^{\infty}(\pi a)^{1/2}.$$

Note that the unit of stress intensity factors is $\text{N m}^{-3/2}$ or $\text{MPa m}^{1/2}$.

Contour integrals

Another way of characterizing the singularity of the stress field in the vicinity of the crack tip consists in studying certain contour integrals deduced on the basis of the law of conservation of energy.

Consider a cracked, plane-elastic medium in mixed mode I and mode II loadings, i.e., a plane problem in which the crack displacements are $u_1 \neq 0$, $u_2 \neq 0$, $u_3 = 0$. Let C_1 be an open contour surrounding the crack tip and let n be its outward normal (Fig. 8.5). If w_{e} is the elastic strain energy density and w_{e}^* the density of the complementary energy, we can write:

$$\sigma_{ij} = \partial w_{\text{e}}/\partial \varepsilon_{ij}^{\text{e}}, \qquad \varepsilon_{ij} = \partial w_{\text{e}}^*/\partial \sigma_{ij}.$$

The Rice integral is defined by

- $$J = \int_C \left(w_{\text{e}} n_1 - \sigma_{ij} n_j \frac{\partial u_i}{\partial x_1} \right) \text{d}s$$

and the Bui integral, the dual of the preceding one, is defined by

$$\bullet \qquad I = \int_C \left(-w_e^* n_1 + u_i n_j \frac{\partial \sigma_{ij}}{\partial x_1} \right) ds.$$

These integrals are independent of the (open) contour of the integration. For the proof of this, it is sufficient to consider a closed contour consisting of two open contours C_1 and C_2 and two segments of the crack lips A_1A_2 and B_1B_2. Integral J and I are zero on the closed contours (by virtue of Stoke's theorem) and they are also zero on the two segments where $n_1 = 0$ and $\sigma_{ij}n_j = 0$ since the stresses are zero on the crack lips.

We may therefore apply the integrals to a contour close to the crack tip with Westergaard's stress and displacement fields for the mixed mode obtained by superposing the two particular fields belonging to the mode I and mode II loadings. We find

$$J = I = \begin{cases} \dfrac{1}{E}(K_I^2 + K_{II}^2) & \text{in plane-stress} \\[2ex] \dfrac{1-\nu^2}{E}(K_I^2 + K_{II}^2) & \text{in plane-strain.} \end{cases}$$

If we define the potential energy $\mathscr{V}$ and the complementary energy $\mathscr{V}^*$ of

Fig. 8.5. Integration contour.

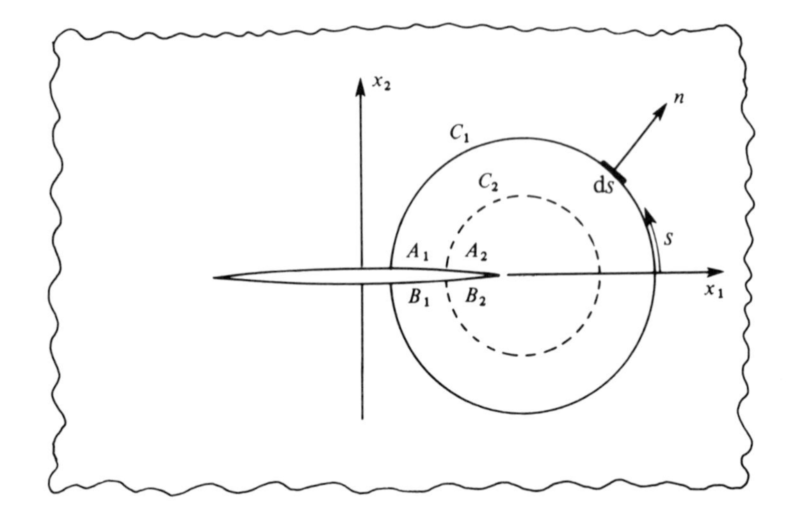

the whole solid (not considering the body forces) by

$$\mathscr{V} = \int_{\mathscr{S}} w_e \, dV - \int_{\partial\mathscr{S}_F} \vec{F}^d \cdot \vec{u} \, dS$$

$$\mathscr{V}^* = -\int_{\mathscr{S}} w_e^* \, dV + \int_{\partial\mathscr{S}_u} \vec{u}^d \cdot \boldsymbol{\sigma} \cdot \vec{n} \, dS$$

then we can show that J and I are respectively equal to the rates of change of the potential and complementary energies with respect to the crack growth (in terms of its prolongation):

$$J = -\,d\mathscr{V}/da, \qquad I = -\,d\mathscr{V}^*/da.$$

These two quantities are equal for the exact solution, but their duality bestows upon them the property of providing bounds to the exact solution if they are obtained in an approximate manner, by a kinematic method for the J integral and by a static method for the I integral. For the finite element method, for example, we obtain (see Section 8.6.1):

$$J_{\text{approx}} \leqslant J = I \leqslant I_{\text{approx}}.$$

Strain energy release rate

Yet another global approach consists in studying the balance of energies taking part in the crack growth process. We have just seen that contour integrals, characteristics of the singularity of the stress field, also characterize the rate of change of the potential energy of the whole solid.

Consider as usual a plane-elastic medium of thickness e with a crack as shown in Fig. 8.3. Let the length of the crack grow at a rate $\dot{a}$ and its area at a rate $\dot{A} = e\dot{a}$. Then, applying the first principle of thermodynamics (stated in Chapter 2) to the whole solid, within the assumption of small perturbations, we have

$$\dot{E} + \dot{K} = P_{(x)} + Q - 2\gamma\dot{A}$$

where E is the internal energy, K is the kinetic energy, Q is the rate at which heat is received by the solid, and $P_{(x)}$ is the power of the external forces which (in the absence of body forces) can be written as

$$P_{(x)} = \int_{\partial\mathscr{S}} \vec{F} \cdot \dot{\vec{u}} \, dS.$$

With reference to the first principle stated in Chapter 2, the presence of the additional term $-2\gamma\dot{A}$ is due to the fact that the boundary of the volume under consideration is evolving with the crack (assuming the crack surface forms part of the boundary of the solid). Thus $2\gamma\dot{A}$ is the power dissipated in the process of decohesion, γ being a characteristic constant of the material. The rate of change in the internal energy, expressed as in Section 2.3.1, is

$$\dot{E} = -P_{(\mathrm{i})} + Q = \int_{\mathscr{S}} \boldsymbol{\sigma}:\dot{\boldsymbol{\varepsilon}}\,\mathrm{d}V + Q = \dot{W}_{\mathrm{e}} + Q.$$

Hence, the first principle may be written in the form:

$$\dot{W}_{\mathrm{e}} + \dot{K} = \int_{\partial\mathscr{S}} \vec{F}\cdot\dot{\vec{u}}\,\mathrm{d}S - 2\gamma\dot{A}.$$

The condition for the stability of the crack growth process is that the kinetic energy of the solid is not increasing, i.e.,

$$\dot{K} = \int_{\partial\mathscr{S}} \vec{F}\cdot\dot{\vec{u}}\,\mathrm{d}S - \dot{W}_{\mathrm{e}} - 2\gamma\dot{A} \leqslant 0.$$

Considering A as the only variable of the global balance, the above is reduced to:

$$\dot{K} = \left(\int_{\partial\mathscr{S}} \vec{F}\cdot\frac{\partial\vec{u}}{\partial A}\mathrm{d}S - \frac{\partial W_{\mathrm{e}}}{\partial A} - 2\gamma\right)\dot{A} \leqslant 0$$

from which (since $\dot{A} \geqslant 0$) it follows that:

$$\int_{\partial\mathscr{S}} \vec{F}\cdot\frac{\partial\vec{u}}{\partial A}\mathrm{d}S - \frac{\partial W_{\mathrm{e}}}{\partial A} \leqslant 2\gamma.$$

The left hand side of the above relation represents the energy which is available during crack growth and which may be utilized for generating this mechanism. It is, by definition, the energy release rate G. Denoting by W_{x} the work of external forces (considered as constant forces), we can write

$$\bullet \qquad G = \int_{\partial\mathscr{S}} \vec{F}\cdot\frac{\partial\vec{u}}{\partial A}\mathrm{d}S - \frac{\partial W_{\mathrm{e}}}{\partial A} = \left.\frac{\mathrm{d}W_{\mathrm{x}}}{\mathrm{d}A}\right|_{F} - \frac{\partial W_{\mathrm{e}}}{\partial A}.$$

Then, the Griffith criterion of fracture by instability of the brittle-elastic media is

$$G \geqslant 2\gamma.$$

When $G = 2\gamma$, we may say that the fracture is 'controlled'; the kinetic energy is no longer increasing.

In Section 8.4 we will see that it is possible to introduce directly, by means of thermodynamic formalism, the energy release rate for three-dimensional elastoplastic, viscoplastic media and thus avoiding the assumption of linear elastic behaviour. In the present case, Fig. 8.6(*a*) schematically illustrates the definition of the energy release rate.

In the plane-elastic medium, at least, the energy release rate is simply related to the stress intensity factors K_{I}, K_{II}, K_{III} and to the contour integrals J or I. For this relation, we express G in the form of a contour integral by first transforming the volume integral of W_{x} to an integral over the external surface of the solid. The two crack surfaces do not contribute to the integral because of the free surface conditions consisting of zero tractions and unprescribed displacements.

Applying the principle of virtual work to W_{e} (using here the real displacements) and neglecting the dynamic effects we have

$$\int_{\mathscr{S}} \tfrac{1}{2}\boldsymbol{\sigma}:\boldsymbol{\varepsilon}\,\mathrm{d}V = \tfrac{1}{2}\int_{\partial\mathscr{S}} \vec{F}\cdot\vec{u}\,\mathrm{d}S$$

$$\frac{\partial W_{\mathrm{e}}}{\partial A} = \frac{\partial}{\partial A}\int_{\mathscr{S}} \tfrac{1}{2}\boldsymbol{\sigma}:\boldsymbol{\varepsilon}\,\mathrm{d}V = \frac{1}{2}\int_{\partial\mathscr{S}}\left(\frac{\partial \vec{F}}{\partial A}\cdot\vec{u} + \vec{F}\cdot\frac{\partial \vec{u}}{\partial A}\right)\mathrm{d}S$$

Fig. 8.6. (*a*) Schematic diagram illustrating the energy balance in linear elastic case. (*b*) Local definition of the energy release rate (after H. D. Bui).

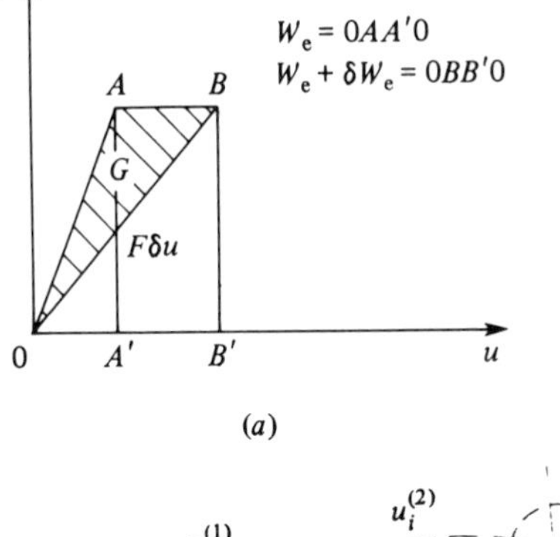

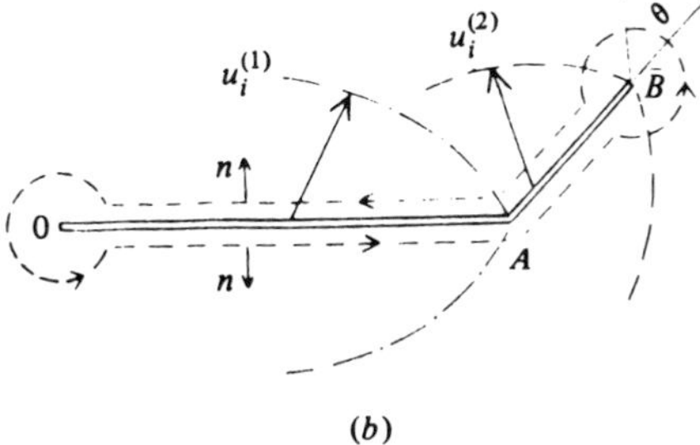

so that we obtain

$$G = \frac{1}{2}\int_{\partial\mathscr{S}} \left(\vec{F} \cdot \frac{\partial \vec{u}}{\partial A} - \vec{u} \cdot \frac{\partial \vec{F}}{\partial A} \right) \mathrm{d}S.$$

Betti's theorem can be used to prove that this integral is independent of the closed contour of integration. Since e is the thickness of the plane medium, $A = ea$ and $\partial\mathscr{S} = eC$, we can therefore, for any closed contour write

- $$G = \frac{1}{2}\int_{C} \left(\vec{F} \cdot \frac{\partial \vec{u}}{\partial a} - \vec{u} \cdot \frac{\partial \vec{F}}{\partial a} \right) \mathrm{d}C$$

and calculate this integral on a contour flattened along the crack which then defines it as the energy rate of crack closure. We thus arrive at a local definition of the dissipated energy rate. Denoting by $\vec{u}^{(1)}$ and $\vec{u}^{(2)}$ the displacement fields before and after the progression from A to B (Fig. 8.6(*b*)), and by $\vec{F}^{(1)} = \boldsymbol{\sigma}^{(1)} \cdot \vec{n}$ the bond forces on the portion AB in state (1), we find

$$G\,\mathrm{d}a = \tfrac{1}{2}\int_{AB^+} \vec{F}^{(1)} \cdot \vec{u}^{(2)}\,\mathrm{d}S + \tfrac{1}{2}\int_{AB^-} \vec{F}^{(1)} \cdot \vec{u}^{(2)}\,\mathrm{d}S.$$

In the symmetric case (of $0AB$ being a straight crack) we have:

$$G\,\mathrm{d}a = \int_{AB} \vec{F}^{(1)} \cdot \vec{u}^{(2)}\,\mathrm{d}S.$$

These quantities can be interpreted as the work required to close the crack AB (of length da).

In the expression for the energy release rate G established above, we may, by considering the given quantities as independent of the crack, effect a partition of $\partial\mathscr{S}$ into $\partial\mathscr{S}_{\mathrm{F}}$ and $\partial\mathscr{S}_u$ and write

- $$G = \tfrac{1}{2}\int_{\partial\mathscr{S}_{\mathrm{F}}} \vec{F}^{\mathrm{d}} \cdot \frac{\partial \vec{u}}{\partial A}\,\mathrm{d}S - \tfrac{1}{2}\int_{\partial\mathscr{S}_{\mathrm{u}}} \vec{u}^{\mathrm{d}} \cdot \frac{\partial \vec{F}}{\partial A}\,\mathrm{d}S.$$

By using Westergaard's equations to define the two fields, we finally obtain Irwin's formulae:

$$G = \frac{1}{E}(K_{\mathrm{I}}^2 + K_{\mathrm{II}}^2) + \frac{1+\nu}{E}K_{\mathrm{III}}^2 \qquad \text{in plane-stress,}$$

$$G = \frac{1-\nu^2}{E}(K_{\mathrm{I}}^2 + K_{\mathrm{II}}^2) + \frac{1+\nu}{E}K_{\mathrm{III}}^2 \quad \text{in plane-strain.}$$

By comparing them with the formulae for J and I obtained previously, we find that for the plane-elastic problems, the energy release rate is exactly equal to these integrals:

$$G = J = I.$$

Three-dimensional problems

Three-dimensional elasticity problems present great difficulties, and only a few explicit solutions exist for characterizing the stress and displacement fields in the vicinity of crack surfaces in three-dimensional media.

Even though the global approach presented in Section 8.4.2 appears more appropriate for the study of three-dimensional crack phenomena, we give here definitions of the stress intensity factors which may prove of interest in some cases.

The plane crack is defined by its surface A and by Γ the contour of its front in a coordinate system $(0, x_1, x_2, x_3)$ with the x_3 axis normal to the crack surface (Fig. 8.7). Let (M, ntv) represent the local coordinate system at point M of the crack front defined by the normal $\vec{n}$ parallel to x_3 and the tangent t to Γ at M, and let (M, r) be any direction in the plane (M, n, v) with abscissa r measured from M.

For the static definition, the stress intensity factors at M are expressed as functions of the limits of the antiplane components of the stress tensor in the local coordinate system. Or, for the kinematic definition, they are expressed as functions of the discontinuities of the components of the relative crack lip

Fig. 8.7. Crack in a three-dimensional medium.

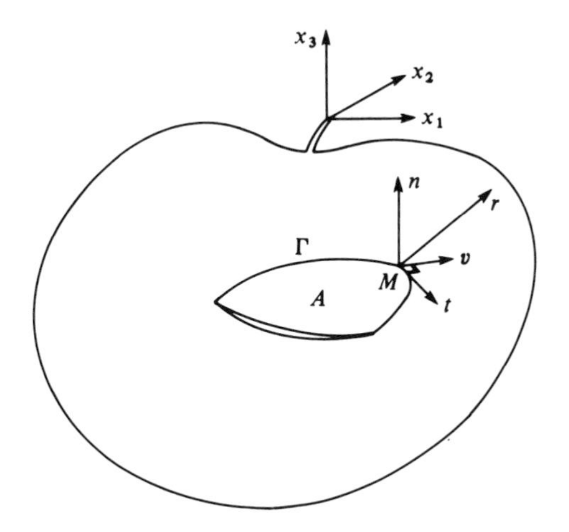

displacements.

- $$K_{\mathrm{I}}(M) = \lim_{r\to 0} [\sigma_{33}(2\pi r)^{1/2}] = \lim_{r\to 0}\left[\frac{E}{8(1-\nu^2)}[\![u_3]\!]\left(\frac{2\pi}{r}\right)^{1/2}\right]$$
- $$K_{\mathrm{II}}(M) = \lim_{r\to 0} [\sigma_{3v}(2\pi r)^{1/2}] = \lim_{r\to 0}\left[\frac{E}{8(1-\nu^2)}[\![u_v]\!]\left(\frac{2\pi}{r}\right)^{1/2}\right]$$
- $$K_{\mathrm{III}}(M) = \lim_{r\to 0} [\sigma_{3t}(2\pi r)^{1/2}] = \lim_{r\to 0}\left[\frac{E}{8(1+\nu)}[\![u_t]\!]\left(\frac{2\pi}{r}\right)^{1/2}\right].$$

We note that these expressions are identical to those which define the stress intensity factors for plane media in plane strain. This is a quite general result: the asymptotic values of the discontinuities of the relative crack lip displacements are identical for the two cases.

8.2.3 *Elastoplastic analyses*

Linear elastic analysis comes close to representing the physical reality for brittle or only slightly ductile elastic materials, but it is inadequate when plastic deformations occur in a volume with a characteristic dimension of more than 5–20% of the crack length. Knowledge of local stresses and strains in the case of nonconfined plasticity requires rather complex elastoplastic analysis. In practice we use either the plastic zone correction factors, or global parameters derived from a nonlinear elastic analysis.

Estimation of the plastic zone at the crack tip

A rough estimate of the dimensions of the plastic zone of a crack can be obtained by using Westergaard's asymptotic solution. For this we again consider the plane medium of Fig. 8.3, loaded in mode I. We then search for the points where the state of stress satisfies the von Mises yield criterion (see Chapter 5)

$$\tfrac{1}{2}[(\sigma_{11}-\sigma_{22})^2 + (\sigma_{22}-\sigma_{33})^2 + (\sigma_{33}-\sigma_{11})^2 + 6(\sigma_{12}^2 + \sigma_{23}^2 + \sigma_{13}^2)] = \sigma_{\mathrm{Y}}^2.$$

By introducing Westergaard's solution for stresses in polar coordinates, the above criterion becomes:

in plane-stress:

$$\frac{K_{\mathrm{I}}^2}{2\pi r}\cos^2\frac{\theta}{2}\left(1 + 3\sin^2\frac{\theta}{2}\right) = \sigma_{\mathrm{Y}}^2$$

in plane-strain:

$$\frac{K_I^2}{2\pi r}\cos^2\frac{\theta}{2}\left[(1-2\nu)^2+3\sin^2\frac{\theta}{2}\right]=\sigma_Y^2$$

where r is the polar distance and θ is the polar angle (see Fig. 8.3). From these expressions it follows that r_Y, the polar distance at which the criterion is satisfied is

$$\bullet \qquad r_Y=\begin{cases}\dfrac{K_I^2}{2\pi\sigma_Y^2}\cos^2\dfrac{\theta}{2}\left(1+3\sin^2\dfrac{\theta}{2}\right) & \text{in plane-stress;}\\[2ex] \dfrac{K_I^2}{2\pi\sigma_Y^2}\cos^2\dfrac{\theta}{2}\left[(1-2\nu)^2+3\sin^2\dfrac{\theta}{2}\right] & \text{in plane-strain.}\end{cases}$$

If r_Y is small with respect to the length of the crack ($r_Y/a<0.2$) we may assume that the stress field calculated on the basis of elastic behaviour is perturbed only slightly by the presence of the plastic deformations (except, of course, for $0<r<r_Y$) and the expression for r_Y constitutes an approximation to the boundary of the plastic zone. The values ρ of r_Y on the crack axes are

in plane-stress:

$$r_Y(\theta=0)=\rho=K_I^2/2\pi\sigma_Y^2,$$

in plane-strain:

$$r_Y(\theta=0)=\rho=\frac{K_I^2}{2\pi\sigma_Y^2}(1-2\nu)^2.$$

The maximum values of r_Y as a function of the polar angle θ are:

in plane-stress:

$$r_Y(\theta=70.5^\circ)=K_I^2/1.5\pi\sigma_Y^2,$$

in plane-strain (for $\nu=0.3$):

$$r_Y(\theta=86.9^\circ)=K_I^2/2.4\pi\sigma_Y^2.$$

Fig. 8.8 shows the plastic zone in a thick metal sheet for which the plane-stress approximation is assumed to hold at the surface, but the plane-strain one in the median plane. These estimates do not take into account stress redistributions. Irwin's plasticity correction, justified by exact analysis in mode III, assumes the existence of a small plastic zone of length ρ. In the interval $0<x_1<\rho$ where the elastic solution has exceeded the elastic limit, σ_{22} is restricted to σ_Y, the excess stress having to be distributed further, with $\sigma_{22} = \sigma_Y$ for $\rho < x_1 < 2\rho$ and then beyond this range varying according to $K_I'(x_1 - r_Y)^{-1/2}$. This corrected field is shifted by ρ without any change in the intensity (see Fig. 8.9(*a*)): confined plasticity models admit the existence of the elastic field outside the critical plastic zone in the form

$$\sigma_{ij} = \frac{K_I'}{(2\pi r)^{1/2}} g_{ij}(\theta),$$

with r and θ measured from the centre of the plastic region. For an infinite plate in mode I with a central crack of length $2a$ the situation is exactly as if the effective crack length were $2a' = 2(a + \rho)$; the stress intensity factor is expressed by

$$K_I' = \sigma_{22}^{\infty}[\pi(a + \rho)]^{1/2}.$$

We also note the more rigorous model of Dugdale and Barenblatt. This model introduces the concept of cohesive forces and leads to a complete solution in plane-plasticity on the basis of the Tresca criterion. The plastic

Fig. 8.8. Plastic zone in plane-stress and plane-strain.

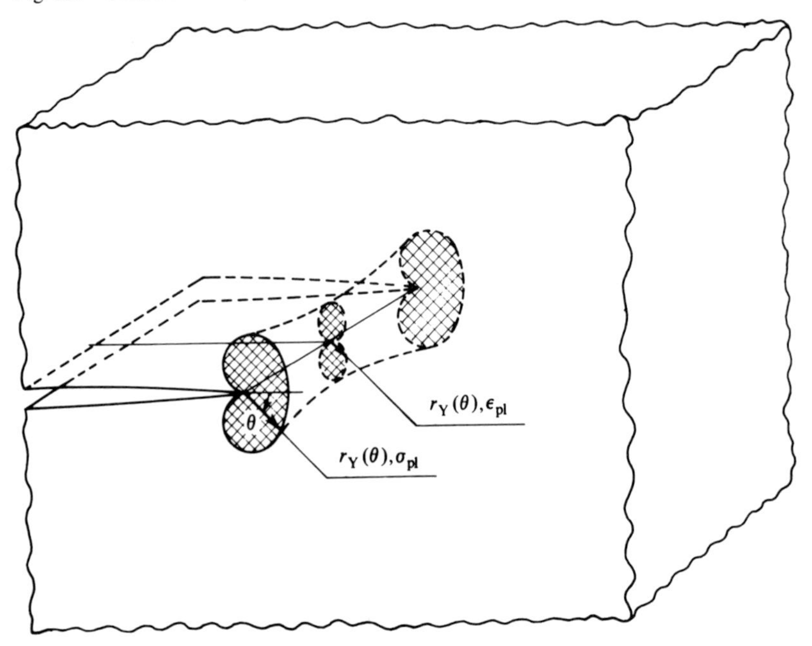

half-zone is then expressed in plane-stress by

$$\rho = \frac{\pi}{16} \frac{K_{\mathrm{I}}^2}{\sigma_{\mathrm{Y}}^2}$$

which is only slightly different from $K_{\mathrm{I}}^2/2\pi\sigma_{\mathrm{Y}}^2$. This model can be used to calculate the opening of the crack, defined by the discontinuity in

Fig. 8.9. Corrected stress fields as functions of the plastic zone.

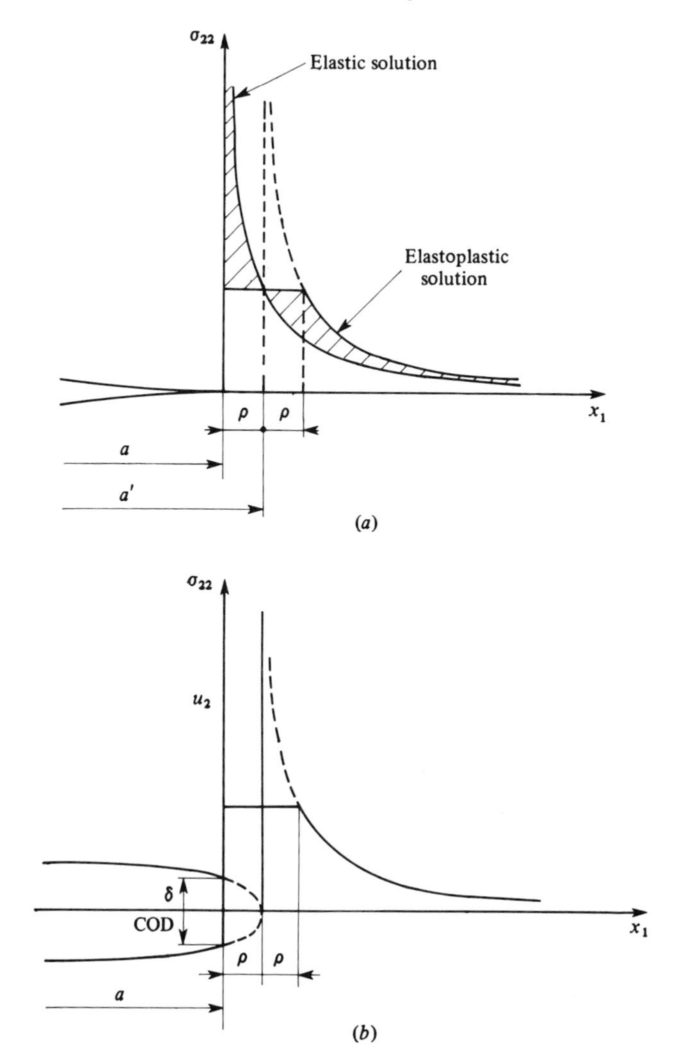

displacement u_2 at the point $x_1 = a$ (Fig. 8.9(*b*). This quantity is often called COD (crack opening displacement) and is given approximately by

$$\bullet \qquad \delta = [\![u_2(x_1 = a)]\!] = \sqrt{2}\frac{K_I^2}{E\sigma_Y}.$$

The same concept when introduced in Irwin's model leads to

$$\delta = \frac{8}{2\pi}\frac{K_I^2}{E\sigma_Y}.$$

Contour integrals in plasticity and viscoplasticity

Rice integral in plasticity

By assuming Hencky type relations (Chapter 5), plasticity is treated as nonlinear elasticity. This scheme is therefore useful only for monotonic proportional loading, since neither unloading nor the presence of residual stresses is taken into consideration. The nonlinear elastic material is described by the strain energy density:

$$W(\boldsymbol{\varepsilon}) = \int_0^{\boldsymbol{\varepsilon}} \sigma_{ij}\,d\varepsilon_{ij}, \quad \sigma_{ij} = \partial W/\partial\varepsilon_{ij}$$

and the Rice integral is defined, as in the linear case, by

$$\bullet \qquad J_p = \int_C [W(\boldsymbol{\varepsilon})n_1 - \sigma_{ij}n_j(\partial u_i/\partial x_1)]\,ds$$

and it still is independent of the choice of the arbitrary contour encircling the crack tip. We can express this quantity in terms of the potential energy as

$$\mathscr{V} = \int_{\mathscr{S}} W(\boldsymbol{\varepsilon})\,dV - \int_{\partial\mathscr{S}_F} F_i^d u_i\,dS$$

$$J_p = -d\mathscr{V}/dA.$$

In plasticity without unloading J_p can thus be considered as the change in the potential energy of two structures whose cracks differ by a length da as illustrated in Fig. 8.10 for the controlled load case.

The variable J_p is used for unconfined plasticity with two complementary aims:

to act as a global parameter associated with the fracture process when linear fracture mechanics proves to be in error, to allow a determination of the stress and strain fields in the vicinity of the crack tip.

When the tension curve can be expressed as a power function

$$\sigma \sim \varepsilon^{1/M}$$

the stress field can be put in the following form (where α is a coefficient):

$$\sigma_{ij} = \alpha(J_p/r)^{1/(M+1)}\sigma_{ij}(\theta, M).$$

When $M = 1$ we recover the elastic case (with $J_p = K_I^2/E$ in plane-stress, for example). With this law, the expression for J_p is always given by

$$J_p = h(a, M)\sigma_\infty^M$$

where σ_∞ is the stress 'at infinity' (far from the crack) and the function h may be tabulated as a function of the crack length and of the hardening exponent (we may use finite elements for these calculations). Note that this formulation is valid only for generalized plasticity, that it presents uncertainties in the transition regime, and that, at least in the case of the power function – in terms of total strain – it neglects the elasticity effects. Moreover, it does not correspond to a correct physical description since the unloading has no meaning.

Contour integral in viscoplasticity

Here again, this integral is introduced for the case of generalized viscoplasticity. In principle, it cannot be justified except in the stationary case. Elasticity is neglected and it is assumed that a power law can be used to

Fig. 8.10. Definition of the Rice integral in plasticity.

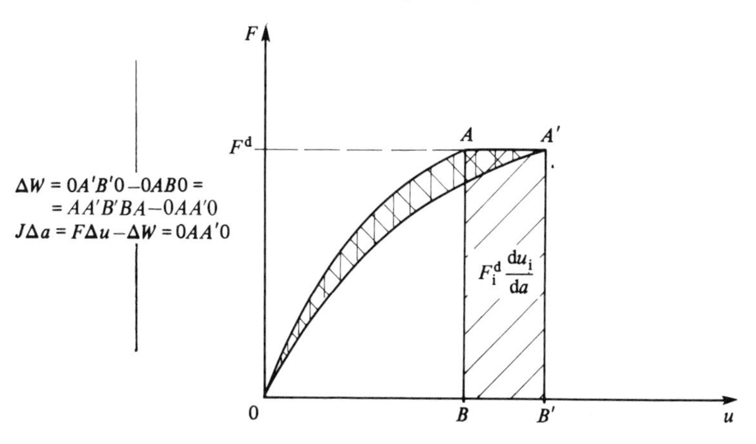

describe the secondary creep, for example, Odqvist's law (Chapter 6):

$$\dot{\varepsilon}_{ij} = \frac{3}{2}\left(\frac{\sigma_{\text{eq}}}{\lambda}\right)^N \frac{\sigma'_{ij}}{\sigma_{\text{eq}}}$$

where σ'_{ij} is the deviator of σ_{ij} and σ_{eq} is the von Mises equivalent stress. Analogous to the nonlinear elastic case, the integral C^* (sometimes denoted by $\dot{J}$) is defined by

• $$C^* = \int_C (W^*(\dot{\boldsymbol{\varepsilon}})n_1 - \sigma_{ij}n_j(\partial \dot{u}_i/\partial x_1))\,\mathrm{d}s$$

with

$$W^*(\dot{\boldsymbol{\varepsilon}}) = \int_0^{\dot{\varepsilon}} \sigma_{ij}\,\mathrm{d}\dot{\varepsilon}_{ij}.$$

This integral may be written in yet another form:

$$C^* = h_1(a, N)\sigma_\infty^{N-1}.$$

The parameter C^* may be used to correlate creep crack growth tests. Note that a correct use of this integral must include tables of the function $h_1(a, N)$ obtained by calculations for each type of specimen. Sometimes the integral is misused, for example by using values of C^* obtained from direct measurements of the COD during the test. The only valid correlations are those where at least one verification is done by means of calculations (or with precalculated tables). In addition, the domain of validity of this integral is very limited: essentially, to (secondary) creep when the viscoplastic zone is not confined in any way. For example, it is not applicable to situations involving relaxation.

8.3 Phenomenological aspects

8.3.1 *Variables governing crack behaviour*

Strain energy release rate, contour integrals and stress intensity factors

Three types of variables characterize the perturbation of the stress field due to the presence of a crack: the stress intensity factors, the contour integrals and the strain energy release rate G. We have seen in Section 8.2.2 that G defines the stable or unstable character of a crack. If $G \geqslant 2\gamma$, the crack grows suddenly. We shall see in Section 8.4.2 that G is a thermodynamic variable associated to the area A of the crack, and this justifies a relationship between G and A. Such phenomenological relations have been demonstrated empirically for ductile fracture and for fatigue crack growth. Since the variables G, J and K are related to each other, at least for plane, linear elastic media, crack behaviour can be described by any one of them.

Historically, the stress intensity factors are the ones which have been most often used. We shall therefore use these variables to present basic experimental results for modelling purposes. Their analytical expressions for some simple situations will be found below.

Usual values for stress intensity factors (After Bui and Barthelemy)

Crack in an infinite medium:

$$K_{\mathrm{I}} = \sigma_\infty (\pi a)^{1/2} \cos^2 \alpha$$
$$K_{\mathrm{II}} = \sigma_\infty (\pi a)^{1/2} \cos \alpha \sin \alpha.$$

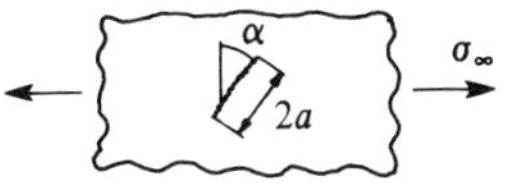

Edge crack in a semiinfinite medium:

$$K_{\mathrm{I}} = 1.122\sigma_\infty (\pi a)^{1/2}.$$

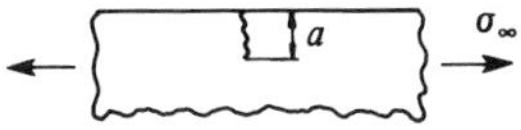

Central crack in a plate of finite width:

$$K_{\mathrm{I}} \approx \sigma_\infty (\pi a)^{1/2} \left[1 - 0.025\left(\frac{a}{b}\right)^2 + 0.06\left(\frac{a}{b}\right)^4\right]\left[\cos\frac{\pi a}{2b}\right]^{-1/2}$$

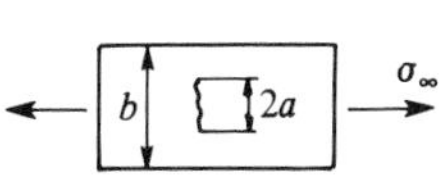

Edge crack in a plate of finite width:

$$K_{\mathrm{I}} \approx \sigma_\infty (\pi a)^{1/2} \left(\frac{2b}{\pi a}\tan\frac{\pi a}{2b}\right)^{1/2} \times \frac{0.752 + 2.02(a/b) + 0.37[1 - \sin(\pi a/2b)]^3}{\cos(\pi a/2b)}.$$

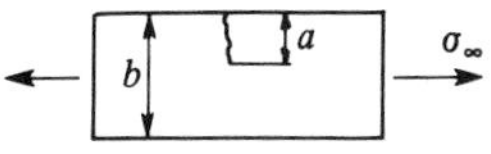

Double edge cracks in a plate of finite width:

$$K_{\mathrm{I}} \approx \sigma_{\infty}(\pi a)^{1/2}\left[1 + 0.122\cos^4\left(\frac{\pi a}{2b}\right)\right]\left[\frac{2b}{\pi a}\tan\frac{\pi a}{2b}\right]^{1/2}.$$

Crack in a bar subjected to bending:

$$K_{\mathrm{I}} \approx \sigma_{\mathrm{Max}}(\pi a)^{1/2} \times\left[1.122 - 1.4\frac{a}{b} + 7.33\left(\frac{a}{b}\right)^2 - 13.08\left(\frac{a}{b}\right)^3 + 14\left(\frac{a}{b}\right)^4\right].$$

Crack in a bar subjected to three point bending ($L/b = 8$):

$$K_{\mathrm{I}} \approx \frac{3PLa^{1/2}}{2b^2e} \times\left[1.96 - 2.75\frac{a}{b} + 13.66\left(\frac{a}{b}\right)^2 - 23.98\left(\frac{a}{b}\right)^3 + 25.22\left(\frac{a}{b}\right)^4\right].$$

CT 15 specimen:

$$K_{\mathrm{I}} \approx \frac{Pa^{1/2}}{be} \times\left[29.6 - 185.5\frac{a}{b} + 655.7\left(\frac{a}{b}\right)^2 - 1.017\left(\frac{a}{b}\right)^3 + 638.6\left(\frac{a}{b}\right)^4\right].$$

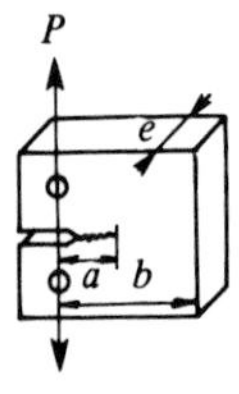

Semielliptical crack in a shell:

$$K_{\mathrm{I}}(M) \approx 0.5(1.2\pi a)^{1/2}\sigma_{\infty}\left[2+\frac{0.731e}{R(1-\nu^{2})}\left(\frac{b}{a}\right)^{0.043}\left(\frac{e}{a}\right)^{0.047}\right]$$

Elliptical crack in an infinite medium:

$$K_{\mathrm{I}}(\theta)=\frac{\sigma_{\infty}}{H}\left(\frac{b}{a}\right)^{1/2}(a^{2}\sin^{2}\theta+b^{2}\cos^{2}\theta)^{1/4}$$

with

$$H=\int_{0}^{\pi/2}\left\{1-\left[1-\left(\frac{b}{a}\right)^{2}\right]\sin^{2}\theta\right\}^{1/2}\mathrm{d}\theta.$$

Crack emanating from a hole:

$$K_{\mathrm{I}} \approx f\left(\frac{a}{R}\right)\sigma_{\infty}(\pi a)^{1/2}.$$

a/R	0	0.1	0.2	0.5	1	2	5
$f(a/R)$	3.39	2.73	2.30	1.73	1.37	1.06	0.81

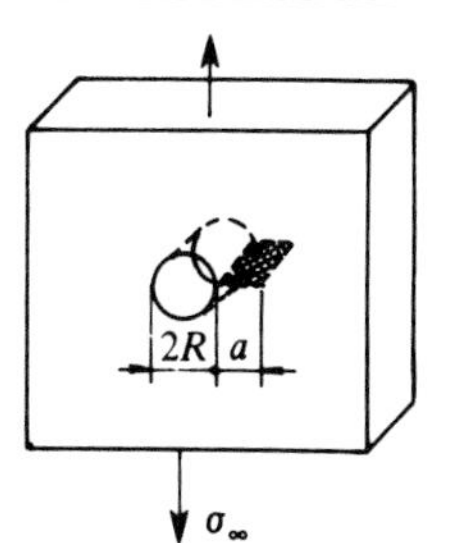

Nontransverse crack emanating from a hole:

$$K_{\mathrm{I}}(M) \approx 3.36\sigma_{\infty}\left(\frac{\pi a}{H}\right)^{1/2}\left(\frac{2e}{\pi a}\tan\frac{\pi a}{2e}\right)^{1/2}$$

$$H = \int_0^{\pi/2} \{1 - [1 - (a/c)^2]\sin^2\theta\}^{1/2}\,\mathrm{d}\theta.$$

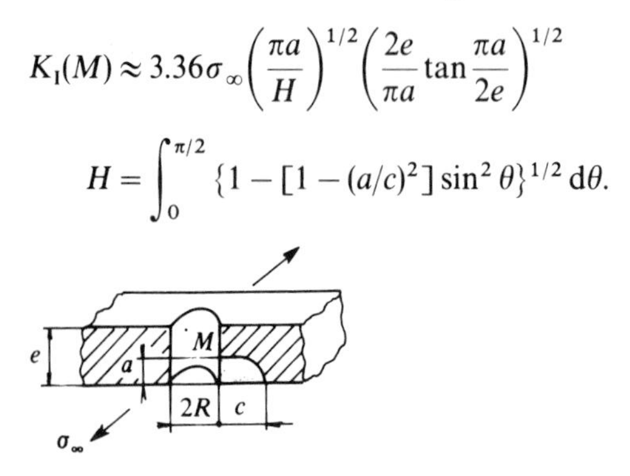

Crack in a notch:

$K_{\mathrm{I}} \approx 1.1215\sigma_{\infty}(\pi a)^{1/2}(1 + c/a)^{1/2}$ for $a/R \geqslant 0.4$.

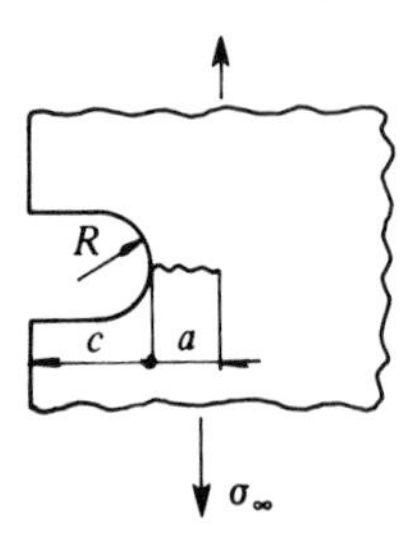

8.3.2 ***Elementary experimental results***

Brittle fracture. Toughness of common materials

If a cracked specimen of a brittle-elastic material is subjected to monotonically increasing loads in mode I, its response is initially elastic without any

Fig. 8.11. Critical force for brittle fracture.

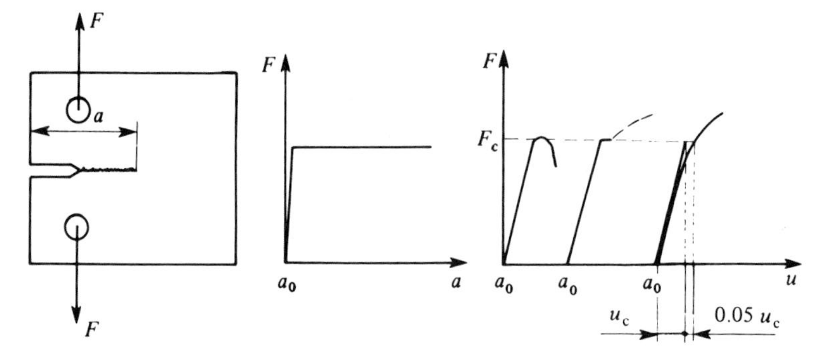

Table 8.1. *Toughness of common materials at room temperature (after Barthelemy and Kausch).*

Materials		Treatment	K_{Ic} (MPa m$^{1/2}$)	K_{IScc} (MPa m$^{1/2}$)
Aluminium alloys	2024	T351	≈ 35	
		T851	≈ 25	
	7075	T651	≈ 27	
		T351	≈ 35	
Titanium alloys	T21		≈ 80	
	T68		≈ 130	
Steels	30 CD 2 B		≈ 110	≈ 30
	30 CND 8		≈ 120	≈ 20
	45 CD 4		≈ 90	≈ 30
	18 Ni(300) Maraging		≈ 65	≈ 8
Polymers			≈ 3	
Concrete			≈ 1	
Wood			≈ 2	

crack growth. However, when conditions of instability are reached, there is a sudden growth of the crack until the specimen fractures. The corresponding graphs are shown in Fig. 8.11. Since common materials are never perfectly brittle, the practical determination of the critical fracture force F_c presents some difficulties when the curve displays a step or is significantly nonlinear. The conventional method is shown in Fig. 8.11.

Knowing the geometry of the specimen, we can calculate the stress intensity factor in mode I corresponding to the critical force:

$$K_{Ic} = K_I(F_c).$$

Experiment shows that for a given thickness of the specimen, the stress intensity factor value is almost independent of the length of the crack and the specimen shape. Some experimental values are given in Table 8.1.

In the state of plane-stress, we obtain the critical stress intensity factor for brittle fracture of thin metal sheets: K^{σ}_{Ic}.

In the state of plane-strain, we obtain the critical stress intensity factor which characterizes the toughness of the material, and which can be 10–20% lower than K^{σ}_{Ic}, sometimes by as much as a factor of 2–3. It is found that this incompatibility stems from the assumption of plane-stress (see Section 8.2.2). In order to realize the condition of plane-strain over 90–95% of the thickness of the specimen, the latter must be such that

$$e \geqslant 2.5(K_{Ic}/\sigma_Y)^2.$$

The action of a hostile environment decreases the toughness of materials. In stress corrosion, for example, the concept of brittle fracture K_{Ic} is replaced by a long term critical value $K_{\mathrm{IScc}} < K_{\mathrm{Ic}}$. For strongly anisotropic media, we may also define toughness in mode II, K_{II}; and in mode III, K_{III}. Generaly, however, pure mode II and mode III do not exist; they are unstable and the crack deviates towards the mode I type.

Ductile fracture. Resistance curves or 'R' curves

The experiment, described in the preceding section, when performed on a ductile material, shows the possibility of stable crack growth before fracture by instability as indicated in Fig. 8.12. The term stable is related to the fact that the crack does not propagate if the external loads are kept constant.

Knowing the loading force F and the length a of the crack, either by calculation or by measuring the stiffness of the specimen under consideration (see Section 8.4.2.), we can transform the graph $F(a)$ to that of $K_{\mathrm{I}}(a)$ or $G(a)$.

For a given thickness the experiment shows that the curve obtained is almost independent of the geometry of the specimen; it is therefore characteristic of the material and the thickness. In contrast, the point of instability depends on the crack length. This is easily shown by using again the energy-based analysis of Section 8.2.2. The condition of brittle fracture, $G \geqslant 2\gamma$, must now be replaced by a condition of ductile fracture where the resistance $R(a)$ of a cracked medium is a function of the crack length. During stable growth we have

$$G(a, F) = R(a - a_0).$$

Fig. 8.12. Graphs of ductile fracture.

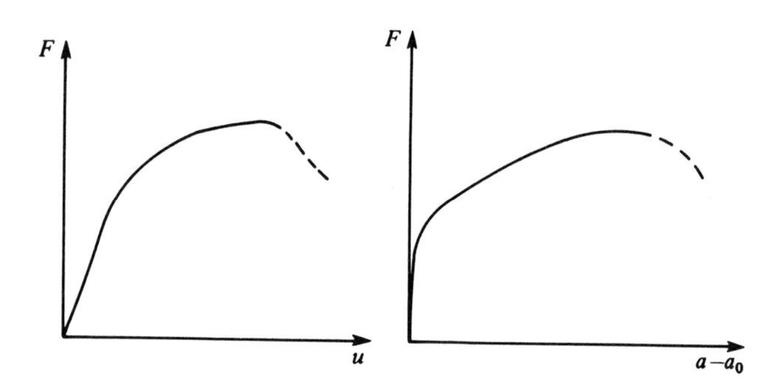

This case is represented by the point P_s in Fig. 8.13. The graph of the function $G(a)$ (approximately linear in a) is such that for any virtual growth of the crack under constant load, we have:

$$G < R.$$

On the other hand, when the load is such that the line $G(a, F = \text{constant})$ is a tangent to the curve $R(a - a_0)$ at point P_i with abscissa a_c, any virtual growth of the crack leads to:

$$G > R.$$

This is the instability condition; The critical crack length is defined by

$$G = R \quad \text{and} \quad \partial G/\partial a = \mathrm{d}R/\mathrm{d}a.$$

From the shape of the 'R' curve, we observe that a_c increases with a_0. Therefore, for the experimental determination of the 'R' curves, it is better to perform experiments on specimens with large initial cracks, but with a sufficiently large 'ligament'.

Fatigue fracture

Let us consider the situation in which a single specimen of a brittle or a ductile material is loaded in mode I, but this time by a periodic load defined by its range ΔF and the ratio R of its minimum value F_m to its maximum F_{Max}. In general the crack propagates with a per cycle growth rate $\delta a/\delta N$ which increases with crack length. If the initial crack is an artificial one, or is incompatible with the applied load, a certain number of cycles N_i is necessary to initiate the propagation (or to produce a deviation of the

Fig. 8.13. The 'R' curve.

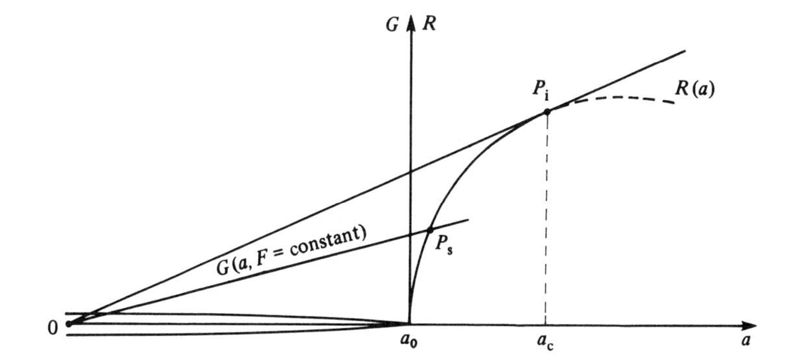

crack). When the crack length associated with the applied load reaches the instability condition, catastrophic failure takes place (Fig. 8.14).

Paris' law

The same type of experiments conducted with specimens of different geometries, but always in mode I loading and with almost zero minimum force, $R \approx 0$, show that a simple relation exists between the crack growth rate and the range of the stress intensity factor defined by

$$\Delta K = K_r(a)\Delta F$$

where K_r is the reduced stress intensity factor corresponding to a unit load

$$K_r = K_I(a, F = 1).$$

In 1962, Paris proposed the law

● $$\delta a/\delta N = C\Delta K^\eta$$

where C and η are constants which are characteristic of the material. Fig. 8.15 compares the results of different experiments performed on cracked panels under cyclic tension. Table 8.2 gives some typical values for C and η.

The influence of the mean loading, i.e., the load ratio, is not reproduced correctly by Paris' law. It may be corrected by a multiplicative influence function:

$$\delta a/\delta N = C(K_{Max} \cdot f(R))^\eta.$$

Fig. 8.14. Growth of the crack by fatigue.

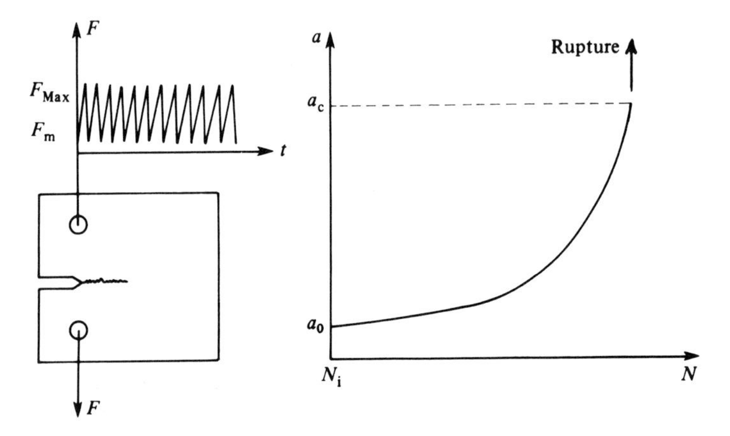

Paris' law corresponds to $f(R) = 1 - R$. Different expressions may be chosen for f; they are illustrated in Fig. 8.16 with some experimental results for the light alloy AU4Gl T3.

Forman's law

If we examine the typical curve of Fig. 8.15 closely, we find that Paris' law does not represent the existence of two asymptotes which define fracture by

Fig. 8.15. Curve of fatigue induced crack growth: AU2GN T6 alloy, thickness 1.6 mm, T = 20 °C (after G. Baudin).

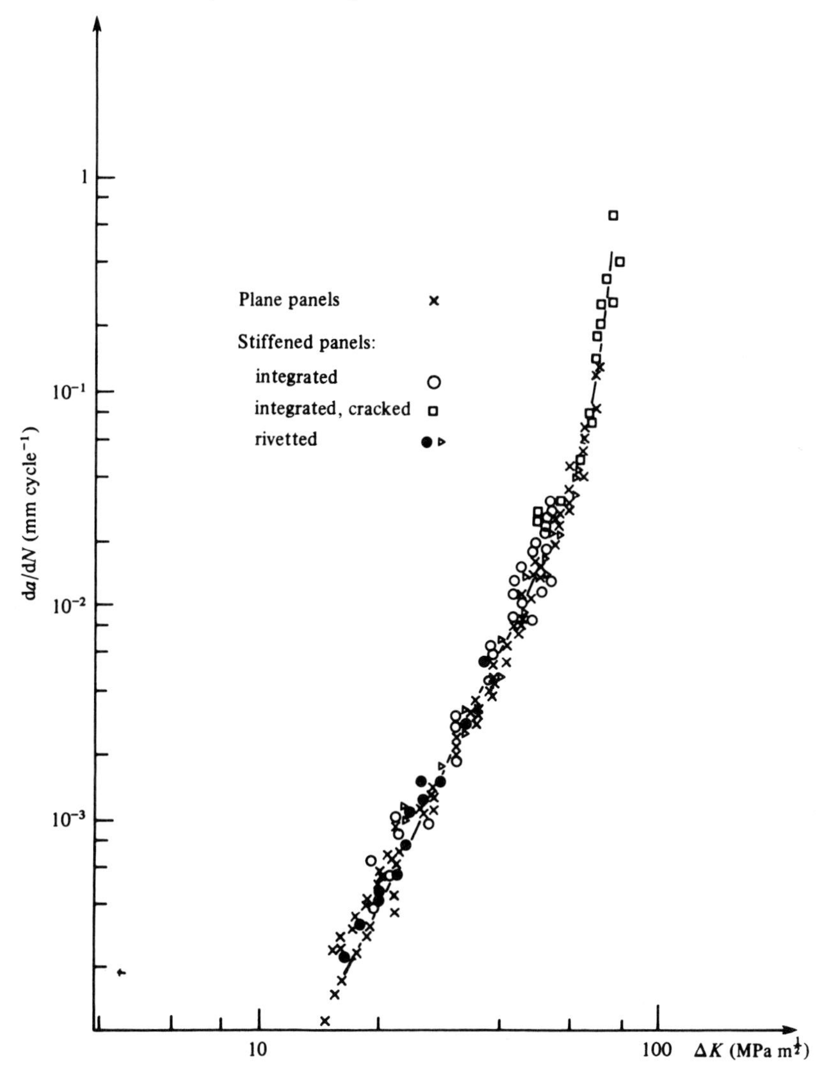

Table 8.2. *Examples of the values of the coefficients of Paris' law*

	Materials	C (mm cycle^{-1} (MPa m$^{1/2}$)$^{-\eta}$)	η
Aluminium alloys	AU4G T3	10^{-7}	2.9
	AU2GN	1.2×10^{-8}	4.2
Steels	NiCrMoV	6.4×10^{-14}	1.4
	35 CD 4	5.2×10^{-9}	2.9
	CrMoV	1.3×10^{-10}	6.7
	32 CDV 13	5.01×10^{-8}	2.3
	TA 6Zr 5D	1.04×10^{-12}	5.7
	NK 17 CDAT (600 °C)	2.27×10^{-8}	2.8
	NC 20 K 14 V (700 °C)	1.8×10^{-7}	2.2

Fig. 8.16. Load ratio influence functions (a_s is the crack length at the instant of the overload).

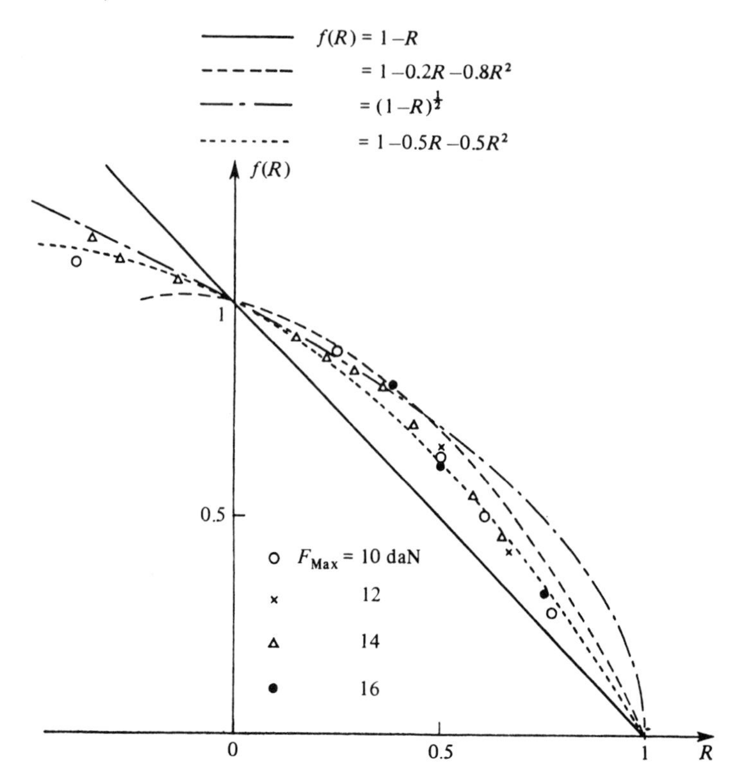

instability and the existence of the threshold below which there is almost no crack growth:

$$K_I \to K_{Ic} \Rightarrow \delta a/\delta N \to \infty$$
$$K_I \to K_{th} \Rightarrow \delta a/\delta N \to 0.$$

Among the hundreds of mathematical models originating from the wild imagination of researchers during the 1970s to represent these two asymptotes, we mention the model of Forman:

$$\frac{\delta a}{\delta N} = \frac{C_1 \Delta K^{\eta_1}}{(1-R)(K_{Ic} - K_{Max})}.$$

This can be improved by introducing K_{th}, the crack growth threshold and a coefficient m which can be used to adjust the effect of the mean force better, so that

$$\bullet \qquad \frac{\delta a}{\delta N} = C_2 \left(\frac{K_M \dfrac{1-R}{1-mR} - K_{th}}{K_{Ic} - K_{Max}} \right)^{\eta/2}.$$

The coefficient m is usually of the order of 0.5.

Effects of the loading history

The plastic zone at the crack tip induces a field of residual stresses which tends to delay the opening of the crack under increasing load. When the load is no longer periodic this effect depends on the history of the local plastic deformations. The most important characteristic effect is the crack propagation delay induced by overloads.

If a periodic load is interrupted by one cycle and replaced by a load with significantly greater amplitude, the ensuing crack growth rate, after a short acceleration, diminishes rapidly and returns to the value it would have acquired without the oberload only after a number of cycles, called the delay to crack growth. This effect is displayed qualitatively in Fig. 8.17 according to the schematic representation due to Wheeler.

If there had been no overload, the crack would have progressed with a plastic zone of size equal to

$$\rho_{SS} = K_{Max}^2/2\pi\sigma_Y^2;$$

at the moment of overload $F_{Max\,S}$ the plastic zone becomes

$$\rho_{AS}^0 = K_{Max\,S}^2/2\pi\sigma_Y^2,$$

then the crack progresses with a plastic zone ρ_{AS} greater than ρ_{SS}

$$\rho_{\text{AS}} = \rho_{\text{AS}}^0 - (a - a_{\text{S}})$$

until the moment when the crack length is such that

$$\rho_{\text{AS}} = \rho_{\text{SS}}(K_{\text{Max}})$$

i.e., when

$$a^* = a_{\text{S}} + \frac{K_{\text{Max S}}^2 - K_{\text{Max}}^2}{2\pi\sigma_{\text{Y}}^2}.$$

This is the crack length for which the growth rate no longer depends on the history of the overload. It is understood that the existence of a larger plastic zone modifies the residual stresses in front of the crack tip and also the conditions of closure when the crack has grown: it closes for greater values of the load than it does without overload and thereby the effective amplitude of the stress intensity factor is reduced.

The Wheeler model consists in introducing a power function of the ratio

Fig. 8.17. Effect of an overload on the crack growth.

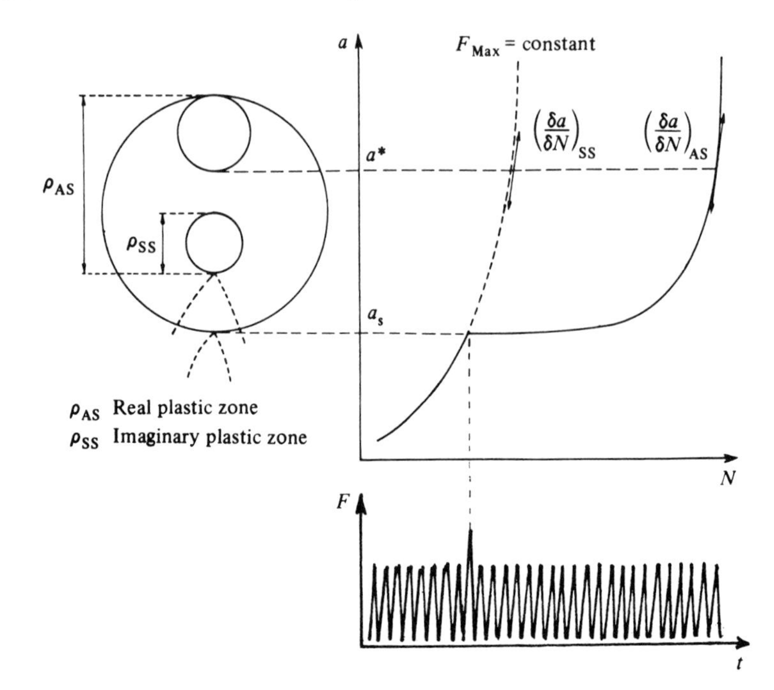

ρ_{SS}/ρ_{AS} in Paris' law

$$\delta a/\delta N = C\Delta K^{\eta}(\rho_{SS}/\rho_{AS})^{\beta}.$$

Another history effect correlated with the previous one is the influence of overloads on the crack growth threshold. This threshold introduced in Forman's law is not a characteristic property of the material except in the case of a virgin crack. Since it is related to the opening of the crack lips, it depends on the size of the plastic zone that borders the crack and the residual stress field. Technical details regarding its detection and measurement will be given in Section 8.5.4.

8.4 Thermodynamic formulation

In Sections 8.2 and 8.3, we have essentially dealt with linear fracture mechanics, restricted to plane media in plane-stress or plane-strain. It is valid as long as the size of the plastic zone remains small compared to the length of the crack. However, generalization of the concepts of stress intensity factors and contour integrals to three-dimensional problems with large scale plasticity and viscoplasticity is a difficult matter. The thermodynamic global approach such as that employed in the preceding chapters which generalizes the concept of the energy release rate, seems more accurate although it masks certain local effects which it cannot take into account. For these local effects, the natural complementary approach is that of damage theory applied to crack phenomena as described in Section 8.6.4.

The approach followed here is valid for a brittle-elastic solid, and by extension, for very localized plastic deformations at the crack tip. It will be seen that the concept of the rate of energy released by crack growth can be generalized to cases where plastic or viscoplastic deformations are important in a large part of the structure. This can only be done by a global method by considering the fact that the energy available to propagate a crack is an elastic energy and not that which is stored in dislocations or dissipated as heat. The approach developed here cannot be used to find local stresses and strains due to plasticity, but it does complement the approach based on using semiglobal parameters, such as J or C^*, as described in Section 8.2.3. More precisely, crack evolution can be described with the help of an intensity parameter, suitable for elasticity (the stress intensity factor or the elastic energy release rate) and incorporating the plasticity effects through a set of internal variables.

8.4.1 *Choice of variables. Thermodynamic potential*

Let us consider a three-dimensional solid $\mathcal{S}$ which may be elastic, elastoplastic, or viscoplastic. To simplify the presentation let the temperature T be uniform. Let the solid contain a crack of area A and contour $\Gamma(x_1, x_2, x_3)$ and let it be subjected to external forces (and moments) represented by static parameters F_q in a finite dimension vector space (Fig. 8.18(*a*)). This notation implicitly includes the case of distributed loads assumed to evolve as a function of one or several of these discrete parameters F_q. The general description without discretization does not pose any problem of principle.

We have seen in Section 8.3 that from a phenomenological view point, the action of the loads F_q can result in:

> a displacement, eventually elastoplastic, represented here by the kinematic parameters u_q; each displacement may be decomposed into a part called permanent u_q^p, defined as the displacement corresponding to the relaxed configuration ($F_q = 0$) and a part called 'elastic' u_q^e so that
>
> $u_q = u_q^e + u_q^p$;

Fig. 8.18. Three-dimensional cracked media: (*a*) geometry of the crack; (*b*) load–displacement curve.

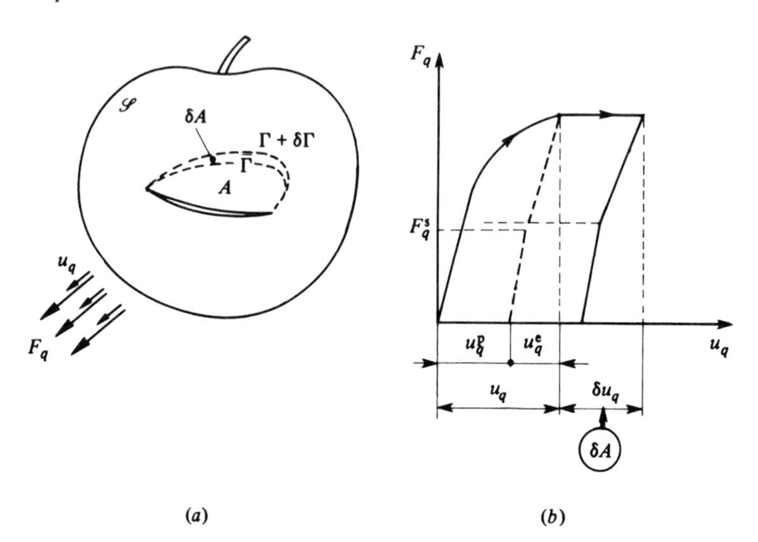

Table 8.3. *Crack variables*

Observable variables	Internal variables	Associated variables
u_q		F_q
T		S
	u_q^{e}	F_q
	u_q^{p}	F_q
	A	G^*
	h_{S}	H_{S}
	h_{p}	H_{p}

a growth of the crack compatible with the loading, manifested by a change in the crack front contour from Γ to $\Gamma + \delta\Gamma$;
a change in the crack growth threshold, represented here by the totality of the static parameters F_q^{S} whose values depend on the mode of the previous loading. This threshold can be associated with the phenomena of opening and closing of the crack, i.e., to a change in the stiffness of the structure (see Fig. 8.18(*b*)).

Consequently, with reference to Chapter 2, the variables for a global thermodynamic analysis of a cracked solid $\mathscr{S}$, isolated as a whole like a volume element in the previous chapters, are the following (see also Table 8.3):

(*a*) The observable variable T, the temperature assumed uniform throughout the solid, and the associated variable S, the total entropy

$$S = \int_{\mathscr{S}} s \, \mathrm{d}V;$$

(*b*) Observable variables u_q, the displacements associated with forces F_q.

(*c*) Internal variables u_q^{p}, the permanent displacements introduced in the same way as plastic strains in plasticity (Chapter 5), the associated dissipation being

$$\Phi = F_q \dot{u}_q^{\mathrm{p}};$$

(*d*) The elastic displacements, derived as $u_q^{\mathrm{e}} = u_q - u_q^{\mathrm{p}}$, define the reversible power of external forces for the whole solid

$$P_{\mathrm{e}} = F_q \dot{u}_q^{\mathrm{e}};$$

(*e*) A variable that should be included naturally as an internal variable is the crack configuration represented by its contour $\Gamma(x_1, x_2, x_3)$. However, in Sections 8.4.2–8.4.4 it will be sufficient to represent the geometry of the crack by its area A. Of course, in plane problems, the configuration of the crack is quite measurable and is an internal variable only in the thermodynamic sense. It will be shown that in cases where there is no threshold (see below), the variable associated with the crack area is the elastic energy release rate G, thus generalizing the definition of Section 8.2.2. We should note this quantity is not expected to represent the local stress field (in nonconfined plasticity) exactly; it should be considered as a global variable associated with the crack in the same way as in continuum mechanics the stress is associated with the strain in a volume element. By definition the dissipation associated with the crack propagation process is $G\dot{A}$;

(*f*) To represent all phenomenological aspects of the crack propagation process, it is necessary to introduce an internal variable which characterizes the change in crack propagation threshold represented by the parameters F_q^{S}. If we denote this internal variable by h_{S} and the associated variable by H_{S}, then $H_{\mathrm{S}}\dot{h}_{\mathrm{S}}$ is the energy, related to the residual stresses, which plays a part in varying the crack propagation threshold. In the presence of this threshold, the variable associated to A is $G^* = \langle G^{1/2} - G_{\mathrm{S}}^{1/2} \rangle^2$ as shown in Section 8.4.3, where G_{S} depends on A and h_{S};

(*g*) To represent the process related to plastic flow, whether or not confined to the crack tip, we use the additional internal variables h_{p} and the associated variables H_{p}. The physical significance of h_{p} is not clear; we may consider them as representing plastic displacements u_q^{p}. In practice, it will be sufficient to consider the crack opening displacement to be such a variable.

Let Ψ be the thermodynamic potential associated with the whole solid. According to the method of Chapter 2, it is a function of all the observable and internal state variables

$$\Psi = \Psi(u_q, u_q^{\mathrm{e}}, u_q^{\mathrm{p}}, T, A, h_{\mathrm{S}}, h_{\mathrm{p}})$$

Since the displacements appear only as $u_q^{\mathrm{e}} = u_q - u_q^{\mathrm{p}}$, we have

$$\Psi = \Psi(u_q^{\mathrm{e}}, T, A, h_{\mathrm{S}}, h_{\mathrm{p}}).$$

As for damage, studied in Chapter 7, we assume that there is a decoupling between the plastic and the thermoelastic effects associated with the crack phenomena. Thus

$$\Psi = \Psi_{\mathrm{e}}(u_q^{\mathrm{e}}, T, A, h_{\mathrm{S}}) + \Psi_{\mathrm{p}}(T, h_{\mathrm{S}}, h_{\mathrm{p}}),$$

and by definition, we have

$$F_q = \partial \Psi / \partial u_q^{\mathrm{e}}$$

$$S = -\partial\Psi/\partial T$$
$$G^* = -\partial\Psi/\partial A$$
$$H_S = \partial\Psi/\partial h_S$$
$$H_p = \partial\Psi/\partial h_p.$$

8.4.2 *Elastic strain energy release rate*

If the evolution of the crack front depends on only one parameter (e.g. a symmetric line crack in a two-dimensional medium, a circular crack in a three-dimensional medium, etc.), then the crack is completely characterized by its area A, with its evolution dependent on the associated variable G. We limit ourselves here to the case with no threshold, which means $G^* = G$. The effect of threshold is considered later.

Definition and mechanical meaning

In the framework of elastic, elastoplastic, or viscoplastic materials with linear elastic behaviour, we may postulate that the thermodynamic potential is a positive definite quadratic form in elastic displacements. This assumes that the elastic unloading defining the relaxed configuration is linear, which amounts to neglecting the influence of the residual stress field on the thermodynamic potential. This is the extension of the classical elastoplastic assumption for the volume element (with or without damage) to the whole solid. It will be seen that, according to this scheme, the residual stresses are considered through the internal variables (cf. Section 8.4.3).

Denoting by $[R]$, the generalized stiffness matrix of the cracked solid $\mathscr{S}$, we have

$$F_q = R_{qr}u_r^e$$
$$\Psi = \tfrac{1}{2}R_{qr}(T, A)u_q^e u_r^e + \Psi_p$$

which in simplified notation can be written as

$$\Psi = \tfrac{1}{2}\{u_e\}^T[R]\{u_e\} + \Psi_p.$$

The variable G is derived as

$$G = -\partial\Psi/\partial A = -\tfrac{1}{2}\{u_e\}^T[\partial R/\partial A]\{u_g\}$$

which, by virtue of the load–displacement relation (and symmetry of $[R]$), can be written as

$$G = -\tfrac{1}{2}\{F\}^T[R]^{-1}[\partial R/\partial A][R]^{-1}\{F\}.$$

Note that this definition implies no hypothesis on the geometry of the cracked medium nor on its elastic, plastic, or viscoplastic nature except the existence of a relaxed configuration obtained by linear elastic unloading. It generalizes the expression for the elastic energy release rate introduced in Section 8.2.2.

For loading by prescribed forces the expression for G is simplified upon using

$$\{\mathrm{d}F\} = 0 = [\partial R/\partial A]\{u_\mathrm{e}\}\,\mathrm{d}A + [R]\left\{\frac{\partial u_\mathrm{e}}{\partial A}\right\}_F \mathrm{d}A.$$

The displacements $\{u_\mathrm{e}\}$ depend on A by virtue of equilibrium and we have

$$[\partial R/\partial A] = -[R]\{\partial u_\mathrm{e}/\partial A\}_F\{u_\mathrm{e}\}^{-1}$$
$$G = \tfrac{1}{2}\{u_\mathrm{e}\}^\mathrm{T}[R]\{\partial u_\mathrm{e}/\partial A\}_F = \tfrac{1}{2}\{F\}^\mathrm{T}\{\partial u_\mathrm{e}/\partial A\}_F.$$

For prescribed displacements, we similarly obtain

$$G = -\tfrac{1}{2}\{u_\mathrm{e}\}^\mathrm{T}\{\partial F/\partial A\}_u$$

or more generally,

- $$G = \tfrac{1}{2}\left(\{F\}^\mathrm{T}\left\{\frac{\partial u_\mathrm{e}}{\partial A}\right\}_F - \{u_\mathrm{e}\}^\mathrm{T}\left\{\frac{\partial F}{\partial A}\right\}_u\right) = \tfrac{1}{2}\left(F_q\frac{\partial u_q^\mathrm{e}}{\partial A} - u_r^\mathrm{e}\frac{\partial F_r}{\partial A}\right)$$

where subscripts q and r distinguish the zones where forces and displacements respectively are prescribed. We also recover a relation analogous to that obtained in Section 8.2.2. namely

$$G = \frac{1}{2}\left[\int_{\partial\mathscr{S}_F} \vec{F}^\mathrm{d}\cdot\frac{\partial\vec{u}}{\partial A}\,\mathrm{d}S - \int_{\partial\mathscr{S}_u} \vec{u}^\mathrm{d}\cdot\frac{\partial\vec{F}}{\partial A}\,\mathrm{d}S\right].$$

In particular, in the simple case of a single prescribed force

- $$G = \frac{1}{2}F\frac{\mathrm{d}u_\mathrm{e}}{\mathrm{d}A} = -\frac{1}{2}\frac{F^2}{R^2}\frac{\mathrm{d}R}{\mathrm{d}A}.$$

We thus find an identity of analysis in the treatments of crack phenomena and that of damage in Chapter 7: A and G play the same role in a cracked solid as D and $-\dot{Y}$ do in a damaged volume element.

The elastic energy release rate can be visualized as in Fig. 8.19(a) for the classical case of 'brittle-elastic' material without initial stresses, when the external load is prescribed. $G\delta A$ is equal to half the work of the given external loads through the corresponding displacements (the load being assumed constant during crack growth).

The use of this variable for the cracked solid subjected to unconfined plastic or viscoplastic deformations, is justified, in a purely macroscopic context, by the following hypotheses:

the energy available to effect decohesion at the crack tip is necessarily an elastic energy provisionally stored in the structure;

the fracture process takes place under an increasing external load;

the relaxed configuration is to be considered in a fictitious fashion for an infinitesimal unloading. In this case there is plastic flow in compression, but it remains confined and the usual conditions of linear fracture mechanics continue to apply. G is then defined by the difference between the stiffness existing before and after an infinitesimal increment δa in the crack length. Fig. 8.19 shows this hypothesis for the case where the crack is assumed to occur at the maximum load and where the effects of threshold and flow during the propagation are neglected.

Note that, according to this definition, in plane problems, the variable G is independent of the thickness e in the same way as the stress intensity factor K is. This evidently holds true for a fixed stress value 'at infinity'; in a plate of width b containing a small central crack of length $2a$, we have $K = \sigma_\infty(\pi a)^{1/2}$ regardless of the thickness e. The elastic energy release rate is

Fig. 8.19. Visualization of elastic energy release rate; (*a*) brittle-elastic solid; (*b*) brittle elastoplastic solid without threshold effects: (1) flow during loading; (2) real unloading before and after crack propagation; (3) fictitious elastic unloading.

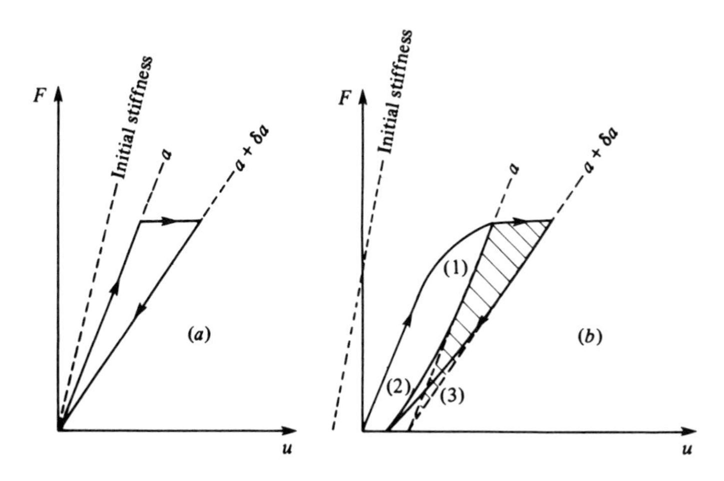

expressed as follows:

$$G = \frac{1}{2} F \frac{du}{dA} = \frac{1}{2} \sigma_\infty be \frac{du}{dA} = \frac{1}{2} \sigma_\infty be \frac{1}{e} \frac{du}{da}$$

$$G = \frac{1}{2} \sigma_\infty b \frac{du}{da}.$$

The unit of G is, for example, $N\,m^{-1}$ or MPa m. It is the same unit as that of K^2/E.

Methods of determining G

We first consider the case in which crack growth depends only on a single parameter: its area A. This condition is approximately realized if the crack is compatible with the external load (Section 8.2.1) and if the latter depends only on one parameter. The case in which there is bifurcation of the crack will be discussed in Section 8.4.5.

Formulae for bounds on G

The formula $G = \frac{1}{2}[F_q(\partial u_q^e/\partial A) - u_r^e(\partial F_r/\partial A)]$ can be exploited to calculate G directly by numerical methods, for example the finite element method (see Chapters 4, 5 and 6 for its use in elasticity, plasticity or viscoplasticity). However, since a change with respect to the crack area is required, at least two calculations must be performed.

The partial derivatives are replaced by their finite difference approximations arising from the calculation performed on two neighbouring configurations A and $A + \delta A$:

$$G = \frac{1}{2}\left(F_q \frac{\delta u_q^e}{\delta A} - u_r^e \frac{\delta F_r}{\delta A} \right)$$

with

$$\delta u_q^e = u_q^e(F_q, A + \delta A) - u_q^e(F_q, A)$$
$$\delta F_r = F_r(u_r^e, A + \delta A) - F_r(u_r^e, A).$$

This method assumes that there is no change in the boundary conditions during the growth δA of the crack. However, in practice the boundary conditions do change in an unknown fashion because of dynamic effects and imperfectly rigid bonds, even though the prescribed forces and displacements are constant. Experiments show that the real displacements are increased ($\delta u_q^{e*} > 0$) and the forces are decreased ($\delta F_r^* < 0$). Two hypotheses lead to an upper and a lower bound on the energy release rate:

the external loads are assumed to be constant everywhere during the growth of the crack ($\delta F_r = 0$) so that

$$G_F = \tfrac{1}{2} F_q (\delta u_q^{\mathrm{e}} / \delta A);$$

the displacements are assumed to be constant everywhere during the growth of the crack ($\delta u_q^{\mathrm{e}} = 0$) so that

$$G_u = -\tfrac{1}{2} u_r^{\mathrm{e}} (\delta F_r / \delta A).$$

These two values are bounds on the real G which correspond to $\delta u_q^{\mathrm{e}*} > 0$ and $\delta F_r^* < 0$

$$G = \frac{1}{2}\left(F_q \frac{\delta u_q^{\mathrm{e}*}}{\delta A} - u_r^{\mathrm{e}} \frac{\delta F_r^*}{\delta A} \right).$$

In fact, the definition of the stiffness R_{qr} before the growth of the crack and R'_{qr} after it, shows that (Fig. 8.20):

$$F_q = R_{qr} u_r^{\mathrm{e}}, \quad F_q = R'_{qr}(u_r^{\mathrm{e}} + \delta u_r^{\mathrm{e}}), \quad F_q + \delta F_q^* = R'_{qr}(u_r^{\mathrm{e}} + \delta u_r^{\mathrm{e}*}),$$
$$F_q + \delta F_q = R'_{qr} u_r^{\mathrm{e}}.$$

Introduction of these relations into the expression for G leads to:

$$G = \tfrac{1}{2}(F_q + \delta F_q^*)(\delta u_q^{\mathrm{e}} / \delta A) = -\tfrac{1}{2}(\delta F_r / \delta A)(u_r^{\mathrm{e}} + \delta u_r^{\mathrm{e}*})$$

or

$$G = G_F + \tfrac{1}{2} \delta F_q^* (\delta u_q^{\mathrm{e}} / \delta A) = G_u - \tfrac{1}{2}(\delta F_r / \delta A) \delta u_r^{\mathrm{e}*}.$$

Fig. 8.20. Crack growth due to prescribed load or prescribed displacement.

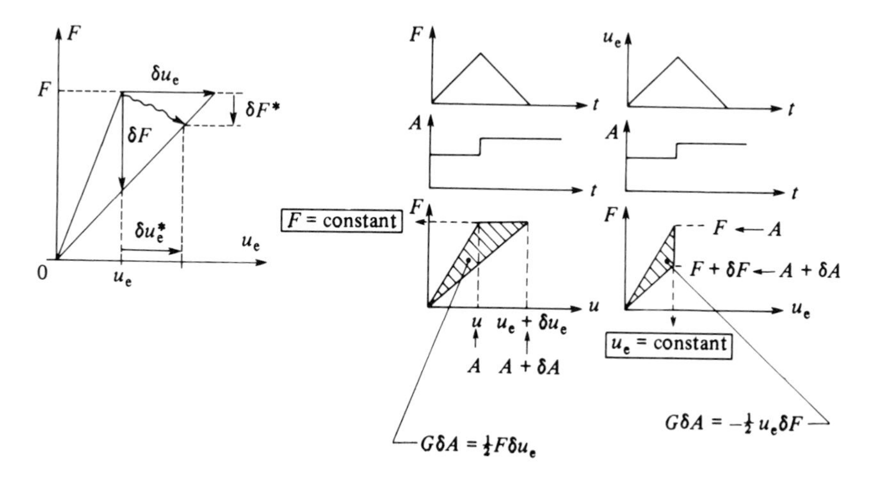

Since δF_q^* and δF_r are negative while δu_q^e and δu_r^{e*} are positive, we have:

$$G_u < G < G_F.$$

This important result is schematically represented in Fig. 8.20. The numerical schemes implementing this method are described in Section 8.6.2.

Experimental stiffness measurement method

Since the definition of the elastic energy release rate is related directly to the change in stiffness, a natural way of evaluating this rate is by stiffness measurements. For this, it is necessary to perform at least two measurements of stiffness R for two neighbouring crack configurations. If the successive configurations are known, they can be artificially realized on a model of the work-piece under study and we thus obtain the evolution of $G(A)$:

$$G = -\frac{1}{2}\frac{F^2}{R^2}\frac{\mathrm{d}R}{\mathrm{d}A}.$$

Only a few measurements are necessary to obtain, through interpolation, a good definition of the graph $R(A)$ since, in general, this function is never far from a linear function. However, one should be careful in measuring R by

$$R = \mathrm{d}F/\mathrm{d}u$$

Fig. 8.21. Determination of G by the stiffness method: (a) in fatigue, (b) in creep.

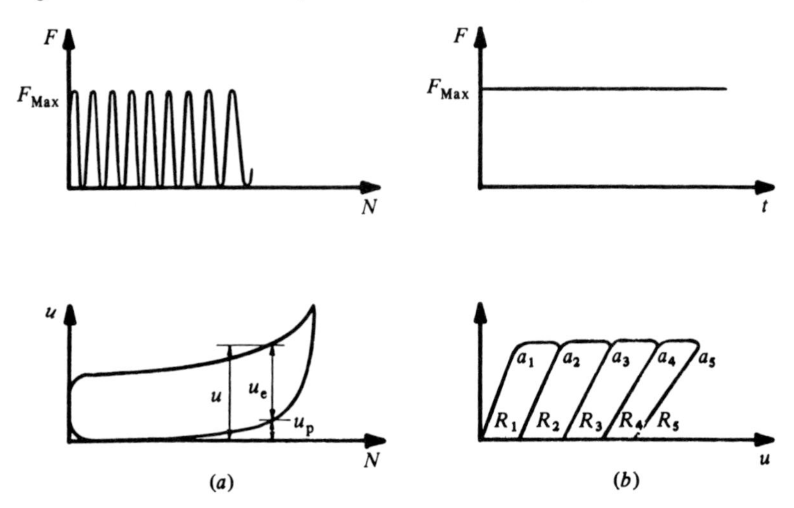

in a load range higher than the one corresponding to the closure effects. A recommended value in practice is $F \approx \frac{1}{2}F_c$ (where F_c is the critical force causing fracture). Fig. 8.21 describes the different operations performed to obtain $G(A)$.

Crack opening method

For an elastic medium subjected to mode I loading, we can determine the energy release rate by a single calculation, or by a single stiffness experiment without the need to study the effect of a crack growth. In this case G is related to the stress intensity factor K_I by

$$G = \begin{cases} \dfrac{K_I^2}{E} & \text{in plane-stress} \\ \dfrac{K_I^2}{E}(1-\nu^2) & \text{in plane-strain} \end{cases}$$

with K_I itself being related to the crack opening displacement by

$$K_I = \begin{cases} \lim\limits_{r\to 0}\left[\dfrac{E}{8}[\![u_2]\!]\left(\dfrac{2\pi}{r}\right)^{1/2}\right] & \text{in plane-stress} \\ \lim\limits_{r\to 0}\left[\dfrac{E}{8(1-\nu^2)}[\![u_2]\!]\left(\dfrac{2\pi}{r}\right)^{1/2}\right] & \text{in plane-strain.} \end{cases}$$

Thus, a calculation or a measurement of $u_2(r)$ can be used to obtain G from

$$G = \begin{cases} \dfrac{E}{64}\left\{\lim\limits_{r\to 0}\left[[\![u_2]\!]\left(\dfrac{2\pi}{r}\right)^{1/2}\right]\right\}^2 & \text{in plane-stress} \\ \dfrac{E}{64(1-\nu^2)}\left\{\lim\limits_{r\to 0}\left[[\![u_2]\!]\left(\dfrac{2\pi}{r}\right)^{1/2}\right]\right\}^2 & \text{in plane-strain.} \end{cases}$$

Fig. 8.22 shows the result of application of this method to a stiffened plane with the crack tip close to the stiffener, the crack openings having been measured photographically with an enlargement $\times$ 100.

Case of proportional loading. Reduced elastic strain energy release rate

An interesting case is that of radial loading where all loads applied to the structure increase proportionally to a single parameter, or where, in the extreme but frequently occurring case, the external load consists of a single load. Let therefore

$$F_q = C_q F(t)$$

be the external forces where $F(t)$ is the load parameter and where C_q are constant. Within the framework of linear elasticity, the energy release rate can be written as

$$G = -\frac{1}{2}\frac{\partial R_{qk}}{\partial A} u_q^e u_k^e = -\frac{1}{2}\frac{\partial R_{qk}}{\partial A} R_{qp}^{-1} F_p R_{lk}^{-1} F_l$$

or

$$G = -\frac{1}{2}\frac{\partial R_{qk}}{\partial A} R_{qp}^{-1} C_p R_{lk}^{-1} C_l F^2(t).$$

The quantity

$$G_r = -\frac{1}{2}\frac{\partial R_{qk}}{\partial A} R_{qp}^{-1} C_p R_{lk}^{-1} C_l$$

is called the reduced strain energy release rate. G_r depends only on the geometry of the cracked structure and the system of external forces but not

Fig. 8.22. Determination of G for a symmetric panel of AU2GN Alloy ($E = 72\,500$ MPa).

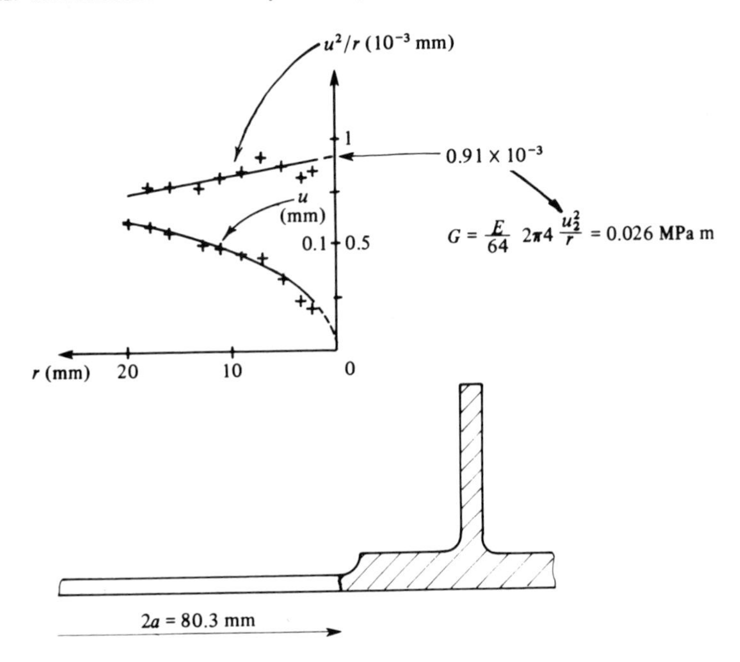

on their intensity. This variable can therefore be calculated as a function of the evolution of the crack configuration, if it is known independently of the evolution of the loading. In the particular case of a single external force

$$\bullet \qquad G_r = -\frac{1}{2R^2}\frac{dR}{dA}, \qquad G = G_r F^2.$$

8.4.3 *The crack growth threshold variable*

Here we again consider the case of a crack described by only one scalar variable. In cases involving significant plastic strains or cyclic loads, the variables A and G are insufficient to describe the current state and the evolution of the cracked solid. In the latter the effect of crack closure during unloading obviously plays an important role. In the former, the effects of nonlinearity induced by unconfined plastic strains require the use of an additional variable; this is defined through the concept of a fictitious relaxed state described in Section 8.4.1.

The crack growth threshold variable was introduced in a purely formal fashion through the internal variable which has yet to be defined. We take it to be related to the value of the external load at which the crack begins to open up. Very schematically we say that:

before this value is reached, the crack is closed and the stiffness of the cracked structure is that of the virgin structure devoid of any crack;
beyond this value the stiffness depends on the crack configuration represented by the variable A.

Let us return to Fig. 8.19(*b*) which concerns the significance of G for an elastoplastic solid without the threshold effect. It is necessary to modify it (Fig. 8.23) so as to take account of:

(1) a coupling between the plasticity and the crack growth, since an irreversible displacement $\delta u_p/\delta A$ can occur during the growth δa of the crack represented by $C'C = E'E$. This displacement is produced by part of the elastic energy released during the growth, but it does not contribute to the energy available at the crack tip;
(2) the effect of the threshold, assumed here for simplicity to be identical for crack opening and closure. The only elastic energy released by the structure and available to propagate the crack is given by the area of the triangle $BC'D'$, the path $C'D'$ corresponding to an instantaneous fictitious unloading 'before' the irreversible

displacement had time to occur. For $F < F_S$ we regain in all cases the initial stiffness of the structure. We note that the concept of crack closure employed here does not correspond to the initiation of the plastic flow in compression at the crack tip during unloading but to a real closing of the crack over a finite length, or more schematically over the whole length of the crack.

The above description is very rough since in Fig. 8.23 we have assumed that crack growth occurs at the maximum load and that the threshold F_S, was constant and identical for crack opening and closure. However, this scheme is sufficient to justify the energy balance and to introduce the concept of a crack growth threshold G_S presented below.

The energy dissipated during the crack growth is therefore $G^* \, dA$, defined as the difference between the variation of the work done by the external loads (considered as constant) and the work of the internal forces during the growth dA of the crack area

$$G^* = \frac{dW_x}{dA}\bigg|_F - \frac{\partial W}{\partial A} = \int_{\partial \mathscr{S}} \vec{F} \cdot \frac{d\vec{u}}{dA} \, dS - \frac{\partial}{\partial A} \int_{\mathscr{S}} W(\varepsilon) \, dV.$$

The quantity $W(\varepsilon)$ results from the integration of $dW = \boldsymbol{\sigma} : d\boldsymbol{\varepsilon}$. From the theorem of virtual work we have

$$\int_{\mathscr{S}} dW \, dV = \int_{\mathscr{S}} \boldsymbol{\sigma} : d\boldsymbol{\varepsilon} \, dV = \int_{\partial \mathscr{S}} \vec{F} \cdot d\vec{u} \, dS$$

in which we allow for the dependence between $\vec{u}$ and $\vec{F}$. From the above

Fig. 8.23. Visualization of the elastic energy released assuming a crack opening threshold: (1) fictitious unloading before the growth; (2) schematic representation of the real unloading after δa.

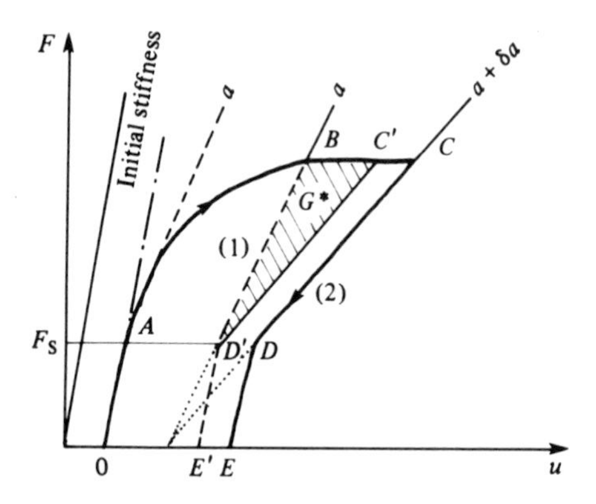

schematic representation it follows that the plastic displacements do not contribute to the cracking: the integration terms $\int \vec{F}\cdot(\mathrm{d}\vec{u}_{\mathrm{p}}/\mathrm{d}A)\,\mathrm{d}S$ and $(\partial/\partial A)\int \vec{F}\cdot\vec{u}_{\mathrm{p}}\,\mathrm{d}S$ correspond to the same area $E'D'C'CDE$ in Fig. 8.23, and cancel each other out. It is therefore sufficient to continue with only the elastic part of the displacement (Fig. 8.24). We have:

$$G^* = \int_{\partial\mathscr{S}} \vec{F}\cdot\frac{\mathrm{d}\vec{u}_{\mathrm{e}}}{\mathrm{d}A}\,\mathrm{d}S - \frac{1}{2}\frac{\partial}{\partial A}\int_{\partial\mathscr{S}} \vec{F}\cdot\vec{u}_{\mathrm{e}}\,\mathrm{d}S.$$

As a simplification let us consider that the loading is dependent only on a single parameter F and consists of prescribed forces. The crack opens when F reaches the value F_{S} and we denote by R_0 and R the initial and current stiffnesses which depend on A. By hypothesis we have (for $\mathrm{d}A = 0$):

$$\mathrm{d}u_{\mathrm{e}} = \mathrm{d}u = \begin{cases} \mathrm{d}F/R_0 & \text{if} \quad F < F_{\mathrm{S}} \\ \mathrm{d}F/R & \text{if} \quad F \geqslant F_{\mathrm{S}}. \end{cases}$$

The quantity W_{e} and the displacement associated to this load are obtained by integrating from 0 to F_{Max}

$$W_{\mathrm{e}} = \frac{1}{2}\left(\frac{F_{\mathrm{S}}^2}{R_0} + \frac{F_{\mathrm{Max}}^2 - F_{\mathrm{S}}^2}{R}\right), \quad u = \frac{F_{\mathrm{S}}}{R_0} + \frac{F_{\mathrm{Max}} - F_{\mathrm{S}}}{R}.$$

By differentiating with respect to A, we obtain

$$\frac{\partial W_{\mathrm{e}}}{\partial A} = -\frac{1}{2R^2}\frac{\mathrm{d}R}{\mathrm{d}A}(F_{\mathrm{Max}}^2 - F_{\mathrm{S}}^2) = \tfrac{1}{2}(F_{\mathrm{Max}} + F_{\mathrm{S}})\left.\frac{\mathrm{d}u}{\mathrm{d}A}\right|_F$$

Fig. 8.24. Schematic illustration of the released elastic energy.

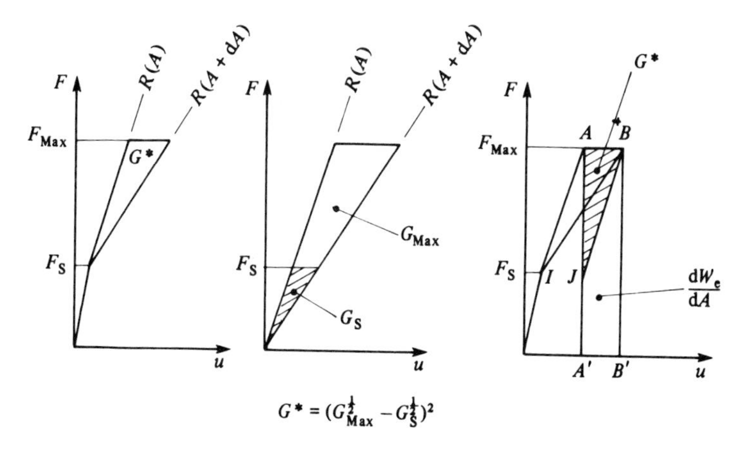

and thus find that

$$G^* = F_{\text{Max}} \frac{\mathrm{d}u}{\mathrm{d}A}\bigg|_F - \frac{\partial W_e}{\partial A} = \tfrac{1}{2}(F_{\text{Max}} - F_S)\frac{\mathrm{d}u}{\mathrm{d}A}\bigg|_F$$

which, in yet another form, can be expressed as

$$G^* = -\frac{1}{2R^2}\frac{\mathrm{d}R}{\mathrm{d}A}(F_{\text{Max}} - F_S)^2.$$

Let us call the elastic energy release rate with no threshold G_{Max} when $F = F_{\text{Max}}$, and G_S when $F = F_S$. We then have

$$G_{\text{Max}} = G_r(A)F_{\text{Max}}^2 = -\frac{1}{2R^2}\frac{\mathrm{d}R}{\mathrm{d}A}F_{\text{Max}}^2, \quad G_S = -\frac{1}{2R^2}\frac{\mathrm{d}R}{\mathrm{d}A}F_S^2$$

so that the elastic energy release rate with a threshold can be expressed, when $F = F_{\text{Max}}$, by

- $$G^* = \langle G_{\text{Max}}^{1/2} - G_S^{1/2}\rangle^2 = G_r(A)\langle F_{\text{Max}} - F_S\rangle^2.$$

This quantity is zero when $G < G_S$, as is indicated by the brackets $\langle\ \rangle$. Fig. 8.24 illustrates the decomposition of G^* and the definition of G_{Max} and G_S, as well as the decomposition of the work of the external force (rectangle $ABB'A'$) into $G^* = \frac{1}{2}(F_{\text{Max}} - F_S)\,\mathrm{d}u/\mathrm{d}A$ (the triangle ABI or ABJ) and

$$\mathrm{d}W_e/\mathrm{d}A = \tfrac{1}{2}(F_{\text{Max}} + F_S)\,\mathrm{d}u/\mathrm{d}A.$$

Proceeding in the same fashion for the case where displacements are prescribed, we obtain

$$G^* = -\frac{\partial W_e}{\partial A} = -\frac{1}{2}\frac{\mathrm{d}R}{\mathrm{d}A}(u_{\text{Max}} - u_S)^2$$

$$= -\frac{1}{2}\frac{\mathrm{d}R}{\mathrm{d}A}\left(\frac{F_{\text{Max}} - F_S}{R}\right)^2 = -\tfrac{1}{2}(u_{\text{Max}} - u_S)\frac{\mathrm{d}F}{\mathrm{d}A}\bigg|_u$$

where $u_S = F_S/R_0$. It can be seen that in the general case, G^* can be expressed in a form similar to that used in Section 8.4.2 by considering the presence of a threshold:

$$G^* = \tfrac{1}{2}(F_{\text{Max}} - F_S)\frac{\mathrm{d}u}{\mathrm{d}A}\bigg|_F - \tfrac{1}{2}(u_{\text{Max}} - u_S)\frac{\mathrm{d}F}{\mathrm{d}A}\bigg|_u.$$

On the other hand, the following expression always remains valid:

$$G^* = \langle G_{\text{Max}}^{1/2} - G_S^{1/2}\rangle^2.$$

This relation demonstrates the validity of the concept of the effective stress intensity factor frequently used in fatigue. For mode I and plane stress, we can replace G by K^2/E and find that

- $$K_{\mathrm{eff}} = K^* = \langle K_{\mathrm{Max}} - K_{\mathrm{S}} \rangle.$$

8.4.4 *Dissipation analysis*

In this section we will consider only the case where the current state of the cracked solid can be identified by a single scalar variable, for example the crack area; the case of a crack which requires more than one kinematic parameter is covered in Section 8.4.6. Let us first of all recall the basic results derived from the second law of thermodynamics, expressed, for simplicity, in terms of a single external load (a scalar notation then suffices):

$$F\dot{u} - \dot{\Psi} - S\dot{T} \geqslant 0$$

where Ψ and S are the free energy and global entropy of the structure respectively. For simplicity, the radiation and heat flux terms have been omitted. The free energy can be expressed as

$$\Psi = \Psi(u_{\mathrm{e}}, A, T, h_{\mathrm{p}}, h_{\mathrm{S}}).$$

Using this expression in the preceding inequality, we get

$$\left(F - \frac{\partial \Psi}{\partial u_{\mathrm{e}}}\right)\dot{u}_{\mathrm{e}} - \left(\frac{\partial \Psi}{\partial T} + S\right)\dot{T} + F\dot{u}_{\mathrm{p}} - \frac{\partial \Psi}{\partial A}\dot{A} - \frac{\partial \Psi}{\partial h_{\mathrm{p}}}\dot{h}_{\mathrm{p}} - \frac{\partial \Psi}{\partial h_{\mathrm{S}}}\dot{h}_{\mathrm{S}} \geqslant 0.$$

By the classical thermodynamic argument the first two terms lead to

$$F = \partial \Psi / \partial u_{\mathrm{e}} \qquad S = -\partial \Psi / \partial T.$$

Finally, the intrinsic dissipation, i.e., the energy dissipated as heat is

$$\Phi_1 = F\dot{u}_{\mathrm{p}} - \frac{\partial \Psi}{\partial A}\dot{A} - \frac{\partial \Psi}{\partial h_{\mathrm{p}}}\dot{h}_{\mathrm{p}} - \frac{\partial \Psi}{\partial h_{\mathrm{S}}}\dot{h}_{\mathrm{S}} \geqslant 0.$$

In the isothermal case, and in the absence of a threshold, the free energy is expressible in the following simple form:

$$\Psi = \tfrac{1}{2} R(A) u_{\mathrm{e}}^2 + \Psi_{\mathrm{p}}(h_{\mathrm{p}}).$$

In order to take into account the crack opening threshold, we choose the following expression:

$$\begin{aligned}\Psi = {} & \tfrac{1}{2} R_0 u_{\mathrm{e}}^2 H(u_{\mathrm{S}} - u_{\mathrm{e}}) + \tfrac{1}{2} R_0 u_{\mathrm{S}}^2 H(u_{\mathrm{e}} - u_{\mathrm{S}}) \\ & + R_0 u_{\mathrm{S}} \langle u_{\mathrm{e}} - u_{\mathrm{S}} \rangle + \tfrac{1}{2} R(A) \langle u_{\mathrm{e}} - u_{\mathrm{S}} \rangle^2 + \Psi_{\mathrm{p}}(h_{\mathrm{p}})\end{aligned}$$

where $h_S = u_S$ is the displacement at the point of application of the force F_S and where the symbol H and $\langle\ \rangle$ are defined by $H(x) = 1$ if $x \geqslant 0$, $H(x) = 0$ if $x < 0$, and $\langle x \rangle = xH(x)$. This choice enables us to write the intrinsic dissipation for all cases in the form

$$\Phi_1 = F\dot{u}_p + G^*\dot{A} - \frac{\partial \Psi}{\partial h_S}\dot{h}_S - \frac{\partial \Psi_p}{\partial h_p}\dot{h}_p$$

in terms of the quantities defined above. If the second term is considered independently, we obtain the trivial condition $\dot{A} \geqslant 0$ in view of the fact that $G^* = (G^{1/2} - G_S^{1/2})^2 \geqslant 0$. The term $(\partial\Psi_p/\partial h_p)\dot{h}_p$ can be used to account for the effects of work-hardening and the term preceding it to describe the evolution of the crack opening threshold.

The evolution laws for the internal variables, u_p, A, h_p and h_S, can be derived from a dissipation potential

$$\varphi(\dot{u}_p, \dot{A}, \dot{h}_S, \dot{h}_p; u_e, T, A, h_S, h_p).$$

The potential obtained by Legendre–Frenchel transformation can be written as

$$\varphi^*(F, G^*, H_S, H_p; u_e, T, A, h_S, h_p).$$

The plastic flow $\dot{u}_p$ is not considered here. The law governing the evolution of the crack is given by

$$\dot{A} = \lambda\, \partial\varphi^*/\partial G^*$$
$$\dot{h}_S = -\lambda\, \partial\varphi^*/\partial H_S.$$

For a crack geometry described by a single parameter, this law is considered to be an intrinsic one: in the case of a crack propagation problem we recall that we have a given structure and a given material with a well-defined loading system; λ is a scalar multiplier which has to be determined for each particular case.

The threshold variable h_S which appears in the above formulation is defined independent of the value of A. Its evolution is governed by the two expressions $\dot{h}_S = -\lambda\, \partial\varphi^*/\partial H_S$ and $H_S = -\partial\Psi/\partial h_S$. Knowing h_S we can then define the threshold value G_S appearing in the expression for the parameter G^*, governing crack growth. In the simple example studied here, we successively take

$$u_S = h_S \qquad F_S = R_0 u_S \qquad G_S = G_r(A)F_S^2.$$

We also note that the generalization of this theory to the case of a nonproportional multiaxial loading does not present any basic problem of principle but only difficulty in notation.

8.4.5 *Bifurcation criteria for crack propagation in plane media*

The criteria of bifurcation, or crack deviation, are global concepts which can be used to determine the orientation of the crack growth for complex nonproportional loadings.

Maximum normal stress criterion

This criterion is used often because it is easy to implement. We assume that the direction of the crack growth which corresponds to the given loading is such that the local stress normal to the plane defined by angle θ is maximum; the analysis is performed with the crack configuration before deviation. Fig. 8.25 shows schematically the case of a straight line crack in a plate subjected to tension in a direction which is not perpendicular to the crack.

In practice we consider points at a constant distance r^* from the crack tip. At each point (r^*, θ) we evaluate the principal directions employing, for example the finite element method. If r^* is sufficiently small, the maximum principal stress is in the direction perpendicular to line $0M$. We then obtain the angle θ which corresponds to the maximum principal stress value.

The practical difficulty lies in the choice of r^*, except when the solutions are available in analytical forms. In the case of Fig. 8.25 the components of the unit normal are: $-\sin\theta$, $\cos\theta$, and the normal and tangential stress

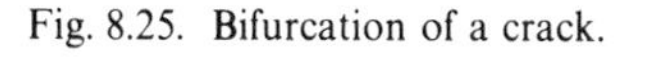
Fig. 8.25. Bifurcation of a crack.

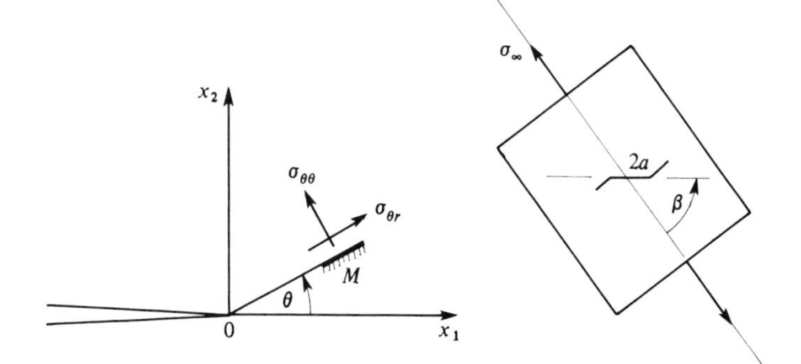

components are:

$$\sigma_{\theta\theta} = \sigma_{11}\sin^2\theta - 2\sigma_{12}\sin\theta\cos\theta + \sigma_{22}\cos^2\theta$$

$$\sigma_{r\theta} = (\sigma_{22} - \sigma_{11})\sin\theta\cos\theta + \sigma_{12}(\cos^2\theta - \sin^2\theta).$$

By using the relations of Section 8.2.2 for plane-stress, we get:

$$4\sigma_{\theta\theta}(2\pi r^*)^{1/2} = K_{\rm I}\left(3\cos\frac{\theta}{2} + \cos\frac{3\theta}{2}\right) - 3K_{\rm II}\left(\sin\frac{\theta}{2} + \sin\frac{3\theta}{2}\right)$$

$$4\sigma_{r\theta}(2\pi r^*)^{1/2} = K_{\rm I}\left(\sin\frac{\theta}{2} + \sin\frac{3\theta}{2}\right) + K_{\rm II}\left(\cos\frac{\theta}{2} + 3\cos\frac{3\theta}{2}\right).$$

We note that the criterion is independent of r^* and that the angle θ which renders $\sigma_{\theta\theta}$ maximum is that for which $\sigma_{r\theta}$ is zero (since $\sigma_{r\theta}$ is proportional to $\partial\sigma_{\theta\theta}/\partial\theta$), and we are interested in maximizing the maximum principal stress. The problem of the direction of the bifurcation is solved as soon as $K_{\rm I}$ and $K_{\rm II}$ are known. In the case of a plate containing a crack of length $2a$ and subjected to tension at an angle β with respect to the crack, the state of stress at infinity is characterized by

$$\sigma_{11}^{\infty} = \sigma_{\infty}\cos^2\beta, \qquad \sigma_{22}^{\infty} = \sigma_{\infty}\sin^2\beta, \qquad \sigma_{12}^{\infty} = \sigma_{\infty}\sin\beta\cos\beta$$

with σ_{11}^{∞} having no influence on the crack. We then know that $K_{\rm I}$ and $K_{\rm II}$ are given by the expressions

$$K_{\rm I} = \sigma_{\infty}\sin^2\beta(\pi a)^{1/2}, \quad K_{\rm II} = \sigma_{\infty}\sin\beta\cos\beta(\pi a)^{1/2}$$

Fig. 8.26. Experimental verification of crack bifurcation criteria.

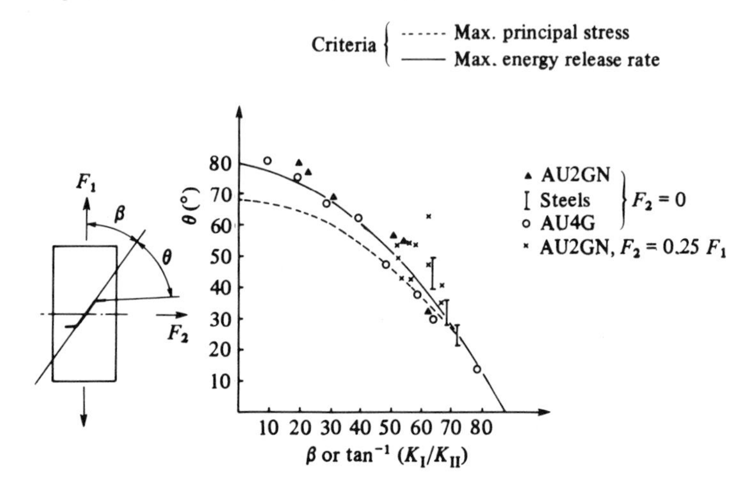

which after simplification (for $\sigma_{r\theta} = 0$) yield the solution

$$\bullet \qquad \frac{1 - 3\cos\theta}{\sin\theta} = \tan\beta = \frac{K_{\mathrm{I}}}{K_{\mathrm{II}}}.$$

Fig. 8.26 shows a comparison between the results obtained from this criterion and those from experiments. The extreme case $K_{\mathrm{I}} = 0$, $K_{\mathrm{II}} \neq 0$ which corresponds to $\beta = 0$ cannot be realized in a plate under uniaxial tension. We can, however, obtain this condition through biaxial loading: the results are then expressed as a function of $\beta = \tan^{-1}(K_{\mathrm{I}}/K_{\mathrm{II}})$.

There are other analogous criteria: those where the maximum strain $\varepsilon_{\theta\theta}$ is considered, or even Sih's criterion in which it is the extremum of the elastic energy density which is considered in determining the direction of crack deviation.

Maximum elastic strain energy release rate criterion

We consider that the direction of crack propagation at any instant is such that it results in a maximum elastic energy release rate G. The latter is defined, for the load under consideration with an already branched crack, to be as small as possible. We can use a local analysis for this purpose starting with the intensity factors K'_{I} and K'_{II} defined for the branched crack. On the other hand, when such a solution is not available, we proceed numerically using, for example, finite elements. The elastic energy release rate is expressed in the general case (see Section 8.2.2) by:

$$G_\theta = \tfrac{1}{2}\int_C \left(\vec{F} \cdot \frac{\mathrm{d}\vec{u}}{\mathrm{d}a} - \vec{u} \cdot \frac{\mathrm{d}\vec{F}}{\mathrm{d}a} \right) \mathrm{d}C$$

where C is a closed contour encircling the tip of the branched crack (see Fig. 8.27). In practice it is simpler to evaluate this quantity globally by

Fig. 8.27. Branched crack.

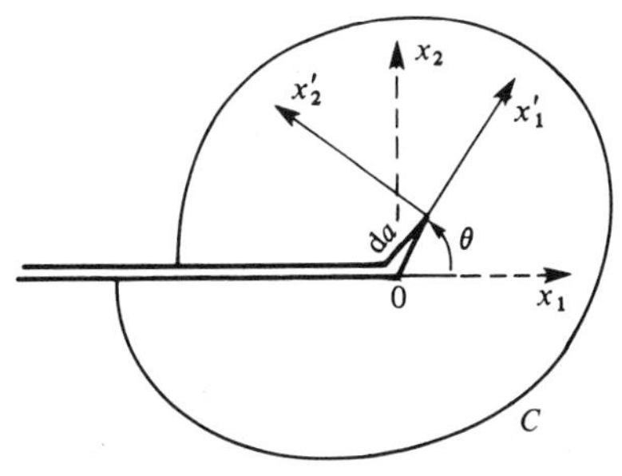

considering the work of external forces. For a prescribed load, for example

$$G = \frac{1}{2}\frac{\mathrm{d}W_x}{\mathrm{d}A}\bigg|_F = \frac{1}{2e}F\frac{\mathrm{d}u}{\mathrm{d}a}.$$

In practice, we can perform a series of calculations for different branch angles θ, keeping the branch length Δa constant and sufficiently small. In each case $G(\theta)$ can be evaluated by the finite difference as follows

$$G(\theta) = \frac{1}{2e\Delta a}F[u(\Delta a, \theta) - u(0,0)].$$

We must, however, note that it is necessary to solve each case of a branched crack independently; because of the change in the geometry of the crack, local perturbation methods cannot be used. The direction of bifurcation is that which renders $G(\theta)$ a maximum. This is determined by smoothing the numerical data.

Although this is a satisfactory criterion from a physical point of view, it is tricky to implement when no analytical solution is available. It gives results which are only slightly different from those obtained from the maximum normal stress criterion, as shown in Fig. 8.26.

Maximum crack growth rate criterion

Again with reference to the analysis of branched cracks, the direction of branching is chosen to be that which maximizes the rate $\dot{a}$ of the crack growth. If the growth law is expressed as a function of G, then there is no difference between the two criteria. On the other hand, this criterion can be used to take account of the effect of the load ratio $R = F_m/F_{Max}$, which, for some materials, quite strongly influences the bifurcation angle. The crack growth laws using this load ratio are mentioned in Section 8.3.2.

8.4.6 *Three-dimensional cracked structures*

The study of crack growth in a three-dimensional structure, even when limited to plane surface cracks, is a difficult problem as much from a stress analysis point of view as from the laws of crack growth and fracture. This is, however, a frequent situation: nontransverse cracks in thin metal sheets, corner cracks in complex parts or in stress concentration zones, etc.

Here, the main principles of a general theory, are examined. This theory will be applied in Section 8.6.2 to solve numerically practical crack growth problems which can be represented by a finite number of parameters.

The crack front Γ is represented by the curvilinear abscissa s of a point $M(s)$. Its evolution is defined by a physically admissible field of crack velocities denoted by $\dot{a}(s)$ (see Fig. 8.28). We restrict ourselves to the case of a brittle-elastic solid and to simplify the presentation, to the case of externally prescribed forces. A point of $\partial\mathscr{S}_F$ is represented by P and $\dot{u}(P)$ denotes the rate of displacement associated with the crack growth rate $\dot{a}(M)$ (it is evident that if F is constant and $\dot{a}(M) = 0$, then $\dot{u}(P) = 0$).

The power of the (constant) external forces in this case can be written as (where F is the surface intensity of the forces):

$$P_{(x)} = \int_{\partial\mathscr{S}_F} F(P)\dot{u}(P)\,\mathrm{d}S.$$

However, since the rate $\dot{u}(P)$ depends only on $\dot{a}(s)$ (at constant F), it can be written, by retaining only the first order term, as

$$\dot{u}(P) = \int_{\Gamma} h(P, M)\dot{a}(M)\,\mathrm{d}s.$$

Substituting this in the expression for $P_{(x)}$ and changing the order of integration we obtain

$$P_{(x)} = \int_{\Gamma} g(M)\dot{a}(M)\,\mathrm{d}s \qquad \text{with} \qquad g(M) = \int_{\partial\mathscr{S}_F} F(P)h(P, M)\,\mathrm{d}S.$$

We note that $h(P, M)$ corresponds to a rate $\mathrm{d}u(P)/\mathrm{d}a(M)$ and that $g(M)$ may be considered as the gradient of the free energy. Special mathematical tools are required to write a formally correct expression of this energy

Fig. 8.28. Velocity field of a crack front in a three-dimensional structure.

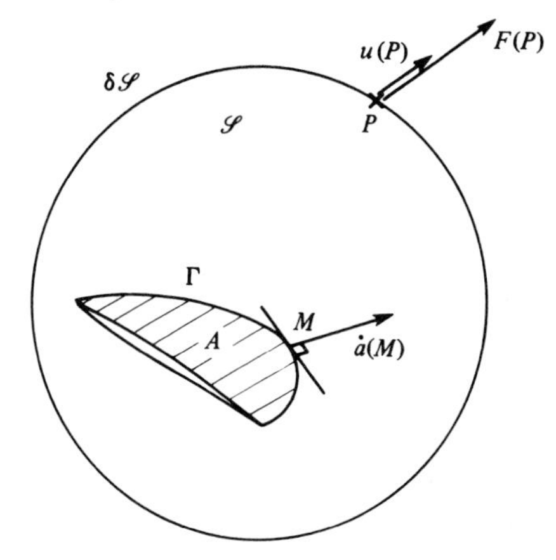

gradient in the infinite dimension case. The free energy Ψ is assumed differentiable in the sense of Gateaux, so that there exists a function $g(M)$ defined on Γ such that

$$D\Psi = -\int_{\Gamma} g(M)\dot{a}(M)\,\mathrm{d}s.$$

By definition, $g(s)$ represents a force distribution along Γ. Using the stress intensity factors as functions of $M(s)$ we can write

$$g(s) = \frac{1-\nu^2}{E}K_{\mathrm{I}}^2(s) + \frac{1-\nu^2}{E}K_{\mathrm{II}}^2(s) + \frac{1}{2\mu}K_{\mathrm{III}}^2(s).$$

This concept of force distribution remains valid in the case of nonlinear elasticity and when the stress singularity changes its nature along Γ, for example in the case of a crack intersection with the external surface.

The power dissipated in the process of crack growth is given by

$$\Phi = \int_{\Gamma} g(s)\dot{a}(s)\,\mathrm{d}s \geqslant 0.$$

This expression generalizes the relation $G\dot{A} = -(\partial\Psi/\partial A)\dot{A} \geqslant 0$ used previously for a crack represented by a single scalar parameter. We assume that this is a normal dissipative mechanism, i.e., there exists at each point of Γ a potential function $\varphi(\dot{a})$, convex in $\dot{a}$, such that

$$g = g(s) = \partial\varphi/\partial\dot{a}.$$

Denoting by $\varphi^*(g)$ the Legendre–Fenchel transform of $\varphi(\dot{a})$, we derive the relation

$$\dot{a}(s) = \partial\varphi^*/\partial g(s)$$

which is valid at any point of the crack front.

Note that the area A of the crack is the surface swept by the front Γ and its growth rate is expressed by

$$\dot{A} = \int_{\Gamma} \dot{a}(s)\,\mathrm{d}s.$$

When the crack can be described by the single variable A, i.e., when $g(s)$ is uniform, $g(s) = G$ and we obviously recover

$$\Phi = \int_{\Gamma} g(s)\dot{a}(s)\,\mathrm{d}s = G\int_{\Gamma} \dot{a}(s)\,\mathrm{d}s = G\dot{A}.$$

The choice of the dissipation potential $\varphi^*(g)$ affects the crack growth law. We may choose a power function

$$\varphi^*(g) = \frac{C'}{\eta/2+1} g^{\eta/2+1}$$

which leads to a local equation of the Paris' law type, since:

$$\dot{a}(s) = \partial\varphi^*/\partial g = C'g(s)^{\eta/2} = C'(K^2(s)/E)^{\eta/2} = CK^\eta(s).$$

If we choose a quadratic potential, $\eta = 2$, we obtain a law called the gradient law, which is well suited to numerical calculations:

$$\dot{a}(s) = \alpha g(s).$$

In this case the growth of the crack takes place so as to maximize the energy release rate, which corresponds to one of the criteria studied in connection with the bifurcation processes. The practical application of this approach is indicated in Section 8.6.2 within the framework of the finite element method.

8.5 Particular crack propagation models

According to the formalism introduced so far, there are three complementary laws governing crack evolution:

- a relation governing $\dot{A}$, the rate of growth of the crack area;
- a relation governing the form of the crack front. In fact, it is not necessary to state this law explicitly since it is advantageously replaced by the bifurcation criteria. In each particular problem, the form of the crack front can always be determined by applying one of the criteria described in Section 8.4.5;
- a relation expressing the variation in the crack growth threshold which should be taken into account when plastic or viscoplastic strains are involved. We will give an example of fatigue fracture modelling, but we should note, as has been seen in Section 8.4.3, that this threshold can also be calculated by solving an elastoplastic problem.

Retaining the initial idea according to which crack behaviour is governed by the energy release rate, the essential variable related to the growth rate, $\dot{A}$, is G in elastic problems or where plastic dissipation is very low, and $G^* = (G^{1/2} - G_S^{1/2})^2$ in the problems where inelastic deformations lead to the existence of a crack growth threshold G_S (see Section 8.4.3).

8.5.1 Cracking by brittle fracture

Critical energy release rate

The modelling of a brittle-elastic cracked solid is amenable to a representation based on perfect plasticity. As long as the energy release rate G is lower than a critical value G_c, the crack area remains unaltered. When G attains the value G_c there is a sudden fracture (Fig. 8.29)

- $G < G_c \rightarrow \dot{A} = 0, \quad G = G_c \rightarrow \dot{A}$ indeterminate.

Fig. 8.29. Diagram illustrating brittle fracture.

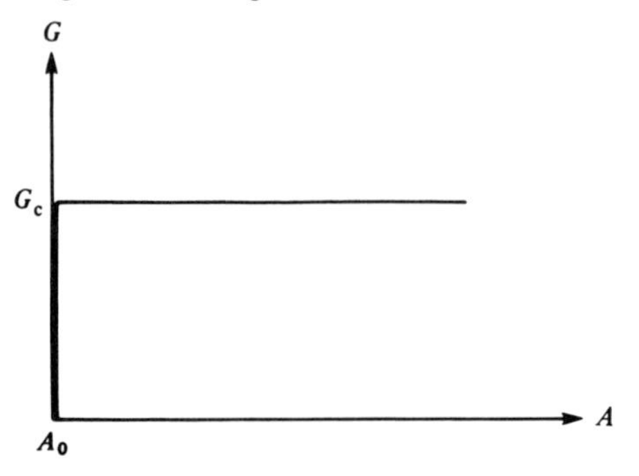

Fig. 8.30. 'R' curve for ductile fracture, INCO 718 alloy: thickness of the sheet $e = 4$ mm, $a_0 = 105$ mm (after Marandet, IRSID).

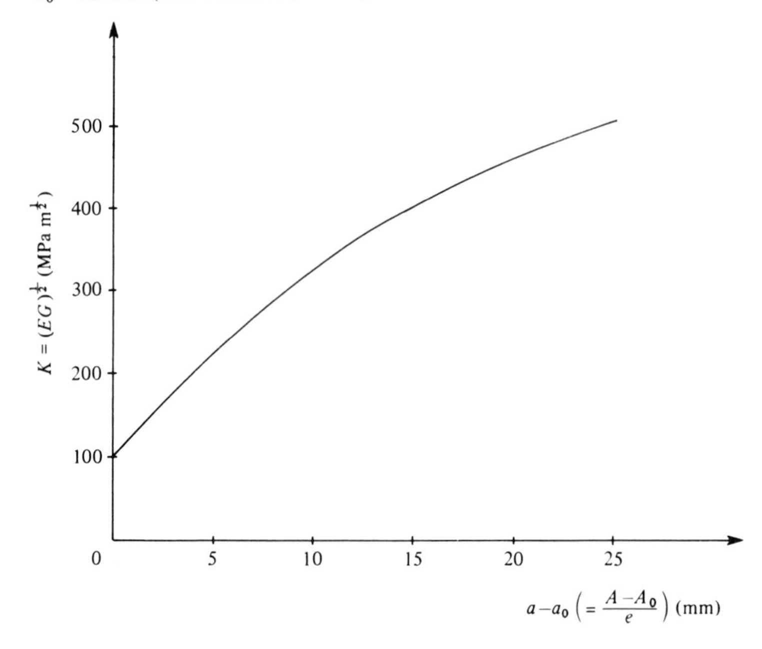

G_c is equal to twice the surface energy of cohesion of the material, and generalizes to three dimensions the concept of toughness K_{Ic} discussed in Section 8.3.2 for plane problems.

The identification of G_c consists in performing a fracture test on a cracked specimen and recording the critical fracture load, taking the precautions mentioned in Section 8.3.2. An elastic analysis or a stiffness measurement can be used to obtain the reduced energy release rate G_r for the geometry considered (see Section 8.4.2). We then have

$$G_c = G_r F_c^2.$$

8.5.2 *Cracking by ductile fracture*

Modelling of cracking by ductile fracture is equivalent to the modelling of the 'R' curve described in Section 8.3.2. The crack grows in a stable manner accompanied by significant plastic strains. The 'R' curve for ductile fracture with variables G and A plays the same role as the hardening curve does for plasticity with stress and strain as variables (Fig. 8.30).

For a modelling of this curve which is consistent with respect to thermodynamics and physics, the dissipation potential is considered to be expressible as a threshold function in the following form

$$\varphi^* = f = G - H_p = 0$$

with a loading–unloading criterion of the plasticity type:

$$f < 0 \rightarrow \text{no crack propagation}$$
$$f = 0 \rightarrow \text{propagation, i.e., stable fracture.}$$

Thus, H_p is a variable which corresponds to an evolutive critical value of G. In fact, it is the R function. To obtain simple expressions we assume that the free energy is a power function, with the plastic part expressed by

$$\Psi_p = G_0 h_p + \frac{Vv}{v+1} h_p^{(v+1)/v}$$

where V, v and G_0 are constants. The evolution equations can be written as

$$\dot{A} = \dot{\lambda}(\partial f/\partial G) = \dot{\lambda}, \quad \dot{h}_p = -\dot{\lambda}(\partial f/\partial H_p) = \dot{\lambda}$$

and the threshold variable H_p as

$$H_p = \partial\Psi/\partial h_p = G_0 + V h_p^{1/v}.$$

The consistency condition, as in plasticity, gives the following expression for the multiplying factor

$$\lambda = \dot{A} = H(f)\frac{v}{V}\frac{\langle \dot{G} \rangle}{h_{\mathrm{p}}^{(1-v)/v}}$$

where the symbols $H(f)$ and $\langle \quad \rangle$ are are used to denote the fact that λ is zero for $f < 0$ or when $\dot{G} < 0$. The crack growth rate, taking into account $f = G - H_{\mathrm{p}} = 0$, is then

$$\dot{A} = H(f)\frac{v}{V}\frac{\langle \dot{G} \rangle}{h_{\mathrm{p}}^{(1-v)/v}}$$

or

• $$\dot{A} = H(f)\frac{v}{V}\left\langle \frac{G - G_0}{V} \right\rangle^{v-1} \langle \dot{G} \rangle.$$

The integration is obvious, and we get

$$A = A_0 + \left\langle \frac{G - G_0}{V} \right\rangle^{v}.$$

This relation governs the crack growth insofar as fracture takes place in a stable fashion. It describes the '*R*' curve by

$$G = R(A - A_0) = G_0 + V(A - A_0)^{1/v}.$$

G depends on the prescribed load F and on A. For example, for a small crack in a large plate we have

$$G = G_{\mathrm{r}}(A)F^2 = kF^2A$$

where k is a constant, dependent on the geometry and the modulus of elasticity. The two equations above can be used to find F as a function of A.

When the prescribed load F is increasing, an instability occurs if, for the same value of A, the rate of change of $G = kF^2A$ becomes equal to that of R (see Fig. 8.13):

$$\partial G/\partial A = \mathrm{d}R/\mathrm{d}A$$

or, when

$$kF^2 = \frac{V}{v}(A - A_0)^{(1-v)/v}.$$

In view of the equality $G = R(A - A_0)$, we find that the critical value A_{c} of A is given by

$$G_0 + V(A_{\mathrm{c}} - A_0)^{1/v} = A_{\mathrm{c}}\frac{V}{v}(A_{\mathrm{c}} - A_0)^{(1-v)/v}.$$

The critical values of G and F can be obtained from the above relations. An explicit solution is easily obtained when the threshold G_0 is neglected. We find

$$\bullet \qquad A_c = A_0 \frac{v}{v-1}, \qquad G_c = V\left(\frac{A_0}{v-1}\right)^{1/v}.$$

The curves $G_c(A_0)$ and $R(A - A_0)$ are therefore homothetic to each other. The critical value G_c depends on A_0. In the particular case of brittle fracture ($v \to \infty$), we recover the results $A_c = A_0$ and $G_c = G_0$, independent of A_0.

The identification of the coefficients V and v by means of experiments in metal sheets shows that V is sensitive to the thickness e.

8.5.3 *Creep crack growth*

Creep cracking is the result of the decohesion of metals at medium and high temperatures at which viscoplastic deformations take place. The crack propagates even if the loading is constant (Fig. 8.31). When viscoplastic strains are small within the total structure (e.g., for materials of low ductility), the phenomenon can be treated in the same way as in viscoplasticity. A dissipation potential as a power function of G^* leads to a good model:

$$\varphi^* = \frac{W}{w+1}\left(\frac{G^*}{W}\right)^{w+1}$$

Fig. 8.31. Creep crack growth tests (after H. Policella).

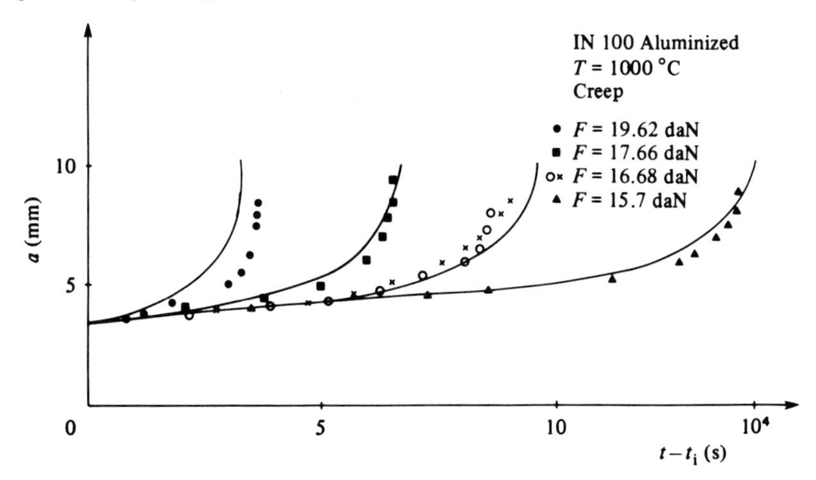

from which

$$\dot{A} = \frac{\partial \varphi^*}{\partial G^*} = \left(\frac{G^*}{W}\right)^w$$

so that

$$\bullet \quad \dot{A} = \left(\frac{\langle G^{1/2} - G_S^{1/2} \rangle^2}{W}\right)^w.$$

W and w are constants (possibly functions of the temperature) which are determined for each material by identification with crack growth experiments (W in this case is also sensitive to the thickness of the metal sheets).

Fig. 8.31 illustrates the case of the IN 100 refractory alloy, a slightly ductile material, with, in this case, $G_S = K_S = 0$. For more ductile materials it would be necessary to consider the evolution of the threshold G_S. The equation can also be written in terms of the stress intensity factor by introducing $K_{eff} = K - K_S$:

$$\dot{A} = \left(\frac{K_{eff}}{W'}\right)^w.$$

8.5.4 *Fatigue crack growth*

The phenomena linked to fatigue crack growth are so complex that it is not possible to formulate representative models of all the essential properties by thermodynamics. We will therefore limit ourselves to a thermodynamic generalization of the simplest law and to the description of a phenomenological model with an evolving threshold.

Generalization of Paris' law

Let us consider a cracked three-dimensional medium subjected to a force F which leads to a crack evolution dependent only on the single variable A. The medium is linear elastic and the energy dissipated in plastic deformation is small compared to the elastic energy. The threshold is G_S.

As for creep induced crack growth, we assume that the crack growth rate $\dot{A}$ is a function of $G^* = (G^{1/2} - G_S^{1/2})^2$ through the intermediary of a dissipation potential. Moreover, we use the same formalism as for ductile fracture. With variables h_p and H_p we choose the potential φ^* in the form

$$\varphi^* = f = G^* - H_p$$

and the part Ψ_p of the thermodynamic potential in the form

$$\Psi_p(h_p) = \frac{\eta C'}{\eta + 2} h_p^{2/\eta + 1}$$

where η is the exponent in Paris' law and C' is a constant, characteristic of the material. The evolution laws are given by expressions similar to those of Section 8.5.2:

$$\dot{A} = \frac{\eta}{2C'}\left(\frac{G^*}{C'}\right)^{\eta/2 - 1} \langle \dot{G}^* \rangle.$$

In practice the fatigue laws are expressed in terms of cycles. It is sufficient to integrate over a cycle of period Δt defined by G^*_{Max} and G^*_{m}, the maximum and minimum values of G^*:

$$\frac{\delta A}{\delta N} = \int_t^{t+\Delta t} \dot{A}\,\mathrm{d}t = \int_{G^*_{\text{m}}}^{G^*_{\text{Max}}} \frac{\eta}{2C'}\left(\frac{G^*}{C'}\right)^{\eta/2 - 1} \mathrm{d}G^*$$

- $$\frac{\delta A}{\delta N} = \frac{1}{C'^{\eta/2}}(G^{*\eta/2}_{\text{Max}} - G^{*\eta/2}_{\text{m}})$$

with

$$G^*_{\text{Max}} = \langle G^{1/2}_{\text{Max}} - G^{1/2}_{\text{S}} \rangle^2, \quad G^*_{\text{m}} = \langle G^{1/2}_{\text{m}} - G^{1/2}_{\text{S}} \rangle^2.$$

When G_{m} is smaller than the threshold G_{S} we recover Paris' law exactly, with $\Delta K_{\text{eff}} = K_{\text{Max}} - K_{\text{S}}$ and $G = K^2/E$:

$$\frac{\delta A}{\delta N} = \frac{1}{C'^{\eta/2}} \langle G^{1/2}_{\text{Max}} - G^{1/2}_{\text{S}} \rangle^{\eta} = C(K_{\text{Max}} - K_{\text{S}})^{\eta}.$$

Moreover, experiments show that C' is sensitive to the thickness of the metal sheets. The threshold G_{S} (or K_{S}) can be of an evolving type. An example of a model with a variable threshold is studied below.

Model with a variable threshold

Wheeler's model presented in Section 8.3.2 introduces the effects of the loading history through the actual size of the plastic zone. It correctly takes into account the effects of spaced overloads but gives incorrect results in other cases. Fig. 8.32 illustrates this for AU4G1 T3 alloy by comparing the case of an overload (ratio $R_{\text{S}} = F_{\text{MaxS}}/F_{\text{Max}} = 1.5$) with that of a loading sequence with two steps (still with $R_{\text{S}} = 1.5$); the delay effect is much more important in the second case, although the Wheeler model furnishes identical results.

Fig. 8.32. Demonstration of the inadequacy of Wheeler's model (compare the experiments (dotted line) for $R_s = 1.5$). The full lines are simulations using the model with a variable threshold.

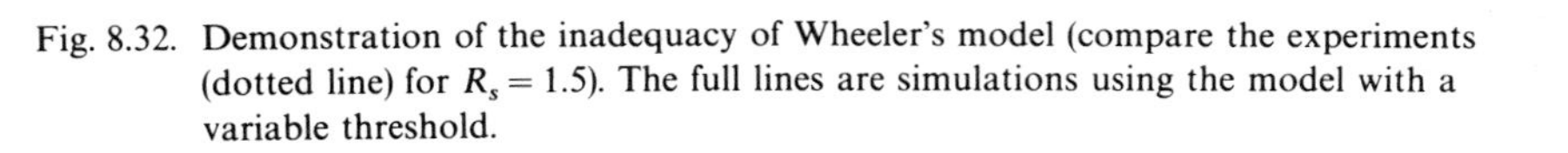

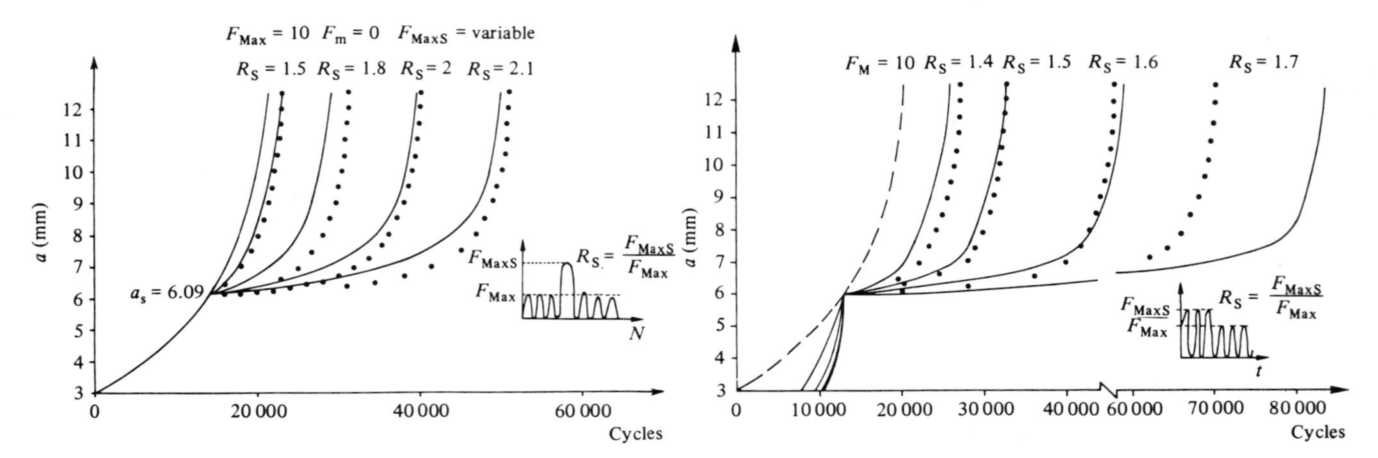

This defect can be remedied by considering a variable crack threshold. The evolution of the threshold as a function of the load history can be demonstrated experimentally by considering two extreme cases (Fig. 8.33):

the case of an overload; for a critical ratio $F_{\text{MaxS}}/F_{\text{Max}}$ denoted by R_{c_1}, there is a complete blockage of the crack growth;
the case of two-step loading; the blockage occurs when the ratio attains R_{c_2}, a critical value lower than R_{c_1}.

The experimental values obtained are independent of the applied load levels and the crack length. For AU4G1 light alloy, we have, for example $R_{c_1} = 2.7$ and $R_{c_2} = 1.9$. In the second case, we can indeed observe the effect of the threshold: during the load at level F_{MaxS}, it attains the value $F_S = F_{\text{MaxS}}/R_{c_2}$. In the first case, it is seen that the overload shifts the threshold from $F_S = F_{\text{Max}}/R_{c_2}$ to F_{Max}, and this proves its evolution.

For an arbitrary load, evolution laws can be developed in the form

$$\delta A/\delta N = f(G^*_{\text{Max}}, G_S, \ldots)$$
$$\delta G_S/\delta N = g(G^*_{\text{Max}}, G_S, \ldots).$$

Although they are not included in the general scheme of Section 8.4, the expressions below furnish a good model for a wide range of loading. We

Fig. 8.33. Demonstration of the loading history effect on the threshold.

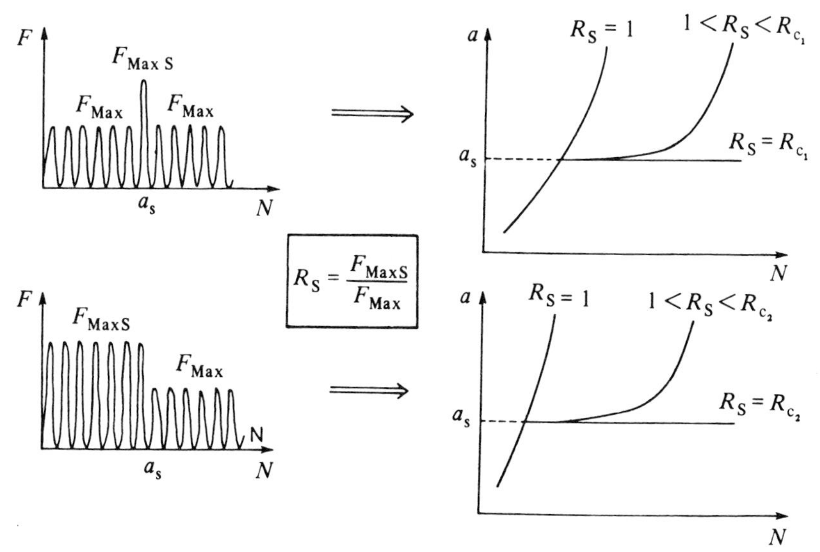

express them in terms of the stress intensity factors as

- $$\frac{\delta A}{\delta N} = C_3 \langle K_{\text{Max}} - K_S \rangle^{\eta_3},$$
- $$\frac{\delta K_S}{\delta N} = \alpha (K_{\text{Max}} K_S)^{1/2} \left(\frac{1}{R_{c_2}} - \frac{K_S}{K_{\text{Max}}} \right) \left\langle 1 - \frac{K_S}{K_{\text{Max}}} \right\rangle^{\eta_3} \left(\frac{\rho_{SS}}{\rho_{AS}} \right)^{\delta},$$
- $$\frac{\rho_{SS}}{\rho_{AS}} = \frac{K_{\text{Max}}^2}{K_{\text{MaxS}}^2 - 2\pi\sigma_Y^2 (a - a_S)},$$

where ρ_{SS}/ρ_{AS} represents the ratio of the fictitious plastic zone corresponding to a periodic load to the current plastic zone (the introduction of ρ_{SS}/ρ_{AS} has been discussed in Section 8.3.2).

The term $(1/R_{c_2} - K_S/K_{\text{Max}})$ is introduced so that the threshold tends to its normal saturation value in a periodic loading; δ is a coefficient intrinsic to the material; the function $(\rho_{SS}/\rho_{AS})^{\delta}$ is used to represent the differences in the threshold evolutions during and after the overload; α is a constant expressible as a function of R_{c_1} and R_{c_2}:

$$\alpha = (R_{c_1} R_{c_2})^{1/2} \frac{R_{c_2} - 1}{R_{c_1} - 1} \left(\frac{R_{c_1} R_{c_2}}{R_{c_1} R_{c_2} - 1} \right)^{\eta_3}.$$

The computed curves of Fig. 8.32 allow a verification of this model on an example chosen from many possible ones (light alloy AU4G1).

8.6 Elements of the crack analysis of structures by the global approach

The global approach to crack growth phenomena uses the crack variables K_I, J or G. Its advantage lies in the decoupling of the calculation steps in the analysis of the structure itself (Section 8.6.1 and 8.6.2) and in integrating the propagation model (Section 8.6.3).

8.6.1 *Elastic analysis by finite elements (two-dimensional media)*

Analytical values of stress intensity factors are available only for structures with very simple geometries and under the action of special loads (see Section 8.3.1). In practice, it becomes necessary to resort to finite element analysis. The setting up of a mesh constitutes a difficult problem since it must represent the singular stress field in the vicinity of the crack tip and must permit an economical calculation of the reduced energy release rate or the stress intensity factors, as a function of the area or length of the crack:

$G_r(A)$ or $K_{Ir}(a)$. For a plane medium of thickness e, we denote the crack length by a, so that $A = ea$.

In what follows, the crack is described by the constraint conditions imposed at the nodes of the finite element mesh. In the symmetric case, for example, for the mesh shown in Fig. 8.34, the crack tip is at node P; the lips are free (neither forces nor constraints are present), and the growth PQ of the crack is prevented by restraining the nodes in direction 2 (the symmetry condition). The crack tip is thus the first node which is restrained.

Local analysis and associated meshes

The stress intensity factor is determined from the computed stresses, or better from the displacements, in the vicinity of the crack tip. This technique requires a high accuracy in calculations, which, due to the presence of the singularity, calls for adapted meshes:

either a very fine mesh converging at the crack tip: common isoparametric six node finite elements for example (Fig. 8.35(a)); a sufficient number of sectors (at least six) must converge towards the tip and the radial dimension of the elements must become refined towards the tip in a geometric progression; the size of the first triangle at the tip must lie between 1/1000 and 1/100 of the crack length depending on the desired precision;

or special elements in the vicinity of the tip; the simplest way is to proceed by shifting the side nodes of an isoparametric element (Fig. 8.35(b)); this induces at the vertex of the triangle a singular field similar to that of the desired solution.

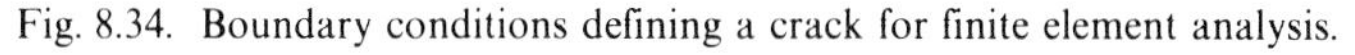

Fig. 8.34. Boundary conditions defining a crack for finite element analysis.

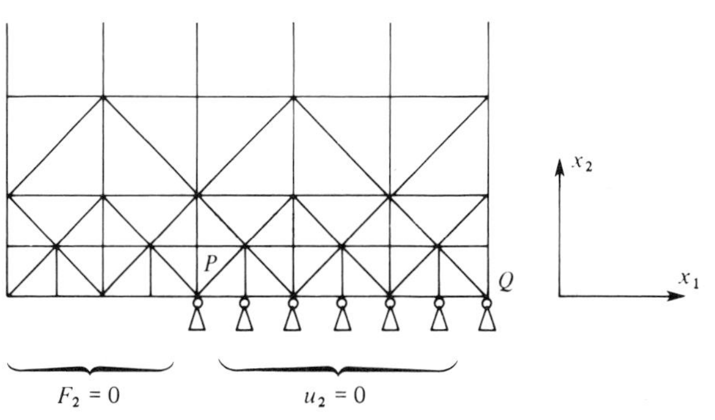

Fig. 8.35. Examples of finite element meshes.

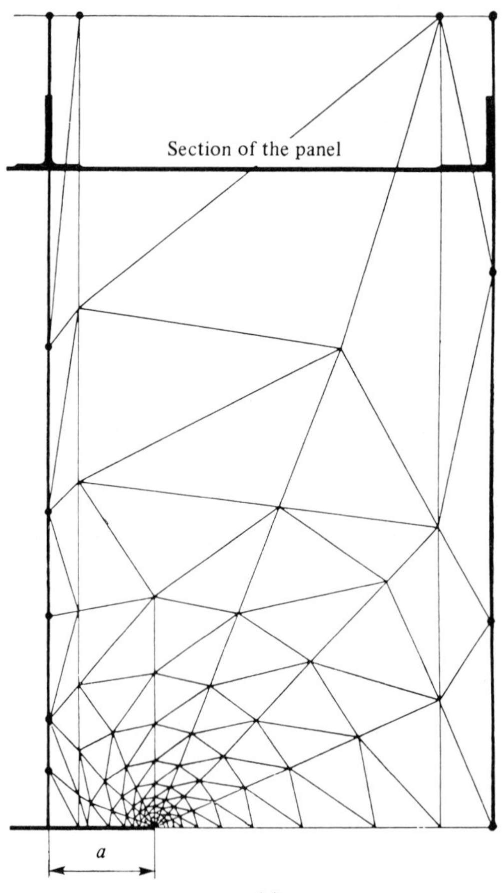

(*a*)

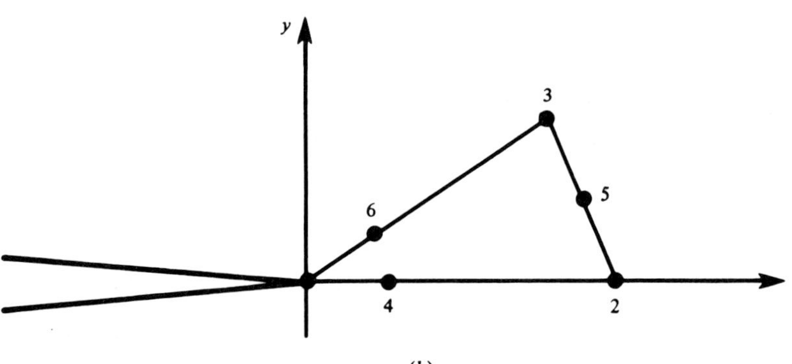

(*b*)

Once the displacement field has been determined, a simple way of obtaining the stress intensity factors is by using the relations of Section 8.2.2, for example for mode I loading in plane stress:

$$K_{\mathrm{I}} = \lim_{\substack{r \to 0 \\ \theta = \pi}} \left[\frac{E}{8} \left(\frac{2\pi}{r} \right)^{1/2} [\![u_2]\!] \right].$$

In practice, a linear extrapolation of the values of u_2^2/r near $r = 0$ provides the value of K_{I}.

Energy based methods

The methods recommended here are based on the calculation of the elastic energy release rate during a unit growth of the crack. Knowing G it is easy to find an equivalent value of the stress intensity factor. In plane-stress and for mode I for example, $K_{\mathrm{I}} = (EG)^{1/2}$.

The direct method consists in calculating the work of the external forces for different crack lengths (for the same external forces F_q):

$$W_{\mathrm{x}} = \sum F_q u_q$$

and in smoothing the obtained relation between W_{x} and A in order to compute a derivative. We then obtain G by

$$G = \frac{1}{2} F \left. \frac{\mathrm{d}u}{\mathrm{d}A} \right|_F = \frac{1}{2} \frac{\mathrm{d}W_{\mathrm{x}}(A)}{\mathrm{d}A}.$$

This method is accurate and does not require a very fine mesh in the vicinity of the crack tip (see Fig. 8.36). However, it is necessary to obtain a complete solution for each crack length considered (generally a dozen points are necessary to obtain a correct smoothing over an interval length of practical interest).

Method of bounds

Let us recall the bracketing formulae derived in Section 8.4.2 for structures subjected to a load F_q and the corresponding elastic displacement u_q^{e}:

$$G_u = -\frac{1}{2} u_q^{\mathrm{e}} \frac{\delta F_q}{\delta A} < G < G_F = \frac{1}{2} F_q \frac{\delta u_q^{\mathrm{e}}}{\delta A}.$$

In practice this method is implemented in the following manner:

a first analysis of the structure corresponding to a crack configuration is used to obtain the displacement u_q^{e} at points where loads

are prescribed as well as the forces F_q at points where displacements are prescribed;
the configuration of the structure is changed to $A + \delta A$. In practice if the analysis is performed by finite elements, the crack growth may be of the same order as the smallest element. In contrast to the local approach of stress intensity factors, this method does not require a particularly refined mesh;
a second analysis of the structure is performed with the new configuration and boundary conditions consisting of all the given and also the computed forces from the first analysis. The computed displacements furnish δu_q^e and

$$G_F = \tfrac{1}{2} F_q(\delta u_q^e / \delta A);$$

Fig. 8.36. Mesh for one-quarter of a plate with a symmetric crack in its centre for use in the energy method.

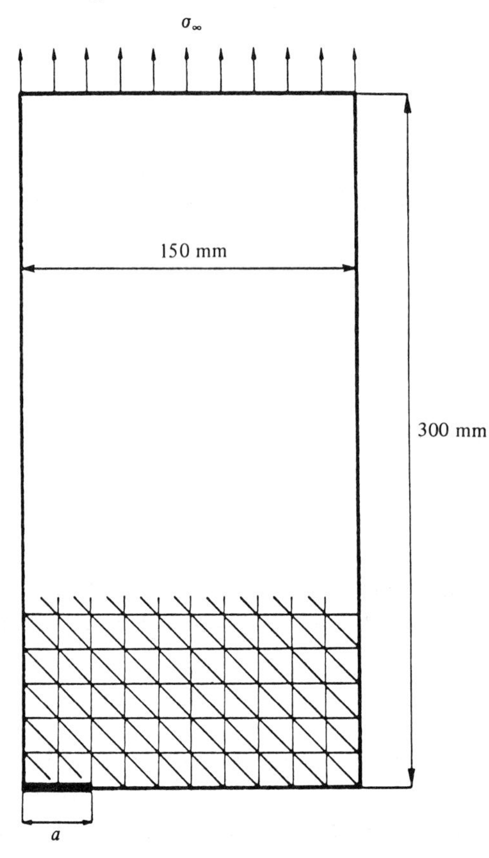

a third analysis of the structure is performed without changing the configuration but with boundary conditions consisting of all the given displacements and those computed in the first analysis. This analysis furnishes new forces from which we find δF_q and then

$$G_u = -\tfrac{1}{2}u_q^{\mathrm{e}}(\delta F_q/\delta A).$$

This method is relatively expensive due to the fact that three successive analyses are required to obtain one point. On the other hand, it is interesting to obtain these bounds, which are not given by other methods.

Perturbation method

This method is applicable only when a single crack length is to be studied; however, its generalization to the three-dimensional case is very useful. With reference to the direct method defined above, it consists of

performing a complete analysis of the structure for the considered crack length;
slightly displacing the node representing the crack tip by a small amount δa by disturbing only the elements containing this node.

This perturbation produces only very local modification δK to the stiffness matrix $[K]$. By perturbing the equation $[K]\{q\} = \{F\}$ corresponding to the first analysis, we find (omitting the matrix symbols) that

$$\delta K\, q + K\, \delta q = \delta F = 0$$

or

$$\delta q = -K^{-1}\delta K\, q.$$

The increment of work at constant force is expressed by

$$\delta W_{\mathrm{x}} = \delta q^{\mathrm{T}} F = -q^{\mathrm{T}}\delta K\, K^{-1}\, F = -q^{\mathrm{T}}\delta K\, K^{-1} K\, q$$

$$\delta W_{\mathrm{x}} = -q^{\mathrm{T}}\delta K\, q.$$

Knowing the nodal displacements q (only their local values are required) and the change in the stiffness matrix δK, δW_{x} can easily be calculated, and then

$$G = \tfrac{1}{2}(\delta W_{\mathrm{x}}/\delta A).$$

This method has the advantage of not requiring a smoothing of the calculation results followed by a calculation of the derivative as in the direct method. However, because of the local aspect, it is necessary to use a sufficiently fine mesh. The increment δa must be chosen to lie between 1/100 and 1/10 of the dimension of the tip element.

Inverse flexibility method

This method is of interest when the entire path of the crack must be calculated (assuming the path to be known). It consists in a single triangular decomposition of the stiffness matrix for a maximum number of unknowns, i.e., when the crack has reached its maximum length. A numerical treatment corresponding to successive reclosing of the crack then provides the values of G in the middle of the interval between the successive nodes. With this method we can still work with a regular mesh such as that of Fig. 8.36. We restrict ourselves to the simplified case where there is symmetry. Let us denote the compliance matrix by $[C]=[K]^{-1}$. The equilibrium condition for an intermediate crack length can be written in the form

$$\begin{Bmatrix} q_V \\ \hline q_F \\ \hdashline q_M \end{Bmatrix} = \left[\begin{array}{c|c:c} C_V & C_{VF} & C_{VM} \\ \hline C_{VF}^T & C_F & C_{FM} \\ \hdashline C_{VM}^T & C_{FM}^T & C_M \end{array}\right] \begin{Bmatrix} F_V \\ \hline F_F \\ \hdashline F_M \end{Bmatrix}$$

$$\underbrace{\qquad\qquad\qquad}_{n_F}$$

with the following meanings for the subscripts:

V: degree of freedom not on the crack path
F: degree of freedom on the (still open) lips of the crack
M: degree of freedom already blocked against crack growth.

Obviously $F_F = q_M = 0$. It is easy to show that

$$F_M = -C_M^{-1} C_{VM}^T F_V$$
$$q_V = (C_V - C_{VM} C_M^{-1} C_{VM}^T) F_V.$$

The work is done only through the displacements q_V. The work of external forces is expressed by

$$W = q^T F = q_V^T F_V = F_V^T (C_V - C_{VM} C_M^{-1} C_{VM}^T) F_V.$$

If all of the prescribed displacements are equal to zero, $\delta F_V = 0$ during crack growth, and we have

$$\delta W = -F_V^T \delta (C_{VM} C_M^{-1} C_{VM}^T) F_V.$$

The method is then applied in three steps:

(1) triangular decomposition of the complete K matrix.
(2) Solution of n_F problems to obtain the n_F columns of the K^{-1} matrix which correspond to the degrees of freedom associated with the crack length (we use unity on the right hand side).
(3) n_F calculations of δW; the matrix C_M^{-1} for the length a_j is easily obtained from that for length $a_j + 1$ by a method of inversion by blocks.

This method is very economical since the major part C_V of the C matrix is not involved and only one triangular decomposition is needed to treat all the crack lengths. The computation time saved varies between 8 and 20 times (compared to the classical method) depending upon the type of example. However, the method is slightly less accurate than the preceding one because of the finite increments that are used (the obtained values of $\delta W/\delta a$ are not as close to the derivative as before). It can be implemented in a finite element program without difficulty. It has been validated with respect to several examples. Fig. 8.37 shows the case of a plane plate loaded in tension perpendicular to the crack and also illustrates the edge effect (by comparison with Dixon's correction).

Fig. 8.37. Calculation of $K(a)$ for a cracked plate in tension ($w = 150$ mm) (after Labourdette).

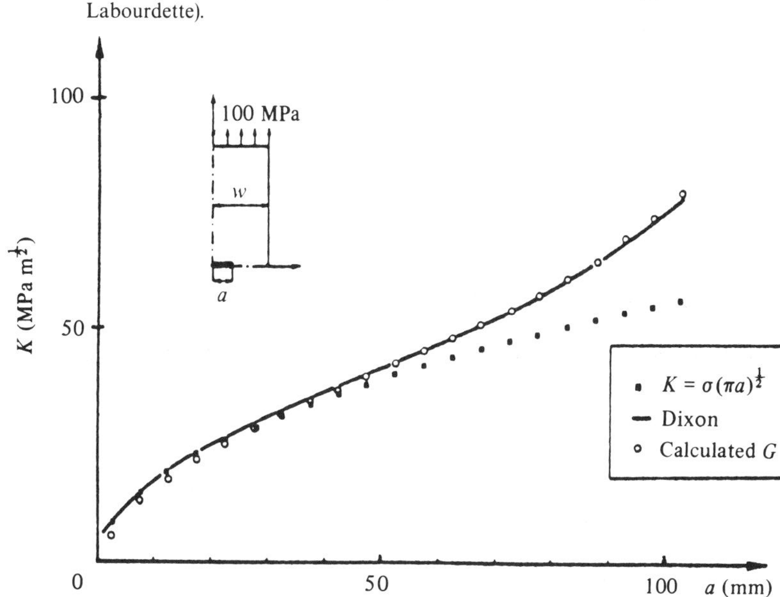

J integral method

This is a different method from the preceding one since it requires the calculations to be done only once. We make use of the following relation proved in elasticity:

$$G = J = \int_C [W_e n_1 - \sigma_{ij} n_j (\partial u_i / \partial x_1)] \, ds.$$

The contour C must enclose the crack tip but remain somewhat far from the crack (to avoid the errors caused by the singularity). The difficulty of obtaining d/dA, the derivative with respect to A, is replaced by that of obtaining spatial derivatives such as $\partial u_2/\partial x_1$. The stresses and strains must be evaluated on the contour, which is, in general, formed by the sides of the finite elements; an interpolation process must therefore be used to obtain the edge values from the known values at the Gauss points.

In practice it is advisable to introduce one or several rectangular contours inside the finite element mesh of the structure (the interpolation of stresses is more difficult for the free edges). The operations of interpolation and differentiation are easier if higher order finite elements using displacement gradients as unknowns are employed.

8.6.2 *Three-dimensional cracked structures*

We are now interested in cracks represented by a finite number of variables: this represents a discretization of the continuum formalism presented in Section 8.4.6. Let us consider the crack front Γ decomposed into a certain number of segments and let us suppose that its evolution is entirely defined

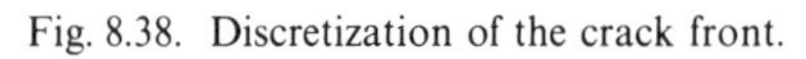
Fig. 8.38. Discretization of the crack front.

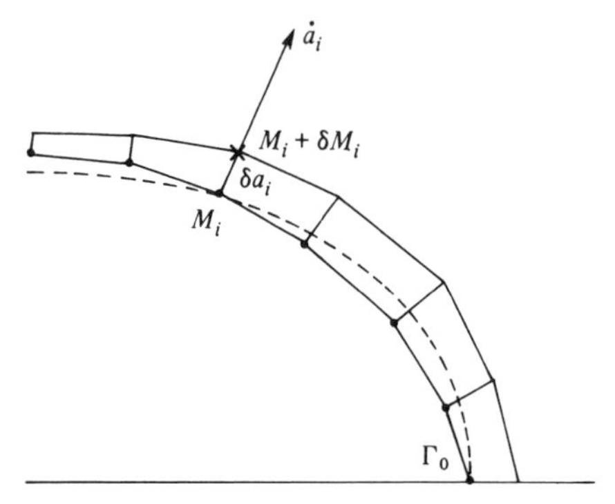

by the evolution of the points M_i. Let $\dot{a}_i$ denote the normal velocity of the point M_i: it is not the velocity of the material points but that of the geometric front (Fig. 8.38).

Let us consider the case of a brittle elastic solid, and to fix the ideas, let the external forces be imposed on the boundary $\partial\mathscr{S}_F$. As a simplification we consider only a single external force, with $\vec{u}$ denoting the displacement of the point of application of this force. The free energy of the system can be written as

$$\Psi(\vec{u}, a_i) = \tfrac{1}{2}R(a_i)\vec{u}\cdot\vec{u},$$

assuming that a_i are reckoned from the front Γ_0 which is considered to be the initial front. To this initial front, there corresponds a certain stiffness of the cracked structure. Its free energy is

$$\Psi(\vec{u}, 0) = \tfrac{1}{2}R(0)\vec{u}\cdot\vec{u}.$$

It should be noted that we can imagine an infinite number of modifications to this front Γ_0 which result in the same stiffness (the dotted line in Fig. 8.38 represents one example). This means that in the space of parameters a_i we can trace an equipotential surface defined by

$$\Psi(\vec{u}, a_i) = \Psi(\vec{u}, 0).$$

The evolution of the crack front Γ is defined through the discretization, by the evolution of points M_i:

$$M_i = M_{i0} + \delta M_i = M_{i0} + \delta a_i = M_{i0} + \dot{a}_i\delta t$$

and also by that of adjacent segments, with a linear approximation. The variables of the evolution problem are, in fact, the areas A_i defined by

$$\delta A_i = \tfrac{1}{2}\delta a_i |M_{i+1} - M_{i-1}|.$$

We note in Fig. 8.39 that the evolution of the segment (M_{i-1}, M_i) is easily decomposed into that of the evolution of points M_{i-1} and M_i, and that the areas δA_i add up in a consistent fashion (neglecting the second order terms when δa_i is small). At the extremity of the front, on a free edge, the evolution $\dot{a}_n$ of point M_n will follow the free edge. In this case $\delta A_n = \tfrac{1}{2}\delta a_n |M_n - M_{n-1}|$. The free energy is now expressible as

$$\Psi(\vec{u}, A_i) = \tfrac{1}{2}R(A_i)\vec{u}\cdot\vec{u}.$$

The thermodynamic variable associated to the variable A_i is defined by

$$g_i = -\partial\Psi/\partial A_i = -\tfrac{1}{2}(\partial R/\partial A_i)\vec{u}\cdot\vec{u}.$$

The dissipated power (plasticity and threshold effects are neglected) is expressed by

$$\Phi_1 = -(\partial\Psi/\partial A_i)\dot{A}_i = g_i\dot{A}_i$$

and we assume the crack phenomenon to be a normal dissipative mechanism. Writing the dissipation potential as $\varphi^*(g_i)$, the crack growth law can be expressed as

$$\dot{A}_i = \partial\varphi^*/\partial g_i.$$

If φ^* is quadratic, crack growth takes place in the direction of the energy gradient

$$\dot{A}_i = \partial\varphi^*/\partial g_i = \alpha g_i = -\alpha\,\partial\Psi/\partial A_i.$$

The above comes down to considering the following criterion for the crack propagation: among all the physically admissible mechanisms of crack evolution, the real mechanism maximizes the released elastic energy.

Fig. 8.39. Definition of perturbation of crack tip nodes and the growth of the corresponding areas.

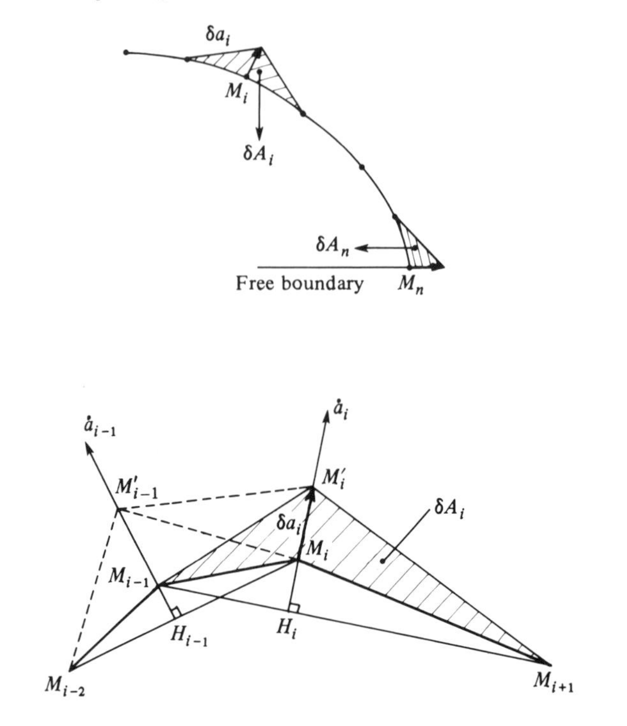

In other words, for an infinitesimal growth of the crack of a fixed magnitude

$$\| \mathrm{d}a \| = \left(\sum_i \mathrm{d}a_i \, \mathrm{d}a_i \right)^{1/2}$$

the evolution $\mathrm{d}a_i$ is such that the decrease in Ψ is maximum, since $\mathrm{d}a_i$ is colinear with the gradient and hence perpendicular to the equipotential $\Psi(u, a_i) = \Psi(u, 0)$ defined above. We note the difference with respect to a maximization with fixed $\mathrm{d}A$ (norm $\sum_i |\mathrm{d}a_i|$).

This law, known as the gradient law, is well supported by experiment. Good correlation has been obtained between the normal $\dot{A}_i$ measured at selected points of the experimental crack front and the quantities g_i calculated by means of finite element analysis from the change in the work of external forces for a number of materials, different geometries, different types of initial cracks, and several types of loadings. Fig. 8.40 shows the example of the AU2GN T6 alloy with g_i expressed as functions of $\alpha = \dot{A}_i/g_i$. Moreover, Fig. 8.40(*a*) shows that reasonable predictions can be made: the final crack front was calculated in a step-by-step fashion (nine successive fronts) each front being deduced from the previous one by means of the crack growth law. The constant α which furnishes the number of cycles is determined from Fig. 8.40(*b*) which regroups a number of tests.

The numerical calculation of the local rate of dissipated energy g_i is done by the perturbation method. For each front under study, the structure of the front is discretized using three-dimensional finite elements. The solution of the elastic problem furnishes the nodal displacements $\{q\}$. Each node of the front is then perturbed in the direction of the normal represented by the quantity δa_i (Fig. 8.39), and the associated quantity g_i is obtained from

$$g_i = \frac{1}{2\delta A_i} (- \{q\}^{\mathrm{T}} [\delta K] \{q\})$$

where $[\delta K]$ is the variation in the stiffness matrix induced by the perturbation (in the interest of economy only the nodes close to the front are considered). This method is sufficiently accurate provided that the finite element mesh is refined enough in the crack front region.

This approach is well suited to problems of propagation of nontransverse cracks. It can be used:

(1) to calculate the local stress intensity factors based on the energy release rate

$$g(s) = \frac{1 - \nu^2}{E} [K_{\mathrm{I}}^2(s) + K_{\mathrm{II}}^2(s)] + \frac{1}{2\mu} K_{\mathrm{III}}^2(s);$$

(2) to calculate the evolution of the crack front according to some growth law (for example, a gradient law or a more general law) established on the basis of a certain number of experiments;

(3) to treat the same type of problem by means of an approximate method using elliptical cracks: the two variables of the problem are then the two semiaxes of the ellipse;

Fig. 8.40. Prediction of the three-dimensional crack propagation: (*a*) successive fronts; (*b*) demonstration of the constant value of α (after Labourdette).

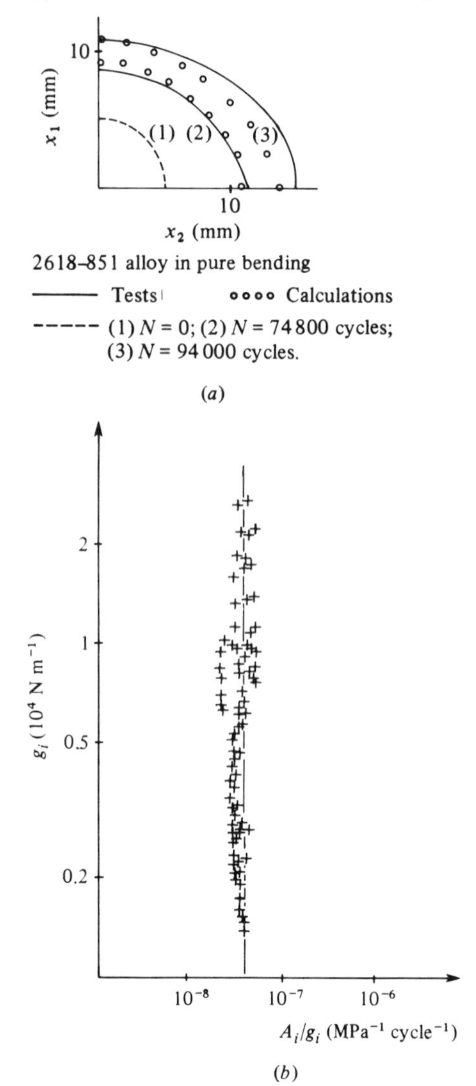

(4) to solve other problems of (plane-stress or plane-strain) transverse cracks in which two crack tips are simultaneously involved, the rates $\dot{a}_1$ and $\dot{a}_2$ are *a priori* taken to be different: e.g., in predicting edge effects (asymmetric cracks), and the influence of stiffness.

8.6.3 *Integration of the models*

As in the prediction of crack initiation by damage mechanics, the prediction of crack propagation by fracture mechanics is carried out in two steps. All models use either the stress intensity factors or the energy release rate as a characteristic variable of the structure and the load. If we restrict ourselves to the simple loading case, the concept of reduced energy release rate can be used to decouple the structural problem from the temporal problem of evolution since (see Section 8.4.2):

$$G(A,t) = G_r(A)F^2(t).$$

Firstly, knowing the path of the crack in the structure (a straight line in a two-dimensional structure, or an ellipse of constant eccentricity in a three-dimensional structure) several analyses of the structure such as those described in Section 8.6.1 can be used to find a set of discrete values $G_r(A_i)$ corresponding to the evolution of the crack starting from its initial configuration A_0 and ending with the fracture of the structure into two parts. Using a smoothing method it is easy to define a function $G_r(A)$ whose graph is a best fit to the numerical values $G_r(A_i)$. A power function facilitates the next step.

Secondly, the nonlinear differential equation or the system of differential equations which represents the model is integrated. For example for a law of crack growth by fatigue such as the one described in Section 8.5.4:

$$\frac{\delta A}{\delta N} = \frac{1}{C'^{\eta/2}}(G^{*\eta/2}_{\mathrm{Max}} - G^{*\eta/2}_{\mathrm{m}})$$

which for $G_m < G_S$ (crack growth threshold) can be put in the form

$$\frac{\delta A}{\delta N} = \left[\frac{G_r(A)}{C'}\right]^{\eta/2}(F_{\mathrm{Max}} - F_S)^{\eta}.$$

This differential equation is separable in its variables:

$$G_r(A)^{-\eta/2}\delta A = C'^{-\eta/2}(F_{\mathrm{Max}} - F_S)^{\eta}\delta N$$

and can be integrated for a given $A(N)$ if knowledge $F(t)$ of the loading history can be used to define two cyclic functions corresponding to the maximum and to the threshold of $F(t)$: $F_{\text{Max}}(N)$, $F_S(N)$

$$\int_{A_0}^{A} G_r(A)^{-\eta/2}\delta A = \int_0^N C'^{-\eta/2}[F_{\text{Max}}(N) - F_S(N)]^{\eta}\delta N.$$

This integration must be continued until the conditions of brittle fracture are reached:

$$G_r^*(A)F_{\text{Max}}^2(N) = G_c \rightarrow \begin{cases} A = A_c & (A_c \text{ unstable}) \\ N = N_R \end{cases}$$

Examples corresponding to crack growth in stiffened panels subjected to periodic loads are described in Fig. 8.41.

When the loading history is known only through the statistical variables, two cases can occur:

either the crack growth model is a differential equation with separable variables, which results in the rule of linear accumulation (see Chapter 3); then knowing the probability density function of the force F_{Max}, one can analytically derive a relation between the probabilistic mean of the crack surface and the number of cycles;

or, the model does not include the rule of linear accumulation. Then, a numerical method must be used. A large number of analyses are performed for different, independent cases of the random loading, defined by the probability density of the maximums for example. An analysis of these calculations then

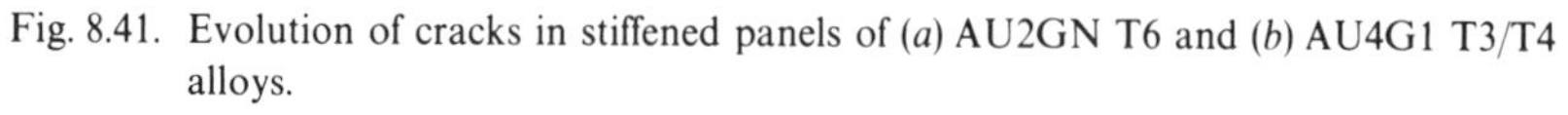

Fig. 8.41. Evolution of cracks in stiffened panels of (*a*) AU2GN T6 and (*b*) AU4G1 T3/T4 alloys.

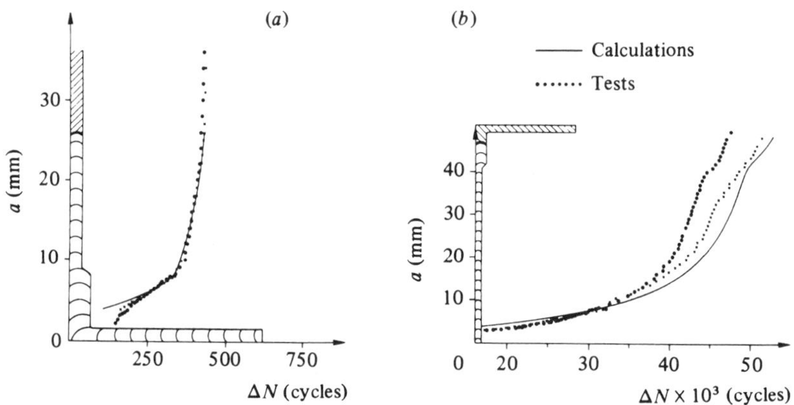

furnishes the desired statistical quantities: a relation between the mean of A and the number of cycles, or its reverse (the mean number of cycles required to attain a given crack surface). This is the Monte Carlo method.

8.7 Crack analysis by the local approach

8.7.1 *Limits and inadequacies of the global fracture mechanics*

The methods of analysis of cracked structures, presented in Section 8.6, are based on variables defined at the global level of the cracked medium. They are applicable to a number of situations in which it is not necessary to know the exact state of stress or of damage in the vicinity of the crack tip: a two-dimensional, almost elastic medium, with only a small plastic zone related to the crack size; a three-dimensional medium subjected to proportional loading, or periodic loading in fatigue corresponding to the organization chart of Fig. 8.42.

On the other hand, in other situations, this approach may prove to be deficient, either because of the size of the cracks (problems of short cracks), or because of a pronounced overall plasticity (ductile fracture, creep crack growth), or because of the history effects (of loading, or temperature), or finally because of the nature of the deterioration mechanisms (multiple crack phenomenon in concrete, delamination of composites, etc.). A few examples are presented below.

Fig. 8.42. Organization chart for classical analysis.

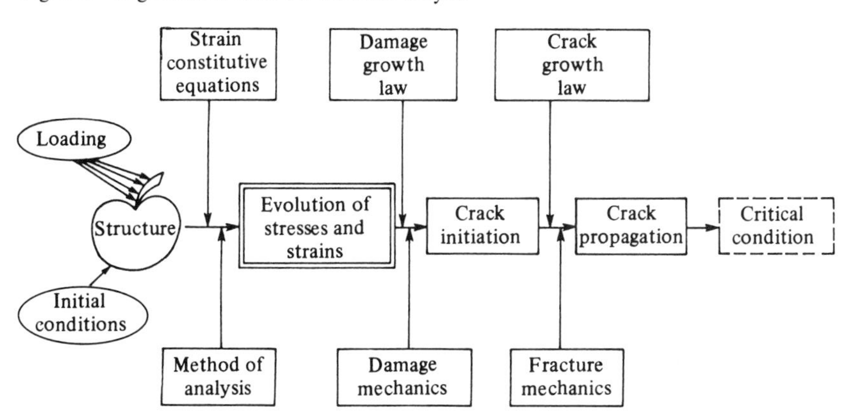

Short cracks

Here we are concerned with defects or cracks whose size, in metals, is of the order of a few crystals (between 0.1 and a few millimetres) and whose evolution can represent an important part in the life of the structure. The crack has a size of the same order as that of the local inhomogenities of the microstructure and may interact with neighbouring microcracks of similar size; the plastic zone at the crack tip is of a different nature than that for long cracks. The states of stress in the vicinity of the crack tip are unknown because they should be defined at the crystallographic scale. In practice, we observe a growth which is somewhat chaotic, with bifurcations and branching, before reaching the 'long crack' regime. Fig. 8.43 illustrates the impossibility of using the usual variable ΔK to correlate fatigue results in relation to cracks of the size of 100 μm to few millimetres since for an increasing ΔK the crack growth rate first decreases and then increases.

Overall plastification

Overall plastification can occur at two levels:

before initiation in that part of the work-piece in which the crack will initiate and grow;

Fig. 8.43. Schematic representation of the domain of propagation of short cracks (after K. Miller).

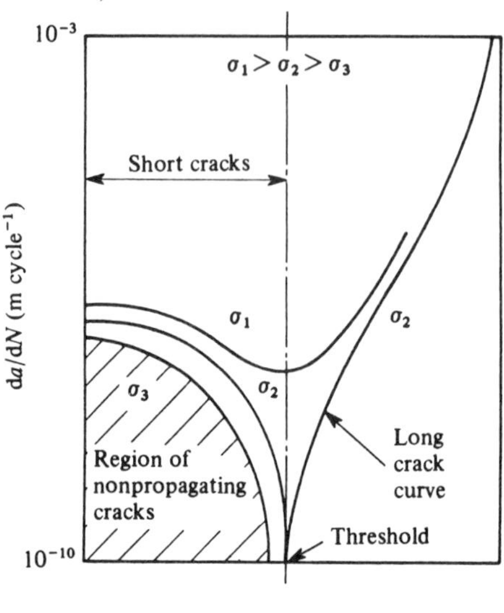

in the region in front of the crack-tip, during growth and during fracture.

In the former, a plastic analysis of the uncracked part of the work-piece is required to evaluate the stress at 'infinity' correctly before applying the methods of fracture mechanics. In the latter, it is the mechanism of the propagation itself which interacts with the plasticity and can render the classically defined global parameters, namely K, G, J, unusable even when plasticity associated singular fields are used (see Section 8.2.3). Ductile fracture and fatigue retardation effects belong to this category.

Creep crack growth

In this case, overall plastification is accompanied by time effects: primary and secondary creep at the crack tip and eventually in the entire work-piece. The global variables are the stress intensity factors K (see Section 8.5.3) and the contour integral C^* that takes secondary creep into account (Section 8.2.3). Fig. 8.44 shows, through tests with constant K, that the variable C^* is not a suitable one (da/dt increasing) for an aluminium alloy and more generally for rather brittle materials. On the other hand, for very ductile materials, neither K nor C^* furnishes an acceptable correlation

Fig. 8.44. Propagation in creep: 2219 alloy (after P. Bensussan).

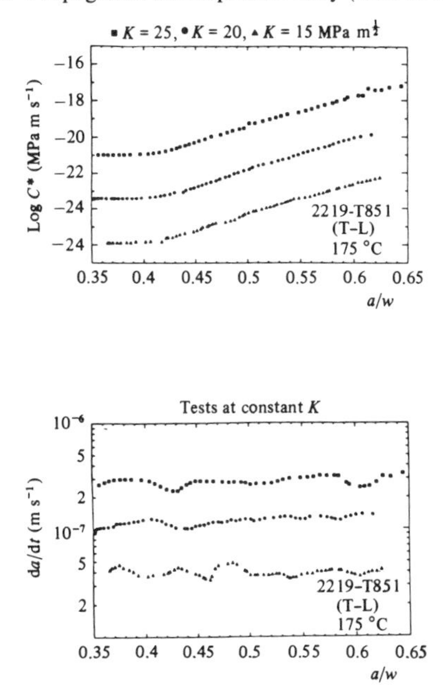

(see e.g., Fig. 8.45). The better correlation with C^* mentioned sometimes in the literature may be due to the experimental method of determining C^* which implicitly incorporates the result da/dt itself.

These few examples show the limits of the global approach of fracture mechanics. It is possible to improve the correlations by introducing additional internal variables to describe other physical processes, effects of crack closing history effects, etc. (Sections 8.4.3 and 8.5.4), but at the cost of additional complications. A simple and often more accurate way is that of the local approach to the crack growth.

8.7.2 *Principles of the local approaches*

These consist in determining as accurately as possible, the stresses, strains, plastic strains and variables which describe the deterioration in the zone in front of the crack, taking into consideration the redistribution of the stresses generated by plastic flow and damage. A local fracture criterion of the volume element is used to increment the crack length.

Different types of local approaches

Schematically, we may distinguish two levels of local approach:

the crack continues to be treated as a discontinuity in the material (with boundary conditions at the free edges corresponding to no

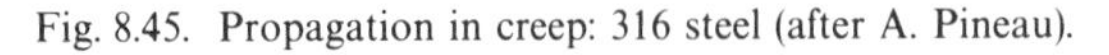

Fig. 8.45. Propagation in creep: 316 steel (after A. Pineau).

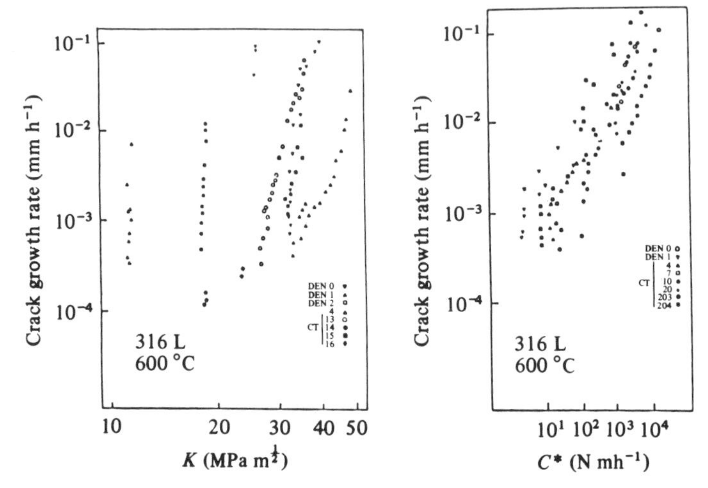

force and unconstrained displacements, and the analysis performed according to the flow chart shown in Fig. 8.46). The crack length grows when the local fracture criterion is met. In practice, in a numerical scheme using the finite element method, the increase in the crack length is accomplished through 'node relaxation': the condition of the nodal connection of the crack tip node (point P in Fig. 8.34) is progressively relaxed. One must take into account the new plastic deformations and the new redistribution produced by this relaxation of a discrete character. Generally, with such a method, one needs the concept of a critical distance at the crack tip because of the fields which remain singular. The physical fracture criterion therefore corresponds to a critical value of a variable (stress, strain, accumulated strain, energy, damage) reached at a critical distance d_c from the crack tip depending on the material. In practice, this distance also depends on the finite element mesh in the local zone at the bottom of the crack. A simplification of this first type of local approach consists in using the stress–strain fields obtained by analytical means on the basis of simple constitutive laws;

we employ damage mechanics (Chapter 7) with the crack represented schematically by a completely damaged zone which remains continuous in the sense of continuum mechanics. A crack is then defined as the neighbourhood of points for which the damage has reached its critical fracture value $D = D_c$ or $D = 1$ (Fig. 8.47). In this way we take into account the progressive deterioration of the material and the corresponding loss in the stiffness. The crack, or

Fig. 8.46. Flow chart for method without local coupling.

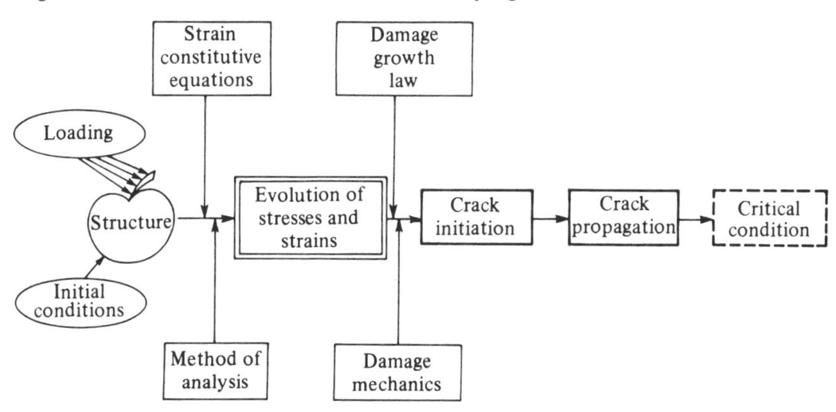

the completely damaged zone, is described by the points which no longer possess any stiffness.

In principle, such a procedure is of interest because one does not need the concept of a critical distance and the discrete aspect of the process of node relaxation. In fact, the redistributions of the stress in the partially damaged zone are such that the singularities disappear: the (principal) stress decreases in reaching the completely damaged zone ($D = 1$). This method which corresponds to the flow chart of Fig. 8.48, was used first by Hayhurst for creep. It is possible to construct a simplified method by considering a sharp variation in damage between $D = 0$ and $D = 1$: analytical solutions may then be obtained in particular cases.

Fig. 8.47. Crack defined by a completely damaged zone.

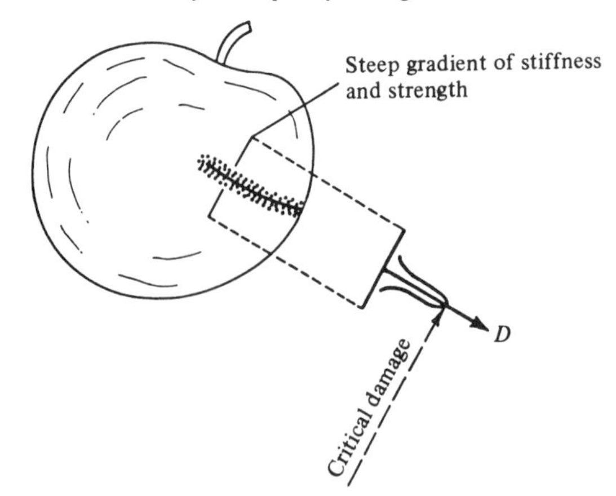

Fig. 8.48. Flow chart for the method with coupling.

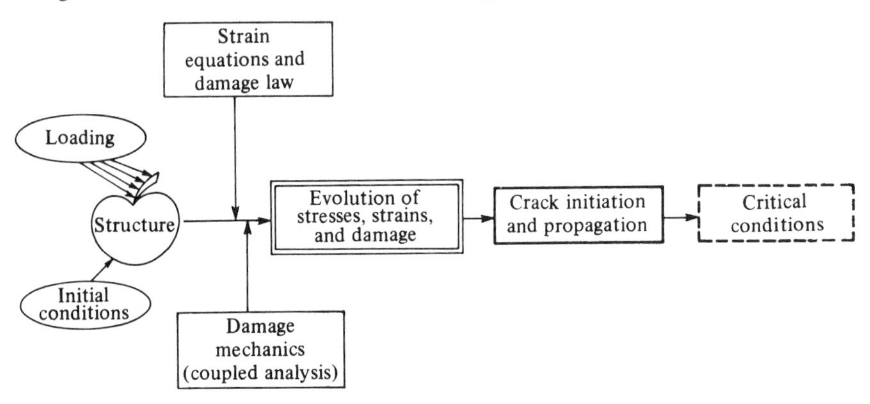

Practical methodology

These local approaches are evidently more complex to apply than the classical approaches of fracture mechanics. In practice, with either of the two methods, it is necessary:

(i) to establish correctly the constitutive law, and also the damage law for the volume element of the material, incorporating the essential physical mechanism of the fracture process, as indicated in Chapters 3–7;

(ii) to devise a 'model' of the cracked part of the work-piece with a schematic representation of the initial crack, the growth zone (mesh discretization), the critical distance (or the fracture criterion of the first element), the material, etc. This step is indispensable for the first approach because the prediction is strongly dependent on the local discretization. It is of interest to use this second method as a verification method. The results of crack growth analysis on this model of the cracked work-piece (for example a CT specimen) must be compared with the results of the corresponding laboratory tests;

(iii) the prediction of other situations for the same material then becomes possible for work-pieces of different geometries, different loadings, different initial crack lengths, etc. A local discretization as close as possible to that calibrated in (ii) is used.

To use the damage mechanics approach, we must simultaneously account for damage–(visco)plasticity and damage–elasticity couplings; the finite element method must be modified to implement an accurate numerical scheme so as to assure spatial as well as temporal convergence within a reasonable computing cost. A practical method of avoiding updating of the stiffness matrix at each time (or load) increment consists in including the variations corresponding to the right hand side of the equation. Thus, the linear system

$$Kq = Q + Q_0$$

of the plastic analysis (Sections 5.6.1 and 6.5) is modified by including the influence of damage on the elastic stiffness (Q_0 corresponds to the plastic strains):

$$K(D)q = Q + Q_0.$$

The practical method consists in writing

$$K^*q = Q + Q_0 + [K^* - K(D)]\tilde{q}$$

where $\tilde{q}$ is a close estimate of q (based on predictions from the previous step). $K^* = K(D^*)$ is the last updated stiffness matrix. A new updating is performed when the error $D - D^*$ exceeds a prescribed value. This method offers a good compromise between cost and accuracy in the situation where time (or load) increments are already small because of plastic flow.

8.7.3 *Examples*

We give below a summary of a few examples of crack growth analysis using the local approach in conjunction with damage mechanics.

Crack propagation in a plane elastic–perfectly brittle medium

In this domain of highly nonlinear problems, it is difficult to find analytical solutions. Nevertheless, by making a highly simplifying assumption regarding material behaviour, one can obtain interesting results. The material is considered as elastic-perfectly brittle, i.e., with an evolution law of the type

$$D = D_c H(\varepsilon - \varepsilon_R)$$

where D_c is the critical damage value, ε_R is the fracture strain, and H is the Heaviside step function.

By analysis, using conformal mapping in two-dimensional problems, we can show that the damage zone has a thickness $(K/E\varepsilon_R)^2$ where K is the stress intensity factor in mode I or in mode III and ε_R is the fracture strain in tension or in shear respectively. In both cases, the damage front is a cycloid whose size depends on the propagation rate (Fig. 8.49).

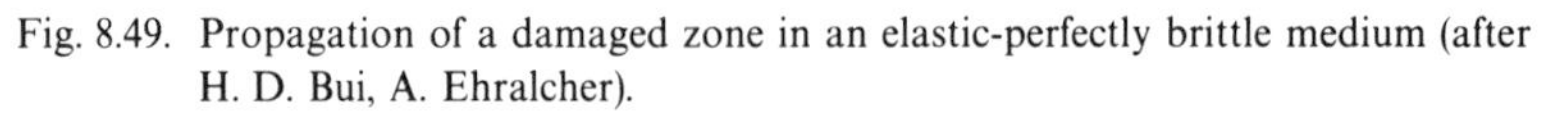

Fig. 8.49. Propagation of a damaged zone in an elastic-perfectly brittle medium (after H. D. Bui, A. Ehralcher).

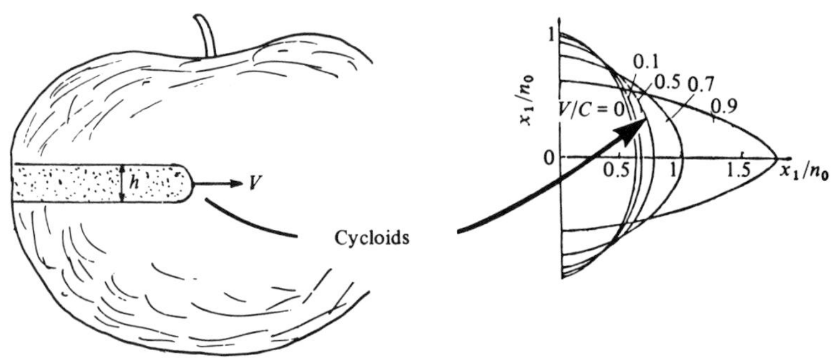

Crack propagation in concrete

The behaviour of concrete can be represented by linear elasticity coupled with a damage law which depends on an equivalent strain defined in Section 7.6.2. An example of the results of a coupled analysis is shown in Fig. 8.50. It is concerned with the plane-strain finite element analysis of a notched concrete specimen. The experiment performed by controlling strains shows a tendency for the development of two cracks, only one of which completely propagates as indicated in Fig. 8.50(*a*).

The analysis of a half-plate gives a damaged zone which progresses as a function of the prescribed strains as shown in Fig. 8.50(*c*). This compares fairly well with Fig. 8.50(*a*). Finally, Fig. 8.50(*b*) gives the comparison between the calculated and measured graphs of the evolution of the crack opening as a function of the applied load.

Bifurcation of cracks

The bifurcation criteria for cracks described in Section 8.4.5 is well suited for application to a plane medium. It is difficult to extend them to the case of a three-dimensional continuum (see Section 8.4.6) and it is possible to think that crack propagation in a three-dimensional medium subjected to nonproportional loading would be better dealt with by the local approach.

As a feasibility study, we describe below a two-dimensional case close to that of Fig. 8.26. This case is concerned with the study of fatigue crack

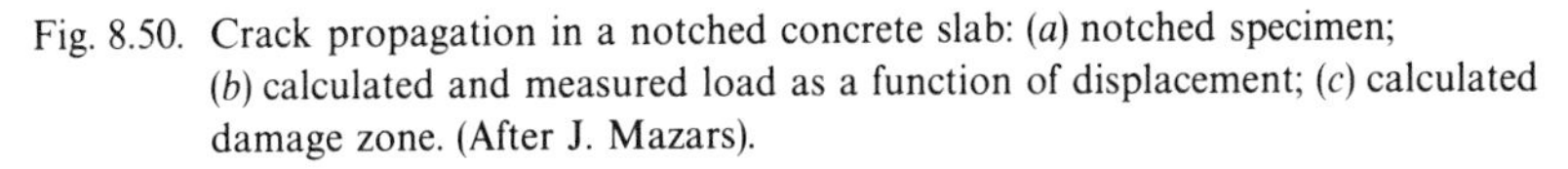

Fig. 8.50. Crack propagation in a notched concrete slab: (*a*) notched specimen; (*b*) calculated and measured load as a function of displacement; (*c*) calculated damage zone. (After J. Mazars).

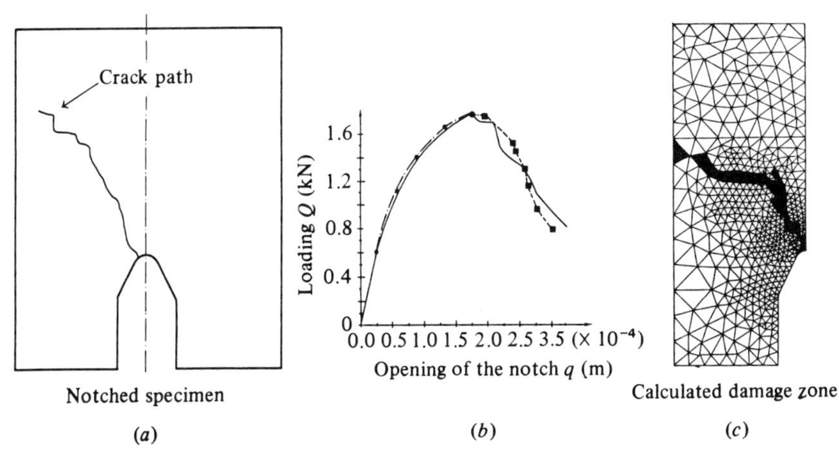

growth in a thin square plate under biaxial loads, F_1 and F_2 (Fig. 8.51). The initial half-crack $0A$ is inclined $-20°$ with respect to F_1. A force ratio of $F_2/F_1 = 1$ generates crack growth characterized by prolongation of quantity AB. Loading with a force ratio $F_2/F_1 = 0$ induces a bifurcation of the crack at an angle $\alpha = 90°$ with the progression of a quantity BC.

The black zone in Fig. 8.51 represents the damaged zone as computed by finite elements using an elasticity law coupled with a fatigue damage law for large numbers of cycles (Section 8.4.3). We consider here that a quarter of an element is completely damaged when the critical damage D_c is reached at the corresponding Gauss point. For the case under consideration (AU4G 2024) this corresponds to a number of cycles of the order of 10^6 which renders it impractical to follow a cyclewise step-by-step procedure; an incremental procedure in terms of damage D is used.

Propagation of fatigue cracks

The case of fatigue is especially difficult because of the effect of crack closure; it requires the consideration of a correct analysis of the stresses and strains in the damaged zone. It is necessary not only to describe the plastic zone at the crack tip and the corresponding redistribution, but also to model the 'plastic trail', i.e., the history of the plastic zone through which the crack has propagated.

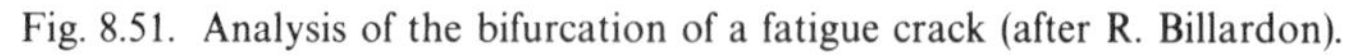

Fig. 8.51. Analysis of the bifurcation of a fatigue crack (after R. Billardon).

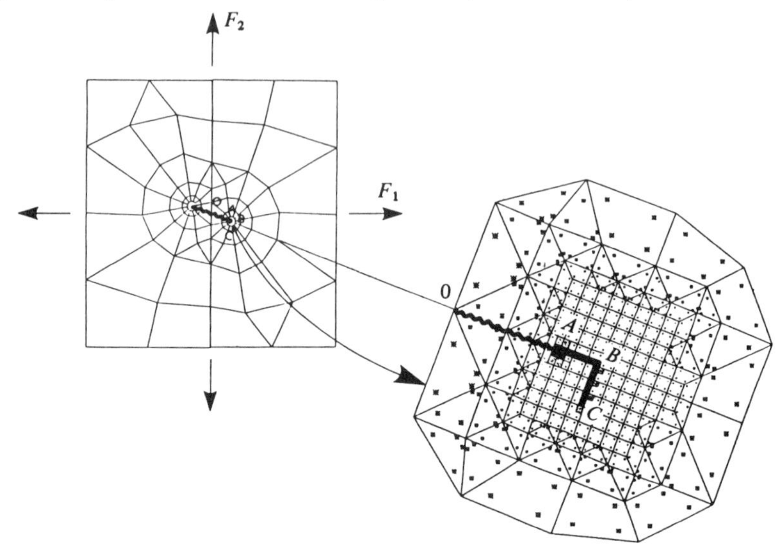

In the case of fatigue, some local approaches restrict themselves to the decoupled analysis without explicitly introducing the damage variable. The approach used by Newman, and generalized by Anquez consists of advancing the crack when the dissipated energy in the first element attains a critical value. The crack growth is realized by relaxing nodes on the width of an element. After such a growth a particular cycle is used to redistribute the stresses induced by the release of the nodal connections. It is easily seen that such a modelling depends on the size and pattern of the local finite element mesh. The model is therefore calibrated by special experiments; but then it can be used for prediction in other situations provided that the same type of mesh is used.

Fig. 8.52(*a*) shows the example of an AU4G1 light alloy bar subjected to bending. We observe marked differences between the plastic strains corresponding to loading and unloading (with components perpendicular to the crack axis).

Fig. 8.52(*b*) shows the distribution of reactions to a minimal external force, after a change in the level of the cyclic load. A growth of $\Delta a =$

Fig. 8.52. (*a*) Normal strain in loading and in unloading for a constant level, (*b*) reactions at contact points for a two level loading (after L. Anquez).

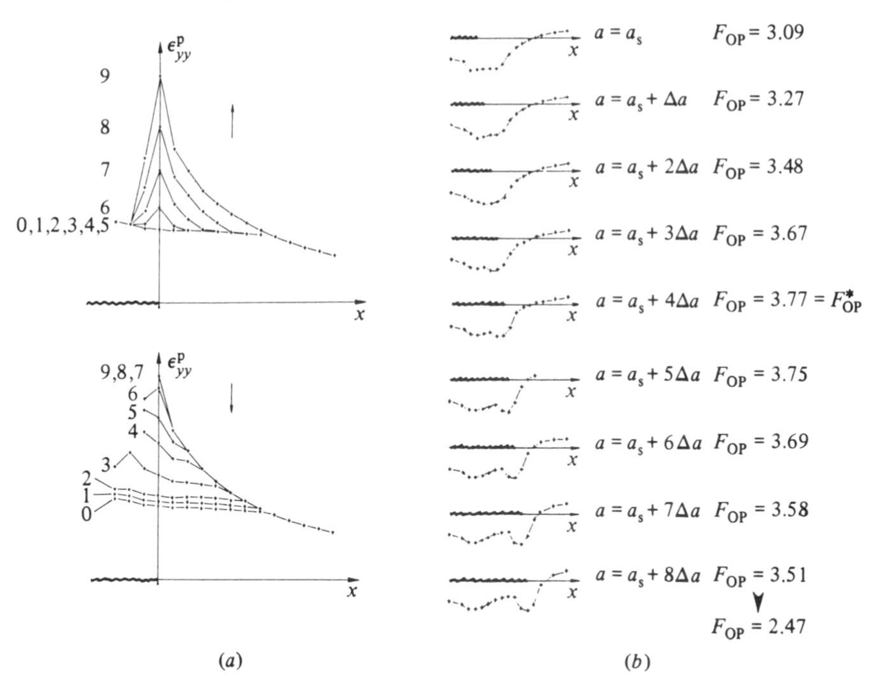

0.01406 mm takes place between each repetition of the cyclic load. We note that the structure retains the effect due to the overload (as can be seen in Fig. 8.52) and that the calculated values of the opening load are first increasing after the overload (applied immediately after the first result) and then decreasing. These effects are indeed seen in the corresponding experiments. They serve as a basis for defining growth models which consider the effects of evolving thresholds such as those described in Section 8.5.4.

Growth of ductile fracture cracks

The growth of cracks in ductile materials requires an analysis which considers coupling of elastoplasticity with ductile damage. In the example presented below the damage law is derived from the models of MacClintock, Rice and Tracey giving the growth of the radii of microcavities. In the finite element analysis performed, the growth of the crack, considered as a completely damaged zone, occurs when the stresses become zero at the Gauss points of the finite elements. The problems of convergence of the calculations are solved by introducing a characteristic element size identified, by comparing the calculated results with experimental ones.

Fig. 8.53 shows some very interesting results of the analysis which by their agreement with experiments on A316 steel, prove that the position of a

Fig. 8.53. Prediction of the initiation of ductile fracture in notched bars: (*a*) identification of the smooth notch case; (*b*) prediction of the sharp notch case (after G. Rousselier).

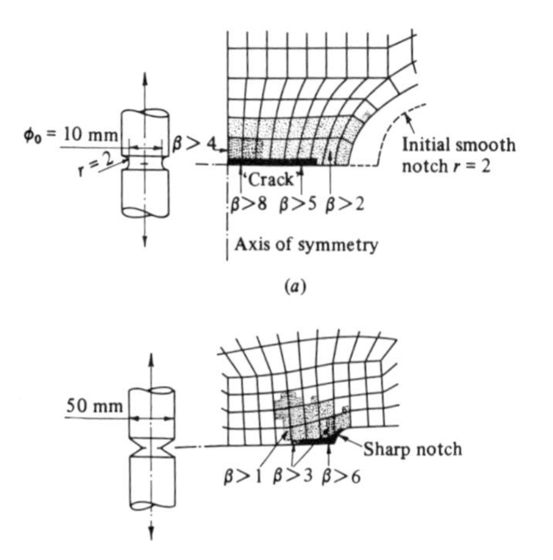

crack in a notched circular specimen subjected to a monotonic axial load depends on the radius at the bottom of the notch:

a crack in the centre of the specimen for a sufficiently large radius (Fig. 8.53(*a*)),
a circumferential crack for a small radius (Fig. 8.53(*b*)).

Creep crack growth

A very localized creep damage in a metallic work-piece can be described by an analysis which takes into account the coupling between the viscoplastic

Fig. 8.54. Calculated and observed path of a crack, 316 steel $T/T_m = 0.6$ (after M. Hayhurst).

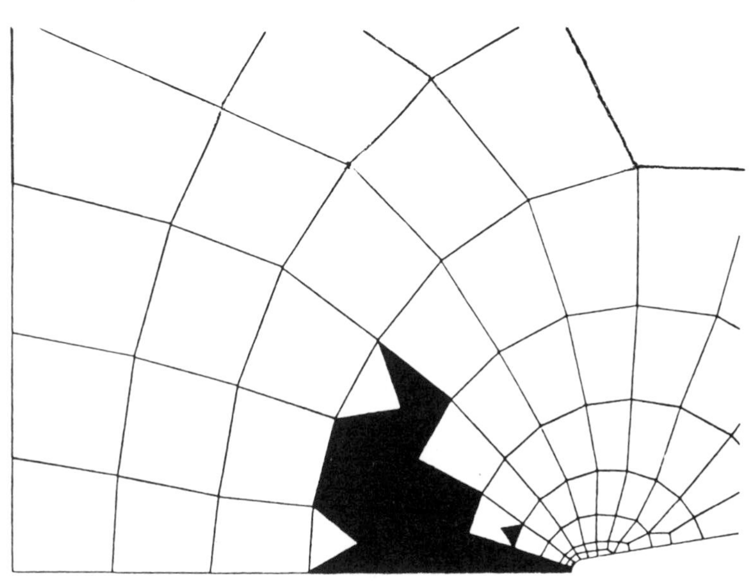

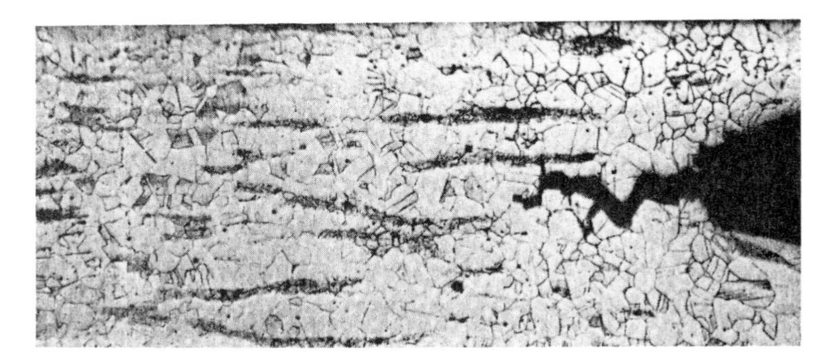

deformations and damage, with isotropic damage and hardening equations, such as the ones described in Section 7.6.2.

A first method consists of continually taking into account the coupling between the viscoplastic law and the creep damage law, but with no coupling at the level of the elastic law. The evolution of the completely damaged zone becomes evident when D is close to 1 ($D \geqslant 0.99$ for example), and the stiffness of the corresponding elements is then eliminated. This simple technique, relatively inexpensive in terms of computation time, can be used to predict the evolution of damaged zones. However, the growth of a crack of small width as the continuation of the initial notch is not correctly described, as is shown by the example of Fig. 8.54.

A more refined but more expensive method consists of adding to the preceding method a continuous coupling with the elasticity law. The stiffness matrix is periodically updated by the practical method indicated in Section 8.7.2. We thus have a continuous evolution of the damage both in space and time which then permits us to eliminate the stress singularity completely and to have a complete redistribution of the stress at the crack 'tip'. The crack is then well described by the completely damaged zone (integration points where D_c attains the critical value were fixed at 0.999). Fig. 8.55 shows the example of a CT specimen of the INCO 718 alloy (used in turbine discs) subjected to creep under constant load at 600 °C. We note that in contrast to the preceding example, the crack propagates along its original direction: this is a consequence of the total coupling.

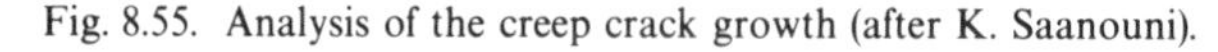

Fig. 8.55. Analysis of the creep crack growth (after K. Saanouni).

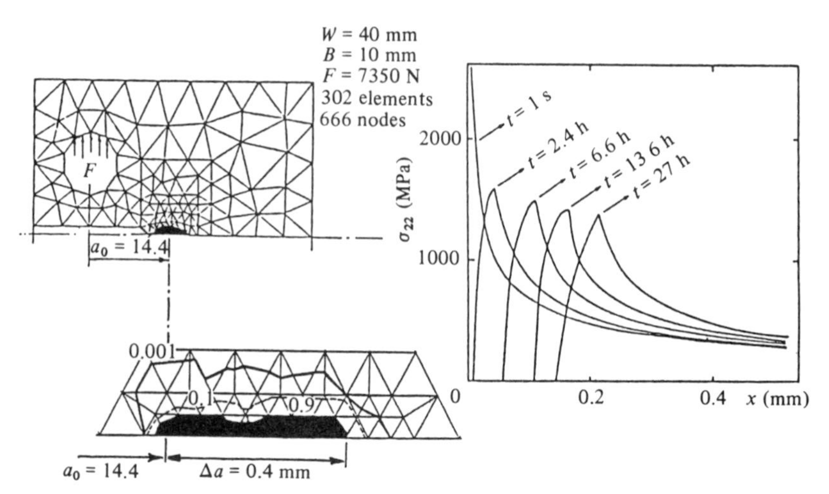

Bibliography

Bui H. D. *Mécanique de la rupture fragile*. Masson, Paris (1978).
Broek D. *Elementary fracture mechanics*. Noordhoff, Groningen (1974).
Liebowitz, *Fracture*. Vols. I, II, III, IV, V. Academic Press, New York (1968–9).
Sih G. *Mechanics of fracture*. Vols. I, II, III, IV, V. Nordhoff, Groningen (1973–7).
Labbens R. *Introduction à la mécanique de la rupture*. Editions Pluralis, Paris (1980).
Barthelemy B. *Notions pratiques de mécanique de la rupture*. Eyrolles, Paris (1980).

INDEX